Y0-BTB-250

BASIC AND APPLIED GENERAL SYSTEMS RESEARCH

A Bibliography
1977–1984

Edited by

Robert Trappl

Department of Medical Cybernetics
University of Vienna Medical School

Werner Horn

Department of Medical Cybernetics
University of Vienna Medical School

George J. Klir

Department of Systems Science
State University of New York (SUNY)
at Binghamton

⬤HEMISPHERE PUBLISHING CORPORATION

Washington   New York   London

INTERNATIONAL FEDERATION FOR SYSTEMS RESEARCH

Laxenburg, Austria

BASIC AND APPLIED GENERAL SYSTEMS RESEARCH:
A BIBLIOGRAPHY 1977–1984

1 2 3 4 5 6 7 8 9   B R B R   8 9 8 7 6 5

Library of Congress Cataloging in Publication Data:

Basic and applied general systems research.

Includes indexes.
1. System theory–Bibliography. I. Trappl, Robert.
II. Horn, Werner. III. Klir, George J., date- .
IV. International Federation for Systems Research.
Z7405.S94B37   1985   [Q295]   016.003   85-8453
ISBN 0-89116-454-5

Q295
A12 B37
1985
MATH

Contents

Preface v

Statistical Information 1

List of Journals 2
List of Publishers 5
List of Institutions 8
List of Types of Entries 10
Publication Years 10

Author Index 11

Key Word in Context (KWIC-) Index 107

Bibliographical List 255

Preface

In May 1977 George J. Klir and Gary Rogers compiled a bibliography on *Basic and Applied General Systems Research*. This bibliography is still available from the International Federation for Systems Research (IFSR) at Hofstrasse 1, A-2361 Laxenburg, Austria, for U.S. $12.00 including postage.

To continue this work, the IFSR has collected material on this topic from the year 1977 onward and 1,944 items have been processed. The result is presented in this bibliography, which consists of four parts:

1. Statistical information, including
   - A list of journals covered, together with the number of papers from each journal.
   - A list of publishers of the books, followed by the number of entries in this bibliography.
   - A list of institutions and the number of papers and reports from these institutions.
   - The number of items of each type of publication. "Collection" denotes

an edited book containing a collection of papers. The bibliography contains all entries of the General Systems Depository of the IFSR, i.e., unpublished papers that are available from the IFSR. These papers are labeled "IFSR-Depository."

- The number of items for each year of publication.

2. An Author Index, listing for each author the titles and the reference numbers in the Bibliographical List for which she/he is one of the authors (first author or coauthor).

3. A KWIC-Index, containing a listing of the title of each item, ordered by the key words in the title. Additionally, the name of the first author and the reference number in the Bibliographical List are given.

4. The Bibliographical List, ordered by the name of the first author and the year of publication. It gives complete bibliographical information for each item. If the language of the reference is not English, an English translation of the title is given and the language is denoted in brackets. Papers and reports available from the General Systems Depository of the IFSR are marked "IFSR-Depository."

The bibliography is implemented in a data base, representing a network of items, authors, journals, publishers, institutions, types, and years. It has been implemented on a Prime computer, running completely in INTERLISP. The data were skillfully input by Ulli Stadler and especially Gabi Wimmer.

This project was supported by the Austrian Federal Ministry for Science and Research, project 11.859/41/31/83. We are especially grateful to the Federal Minister for Science and Research; Dr. Heinz Fischer, the chief of its Department III/1; Dr. Edith Fischer; and the officer-in-charge, Dr. Elisabeth Brandstötter, for this most welcome support.

Although we prepared this bibliography with all possible care, we are sure we have overlooked errors and that we may have unintentionally omitted valuable sources. Your help toward improving this bibliography, especially regarding updates, is therefore greatly appreciated. Please contact us in care of IFSR.

We hope you will find this bibliography a valuable resource.

Robert Trappl
Werner Horn
George J. Klir

STATISTICAL INFORMATION

List of Journals

ACM Communications	8
ACM Computing Surveys	3
ACM Journal	2
ACM Trans. on Database Systems	3
Acta Informatica	5
Acta Polytechnica Scandinavica	4
American Naturalist	2
American Scientist	2
Animal Behavior	1
Annals of Systems Research	6
Annual Review of Ecology and Systematics	1
Applied Mathematical Modelling	2
Artificial Intelligence	3
Automatica	1
Automation and Remote Control	2
Behavioral Science	15
Bell System Technical Journal	1
Bio-Systems	2
Biological Cybernetics	1
Biometrics	1
Biometrika	1
Biosystems	2
Brit. J. for the Philosophy of Science	3
Bulletin of Mathematical Biology	7
Busefal	3
Canadian Psychological Review	1
Chronobiologia	1
Computer Journal	3
Computer-Aided Design	1
Concepts	1
Cybernetica	10
Cybernetics and Systems	80
Dialectica	1
Digital Processes	4
Ecological Modelling	16
Ecology	1
Ekistics	1
Ekonomicko-matematicky obzor	1
Elektronische Informationsverarbeitung u.Kybernetik	2
Engineering Cybernetics	7
Engineering Education	2
Ergonomics	1
Erkenntnis	1
Family Process	2
Fuzzy Sets and Systems	23
General Systems Bulletin	1
General Systems Yearbook	7
Heart and Lung	1
Human Systems Management	4
IBM Journal of Research and Development	2
IEEE Proc.	3
IEEE Trans. on Automatic Control	5
IEEE Trans. on Circuits and Systems	1
IEEE Trans. on Computers	3
IEEE Trans. on Information Theory	7
IEEE Trans. on Pattern Analysis and Machine Intelligence	4
IEEE Trans. on Reliability	1
IEEE Trans. on Systems, Man, and Cybernetics	53
Idealistic Studies	1
Information Processing Letters	2
Information Sciences	12

Information Systems	1
Information and Control	3
Inquiry	1
Int. J. of Bio-Medical Computing	2
Int. J. of Computer and Information Sciences	2
Int. J. of General Systems	87
Int. J. of Man-Machine Analysis	1
Int. J. of Man-Machine Studies	22
Int. J. of Man-Machine Systems	1
Int. J. of Systems Science	23
Interdisciplinary Science Reviews	2
Interfaces	2
J. for General Philosophy of Science	2
J. of Aircraft	1
J. of Applied Systems Analysis	18
J. of Computational and Applied Mathematics	1
J. of Computer and Systems Sciences	4
J. of Cybernetics	80
J. of Enterprise Management	2
J. of Graph Theory	2
J. of Information Processing	1
J. of Management Studies	1
J. of Mathematical Analysis and Applications	2
J. of Mathematical Biology	3
J. of Mathematical Psychology	1
J. of Mathematical Sociology	2
J. of Philosophy	1
J. of Social and Biological Structures	3
J. of Structural Learning	1
J. of Symbolic Logic	1
J. of Systems Engineering	1
J. of Theoretical Biology	1
J. of the Franklin Institute	2
J. of the Operations Research Society	2
Jurimetrics Journal	1
Kybernetes	34
Kybernetik	1
Kybernetika	2
Large Scale Systems	5
Management Science	6
Mathematical Biosciences	1
Mathematical Modelling	1
Mathematical Social Sciences	1
Mathematical Sociology on Social Network Research	1
Mathematical Systems Theory	6
Modeling, Identification and Control	1
Nature and Systems	4
Nature	3
Networks	1
Nursing Resources, Wakefield, Mass.	1
Ohio Journal of Science	1
Origins of Life	2
Pattern Recognition	4
Perspectives in Computing	1
Pfluegers Archiv	1
Philosophy of Science	3
Philosophy of Social Sciences	1
Policy Analysis and Information Systems	1
Postepy Cybernetyki	5
Prakseologia	1
Praxiology	3
Psychological Record	1
Radio and Electronic Engineer	2
Regelungstechnik	1

List of Publishers

Abacus Press, Tunbridge Wells, England	3
Ablex, Norwood, N.J.	3
Abt Books, Cambridge, Mass.	2
Academic Press, London, and New York	60
Adam Hilger, Bristol	1
Addison-Wesley, Reading, Mass.	27
Aerial Press, Santa Cruz, Calif.	1
Affiliated East-West Press, New Dehli / Madras	1
Akademie-Verlag, Berlin	4
Akten des 6. Int. Wittgenstein Symposiums. Ä	0
Alfred A. Knopf, New York	1
Ann Arbor Science, Ann Arbor, Mich.	3
Applied Science Publishers, Barking, England	3
Athenaeum, Koenigstein	1
Athlone-Press, London	1
B.G. Teubner, Stuttgart	1
Basic Books, New York	2
Basil Blackwell, Oxford	4
Beacon Press, Boston	1
Birkhaeuser Verlag, Basel and Stuttgart	4
Blackie, Glasgow	1
Blackwell, Oxford	1
Brunner/Mazel, New York	1
Butterworths, London	2
Cambridge Univ. Press, Cambridge, Mass.	12
Campus Verlag, Frankfurt and New York	2
Canadian Futures, Toronto	1
Chapman and Hall, London	1
Charles C. Thomas, Springfield, Ill.	1
Clarendon Press, Oxford	3
Columbia Univ. Press, New York	3
Computer Science Press, Woodland Hills, Calif.	1
Crane Russak, New York	1
D.Reidel, Dordrecht, Holland, and Boston	17
Digital Press, Bedford, Mass.	1
Dowden, Hutchinson and Ross, Stroudsburg, Penn.	3
Dryden-Press, New York	1
Dunker & Humbolt, Berlin	1
Edinburgh Univ. Press, Edinburgh	2
Edward Arnold, London	1
Elsevier / North-Holland, New York	9
Energoizdat, Moscow	1
Enslow, Hillside, N.J.	1
Entropy, Lincoln, Mass.	2
Erlbaum, Hillsdale, N.J.	1
Franklin Institute Press, Philadelphia	1
Franz Steiner Verlag, Wiesbaden	1
Garland Stpm Press, New York	1
Geest & Portig, Leipzig	1
Gordon and Breach, New York, and London	9
Halstead, New York	3
Harper and Row, New York	3
Harvard University Press, Cambridge, Mass.	1
Hayden, Rochelle Park, New Jersey	2
Hemisphere, Washington, D.C.	11
Heyden, Philadelphia	3
Hodder / Stoughton, London	1
Hoelder-Pichler-Tempsky, Wien	1
Holden-Day, San Francisco	3
Humanities Press, Atlantic Highlands, N.J.	1
IEEE, Piscataway, N.J.	9

IPC Science and Technology Press, Guildford, Surrey, England	1
ISEM, Copenhagen and Elsevier, Amsterdam	1
ISI Press, Philadelphia	1
Information Resources Press, Washington, D.C.	1
Infotech International, Maidenhead, Berkshire, UK	1
International Cooperative Publ. House, Burtonsville, Maryland	1
Intersystems, Seaside, Ca.	4
Ist, New York	1
John Hopkins University Press, Baltimore	2
John Wiley, Chichester and New York	78
Jossey-Bass, San Francisco	2
Karger, Basel	1
Kluwer-Nijhoff, Boston and The Hague	1
Kugl. Vetenskapsoch Vitterhets-Samhallet, Goteborg, Sweden	1
Lexington (Heath), Lexington, Mass.	1
Little, Brown and Company, Boston	1
M.I.T. Press, Cambridge, Mass.	10
Maloine, Paris	1
Mansell, London	1
Marcel Dekker, New York	9
Martinus Nijhoff, Boston and The Hague	10
McGraw-Hill, New York	15
Mir Publishers, Moscow	1
National Behavior Systems, Granada Hills, Calif.	1
Nauka, Moscow	3
Naukova Dumka, Kiev	1
Nelson, London	1
New York Univ. Press, New York	1
Noordhoff, Leiden, Holland	1
North-Holland, Amsterdam, and New York	32
Oxford Univ. Press, Oxford and New York	1
Panstwowe Wydawnictwo Naukowe, Warsaw	2
Paul Elek, London	1
Penguin Books, London	2
Pergamon Press, Oxford and New York	19
Peter Peregrinus, London and New York	1
Petrocelli, New York	6
Philosophical Library, New York	1
Physica-Verlag, Wuerzburg, Germany	1
Pion, London and Methuen, New York	1
Pitman, London and Boston	3
Plenum Press, New York and London	19
Polish Sci. Publ., Warsaw	1
Praeger, New York	2
Prentice-Hall, Englewood Cliffs, New Jersey	19
Princelet Editions, London	1
Princeton University Press, Princeton, N.J.	4
Profil, Linkoping, Sweden	1
Progress Publishers, Moscow	1
Quadrangle, New York	1
Research Studies Press (A division of John Wiley), New York	2
Reston/Prentice-Hall, Reston, Virginia	1
Routledge & Kegan Paul, London	2
Sage, Beverly Hills, Ca.	2
Science Press, Beijing	2
Sijthoff and Nordhoff, Alphen aan den Rijn (The Nederlands)	1
Simon and Schuster, New York	1
Sovietskoye Radio, Moscow	2
Springer-Verlag, Berlin, FRG, and New York	67
Stanford Univ. Press, Stanford, Ca.	1
Tavistock, London	1
Taylor & Francis, London	2
Tioga Publishing Co., Palo Alto, Calif.	1
Uitgewejij Stabo/All Round B.V., Groningen	1

Univ. of Chicago Press, Chicago	1
Univ. of Illinois Press, Urbana, Ill.	1
Univ. of Michigan Press, Ann Arbor, Mich.	1
Univ. of Minnesota Press, Minneapolis	2
Univ. of Texas Press	1
Univ. of Washington Press, Seattle, Wash.	1
University Park Press, Baltimore	1
University Press of Hawai, Honolulu	1
University Science Books, Mil Valley, Calif.	1
Van Nostrand Reinhold, New York	8
Vanderhoek-Ruprecht, Goettingen	1
Vantage Press, New York	1
Viking Press, New York	1
W.H. Freeman, San Francisco	7
Western Periodicals, North Hollywood, Ca.	1
Westview Press, Boulder, Colorado	3
William Kaufmann, Los Altos, Calif.	3
Yale Univ. Press, New Haven, Conn.	2
Yourdon, New York	2
Znaniye Press, Moscow	1

List of Institutions

Addiction Research Foundation, Toronto	1
Akademiai Kiado, Budapest	2
Asian Institute of Technology, Bangkok	2
Austrian Society for Cybernetic Studies, Vienna	22
Calif. State Univ., Long Beach, Calif.	9
California Polytechnic State Univ., San Luis Opisbo, Calif.	2
Centre for Agricultural Publishing and Documentation, Wageningen, Neth.	1
Computing and Systems Consultants, Binghampton, NY	1
Dep. of Psychiatry a. Human Behavior, Univ.of Calif. at Irvine	1
Department of Business and Economics, Katholieke	1
Dept. El. Eng., Univ. of Canterbury, Christchurch, New Zealand	1
Duke University, Durham, N.C.	1
Education and Training Consultants Co., Los Angeles, Box 49899, Ca. 90049	2
Eindhoven Univ. of Technology, Eindhoven, Netherlands	1
Electr. Inst., Techn. Univ. of Denmark, Lyngby	1
George Washington Univ., Maryland	1
Georgia State Univ., Atlanta	2
IIASA, Laxenburg, Austria	9
Indian Institute of Advanced Study, Simla, India	1
Inst. for the Information Society, Tokio, Japan	1
Institute of Sociology, Hungarian Academy of Science	1
Instituto Nacional De Estadistica, Madrid	1
Int. Association for Cybernetics, Namur	1
Int. Society for Ecological Modelling, Copenhagen	2
Johannes Kepler Univ., Linz, Austria	1
Johns Hopkins Univ., Baltimore	1
Katholieke Hogeschool, Tilburg, Netherlands	1
L. Etoevoes Univ. Dept. of Behavior Genetics, Goed, Hungary	1
Laboratoire de Biomathematiques, Faculte de Med. d'Angers, Angers, France	1
Laboratory for Computer Graphics and Spatial Analysis, Harvard Univ.	1
Mathematisch Centrum, Amsterdam	1
Nat.Res.Inst. for Mathematical Sciences, Pretoria, SA	1
Naval Research Lab. Memo.	1
New York Academy of Sciences, New York	1
Open Univ., Milton Keynes, England	1
Oregon State Univ., Corvallis, Oregon	3
Philosophy of Science Asscociation, East Lansing, Mich.	1
Polish Academy of Sciences, Warsaw	9
Regional Engineering College, Kurukshetra, India	1
Rozvojni Center Celje, Celje, Yugoslavia	1
SGSR, Louisville, Kentucky	14
SIAM, Philadelphia	1
SNTL, Prague	1
School of Advanced Technology, SUNY-Binghampton, Binghampton, NY	1
Simulation Council, La Jolla, Ca.	1
State University of New York, Binghamton, New York	1
Systems Science Institute, Univ. of Louisville, KY	1
Technical Univ. of Wroclaw, Poland	2
Univ. De Droit, Aix-en-Provence, France	2
Univ. della Calabria, Italy	1
Univ. of Alberta, Edmonton, Canada	1
Univ. of Calif. at Santa Cruz	1
Univ. of California at Berkeley, Berkeley, California	1
Univ. of Essex, Colchester, England	1

Univ. of Helsinki	1
Univ. of Illinois At Chicago Circle, Dept. of Information Engineering	1
Univ. of Lancaster, Centre in Simulation	1
Univ. of Louisville, Louisville, Kentucky	1
Univ. of Tampere, Finland	1
Univ. of Technology, Delft, Netherlands	1
Univ. of Texas at San Antonio	1
Univ. of Vienna, Dept. of Medical Cybernetics and AI	1
Vrije Universiteit, Amsterdam, Netherlands	2
Yale University	1
Yeshiva Univ., Manhattan, NY	1

List of Types of Entries

Paper	584
Bibliography	8
Book	461
Collection	189
Dissertation	1
Overview	3
Proceedings	26
Report	35
Series	11
Special-Issue	7
IFSR-Depository	55

Publication Year

1963	1
1964	1
1965	2
1968	1
1969	6
1970	3
1971	2
1972	8
1973	10
1974	15
1975	37
1976	52
1977	199
1978	204
1979	239
1980	288
1981	268
1982	353
1983	62
1984	175
1985	18

AUTHOR INDEX

Aarnes, E.:
    On the Problem of Identification in Compartment
        Analysis.   [1]
Abe, J.:
    Schema Representation of 'Speakers-World' in Natural
        Language Processing.   [2]
Abraham, R.:
    Dynamics: The Geometry of Behavior.   [3]
Acalugaritei, G.:
    Hierarchies of Evolutions and Stabilities.   [4]
Acar, W.:
    Bureaucracy as Organizational Pathology.   [5]
Achinstein, P.:
    Can There Be a Model of Explanation?   [6]
Ackerman, E.:
    Mathematical Models in the Health Sciences: a Computer
        Approach.   [7]
Ackoff, R.L.:
    Resurrecting the Future of Operational Research.   [8]
    The Future of Operations Research is Past.   [9]
    The Art and Science of Mess Management.   [10]
Adam, A.:
    Discrete Systems Methodology.   [11]
Adam, N.R.:
    Current Topics in Computer Simulation.   [12]
Adams, W.M.:
    Hiragana and Katakana Syllabaries: Relative
        Information Content.   [13]
Adulbhan, P.:
    Proceedings of the Int. Conference on Systems
        Modelling in Developing Countries.   [1536]
    Systems Models for Decision Making.   [1537]
Agaian, S.:
    On an Algorithm of Spectral Analysis.   [45]
Agassi, J.:
    Scientific Philosophy Today: Essays in Honour of Mario
        Bunge.   [14]
Agazzi, E.:
    System Theory and the Problem of Reductionism.   [15]
Aggarwal, J.K.:
    Nonlinear Systems: Stability Analysis.   [16]
Aggarwal, K.K.:
    An Event Expansion Algorithm for Reliability
        Evaluation in Complex Systems.   [643]
    An Overview of Computer-Orientied Learning Processes. [1597]
Aggarwal, S.:
    Deterministic Representation of Probabilistic Systems
        by Ergodic Machines.   [17]
Agrawal, V.D.:
    An Information Theoretic Approach to Digital Fault
        Testing.   [18]
Aguilo, J.:
    Modularity Grade of the Neural Networks Synthesized
        with Z-Fivex Formal Neurones.   [19]
Agusti, J.:
    Modularity Grade of the Neural Networks Synthesized
        with Z-Fivex Formal Neurones.   [19]
Aho, A.V.:
    The Theory of Joins in Relational Data Bases.   [20]
Airaksinen, T.:
    Systems Theory in Social Sciences.   [21]
    Normative Aspects of Some Stabilizing Social Systems.   [22]
Airenti, G.:
    A Two Level Model of Knowledge and Belief.   [23]

Albegov, M.:
  An Approach to Distributed Modeling.   [1769]
Aleo, F.:
  Analysis of the Purine Biosynthesis by Mathematical
    Logic.   [1405]
Allen, P.M.:
  Self-Organization and Dissipative Structures:
    Applications in the Physical and Social Sciences. [1507]
Allen, R.:
  Signal Analysis of Visual Evoked Potentials.   [838]
Allen, T.F.H.:
  Hierarchy: Perspectives for Ecological Complexity.   [24]
Alspach, B.:
  Algorithmic Aspects of Combinatorics.   [25]
Alter, S.L.:
  Decision Support Systems: Current Practice and
    Continuing Challenges.   [26]
Altmann, A.:
  Interdisciplinary System Analysis.   [27]
Amagasa, M.:
  Structural Modeling in a Class of Systems Using Fuzzy
    Sets Theory.   [1675]
  Structural Modeling with Consensus for Societal
    Systems Using Fuzzy Set Theory.   [1676]
Ambrus, S.Z.:
  Runoff Simulation with a Random Walk Model.   [967]
Amey, G.X.:
  Information Systems and Networks.   [1486]
Andersen, E.B.:
  Discrete Statistical Models with Social Science
    Applications.   [28]
Anderson, B.D.O.:
  Foundations of Systems Theory: Finitary and Infinitary
    Conditions.   [29]
  Partial Prior Information and Decision-Making.   [1355]
Anderson, J.:
  Construction of an Expert Therapy Adviser as a Special
    Case of a General System Protection Design.   [966]
Anderson, R.B.:
  Proving Programs Correct.   [30]
Anderson, T.W.:
  A Bibliography of Multivariate Statistical Analysis. [31]
Andersson, H.:
  Analysis of an Inventory Control System Using the
    Theory of Relatively Closed Systems.   [32]
Anderton, R.:
  Systems in the Seventies: an Emerging Discipline?   [33]
Andre, C.:
  Systematic Method of Realization of Interpreted Petri
    Nets.   [70]
Andreae, J.H.:
  A Dual Model of the Brain.   [34]
  The Competence of a Multiple Context Learning System. [1078]
Andrew, A.M.:
  Autopoiesis and Self-Organization.   [35]
  Logic and Continuity - a Systems Dichotomy.   [36]
  Information Transmission on Nerves.   [37]
Andrews, J.G.:
  Mathematical Modelling.   [38]
Aplin, J.C.:
  An Evaluation of the Effectiveness of Dialectical
    Inquiry Systems.   [349]
  A Cybernetic Model of the MBO Process.   [39]
Apostolakis, G.:

      Synthesis and Analysis Methods for Safety and
         Reliability Studies (NATO Advanced Study Institute).   [40]
Apostolico, A.:
      On the Role of Information and Hierarchy in the
         Modeling of Biomolecular Systems.   [41]
Apple, A.:
      Megasynthesis.   [42]
Apter, M.J.:
      On the Concept of Bistability.   [43]
Aracil, J.:
      Structural Stability of Low-Order System Dynamics
         Models.   [44]
Arakelian, A.:
      On an Algorithm of Spectral Analysis.   [45]
Arbel, A.:
      Aggregation and Information Structuring in Large-Scale
         Dynamic Systems.   [46]
Arbib, M.A.:
      Foundations of Systems Theory: Finitary and Infinitary
         Conditions.   [29]
      Computers and the Cybernetic Society.   [47]
      Foundations of System Theory: The Hankel Matrix.   [48]
      Cooperative Computation and the Cybernetic Society.   [49]
Archer, S.M.:
      Influence of Sensor Failures on Log Regulators.   [1647]
Arhire, R.I.:
      System of Indicators for Estimating the Software
         Performances.   [817]
Arigoni, A.C.:
      Algebraic Structure of Property Spaces - On the
         Synthesis of Formal Properties.   [50]
      Correctness of the Semiotic Transformation of
         Information.   [51]
Aripov, K.N.:
      Medical Monitoring of Man's Reliability during Cosmic
         Flights.   [1929]
Aris, R.:
      Mathematical Modelling Techniques.   [52]
Arora, P.N.:
      Shannon's Entropy and Cyclicity.   [53]
      Directed Divergence and Cyclic Symmetry.   [54]
Asai, K.:
      A Formulation of Fuzzy Decision Problems with Fuzzy
         Information Using Probability Measures of Fuzzy
         Events.   [1261]
Ashby, R.:
      Mechanisms of Intelligence: Ross Ashby's Writings on
         Cybernetics.   [55]
Asquith, P.D.:
      PSA 1976.   [1639]
Astola, J.:
      The Lee-Scheme and Bounds for Lee-Codes.   [56]
Atanassov, K.:
      Intuitionstic L-Fuzzy Sets.   [57]
Athanassov, A.:
      A System for Training of EDP-Specialists for the
         National Economy.   [58]
Atherton, D.P.:
      Stability of Nonlinear Systems.   [59]
Atherton, D.R.:
      Nonlinear Control Engineering.   [60]
Atkin, R.:
      Polyhedral Dynamics and the Geometry of Systems.   [61]
      Multidimensional Man.   [62]

Atkin, R.H.:
    Time as a Pattern on a Multi-Dimensional Structure.   [63]
Atlan, H.:
    Random Boolean Networks.   [64]
    Information Theory.   [65]
Atre, S.:
    Data Base: Structured Techniques for Design,
        Performance and Management.   [66]
Auble, G.T.:
    Systems Approach to the Concept of Niche.   [1293]
    System Theory and the Ecological Niche.   [1294]
Auger, P.:
    Coupling between N Levels of Observation of a System
        (Biological or Physical) Resulting in Creation of
        Structures.   [67]
    Interactions Between Collective and Individual Levels
        of Organization in a Hierarchically Organized
        System.   [68]
    Order, Disorder in Hierarchically Organized Systems.   [69]
Auguin, M.:
    Systematic Method of Realization of Interpreted Petri
        Nets.   [70]
Augustine, N.R.:
    Augustine's Laws and Major System Development Programs.   [71]
Aulin, A.:
    On the Foundations of Systems Theory.   [72]
    The Cybernetic Laws of Social Progress: Towards a
        Critical Social Philosophy and a Criticism of
        Marixsm.   [73]
Aulin-Ahmavaara, A.:
    The Impossibility of Genuinly Self-Steering Machines:
        A Fundamental Theorem of Actor-Systems.   [74]
Aupperle, K.E.:
    Bureaucracy as Organizational Pathology.   [5]
Averkin, A.N.:
    Generalized Strategies in Problem Solving.   [75]
Baase, S.:
    Computer Algorithms.   [76]
    Computer Algorithms: Introduction to Design and
        Analysis.   [77]
Babloyantz, A.:
    Self-Organization in Biological Systems with Multiple
        Cellular Contacts.   [78]
Bahm, A.J.:
    Five Types of Systems Philosophy.   [79]
    Methodologies of Five Types of Philosophy.   [80]
    Holons: Three Conceptions.   [81]
Bailey, K.D.:
    Equilibrium, Entropy and Homeostasis: A
        Multidisciplinary Legacy.   [82]
Bailey, N.T.J.:
    Some Issues in the Cybernetics of Disease Control.   [83]
Baldwin, J.F.:
    Modelling Controllers Using Fuzzy Relations.   [84]
    Dynamic Programming for Fuzzy Systems with Fuzzy
        Environment.   [85]
Balkus, K.:
    Systems in Sociocultural Development.   [86]
    Organizational Determinants of National Development.   [87]
    Necessities as the Basis of the Social
        Self-Organization.   [88]
Ball, M.O.:
    Complexity of Network Reliability Computations.   [89]
Ballance, R.A.:

Functional Dependence and its Application to System
    Evaluation.   [90]
Ballard, D.H.:
    Computer Vision.   [91]
Ballonoff, P.A.:
    Mathematical Demography of Social Systems, II.   [92]
Balossino, N.:
    ECG Map Filtering by Means of Spherical Harmonics: A
    Simple Approximation and Results.   [93]
Baltes, P.B.:
    Life Span Development and Behavior, Vol.I.   [94]
Banathy, B.H.:
    Developing a Systems View of Education.   [95]
    Systems Science and Science. (Proc. 24th Annual
    Meeting of the Society for General Systems
    Research).   [96]
    Proceedings of the 29th Annual Meeting.   [97]
Bandler, W.:
    On the Theory of Lorrain-Categories.   [98]
    Fuzzy Power Sets and Fuzzy Simplification Operators.   [99]
    Hi & Low - In & Out: Approaches to Social Status. [1512]
    Semantics of Implication Operators and Fuzzy
    Relational Products.   [100]
    The Valuton: An Automation Model of a Sociological
    Theory of Value Genesis.   [101]
    Construction of an Expert Therapy Adviser as a Special
    Case of a General System Protection Design.   [966]
    The Four Modes of Inference in Fuzzy Expert Systems.   [102]
Banerji, R.B.:
    Theoretical Approaches to Non-Numerical Problem
    Solving.   [103]
Baptistella, L.F.B.:
    Fuzzy Methodologies for Interactive Multicriteria
    Optimization.   [104]
Bara, B.G.:
    A Two Level Model of Knowledge and Belief.   [23]
Barber, M.C.:
    A Markovian Model for Ecosystem Flow Analysis.   [105]
Barenblatt, G.I.:
    Similarity, Self-Similarity and Intermediate
    Asymptotics.   [106]
Barfoot, C.B.:
    Aggregation of Conditional Absorbing Markov Chains.   [107]
Barmark, J.:
    Perspectives in Metascience.   [108]
Barnes, B.H.:
    Decision Table Languages and Systems.   [1155]
Barnett, S.:
    Categories in Systems and Control Theory.   [422]
Barr, A.:
    The Handbook of Artificial Intelligence (Vol.1).   [109]
    The Handbook of Artificial Intelligence (Vol.2).   [110]
Barrett, M.:
    A System Approach to Alcohol Treatment.   [622]
Barto, A.G.:
    A Note on Pattern Reproduction in Tessellation
    Structures.   [111]
    Discrete and Continuous Models.   [112]
Bartolini, G.:
    Development of Performance Adaptive Fuzzy Controllers
    with Application to Continuous Casting Plants.   [113]
Basta, D.J.:
    Analyzing Natural Systems: Analysis for Regional
    Residuals-Environmental Quality Management.   [114]

Batchelder, W.H.:
    A Critical Examination of the Analysis of Dichotomous
        Data.   [115]
Batchelor, B.G.:
    Experimental and Programmatic Approaches to Pattern
        Recognition.   [116]
    Pattern Recognition: Ideas in Practice.   [117]
    The Classification of Pole Maps in the Complex Plane.   [118]
    Hierarchical Shape Description Based Upon Convex Hulls
        of Concavities.   [119]
    Shape Description for Labeling Concavity Trees.   [120]
    Two Methods for Finding Convex Hulls of Planar Figures.   [121]
    Converting Run Code to Chain Code.   [122]
Bathe, K.J.:
    Formulations and Computational Algorithms in Finite
        Element Analysis: U.S.-German Symposium.   [123]
Battista, J.R.:
    The Holistic Paradigm and General Systems Theory.   [124]
Bauer, M.A.:
    Defining Structural Descriptions.   [125]
Bavel, Z.:
    Math Companion for Computer Science.   [126]
Bayraktar, B.A.:
    Education in Systems Science.   [127]
Beaumont, M.J.:
    AESOP: an Architectural Relational Data Base.   [1333]
Beck, M.B.:
    Hard or Soft Environmental Systems?   [128]
    Uncertainty and Arbitrariness in Ecosystem Modelling:
        A Lake Modelling Example.   [515]
Becker, P.W.:
    On Systems with Redundancy and their Inherent Bounds
        on Achievable Reliability.   [129]
    On Systems with Redundancy and Their Inherent Bounds
        on Achievable Reliability.   [130]
    On Systems' Reliability and Redundancy.   [131]
Beckett, J.A.:
    Management: The Case of the Disappearing Model.   [132]
Bednarek, A.R.:
    Dynamical Systems II.   [133]
Beer, S.:
    Brain of the Firm (second edition).   [134]
    Death is Equifinal: Eighth Annual Ludwig Von
        Bertalanffy Memorial Lecture.   [135]
    The Heart of Enterprise.   [136]
    Introduction: Questions of Quest.   [137]
    Recursions of Power.   [138]
Beeri, C.:
    The Theory of Joins in Relational Data Bases.   [20]
Beights, H.W.:
    An Algorithm for Converting Signed Digraphs to
        Forrester Schematics.   [249]
Beishon, J.:
    Learning about Systems.   [139]
Bekey, G.A.:
    Models and Reality: Some Reflections on the Art and
        Science of Simulation.   [140]
Bekjarov, E.:
    A System for Training of EDP-Specialists for the
        National Economy.   [58]
Bell, D.A.:
    Physical Record Clustering in Databases.   [141]
Bell, D.E.:
    Conflicting Objectives in Decisions.   [142]

Bellman, R.:
  Introduction to Mathematical Theory of Control
    Processes: Nonlinear Processes.   [143]
Bello, J.A.:
  Leadership as a Task Process Uncertainty Control
    Process.   [1084]
Belnap, N.P.:
  The Logic of Questions and Answers.   [144]
Beloglavec, E.:
  A General Systems Theory Approach to the Properties of
    a Periodic Discrete String.   [145]
Belov, K.:
  Valuation and Control.   [146]
Belova, N.:
  Valuation and Control.   [146]
Beltrami, E.J.:
  Models for Public Systems Analysis.   [147]
Ben-Dov, Y.:
  Optimal Testing Procedures for Special Structures of
    Coherent Systems.   [148]
Ben-Eli, M.U.:
  Amplifying Regulation and Variety Increase in Evolving
    Systems.   [149]
Bender, E.A.:
  An Introduction to Mathematical Modelling.   [150]
Benedikt, S.:
  Non-Repetitive Decision Making under Risk.   [151]
Bennett, P.G.:
  Hypergames as a Soft Systems Methodology.   [800]
Bennett, R.J.:
  Environmental Systems: Philosophy, Analysis and
    Control.   [152]
Bennett, S.:
  An Introduction to Multivariate Techniques for Social
    and Behavioral Sciences.   [153]
Bennett, T.J.A.:
  Language as a Self-Organizing System.   [154]
Benningfield, L.M.:
  Circuit and Systems Theory.   [155]
Benoit, B.M.:
  Fractals: Form, Chance, and Dimension.   [156]
Benseler, F.:
  Autopoiesis, Communications, and Society: The Theory
    of Autopoietic System in the Social Sciences.   [157]
Bensoussan, A.:
  New Trends in Systems Analysis: International
    Symposium, Versailles 1976.   [158]
  International Symposium on Systems Optimization and
    Analysis.   [159]
  Analysis and Optimization of Systems. (Proc. of a
    Conference in Versailles, France, Dec.1980).   [160]
Beran, H.:
  Discrete Systems Methodology.   [11]
Bergeron, R.J.:
  Imaginary, Fluid, and Dancing Systems.   [161]
Bernard, N.:
  Multiensembles, a Contribution to the Search of Fuzzy
    Sets Theory.   [162]
Bertoni, A.:
  Analysis and Compacting of Musical Texts.   [163]
Best, D.P.:
  A Case Study in the Implementation of a Cybernetic
    Organisation Structure in a Medium Sized UK Company. [164]
  A Cybernetic Approach to the Application of

Information Technology.   [165]
Bezdek, J.C.:
  Fuzzy Partitions and Relations: an Axiomatic Basis for
    Clustering.   [166]
  Patterns Recognition with Fuzzy Objective Function
    Algorithms.   [167]
Bibbero, R.J.:
  Microprocessor System, Interfacing and Applications.   [168]
Bilcio, C.:
  Modern Trends in Cybernetics and Systems (3 Vols.). [1436]
Binder, D.A.:
  Bayesian Cluster Analysis.   [169]
Binder, Z.:
  Stabilization, Co-ordination and Structuralization in
    the Integrated Automation of Complex Systems. [170]
Biswas, A.K.:
  Mathematical Modelling and Environmental
    Decision-Making.   [171]
Blackie, M.J.:
  Systems Simulation in Agriculture.   [410]
Blake, I.F.:
  Computer Security and Protection Structures.   [1829]
Blaquire, A.:
  Dynamical Systems and Microphysics. (Seminar in
    Udine/Italy, Sept.1979).   [172]
Blauberg, I.V.:
  Systems Theory: Philosophical and Methodological
    Problems.   [173]
  The Systemic Approach: Prerequisites, Problems and
    Difficulties.   [174]
Blomberg, H.:
  Algebraic Theory for Multivariable Linear Systems. [175]
Blunden, M.:
  Military Technology and Systemic Interaction.   [176]
  Systems and Management - The Political Dimension. [177]
Bluth, B.J.:
  Parson's General Theory of Action. A Summary of the
    Basic Theory.   [178]
Bobrow, D.G.:
  Representation and Understanding: Studies in Cognitive
    Science.   [179]
Bocklisch, S.F.:
  Data- and Behaviour Analysis with Classification
    Models.   [1321]
Boden, M.:
  Artificial Intelligence and Natural Man.   [180]
Boeri, F.:
  Systematic Method of Realization of Interpreted Petri
    Nets.   [70]
Bogart, D.H.:
  Feedback, Feedforward, and Feedwithin: Strategic
    Information in Systems.   [181]
Bogdanski, Ch.:
  Basic Elements of Cybernetic Physics.   [182]
Bogner, S.:
  A Cybernetic Model of Cognitive Processes.   [183]
Bohm, D.:
  Wholeness and the Implicate Order.   [184]
Bolter, M.:
  A Statistical Scheme for Structural Analysis in Marine
    Ecosystems.   [185]
Boltyanskii, V.G.:
  Optimal Control of Discrete Systems.   [186]
  Optimal Control of Discrete Systems. (Translated From

        Russian by Ron Hardin.)   [187]
Boncina, R.:
    An Open System Composed of Non-Active Monotonous Parts
        in Passive Surroundings.   [188]
Bonczek, R.H.:
    Foundations of Decision Support Systems.   [189]
Bonisonne, P.P.:
    A Linguistic Approach to Decision Making with Fuzzy
        Sets.   [1701]
Boots, B.:
    Models of Spatial Processes.   [601]
Borisov, A.N.:
    Analysis of Fuzzy Initial Information in
        Decision-Making Models.   [1582]
Borisyuk, G.N.:
    Short-Term Memory as a Metastable State. II.
        Simulation Model.   [974]
Borisyuk, R.M.:
    Short-Term Memory as a Metastable State. II.
        Simulation Model.   [974]
Borko, H.:
    Information Systems and Networks.   [1486]
Bornschein, H.:
    A Second Order Model of the Optic Generator Potential
        and its Relation to Stevens' Power Function.   [1710]
Borodin, A.:
    The Computational Complexity of Algebraic and Numeric
        Problems.   [190]
Boroush, M.A.:
    Technology Assessment: Creative Futures.   [191]
Bose, N.K.:
    Multidimensional Systems: Theory and Applications.   [192]
    Applied Multidimensional Systems Theory.   [193]
Bosserman, R.W.:
    The Role of Mathematics in Systems Science.   [194]
    Internal Security Processes and Subsystems in Living
        Systems.   [195]
    Internal Security Processes in General Living Systems. [196]
Botez, M.C.:
    Toward a Systemic Representation of General Systems:
        Open Multimodelling.   [197]
Bottaci, L.:
    A Relational Scheme for the Abstract Specification of
        Production System Programs Used in Expert Systems. [198]
Bottini, E.:
    Selective Polymorphism of the Erythrocyte Acid
        Phosphate Locus in Humans.   [625]
Boulding, K.E.:
    The Universe as a General System: Fourth Annual Ludwig
        von Bertalanffy Memorial Lecture.   [199]
    Ecodynamics: a New Theory of Societal Evolution.   [200]
    Human Knowledge as a Special System.   [201]
Bowen, B.D.:
    An Introduction to Data Analysis.   [202]
Bower, B.T.:
    Analyzing Natural Systems: Analysis for Regional
        Residuals-Environmental Quality Management.   [114]
Bowers, D.:
    An Introduction to Multivariate Techniques for Social
        and Behavioral Sciences.   [153]
Bowler, T.D.:
    General Systems Thinking: Its Scope and Applicability. [203]
Boyce, W.E.:
    Case Studies in Mathematical Modeling.   [204]

Boyd, G.M.:
  Cybernetic Aesthetics: Key Questions in the Design of
    Mutual Control in Education.   [205]
Bozic, S.M.:
  Digital and Kalman Filtering.   [206]
Bozinovski, S.:
  A Self-Learning System Using Secondary Reinforcement. [207]
  A Representation of Pattern Classification Teaching.   [208]
Braae, M.:
  Selection of Parameters for a Fuzzy Logic Controller. [209]
Bradley, R.:
  Case Studies in Mathematical Modelling: A Course Book
    for Engineers and Scientists.   [210]
Brams, S.J.:
  Applied Game Theory.   [211]
Brandt, A.:
  Using Nursing Research: Discovery, Analysis, and
    Interpretation.   [1755]
Branin, F.H.:
  Problem Analysis in Science and Engineering.   [212]
Braten, S.:
  Frontiers in Systems Research: Implications for the
    Social Sciences (a New Book Series) .   [941]
Bratko, I.:
  A Circuit Analysis Program that Explains its Reasoning. [565]
Brauer, W.:
  Net Theory and Applications. (Proc. of the Advanced
    Course on General Net Theory of Processes and
    Systems, Hamburg, 1979).   [213]
Brayton, R.K.:
  Stability of Dynamical Systems: a Constructive
    Approach.   [214]
Breitenecker, F.:
  Simulation and Analysis of Pathological Blood Pressure
    Behaviour after Treadmill Test in Patients with
    Coarctation of the Aorta.   [215]
  Socio-Economic Consequences of the Volterra Approach
    for Nonlinear Systems.   [1325]
Brewer, J.N.:
  Structure and Cause and Effect Relations in Social
    System Simulations.   [216]
Bridge, J.:
  Beginning Model Theory.   [217]
Brigham, E.O.:
  The Fast Fourier Transform.   [218]
Brightman, H.J.:
  Problem Solving: A Logical and Creative Approach.   [219]
Brightman, H.L.:
  Problem Solving: A Logical and Creative Approach.   [220]
Brillinger, D.R.:
  Time Series: Data Analysis and Theory.   [221]
Brinkmann, K.:
  Identification of the Temperature Masking of the
    Circadian System of Euglena Gracilis.   [847]
Brockett, P.L.:
  Information Theoretic Analysis of Questionnaire Data.   [222]
Broekstra, G.:
  Constraint Analysis and Structure Identification II.   [223]
  On the Representation and Identification of Structure
    Systems.   [224]
  On the Foundations of GIT (General Information Theory). [225]
  C-Analysis of C-Structures: Representation and
    Evaluation of Reconstruction Hypotheses by
    Information Measures.   [226]

MAMA: Management by Matching. A Consistency Model for
    Organizational Assessment and Change.   [227]
Brown, C.M.:
  Computer Vision.   [91]
Brown, D.B.:
  Systems Analysis and Design for Safety.   [228]
Brown, D.J.H.:
  A Task-Free Concept Learning Model Employing
    Generalization an Abstraction Techniques.   [229]
Brown, J.H.U.:
  Integration and Coordination of Metabolic Processes: a
    Systems Approach to Endocrinology.   [230]
Brown, K.C.:
  Role and Effectiveness of Theories of Decision in
    Practice.   [1862]
Brown, R.H.:
  Artificial Intelligence: An MIT Perspective (2 Vols.). [1871]
Bruckmann, G. (ed.):
  Input-Output Approaches in Global Modelling. (Proc.of
    the Fifth IIASA Symposium on Global Modelling).   [231]
Brugger, K.:
  A Simple Mathematical Model of Biphasic Insulin
    Secretion.   [232]
Bruni, C.:
  Systems Theory in Immunology.   [233]
Bryant, D.T.:
  Manpower Planning and Organization Design.   [234]
  Current Status of Hospital Management Information
    Systems.   [793]
Bubnicki, Z.:
  Global and Local Identification of Complex Systems
    with Cascade Structure.   [235]
  Systems Science. (A Journal Published Quarterly.)   [236]
Buchberger, B.:
  Input/Output Codings and Transition Functions in
    Effective Systems.   [237]
Buchberger, E.:
  VIE-LANG: A German Language Understandig System.   [238]
  VIE-LANG: A German Language Dialogue System.   [1753]
  On Generation of Anaphora and Gapping in German.   [781]
Buchstaller, W.:
  Interactive Computer-Aided Planning: Estimating
    Health-Care Resource Requirements.   [782]
  Health Care in Austria: 1980-2000.   [783]
  The Structure of Manifestations in a Medical
    Consultation System.   [785]
Bunge, M.:
  General Systems and Holism.   [239]
  A Systems Concept of Society: Beyond Individualism and
    Holism.   [240]
  Analogy between Systems.   [241]
Bunn, D.W.:
  Formal Methods in Policy Analysis.   [242]
Burghes, D.N.:
  Mathematical Models in the Social, Management and Life
    Sciences.   [243]
Burgin, M.S.:
  Products of Operators of Multidimensional Structured
    Models of Systems.   [244]
Burkhardt, H.:
  The Real World and Mathematics.   [245]
Burkheiner, G.J.:
  Response Models for Detection of Change.   [1382]
Burks, A.W.:

Models of Deterministic Systems.   [246]
Burns, D.W.:
    The Synthesis of Forecasting Models in Decision
        Analysis.   [247]
Burns, J.R.:
    A Sector Approach to the Formulation of Systems
        Dynamics.   [248]
    An Algorithm for Converting Signed Digraphs to
        Forrester Schematics.   [249]
    An Input/Output Approach to the Structural Analysis of
        Digraphs.   [250]
Burton, D.J.:
    Methodology and Epistemology for Second-Order
        Cybernetics.   [251]
Busch, G.M.:
    Cybernetics IV: A System-Type Applicable to Human
        Groups.   [252]
Busch, J.A.:
    Cybernetics IV: A System-Type Applicable to Human
        Groups.   [252]
Buss, A.R.:
    The Role of General Systems and Information Theory in
        Multifactor Individuality Theory.   [1453]
Butzer, K.W.:
    Civilizations: Organisms or Systems?   [253]
Byatt, W.J.:
    Fuzzy Sets, Fuzzy Algebra, and Fuzzy Statistics.   [873]
    Fuzzy Processes.   [875]
Byerly, H.:
    Teleology and Evolutionary Theory Mechanisms and
        Meanings.   [254]
Caianiello, E.R.:
    A Systemic Study of Monetary Systems.   [255]
    Energetics Versus Communication in the Nervous System. [256]
Calhoun, G.:
    Systemic Inactivism: A Modern Laissez-Faire Thinking. [1469]
Callebaut, W.:
    The Distribution Approach to Problem-Solving in
        Science: Prospects for GSM.   [257]
Camana, P.C.:
    Determination of Feedback for Human Posture Control
        without Physical Intervention.   [258]
Campbell, J.:
    Grammatical Man: Information, Entropy, Language, and
        Life.   [259]
Campbell, J.P.:
    Measurement Theory for the Behavioral Sciences.   [606]
Cannon, R.L.:
    A Generator for Models of Discrete-Event Systems.   [1632]
Capocelli, R.M.:
    A Cybernetic Approach to Population Dynamics Modeling. [260]
Caputo, R.S.:
    Establishing a Rational Energy Policy for Western
        Europe.   [261]
Carlsson, C.:
    A System of Problems and How to Deal with It.   [262]
    A System of Problems and a Way Treating it.   [263]
    An Approach to Handling Fuzzy Problem Structures.   [264]
Carter, C.F.:
    Uncertainty and Expectations in Economics: Essays in
        Honour of G.L.S. Shackle.   [265]
Carvallo, M.E.:
    In Search of the Noetic Structure of Systems.   [266]
    Chance and the Necessity of the Third Classical

    Paradigm.   [267]
Casalino, G.:
    Development of Performance Adaptive Fuzzy Controllers
      with Application to Continuous Casting Plants.   [113]
Caselles, A.:
    A Method to Compare Theories in the Light of General
      Systems Theory.   [268]
Castelfranchi, C.:
    A Lexicon Based Approach to Sentence Understanding.   [269]
Casti, J.:
    Illusion or Reality?: the Mathematics of Attainable
      Goals and Irreducible Uncertainties.   [270]
    Dynamical Systems and Their Applications: Linear
      Theory.   [271]
    Polyhedral Dynamics and the Geometry of Systems.   [61]
    Frontiers in Systems Research: Implications for the
      Social Sciences (a New Book Series) .   [941]
    Connectivity, Complexity and Catastrophe in
      Large-Scale Systems.   [272]
    Lake Ecosystems: a Polyhedral Dynamic Representation. [273]
    Topological Methods for Social and Behavioral Systems. [274]
Cavallo, R.E.:
    General Systems Methodology and Large-Scale Systems. [275]
    A Conceptual Foundation for Systems Problem Solving. [276]
    The Role of Systems Methodology in Social Science
      Research.   [277]
    Recent Developments in Systems Methodology for Social
      Science Research.   [278]
    Supplementary Material to the SGSR Report on Systems
      Research Movement.   [279]
    Sysems Research Movement: Characteristics,
      ¨Accomplishments, and Current Developments. (a
      Report Sponsored by the Society for General Systems
      Research.)   [280]
    Systems Research Movement: Characteristics,
      Accomplishments and Current Developments.   [281]
    The Structure of Reconstructable Relations: a
      Comprehensive Study.   [282]
    Reconstructability Analysis: Evaluation of
      Reconstruction Hypothesis.   [283]
    Reconstructability Analysis: Overview and Bibliography. [284]
    Decision Making in Reconstructability Analysis.   [285]
    Reconstruction of Possibilistic Behavior Systems.   [286]
    Systems Methodology in Social Science Research: Recent
      Developments.   [287]
Caws, P.:
    Coherence, System, and Structure.   [288]
Cellier, F.E.:
    Process in Modelling and Simulation.   [289]
Cerbone, G.:
    Mean Variance and Skewness of the First Passage Time
      for the Ornstein-Uhlenbeck Process.   [290]
Cesari, L.:
    Dynamical Systems II.   [133]
Chaitin, G.J.:
    Algorithmic Information Theory.   [291]
Chandra, A.K.:
    Intrinsically Difficult Problems.   [1621]
Chandrasekaran, B.:
    Natural and Social System Metaphors for Distributed
      Problem Solving.   [292]
Chandy, K.M.:
    Computer Systems Performance Modeling.   [1496]
Chang, L.C.:

A Graph-Structural Approach for the Generalization of
    Data Management Systems.   [293]
Chang, S.K.:
    Fuzzy Systems: Theory and Application to Policy
    Analysis and Information Systems.   [1836]
Chang, Shi-Kou:
    Policy Analysis and Information Systems, (A Journal
    Published Semi-Anually by the Univ. of Ill. at
    Chicago Circle and Tankang College in Taipei,
    Taiwan.)   [294]
Chastang, J.:
    Use of a Partially Self-Referent Formalism for
    Description of Natural Systems.   [1784]
Chatterjee, K.:
    Comparison of Arbitration Procedures: Models with
    Complete and Incomplete Information .   [295]
Checkland, P.B.:
    Using a Systems Approach: the Structure of Root
    Definitions.   [1595]
    The Origins and Nature of 'Hard' Systems Thinking. [296]
    Techniques in 'Soft' Systems Practice. Part 1: Systems
    Diagram - Some Tentative Guidelines. Part 2:
    Building Conceptual Models.   [297]
    The Shape of the Systems Movement.   [298]
    The Systems Movement and the "Failure" of Management
    Science.   [299]
    Rethinking a Systems Approach.   [300]
    Systems Thinking, Systems Practice.   [301]
Chen, C.N.:
    Pattern Recognition and Artificial Intelligence.   [302]
Chen, D.T.-W.:
    Toward Analogical Reasoning in Problem Solving by
    Computers.   [303]
Chen, K.:
    Technology Assessment: Creative Futures.   [191]
    Information Technology in Developing Countries.   [304]
    System Analysis in Different Social Settings.   [305]
Cheng, J.K.:
    A Subgraph Isomorphism Algorithm Using Resolution. [306]
Chikan, A.:
    Analysis of a Model System.   [307]
Chochol, S.:
    Systems of Material Flow.   [923]
Chomsky, N.:
    Rules and Representations.   [308]
Chorley, R.J.:
    Environmental Systems: Philosophy, Analysis and
    Control.   [152]
Chow, G.C.:
    Econometric Analysis by Control Methods.   [309]
Chowdhary, S.:
    Shannon's Entropy and Cyclicity.   [53]
    Directed Divergence and Cyclic Symmetry.   [54]
Christakis, A.N.:
    A Systemic Assessment and Application of the Ekistic
    Typology.   [828]
    Technology Assessment: Creative Futures.   [191]
Christensen, R.:
    Foundations of Inductive Reasoning.   [310]
    Entropy Minimax Sourcebook. Vol.1: General
    Description; Vol.2: Philosophical Origins; Vol.3:
    Computer Implementation; Vol.4: Applications.   [311]
    Contrivedness: The Boundary Between Pattern
    Recognition and Numerology.   [464]

Chroust, G.:
  A Frame-Based Real-Time Graphic Interaction System.   [787]
Churchman, C.W.:
  The Systems Approach and its Enemies.   [312]
  Churchman's Conversations.   [313]
Chyba, M.:
  Processing Sociological Data by the Guha Method: an
    Example.   [737]
Chytil, M.K.:
  Mathematical Methods as Cognitive Problem-Solvers.   [314]
Ciborra, C.:
  A Methodological Inquiry of Organizational Noise in
    Sociotechnical Systems.   [315]
Ciriani, T.A.:
  Mathematical Models for Surface Water Hydrology.
    (Proc. of the Workshop on Mathematical Models in
    Hydrology, Pisa, Italy, Dec. 9-12, 1974)   [316]
Ciupa, M.:
  A Cybernetic Approach to Packet Switched Data
    Communication Systems.   [317]
Clark, W.C.:
  Lessons for Ecology Policy Design: a Case Study of
    Ecosystem Management.   [318]
Cliff, A.D.:
  Elements of Spatial Structure: a Quantitative Approach. [319]
  Spacial Processes: Models and Applications.   [320]
Cliffe, M.J.:
  Cybernetics of Alcoholism.   [321]
  The Contribution of Cybernetics to the Science of
    Psychopathology.   [322]
Coblentz, A.M.:
  Systems Science in Health Care.   [323]
Cochin, I.:
  Analysis and Design of Dynamic Systems.   [324]
Coffman, C.V.:
  Constructive Approaches to Mathematical Models.   [325]
Cohen, J.:
  Non-Deterministic Algorithms.   [326]
Cohen, P.R.:
  The Handbook of Artificial Intelligence (Vol.3).   [327]
Cohen, R.S.:
  Scientific Philosophy Today: Essays in Honour of Mario
    Bunge.   [14]
Colby, J.A.:
  A Hierarchical Model of Lotic Ecosystems.   [1139]
  Mathematical Documentation for a Lotic Ecosystem Model. [328]
Cole, I.D.:
  Naturalized Programming: A Method of Naturalizing
    Programming Languages.   [329]
  Naturalized System Development.   [330]
Collins, A.:
  Representation and Understanding: Studies in Cognitive
    Science.   [179]
Collins, D.M.:
  Electron Density Images from Imperfect Data by
    Iterative Entropy Maximization.   [331]
Collins, L.:
  Use of Models in the Social Sciences.   [332]
Collot, F.:
  A New Mathematical Concept for General Systems Theory.   [333]
Colombano, S.P.:
  Complex Systems Tend to be Unstable. Is this Relevant
    to Ecology?   [334]
Colombetti, M.:

A Two Level Model of Knowledge and Belief.   [23]
Colter, M.A.:
  Advanced System Development/Feasibility Techniques.   [351]
Comstock, F.L.:
  A Systems Approach to Grading of Flight Simulator
    Students.   [335]
Conant, R.C.:
  Estimation of Entropy of a Binary Variable: Satisfying
    a Reliability Criterion.   [336]
  An Informational Analysis of the Inter-Male Behavior
    of the Grasshopper Chortophaga Viridifaciata.   [1618]
  Faculty Observations of Variables and Relations.   [337]
  Structural Modelling Using a Simple Information
    Measure.   [338]
  Detection and Analysis of Dependency Structures.   [339]
  Efficient Proofs of Identities in N-Dimensional
    Information Theory.   [340]
  Identifying Relations from Partial Information.   [341]
  Mechanisms of Intelligence: Ross Ashby's Writings on
    Cybernetics.   [55]
  Set-Theoretic Structure Modelling.   [342]
  Information Loss Due to Data Quantization in
    Reconstructability Analysis.   [1781]
Conrad, M.:
  Adaptability - The Significance of Variability from
    Molecule to Ecosystem.   [343]
Cook, N.D.:
  Stability and Flexibility: an Analysis of Natural
    Systems.   [344]
  An Isomorphism of Control in Natural, Social, and
    Cybernetic Systems.   [345]
Cordier, M.O.:
  Propagation: Another Way for Matching Patterns in KBS.   [346]
Corneil, D.G.:
  The Graph Isomorphism Disease.   [1393]
Cornelis, A.:
  Anthropological Transformations in the Dynamics of
    Scientific Learning.   [347]
  Epistemological Indicators of Scientific Identity.   [348]
Cosier, R.A.:
  An Evaluation of the Effectiveness of Dialectical
    Inquiry Systems.   [349]
Costa, E.:
  Non-Canonical Simplification.   [960]
Cotterman, W.W.:
  Systems Analysis and Design: A Foundation for the
    1980's.   [350]
Couger, J.D.:
  Advanced System Development/Feasibility Techniques.   [351]
Coulter, N.A.:
  A New Look at the Description and Prescription of
    Systems.   [1059]
Courtois, P.J.:
  Decomposability: Queueing and Computer System
    Applications.   [352]
Cover, T.M.:
  Compound Bayes Predictors for Sequences with Apparent
    Markov Structure.   [353]
  Some Equivalences between Shannon Entropy and
    Kolmogorov Complexity.   [1040]
  Maximum Entropy and Conditional Probability.   [1786]
Craine, R.:
  The Optimal Monetary Instrument: An Empirical
    Assessment.   [354]

Cramer, F.:
    Fundamental Complexity: a Concept in Biological
        Science and Beyond.   [355]
Cremers, A.B.:
    Orthogonality of Information Structures.   [356]
Crimmins, E.T.:
    The Art of Abstracting.   [357]
Croskin, C.:
    Ways of Knowing.   [358]
Cross, M.:
    Modelling and Simulation in Practice.   [359]
Cruon, R.:
    Mathematical Models for the Study of the Reliability
        of Systems.   [886]
Csaki, P.:
    An Algebraic Approach to Some General Problems of
        Model Description.   [360]
Csanyi, V.:
    Considerations for the Scientific Formulation of
        Epistemology.   [361]
    General Theory of Evolution.   [362]
Csendes, T.:
    A Simulation Study on the Chemoton.   [363]
Csukas, M.:
    Computer Registration of Patients Waiting for Cardiac
        Operation.   [1388]
Cugno, F.:
    Bellamy's Equalitarian Utopia: A System-Theory
        Approach.   [364]
Cull, P.:
    Flaws of Form.   [365]
Cumani, A.:
    On a possibilistic Approach to the Analysis of Fuzzy
        Feedback Systems.   [366]
Cushing, J.M.:
    Two Species Competition in a Periodic Environment.   [367]
Czogala, E.:
    On Identification of Fuzzy Systems and Its Application
        in Control Problems.   [368]
    Control Problems in Fuzzy Systems.   [369]
    Fuzzy Rules Generation for Fuzzy Control.   [370]
    A Contribution to the Construction of the
        Gamma-Operator.   [371]
D'Angelo, H.:
    System Theoretic Implications of Numerical Methods
        Applied to the Solution of Ordinary Differential
        Equations.   [1866]
Da Silvera e Silva, W.:
    On Developing the Microelectronics Industry: A Systems
        Approach.   [891]
Dale, M.B.:
    Systems Analysis and Ecology.   [372]
Dalenoort, G.J.:
    Modularity of Behavior of Networks.   [373]
    On the Representation of Control in Systems in
        General, and in Neural Networks in Particular.   [374]
Dammasch, I.E.:
    On the Properties of Randomly Connected
        McCulloch-Pitts Networks: Differences between
        Input-Constant and Input-Variant Networks.   [375]
Dang, Y.:
    Application of Fuzzy Decision Making and Fuzzy Linear
        Programming in Production and Planning of Animal
        Husbandry System in Farming Region.   [1837]

Darlington, J.L.:
    Interactive Validation of Program Specifications.   [376]
Darwish, M.:
    On the Stability of Large-Scale Power Systems.   [377]
Das, J.P.:
    Simultaneous and Successive Cognitive Processes.   [378]
Dasiewicz, P.P.:
    Application of Iterative Algorithms to Adaptive
        Predicitive Coding.   [1105]
Dass, B.K.:
    Adjacent-Error Correcting Binary Perfect Codes.   [1538]
David, F.W.:
    Experimental Modelling in Engineering.   [379]
David, P.K.:
    Citizen Participation: A New Factor in Social Systems. [591]
Davies, A.J.:
    The Finite Element Method: A First Approach.   [380]
Davio, M.:
    Discrete and Switching Functions.   [381]
Davis, C.:
    Organization for Program Management.   [382]
Davis, L.S.:
    Reduction Operation for Constraint Satisfaction.   [725]
Davis, R.:
    Frameworks for Cooperation in Distributed Problem
        Solving.   [1593]
    Knowledge-Based Systems in Artificial Intelligence.   [383]
Davoli, F.:
    Development of Performance Adaptive Fuzzy Controllers
        with Application to Continuous Casting Plants.   [113]
Day, J.W.:
    Ecosystem Modeling in Theory and Practice: an
        Introduction with Case Studies.   [713]
Day, R.H.:
    Economic Development as an Adaptive Process.   [384]
Dayal, R.:
    An Integrated System of World Models.   [385]
De Beaufrande,:
    Text, Discourse, and Process: Towards a
        Multidisciplinary Science of Texts.   [386]
De Beaugrande, R.:
    Text, Discourse, and Process.   [387]
De Cleris, M.:
    Certain System Concepts in Law and Politics.   [388]
De Delara, G.K.:
    Computer Simulation of Lateral Inhibition in the
        Retina.   [389]
De Greene, K.B.:
    Force Fields and Emergent Phenomena in Sociotechnical
        Macrosystems: Theories and Models.   [390]
    The Adaptive Organization - Anticipation and
        Management of Crisis.   [391]
De Jong, K.:
    Adaptive System Design: A Genetic Approach.   [392]
De Korvin, A.:
    Switching Mechanisms in a Generalized Information
        System.   [936]
De La Sen, M.:
    Some Aspects of the Modeling of Multivariable
        Aperiodic Sampling Systems.   [393]
    Parameterizations in Single Input-Output Adaptive
        Control Systems to Avoid Delays in the Input
        Generation.   [394]
De Luca, A.:

On the Convergence of Entropy Measures of a Fuzzy Set.   [395]
De Malherbe, M.:
Model Building in Ecology: a Hierarchical Approach.   [396]
De Malherbe, R.:
Model Building in Ecology: a Hierarchical Approach.   [396]
De Mottoni, P.:
Bifurcation of Regular Nonnegative Stationary
Solutions for a Quasilinear Parabolic System.   [397]
De Raadt, J.D.R.:
An Application of Ashby's Law to the Skewness of
Income and Output Distributions.   [398]
De Rosnay, J.:
The Macroscope: a New World Scientific System.   [399]
De Santis, F.:
The Approximation of Factor Scores in a Factor
Analysis Algorithm.   [405]
De Smit, J.:
Systems Thinking: Concepts and Notions.   [981]
De Witt, C.T.:
Simulation of Ecological Processes (2nd Edition,
Revised and Extended) .   [400]
Dedons, A.:
The Information Professional: Survey of an Emerging
Field.   [401]
Degermendzhy, A.G.:
The Reliability of Microevolution Process of the
Stationary and Oscillatory Population in Open
Systems.   [402]
Dekker, L.:
Simulation of Systems.   [403]
Delaney, W.:
A System Theory Based Simulation Language.   [404]
The Role of System Modelling in Natural Language
Processing.   [1779]
Della Riccia, G.:
The Approximation of Factor Scores in a Factor
Analysis Algorithm.   [405]
Demarco, T.:
Structural Analysis and System Specification.   [406]
Demb, A.:
Organization for Program Management.   [382]
Demillo, R.A.:
Social Processes and Proofs of Theorems and Programs.   [407]
Dempster, A.H.:
Deterministic and Stochastic Scheduling (NATO Advanced
Study Institute).   [408]
Denning, P.J.:
Machines, Languages, and Computation.   [409]
Dennis, J.B.:
Machines, Languages, and Computation.   [409]
Dent, J.B.:
Systems Simulation in Agriculture.   [410]
Deschamps, J.-P.:
Discrete and Switching Functions.   [381]
Desouza, G.R.:
System Methods for Socioeconomic and Environmental
Impact Analysis.   [411]
Dessauchoy, A.E.:
Generalized Information Theory and Decomposability of
Systems.   [412]
Destouches, J.-L.:
Basic Concepts of System Theory.   [413]
Detwiler, J.S.:
Stochastic Modeling of Fetal Heart Rate Variability

with Applications to Clinical Risk-Assessment and
Detection of Fetal Respiratory
Movements-Preliminary Results.   [837]
Deutsch, K.W.:
  Toward a Theory of Simplification: Reduction of
    Complexity in Data Processing for World Models.   [414]
Devries, R.P.:
  Systems Theory and the Philosophy of Science.   [415]
Dhillon, B.S.:
  Engineering Reliability: New Techniques and
    Applications.   [416]
Dhrymes, P.J.:
  Mathematics for Econometrics.   [417]
Di Cesare, F.:
  An Input/Output Approach to the Structural Analysis of
    Digraphs.   [1426]
Di Giulio, E.:
  Energetics Versus Communication in the Nervous System. [256]
DiNola, A.:
  On the Fuzziness of Solutions of Alpha-Fuzzy Relation
    Equations on Finite Spaces.   [418]
Dial, G.:
  A Note on Bayes Risk with Utilities.   [419]
  Entropy of Order Alpha and Type Beta and Shannon's
    Inequality.   [420]
Dijkstra, W.:
  Testing Auxiliary Hypothesis Behind the Interview.   [421]
  Modeling Interaction Processes in Interviews; a
    Contribution of Systems Methodology.   [1930]
Dimirivski, G.M.:
  Categories in Systems and Control Theory.   [422]
Dimitrova, L.:
  Cybernetics and the Organization of Material Flow in
    the Industrial Enterprise.   [423]
Dirickx, Y.M.I.:
  Systems Analysis by Multilevel Methods: with
    Applications for Economics and Management.   [424]
Ditleusen, O.:
  Uncertain Modelling.   [425]
Dixon, N.R.:
  Automatic Speech and Speaker Recognition.   [426]
Dogramaci, A.:
  Current Topics in Computer Simulation.   [12]
Dolby, J.L.:
  On the Notions of Ambiguity and Information Loss.   [427]
Dold, A.:
  Lecture Notes in Mathematics, Vol.453   [428]
Dombi, J.:
  Comments on the Gamma-Model.   [429]
Dominique, J.:
  Logical Construction of Systems.   [430]
Dominowski, R.L.:
  Research Methods.   [431]
Domotor, Z.:
  Probabilistic and Causal Dependence Structures.   [432]
Dompere, K.K.:
  On Epistemology and Decision-Choice Rationality.   [433]
Donald, A.:
  Management, Information and Systems (Second Edition). [434]
Donnell, M.L.:
  Fuzzy Decision Analysis.   [1848]
Doorn Van, A.J.:
  On Spotting Three-Dimensional Shape: Rigid Movement,
    Bending and Stretching by Eye: Implications for

Robot Vision.   [962]
Dowd, R.B.:
  A Dual Model of the Brain.   [34]
Dreben, B.:
  The Decision Problem.   [435]
Drechsler, F.S.:
  Computer Integrated Manufacturing (CIM) in Irish
    Industry.   [436]
Dreyfus, S.E.:
  Dynamic Programming and the Calculus of Variations. [437]
  The Art and Theory of Dynamic Programming.   [438]
Driankow, D.:
  An Approach to Multiobjectives Decision Making with
    Ordinal Information.   [439]
Drommenhoek, W.:
  Biological Structures (translated from Dutch).   [440]
Drozen, V.:
  Multilateral Interdependence.   [441]
  Combinatory Spaces - An Approach to Pattern Analysis. [442]
Dubois, D.:
  Fuzzy Sets and Systems: Theory and Applications.   [443]
  A Class of Fuzzy Measures Based on Triangular Norms. [444]
Duda, R.O.:
  Pattern Classification and Scene Analysis.   [445]
Duechting, W.:
  Recent Results in Modelling and Simulation of 3-D
    Tumor Growth.   [446]
Dunn, G.:
  An Introduction to Mathematical Taxonomy.   [447]
Duran, B.S.:
  Cluster Analysis: a Survey.   [448]
Durkin, J.E.:
  Living Groups: Group Psychotherapy and General System
    Theory.   [449]
Dutton, G.:
  Harvard Papers on Geographic Information Systems.   [450]
Dwyer, R.F.:
  Sequential Partition Detectors.   [451]
  Sequential Partition Detectors with Dependent Sampling. [452]
Dym, C.L.:
  Principles of Mathematical Modeling.   [453]
Eckmann, B.:
  Lecture Notes in Mathematics, Vol.453   [428]
Economos, A.C.:
  Thoughts on a Systems View of Life.   [454]
Edmons, E.A.:
  Lattice Fuzzy Logics.   [455]
Edwards, J.B.:
  Analysis and Control of Multipass Processes.   [456]
Efstathiou, J.:
  Multiattribute Decision Making Using a Fuzzy Heuristic
    Approach.   [457]
Egardt, B.:
  Stability of Adaptive Controllers.   [458]
Egiazarian, K.O.:
  The Generalized Hadamard Transforms and Linear Filters. [459]
Ehrenkrantz, E.:
  The Systems Approach to Design.   [460]
Ehrig, H.:
  Universal Theory of Automata.   [461]
Eigen, M.:
  The Hypercycle: a Principle of Natural
    Selforganization.   [462]
  Laws of the Game: How Principles of Nature Govern

Chance.   [463]
Eilbert, R.F.:
  Contrivedness: The Boundary Between Pattern
    Recognition and Numerology.   [464]
Eisen, M.:
  Mathematical Models in Cell Biology and Cancer
    Chemotherapy.   [465]
Eisenstein, B.A.:
  Feature Extraction by System Identification.   [466]
El-Fattah, Y.:
  Learning Systems: Decision, Simulation, and Control.   [467]
El-Sherief, H.:
  Identification and Modeling for Linear Multivariable
    Discrete-Time Systems: A Survey.   [468]
  Identification of Multivariable Systems in the
    Transfer-Function Matrix Form.   [469]
  Identification and Control of Linear Discrete
    Multivariable Systems with Process Noise.   [470]
  Recent Advances in Multivariable System Modeling and
    Identification Algorithms and their Applications.   [471]
El-Shirbeeny, El-H.T.:
  A Sensitivity Augmented Conjugate Gradient Algorithm
    for Solving Constrained Control Design Problems.   [472]
Elandt-Johnson, R.C.:
  Survival Models and Data Analysis.   [473]
Elany, O.A.:
  A Sensitivity Augmented Conjugate Gradient Algorithm
    for Solving Constrained Control Design Problems.   [472]
Elias, D.:
  Procedures for Generating Reconstruction Hypotheses in
    the Reconstructability Analysis: APL Version.   [474]
Elison, D.I.:
  Interactive Simulation on a Computer.   [475]
Ellison, D.:
  Simulation Newsletter.   [476]
Elohim, J.L.:
  Further Steps in our Learning to Organize Educational
    Systems.   [477]
Elstob, C.M.:
  The Formal and the Physical.   [478]
  Emergentism and Mind.   [479]
Elton, M.C.J.:
  Evaluating New Telecommunication Services.   [480]
Engell, S.:
  Variety, Information and Feedback.   [481]
England J.L.:
  Hierarchy Theory and Sociology.   [482]
  An Ashby Hierarchy for Human Action.   [483]
England, J.W.:
  Mathematical Theory of Entropy.   [1110]
Enochson, L.:
  Applied Time Series Analysis. Vol. 1: Basic Techniques. [1273]
Erdi, P.:
  System-Theoretical Approach to the Neural
    Organization: Feed-Forward Control of the
    Ontogenetic Development.   [484]
Erickson, S.W.:
  A Systems View of the Foundations of Education.   [485]
Ericson, R.F.:
  The General Systems Challenge.   [486]
Erlandson, R.F.:
  The Participant-Observer Role in Systems-Methodologies. [487]
  The Satisficing Process. A New Look.   [488]
Espejo, R.:

    Management Information Systems: a System for Design.   [489]
    Organization for Program Management.   [382]
    Cybernetic Praxis in Government: The Management of
       Industry in Chile 1970-1973.   [490]
    Cybernetics in Management and Organization.   [491]
    Management and Information: The Complementarity
       Control-Autonomy.   [492]
Esposito, F.:
    A System Theory Based Simulation Language.   [404]
Estep, M.L.:
    A Systems Integration of Multidisciplinary and
       Interdisciplinary Inquiry in Curriculum Design.   [493]
    Toward Alternative Methods of Systems Analysis: The
       Case of Qualitative Knowing.   [494]
Eugene, J.:
    Aspects of the General Theory of Systems.   [495]
Evans, F.J.:
    Qualitative Aspects of Large Scale Systems: Developing
       Design Rules Using APL.   [541]
Eveland, S.:
    The Use of the QBE Relational Database System for a
       Gothic Ethymological Dictionary.   [496]
Everitt, B.S.:
    Unresolved Problems in Cluster Analysis.   [497]
    An Introduction to Mathematical Taxonomy.   [447]
Eyckhoff, P.:
    Trends and Progress in Systems Identification.   [498]
    Trends and Progress in System Identification.   [499]
Fairhurst, M.C.:
    Cyclic Activity in Information Processing Systems.   [500]
Fairtlough, G.:
    How Systems Thinking Might Evolve.   [501]
Fandel, G.:
    Multiple Criteria Decision Making Theory and
       Application. (Proc. of a conf. in
       Hagen/Konigswinter, Germany, Aug.1979).   [502]
Fantin, J.:
    On the Stability of Large-Scale Power Systems.   [377]
Fararo, T.J.:
    Social Activity and Social Structure: A Contribution
       to the Theory of Social Systems.   [503]
Faribault, M.:
    From Natural Language to a Canonical Representation of
       the Corresponding Semantic Relationships.   [504]
Farinas-del-Cerro, L.:
    Prolegomena for Programming in Modal Logic.   [505]
Farkas, J.:
    Rethinking of Theoretical Presuppositions of Systems
       Theory.   [506]
Farley, A.M.:
    Issues in Knowledge-Based Problem Solving.   [507]
Farreny, H.:
    On the Best Way of Designating Objects in Sentence
       Generation.   [508]
Fatmi, H.A.:
    Parallel Processors for Cybernetic Systems.   [718]
    An Interpolator Based Controller for Cybernetic
       Systems.   [719]
Faucheuz, C.:
    Automation or Autonomy in Organizational Design.   [509]
Favella, L.F.:
    Mathematical Foundations of ECG Map Reconstruction. [510]
    First Passage Time Problems and Some Related
       Computational Methods.   [511]

Application of Loeve-Karhunen Expansion to Acoustic
    Diagnostic Problems for Internal Combustion Engines. [512]
ECG Map Filtering by Means of Spherical Harmonics: A
    Simple Approximation and Results.   [93]
Methodological and Mathematical Problems in ECG
    Analysis.   [513]
Fedanzo, A.J.:
The Origin of Information in Systems.   [514]
Fedra, K.:
Uncertainty and Arbitrariness in Ecosystem Modelling:
    A Lake Modelling Example.   [515]
Feigenbaum, E.A.:
The Handbook of Artificial Intelligence (Vol.1).   [109]
The Handbook of Artificial Intelligence (Vol.2).   [110]
The Handbook of Artificial Intelligence (Vol.3).   [327]
Feintuch, A.:
System Theory: A Hilbert Space Approach.   [516]
Feldman, J.A.:
Total Complexity and the Inference of Bent Programs.   [517]
Fernandez, E.B.:
Database Security and Integrity.   [518]
Ferracin, A.:
Self-Organizing Ability in Living Systems.   [519]
Ferre, M.:
Use of a Partially Self-Referent Formalism for
    Description of Natural Systems.   [1784]
Ferrero, M.:
Bellamy's Equalitarian Utopia: A System-Theory
    Approach.   [364]
Fey, W.R.:
Ecosystem Succession.   [693]
Fiacco, A.V.:
Extremal Methods and Systems Analysis.   [520]
Fick, G.:
Decision Support Systems: Issues and Challenges.   [521]
Field, H.H.:
Science Without Numbers.   [522]
Findeisen, W.:
Control and Coordination in Hierarchical Systems
    (Vol.9 in the Wiley IIASA/ International Series on
    Applied Systems Analysis).   [523]
Findler, N.V.:
Associative Networks: Representation and Use of
    Knowledge by Computers.   [524]
Toward Analogical Reasoning in Problem Solving by
    Computers.   [303]
Aspects of Computer Learning.   [525]
A Multi-Level Learning Technique Using Production
    Systems.   [526]
Progress in Cybernetics and Systems Research, Vol.XI.   [1724]
Fischer, M.M.:
Basic Problems of Modern Econometrics and Some
    Problems of its Acceptability and Interpretation.   [748]
Fishburn, P.C.:
An Analysis of Simple Counting Methods for Ordering
    Incomplete Ordinal Data.   [588]
Fitter, M.:
Towards More 'Natural' Interactive Systems.   [527]
Fitzgerald, G.:
A Taxonomy of Current Approaches to Systems Analysis. [1879]
Fix, G.J.:
Constructive Approaches to Mathematical Models.   [325]
Flaster, P.:
Qualitative Aspects of Large Scale Systems: Developing

      Design Rules Using APL.   [541]
Flavin, M.:
   Fundamental Concepts of Information Modelling.   [528]
Fleddermann, R.G.:
   Teaching the Systems Approach with a Group Thesis.   [529]
Fletcher, R.:
   Practical Methods in Optimization. Vol.1:
   Unconstrained Optimization; Vol.2: Constrained
   Optimization.   [530]
Floyd, R.W.:
   The Paradigms of Programming.   [531]
Flynn, R.:
   On the Strength of Belief of Dialectical Belief
   Systems.   [1173]
Fogel, E.:
   On the Value of Information in System Identification:
   Bounded Noise Case.   [532]
Fogelman-Soulie, F.:
   Random Boolean Networks.   [64]
Fontela, E.:
   Condensed Forms of Large Scale Models.   [533]
Foo, N.Y.:
   Stability Preservation under Homomorphisis.   [534]
   Closure Properties and Homomorphisms of Timevarying
   Systems.   [535]
Ford, J.L.:
   Uncertainty and Expectations in Economics: Essays in
   Honour of G.L.S. Shackle.   [265]
Fortuny-Amat, J.:
   A Solution Procedure for the Hierarchical Coordination
   of Constrained Optimizing Systems.   [536]
Fosko, P.D.:
   Applications of General Systems Theory and Cybernetics
   to the Management of/ Research and Development.   [537]
Foulard, C.:
   Learning Systems: Decision, Simulation, and Control.   [467]
Foulds, L.R.:
   Digraphs: Theory and Techniques.   [1432]
Fowler, T.B.:
   Brillouin and the Concept of Information.   [538]
Fox, M.S.:
   An Organizational View of Distributed Systems.   [539]
Frank, P.M.:
   Introduction to System Sensitivity Theory.   [540]
Frank, W.:
   Flaws of Form.   [365]
Franksen, O.I.:
   Qualitative Aspects of Large Scale Systems: Developing
   Design Rules Using APL.   [541]
Franova, M.:
   Program Synthesis and Constructive Proofs Obtained by
   Beth's Tableaux.   [542]
Franta, W.R.:
   Process View of Simulation.   [543]
Frauenthal, J.C.:
   Mathematical Modelling in Epidemology.   [544]
Freeling, A.N.S.:
   Fuzzy Sets and Decision Analysis.   [545]
Freeman, P.:
   Client-Centered Design: Concepts and Experience.   [1459]
Freire, E.:
   The Structure of Relations in Personnel Management.   [1266]
Freuder, E.C.:
   Synthesizing Constraint Expressions.   [546]

Fritsch, B.:
  Toward a Theory of Simplification: Reduction of
    Complexity in Data Processing for World Models.   [414]
Fu, K.S.:
  Attributed Grammar - A Tool for Combining Syntactic
    and Statistical Approaches to Pattern Recognition.   [1757]
  Pattern Recognition Theory and Applications.   [931]
Fujishige, S.:
  Use of Matroid Theory in Operations Research, Circuits
    and Systems Theory.   [814]
Fukao, T.:
  A Model of Mass and Energy Flow in Ecosystems.   [757]
Fum, D.:
  A Rule-Based Approach to Natural Language Text
    Representation and Comprehension.   [547]
Furukawa, O.:
  Precoordination in Quality Control.   [548]
  Refinement of Quality Control Systems.   [549]
  A Methodology for Quality Goal-Seeking and
    Coordination, and the Practical Application.   [550]
Gabasov, R.:
  The Qualitative Theory of Optimal Processes.
    (Translated From Russian by J. Casti).   [551]
Gage, W.L.:
  Auditorium Management Information Systems.   [552]
  Flexibility in Network Planning.   [553]
  Evidence for Strategic Decisions.   [554]
Gaines, B.R.:
  Fuzzy Automata and Decision Processes.   [685]
  The Fuzzy Decade: a Bibliography of Fuzzy Systems and
    Closely Related Topics.   [555]
  Fuzzy and Probability Uncertainty Logics.   [556]
  General Systems Research: Quo Vadis?   [557]
  General Systems. Yearbook of the Society for General
    Systems Research.   [558]
  Logical Foundations for Database Systems.   [559]
  Sequential Fuzzy System Identification.   [560]
  Fuzzy Reasoning and Its Applications.   [1099]
  The Personal Scientist in the Community of Science.   [1542]
  Methodology in the Large: Modeling All There Is.   [561]
Gal, T.:
  Multiple Criteria Decision Making Theory and
    Application. (Proc. of a conf. in
    Hagen/Konigswinter, Germany, Aug.1979).   [502]
Galitsky, V.V.:
  On Modeling the Plant Community Dynamics.   [562]
Gall, J.:
  Systemantics: How Systems Work and Especially How They
    Fail.   [563]
Gallopin, G.C.:
  The Abstract Concept of Environment.   [564]
Gams, M.:
  A Circuit Analysis Program that Explains its Reasoning. [565]
Ganascia, J.G.:
  Explanation Facilities for Diagnosis Systems.   [566]
Gane, C.:
  Structural Systems Analysis: Tools and Techniques.   [567]
Garbade, K.:
  Time Variation in the Term Structure of Treasury Bills. [568]
Gardiner, P.C.:
  Decision Spaces.   [569]
Gardner, M.R.:
  Predicting Novel Facts.   [570]
Garey, M.R.:

    Computers and Intractability: a Guide to
       NP-Completeness.   [571]
Garfinkel, A.:
    Forms and Explanation.   [572]
Garfolo, B.T.:
    The Renal Dialysis Medical Information System.   [573]
Garofalo, J.:
    Mathematical Problem Solving.   [1038]
Garribba, S.:
    Synthesis and Analysis Methods for Safety and
       Reliability Studies (NATO Advanced Study Institute). [40]
Gasparski, W.W.:
    Problems of Design Methodology.   [574]
    Studies in Design Methodology in Poland (Special Issue
       of "Design Methods and Theories").   [575]
    An Outline of the Logical Theory of Design.   [576]
    Design Methods and Theories: Vol.10, No.2.   [1366]
    Design and System: Methodological Aspects, Vol.3.   [577]
    On Systems Theory and Systems Research and Science and
       Technology.   [578]
    Design Methods and Theories: Vol.15, No.2.   [656]
    Design Methods and Theories; Vol.15, No.3.   [657]
    Papers from the Conference in Design Methods.   [579]
    Praxiology.   [580]
    Science, Technology, and Systems.   [581]
    Si Duo Dicunt Idem Non Est Idem: Or on Design
       Methodology as Seen from Neither an English nor a
       German Speaking Country.   [582]
    What Does it Mean a Notion of "System Modelling".   [583]
    Si duo dicunt idem non est idem: Or on Design
       Methodology as Seen from neither English nor German
       Speaking Country.   [584]
Gatewood, L.C.:
    Mathematical Models in the Health Sciences: a Computer
       Approach.   [7]
Gati, G.:
    Further Annotated Bibliography on the Isomorphism
       Disease.   [585]
Gatlin, L.L.:
    Information Theory and the Living Systems.   [586]
Gause, D.C.:
    Cybernetics and Artificial Intelligence.   [587]
Gauvrit, M.:
    A Two-Level Algorithm for the Simultaneous State and
       Parameter Identification.   [1495]
Gehrlein, W.V.:
    An Analysis of Simple Counting Methods for Ordering
       Incomplete Ordinal Data.   [588]
Gelfand, A.E.:
    A Systems-Theoretic Approach to the Management of
       Complex Organizations: Management by Consensus
       Level and its Interaction with Other Management
       Strategies.   [589]
Gelfand, A.G.:
    A System Theoretic Approach to the Management of
       Complex Organizations: Management by Exception,
       Priority, and Input Span in a Class of
       Fixed-Structure Models.   [1830]
Genesio, R.:
    On the Model Order for Uncertain System Approximation. [590]
Gengoux, K.G.:
    Citizen Participation: A New Factor in Social Systems. [591]
Gentile, S.:
    A Discrete Model for the Study of a Lake.   [592]

George, F.H.:
  Cybernetics and the Environment.   [593]
  The Foundations of Cybernetics.   [594]
  Philosophical Foundations of Cybernetics.   [595]
  The Science of Philosophy.   [596]
Georgiev, A.A.:
  Nonparametric Mathematical Model from Individual Human
    Growth Curve.   [597]
Gerardy, R.:
  Probabilistic Finite State System Identification.   [598]
  Experiments with Some Methods for the Identification
    of Finite-State Systems.   [599]
Gericke, M.:
  Changes of Regional Passenger-Transport Demands by
    Society-Controlled Activities - A Fuzzy Simulation
    Approach.   [600]
Gerlach, J.:
  The Human Brain and its Universe. Vol.I: The World of
    Natural Sciences and its Phenomenology. Vol.II: The
    Brain and its Mind.   [993]
Gesyps, R.G.:
  Basic and Applied General Systems Research: A
    Bibliography.   [937]
Getis, A.:
  Models of Spatial Processes.   [601]
Getz, W.M.:
  Mathematical Modelling in Biology and Ecology.   [602]
Geyer, R.F.:
  Sociocybernetics, Vol.II.   [603]
  Alienation Theories: a General Systems Approach   [604]
Gheorghe, A.:
  Energy Management Planning Tools for the Transition
    towards a Medium Developed Economy.   [605]
Ghiselli, E.E.:
  Measurement Theory for the Behavioral Sciences.   [606]
Ghosal, A.:
  Applied Cybernetics: its Relevance to Operations
    Research.   [607]
  Relevance of Cybernetics in the Study of Developing
    Countries.   [608]
Giere, R.N.:
  Propensity and Necessity.   [609]
Gigley, H.:
  Artificial Intelligence Meets Brain Theory. An
    Integrated Approach to Simulation Modelling of
    Natural Language Processing.   [610]
  How we Name or Misname Objects.   [1019]
Gihman, I.I.:
  Controlled Stochastic Processes.   [611]
Gill, A.:
  Applied Algebra for the Computer Sciences.   [612]
Gill, K.F.:
  Systems Modelling and Control.   [1516]
Gill, P.:
  Practical Optimization.   [613]
Gilmore, R.:
  Catastrophe Theory for Scientists and Engineers.   [614]
Gini, G.:
  Deterministic and Nondeterministic Programming in
    Robot Systems.   [615]
Gini, M.:
  Deterministic and Nondeterministic Programming in
    Robot Systems.   [615]
Ginzberg, M.J.:

A Prescription Model for System Implementation.   [616]
Ginzburg, L.R.:
    Selective Polymorphism of the Erythrocyte Acid
        Phosphate Locus in Humans.   [625]
Gladun, V.P.:
    Adaptive Problem-Solving Systems.   [617]
Glanville, R.:
    Consciousness: And so on.   [618]
    The Model's Dimensions: A Form for Argument.   [619]
    Inside Every White Box there are Two Black Boxes
        Trying to Get Out.   [620]
    Distinguished and Exact Lies.   [621]
Glaser, F.B.:
    A System Approach to Alcohol Treatment.   [622]
Glass, A.L.:
    Cognition.   [623]
Globus, G.G::
    Is There a Ghost in the Machine after all?   [624]
Gloria-Bottini, F.:
    Selective Polymorphism of the Erythrocyte Acid
        Phosphate Locus in Humans.   [625]
Glushkov, V.M.:
    Algebra, Languages Programming.   [626]
Goffman, W.:
    Scientific Information Systems and the Principle of
        Selectivity.   [627]
    Preformation of Autogenetic Patterns.   [884]
Goguen, J.A.:
    The Arithmetic of Closure.   [1798]
Goh, B.S.:
    Stability, Vulnerability and Persistence of Complex
        Ecosystems.   [628]
Gokhale, D.V.:
    The Information in Contingency Tables.   [629]
Gold, H.J.:
    Mathematical Modeling of Biological Systems: an
        Introductory Guidebook.   [630]
Goldfarb, W.D.:
    The Decision Problem.   [435]
Goldkuhl, G.:
    Information Systems Development: A Systematic Approach. [1074]
Golemanov, L.A.:
    On the Coordinability of Hierarchical Systems.   [631]
Golledge, R.G.:
    Proximity and Preference: Problems in the
        Multidimensional Analysis of Large Data Sets. [632]
Gollmann, D.:
    On the Identification of Certain Non-Linear Networks
        of Automata.   [633]
    On the Theory of Bilinear Shift Registers.   [634]
Golubitsky, M.:
    An Introduction to Catastrophe Theory and its
        Applications.   [635]
Gomez, P.:
    Top-Down Versus Bottom-Up Organizational Design: a
        Cybernetic Perspective.   [636]
    Centralization Versus Decentralization in Business
        Organisations: Cybernetic Rules for Effective
        Management.   [637]
    The Systems Approach Problem Solver and a Stockmarket
        Example.   [1778]
    New ('second order') Cybernetics against Mismanagement. [1363]
    Systems-Methodology in Action: Organic Problem Solving
        in a Publishing Company.   [638]

Gonzales, R.C.:
  Digital Image Processing.   [639]
  Syntactic Pattern Recognition.   [640]
Goodman, L.A.:
  Analyzing Qualitative / Categorical Data.   [641]
  Measures of Association for Cross Classification.   [642]
Gopal, K.:
  An Event Expansion Algorithm for Reliability
    Evaluation in Complex Systems.   [643]
Gorbatov, V.A.:
  Theory of Partially Ordered Systems.   [644]
Gottinger, H.W.:
  Systems Approaches and Environmental Problems.   [645]
  Complexity and Catastrophe: Applications of Dynamic
    System Theory.   [646]
  Toward an Algebraic Theory of Complexity in Dynamic
    Systems.   [647]
  Structural Characteristics Economical Models: a Study
    of Complexity.   [648]
Goudriaan, J.:
  Simulation of Ecological Processes (2nd Edition,
    Revised and Extended) .   [400]
Gough, N.E.:
  Categories in Systems and Control Theory.   [422]
Gould, P.:
  Q-Analysis or a Language of Structure: An Introduction
    for Social Scientists, Geographers and Planners.   [649]
Gourlay, A.R.:
  Identification and Analysis of the Subsystems
    Structure of Models.   [650]
Grabowski, J.:
  Sequencing Problem with Resource Constraints.   [651]
Graedel, T.E.:
  Graphical Presentation of Results from Scientific
    Computer Models.   [652]
Graham, J.H.:
  Linguistic Decision Structures for Hierarchical
    Systems.   [653]
Grams, R.R.:
  Systems-Analysis Workbook.   [654]
Granero-Porati, M.I.:
  Random Ecological Systems with Structure:
    Stability-Complexity Relationship.   [655]
Grant, D,P.:
  Design Methods and Theories: Vol.15, No.2.   [656]
  Design Methods and Theories; Vol.15, No.3.   [657]
Grantham, W.J.:
  Optimality in Parametric Systems.   [1818]
Grateloup, G.:
  On the Stability of Large-Scale Power Systems.   [377]
Grauer, M.:
  The Generation of Efficient Energy Supply Strategies
    using Multi-Criteria Optimization.   [658]
  An Evolutionary Analysis of World Energy Consumption
    and World Population.   [982]
Green, M.A.:
  Solar Cells: Operating Principles, Technology, and
    System Application.   [659]
Greenberg, H.J.:
  Computer-Assisted Analysis and Model Simplification.   [660]
  Computer-Assisted Analysis and Model Simplification.   [661]
Greenberg, S.W.:
  A System Approach to Alcohol Treatment.   [622]
Greene, D.H.:

Mathematics for the Analysis of Algorithms.   [662]
Greenspan, D.:
   A Completely Arithmetic Formulation of Classical and
      Special Relativistic Physics.   [663]
   Arithmetic Applied Mathematics.   [664]
   Discrete Modeling in the Microcosm and in the
      Macrocosm.   [665]
   A New Computer Approach to the Modeling of Atoms, Ions
      and Molecules. (Part I: Hydrogen; Part II: The
      Oxygen Molecule).   [666]
Gregory, S.A.:
   Large Technological Projects.   [667]
Greibach, S.A.:
   Theory of Program Structures: Schemes, Semantics,
      Verification.   [668]
Grenander, U.:
   Regular Structures: Lectures in Pattern Theory. Vol.3. [669]
Grindley, K.:
   Systematics: a New Approach to Systems Analysis.   [670]
Groen, J.C.F.L.:
   A Constraint Analytic Measure for the Distance between
      Neighbouring Structures.   [671]
Grogono, P.:
   Problem Solving and Computer Programming.   [672]
Grollmann, J.:
   Description of Feedback Systems via Approximation by
      Delayed Systems.   [673]
Gross, D.:
   Fundamentals of Queueing Theory.   [674]
Grouchko, D.:
   Mathematical Models for the Study of the Reliability
      of Systems.   [886]
Grove, H.D.:
   Measurement, Accounting, and Organizational
      Information.   [1177]
Grubbstrom, R.W.:
   Theory of Relatively Closed Systems and Applications. [675]
   The Axsater Integrated Production-Inventory System
      Interpreted in Terms of the Theory of Relatively
      Closed Systems.   [676]
Gruber, H.:
   Modularization and Abstracting; a Discussion on the
      Basis of Sequential Systems.   [677]
Gruska, J.:
   Mathematical Foundations of Computer Science.   [678]
Guentert, B.:
   The Necessity of Sensitivity Models for Managers,
      Planners and Politicians in Health Care Systems. [1364]
Guida, G.:
   An Integrated Model of Problem Solver.   [679]
   Interactivity and Incrementality in Natural Language
      Understanding Systems.   [680]
   Natural Language Access to Online Data Bases.   [681]
   A Rule-Based Approach to Natural Language Text
      Representation and Comprehension.   [547]
Guidorzi, R.P.:
   Multistructural Model Selection.   [682]
Guild, N.C.F.:
   Modelling Controllers Using Fuzzy Relations.   [84]
Gumovski, I.:
   Recurrences and Discrete Dynamic Systems.   [683]
Gupta, J.S.:
   An Event Expansion Algorithm for Reliability
      Evaluation in Complex Systems.   [643]

Gupta, M.M.:
  Advances in Fuzzy Set Theory and Applications.   [684]
  Fuzzy Automata and Decision Processes.   [685]
  Advances in Fuzzy Set Theory and Applications.   [686]
  On the Design of an Adaptive Controller for Uncertain
    Plants via Liapunov Signal Synthesis Method.   [687]
  Hierarchical Dynamic Optimization for Linear Discrete
    Systems.   [688]
  Feedback Control Applications of Fuzzy Set Theory: A
    Survey.   [689]
Gupta, S.N.:
  Block-Wise Error-Protecting Burst Codes with Weight
    Constraints.   [1539]
Gurdial,:
  A Note on Fano Inequality Involving Distortion.   [690]
Gurel, O.:
  Bifurcation Theory and Applications in Scientific
    Disciplines.   [691]
Gutenbaum, J.:
  Applications of Systems Theory to Economics,
    Management, and Technology.   [692]
Gutierrez, L.T.:
  Ecosystem Succession.   [693]
Gvishiani, J.M.:
  Systems Research: Methodological Problems - 1979
    Yearbook.   [694]
Gyarfas, F.:
  CODEX: Prototypes Driven Backward and Forward Chaining
    Computer-Based Diagnostic Expert System.   [695]
Gyula, P.:
  Common Laws of Sciences and Systems.   [696]
Haack, S.:
  Do We Need 'Fuzzy Logic'?   [697]
Haaland, P.D.:
  Information Theoretic Analysis of Questionnaire Data. [222]
Hafele, W.:
  Modelling of Large Scale Energy Systems. (IIASA
    Proceedings Series, Vol. 12).   [698]
Hagg, C.:
  Possibility and Cost in Decision Analysis.   [699]
Haimes, Y.Y.:
  Hierarchical Analysis of Water Resources Systems:
    Modeling and Optimization of Large-Scale Systems.   [700]
  Hierarchical Holographic Modeling.   [701]
Hajek, P.:
  On Generation of Inductive Hypotheses.   [702]
  Mechanizing Hypothesis Formationmathematical
    Foundations of a General Theory.   [703]
  Special Issue of the International Journal of
    Man-Machine Studies on the Guha Method of
    Mechanized Hypothesis Formation.   [704]
Haken, H.:
  Synergetics - A Workshop.   [705]
  Synergetics.   [706]
  Pattern Formation by Dynamic Systems and Pattern
    Recognition.   [707]
  Dynamics of Synergetic Systems.   [708]
Hakkala, L.:
  An Infeasible Gradient-Type Coordination Algorithm for
    Dynamical Systems.   [709]
  On the Coordinability of Hierarchical Systems.   [631]
Haley, K.B.:
  Applied Operatons Research in Fishing. (Vol. 10 in
    NATO Conference Series.).   [710]

Halfon, E.:
    Adequacy of Ecosystem Models.   [711]
    Theoretical Systems Ecology: Advances and Case Studies. [712]
    Identification of a Dynamic Lake Model by the Method
       of Data Handling: An Application to Lake Ontario. [1666]
Hall, C.A.S.:
    Ecosystem Modeling in Theory and Practice: an
       Introduction with Case Studies.   [713]
Hall, E.L.:
    Computer Image Processing and Recognition.   [714]
Hallam, T.G.:
    On the Intrinsic Structure of Differential Equation
       Models of Cosystems.   [715]
Halldin, S.:
    Comparison of Forest Water and Energy Exchange
       Models.(Proc. of an IUFRO Workshop, Uppsala,
       Sweden, Sept.24-30, 1978).   [716]
Halmes, A.:
    Topics in Systems Theory: Publication in Honour of
       Prof.Hans Blomberg.   [717]
Hanassab, S.:
    Parallel Processors for Cybernetic Systems.   [718]
    An Interpolator Based Controller for Cybernetic
       Systems.   [719]
Hancock, J.K.:
    Applications of General Systems Theory and Cybernetics
       to the Management of/ Research and Development. [537]
Hanika, F.de P.:
    Progress in Cybernetics and Systems Research, Vol.VII. [1335]
    Progress in Cybernetics and Systems Research, Vol.X.   [1723]
Hanken, A.F.G.:
    Cybernetics and Society: An Analysis of Social Systems. [720]
    Social Systems and Learning Systems.   [721]
Hansen, E.B.:
    On Systems' Reliability and Redundancy.   [131]
Hanson, O.J.:
    The Role of Systems Analysis in Designing Systems and
       Methods of Training Using Systems Techniques.   [722]
    Assessing and Modifying an Advanced Business Systems
       Analysis Course Using Feedback from Past Students. [723]
    Review and Re-Design of an Advanced Systems Analysis
       Course.   [724]
Haralik, R.M.:
    Reduction Operation for Constraint Satisfaction.   [725]
    The Characterization of Binary Relation Homomorphisms. [726]
    Digital Image Processing.   [1565]
    Structural Descriptions and Inexact Matching.   [1535]
Hardin, R.C.:
    Political Determinants of Information System
       Configuration.   [1070]
Harleman, D.R.F.:
    Hydrophysical and Ecological Modelling of Deep Lakes
       and Reservoirs.   [858]
Harris, C.M.:
    Fundamentals of Queueing Theory.   [674]
Harris, J.D.:
    Fuzzy Partitions and Relations: an Axiomatic Basis for
       Clustering.   [166]
Harris, L.R.:
    Using the Data Base as a Semantic Component to Aid in
       the Parsing of Natural Language Data Base Queries. [727]
Harrison, M.A.:
    Introduction to Formal Language Theory.   [728]
Hart, P.E.:

    Pattern Classification and Scene Analysis.   [445]
Hartl, R.F.:
    Optimal Seducing Policies for Dynamic Continuous
      Lovers Under Risk of Being Killed a Rival.   [729]
Hartley, R.:
    Recent Developments in Markov Decision Processes.   [730]
Hartley, R.T.:
    Conversation Theory and Cognitive Coherence Theories. [850]
Hartnett, W.E.:
    Systems: Approaches, Theories, Applications.   [731]
Haruna, K.:
    Structural Sensitivity Analysis and its Applications. [732]
Hassebrock, F.:
    Validating Computer Simulation Models of Expert
      Reasoning.   [851]
Hassell, M.P.:
    The Dynamics of Arthropod Predator-Prey Systems.   [733]
Hastings, N.A.J.:
    Decision Networks.   [734]
Hattiangadi, J.N.:
    The Structure of Problems.   [735]
Hatze, H.:
    Myocybernetics - A New Development in Neuromuscular
      Control.   [736]
Haus, G.:
    Analysis and Compacting of Musical Texts.   [163]
Havenner, A.:
    The Optimal Monetary Instrument: An Empirical
      Assessment.   [354]
Havranek, T.:
    On Generation of Inductive Hypotheses.   [702]
    Processing Sociological Data by the Guha Method: an
      Example.   [737]
    Mechanizing Hypothesis Formationmathematical
      Foundations of a General Theory.   [703]
    An Alternative Approach to Missing Information in the
      Guha Method.   [738]
Hawkes, N.:
    International Seminar on Trends in Mathematical
      Modelling.   [739]
    International Seminar on Trends in Mathematical
      Modelling: a UNESCO Report.   [740]
Hawkins, D.M.:
    Topics in Applied Multivariate Analysis (2 volumes). [741]
Haydon, P.G.:
    Coherent and Incoherent Computation in Identified
      Molluscan Neurones.   [742]
Hayes, P.:
    Trends in Artificial Intelligence.   [743]
Hayes-Roth, F.:
    Pattern-Directed Inference Systems.   [1847]
Hazlett, B.A.:
    Quantitative Methods in the Study of Animal Behavior. [744]
Hazony, Y.:
    Problem Solving with Interactive Graphics.   [745]
Heim, R.:
    Theoretical Approaches to Complex Systems.   [746]
Hejl, P.M.:
    Autopoiesis, Communications, and Society: The Theory
      of Autopoietic System in the Social Sciences.   [157]
Hell, P.:
    Algorithmic Aspects of Combinatorics.   [25]
Hellerman, L.:
    Symbolic Initiative and its Application to Computers. [747]

Hellwagner, H.:
  Robotics and Image Processing.   [1338]
Hellwig, Z.H.:
  Basic Problems of Modern Econometrics and Some
    Problems of its Acceptability and Interpretation.   [748]
Hemami, H.:
  Determination of Feedback for Human Posture Control
    without Physical Intervention.   [258]
Hendel, R.J.:
  The Grue Paradox: An Information-Theoretic Solution. [749]
Henize, J.:
  Modelling Large Scale Socio-Economic Systems: Can We
    Begin to Do Better.   [750]
Herman, G.T.:
  Image Reconstruction From Projections: Implementation
    and Applications.   [751]
Hermann, R.:
  Applications of Algebraic Geometry to Systems Theory,
    Part 1   [752]
Herschdorfer, I.:
  Interactive Simulation on a Computer.   [475]
Heymann, J.:
  Microcomputer-Based System for Industrial Energy
    Control.   [952]
Hibbard, T.N.:
  Orthogonality of Information Structures.   [356]
Hickin, J.:
  Applications of Projective Reduction Methods to
    Estimation and Control.   [753]
Higashi, M.:
  Measures of Uncertainty and Information Based on
    Possibility Distributions.   [754]
  On Measures of Fuzziness and Fuzzy Completeness.   [755]
  On the Notion of Distance Representing Information
    Closeness: Possibility and Probability
    Distributions.   [756]
Hill, D.J.:
  Stability Criteria for Large-Scale Systems.   [1192]
Hirata, H.:
  A Model of Mass and Energy Flow in Ecosystems.   [757]
  A Model of Hierarchical Ecosystems with Migration. [758]
  A Model of Hierarchical Ecosystems with Utility
    Efficiency Mass and its Stability.   [759]
Hirzinger, G.:
  Robot-Teaching via Force-Torque-Sensors.   [760]
Hisdal, E.:
  Conditional Possibilities, Independence and
    Noninteraction   [761]
  Generalized Fuzzy Set Systems and Particularization. [762]
Hoaglin, D.C.:
  Data for Decisions: Information Strategies for
    Policymakers.   [763]
Hobitz, H.:
  A Simple Mathematical Model of Biphasic Insulin
    Secretion.   [232]
Hoebeke, L.:
  The Application of Cybernetic Principles for the
    Training of Operator Crews of Nuclear Power Plants. [833]
Hoehle, U.:
  A Remark on Entropies with Respect to Plausibility
    Measures.   [764]
Hoernig, K.M.:
  Aspects of Automatic Program Construction.   [765]
Hoffer, J.A.:

Factors Affecting the Comparative Advantages of
    Sequential and Batched Decision Making.   [766]
Hofferbert, R.I.:
    The Application of General Systems Methodology to the
    Comparative Study of Public Policy.   [767]
Hofmann, Th.R.:
    Qualitative Terms for Quantity.   [768]
Hogeweg, P.:
    Locally Synchronised Developmental Systems: Concrete
    Advantages of Discrete Event Formalism.   [769]
Holden, A.V.:
    Cybernetic Approach to Global and Local Mode of Action
    of General Anestetics.   [770]
    Coherent and Incoherent Computation in Identified
    Molluscan Neurones.   [742]
    Chaotic Activity in Neural Systems.   [771]
Holler, M.J.:
    Markets as Self-Policing Quality Enforcement Systems. [772]
Holling, C.S.:
    Resilience and Stability of Ecological Systems.   [773]
    Adaptive Environmental Assessment and Management.   [774]
    Lessons for Ecology Policy Design: a Case Study of
    Ecosystem Management.   [318]
Holsapple C.W.:
    Foundations of Decision Support Systems.   [189]
Holst, P.:
    Computer Simulation 1951-1976: An Index to the
    Literature.   [775]
Holyoak, K.J.:
    Cognition.   [623]
Honkasalo, A.:
    Design of Work Tasks and Education of Labor Force.   [776]
    Entropic Processes and Economic Systems.   [777]
Hooda, D.S.:
    Two Generalized Measures of "Useful" Information.   [778]
Hooker, C.A.:
    Physical Theory as Logico-Operational Structure.   [779]
Hopcroft, J.E.:
    Introduction to Automata Theory, Languages, and
    Computation.   [780]
Horacek, H.:
    On Generation of Anaphora and Gapping in German.   [781]
Horn, W.:
    Interactive Computer-Aided Planning: Estimating
    Health-Care Resource Requirements.   [782]
    Health Care in Austria: 1980-2000.   [783]
    A Health Hazard Appraisal Program for Austria.   [1717]
    Fast Walsh vs. Fast Fourier Transform: A Comparison of
    Time-Efficiency.   [784]
    Progress in Cybernetics and Systems Research, Vol.XI. [1724]
    The Structure of Manifestations in a Medical
    Consultation System.   [785]
    An Artificial Intelligence System for Medical Decision
    Support.   [786]
    Future, Impacts, and Future Impacts of Artificial
    Intelligence: A Bibliography.   [1616]
    A Frame-Based Real-Time Graphic Interaction System. [787]
    Basic and Applied General Systems Research: A
    Bibliography, 1977-1984.   [1729]
Horowitz, E.:
    Fundamentals of Data Structures.   [788]
Horvath, T.:
    Method and Systematic Reflections.   [789]
Horvitz, J.S.:

A Theoretical Framework for Quantifying Legal
  Decisions.   [844]
Hoskins, R.F.:
  Generalised Functions.   [790]
Hossforf, H.:
  Model Analysis of Structures.   [791]
Hostetter, G.H.:
  Fundamentals of Network Analysis.   [792]
Hoye, R.E.:
  Current Status of Hospital Management Information
  Systems.   [793]
Hsia, P.:
  On the Structure of Directed Graphs with Applications:
  A Rapprochement with Systems Theory (Part I).   [1669]
  On the Structure of Directed Graphs with Applications:
  A Rapprochement with Systems Theory (Part II).   [1670]
Hsu, K.:
  Signal Analysis of Visual Evoked Potentials.   [838]
Hu, N.C.:
  Multi-Dimensional FFT Computation.   [794]
Huang, T.S.:
  A Subgraph Isomorphism Algorithm Using Resolution.   [306]
Huang, Y.F.:
  On the Value of Information in System Identification:
  Bounded Noise Case.   [532]
Huber, O.:
  Nontransitive Multidimensional Preferences:
  Theoretical Analysis of a Model.   [795]
Hull, R.:
  Communication and Information: Their Relationship to
  Computer-Based Information Systems.   [796]
Hunt, E.B.:
  Artificial Intelligence.   [797]
Hunt, G.W.:
  The Instability of Evolving Systems.   [1686]
Hunt, L.R.:
  Controllability of General Nonlinear Systems.   [798]
Husseyin, K.:
  Problem Analysis in Science and Engineering.   [212]
Hutt, A.T.F.:
  A Relational Data Base Management System.   [799]
Huxham, C.S.:
  Hypergames as a Soft Systems Methodology.   [800]
Hwang, C.L.:
  Multiple Objective Decision Making: Methods and
  Applications (a State-of-the-Art Survey).   [801]
  Decomposition of Binary Sequences of Finite Periods
  and a Class of Error Correcting Codes.   [802]
  Multiple Attribute Decision Making: Methods and
  Applications: A State-of-the-Art Survey.   [803]
  System-Reliability Evaluation Techniques for
  Complex/Large Systems: A Review.   [804]
Hyde, M.O.:
  Computers that Think?: The Search for Artificial
  Intelligence.   [805]
Hyvarinen, L.P.:
  Information Theory for Systems Engineers.   [806]
Ibidapo-Obe, O.:
  An Approximate Differential Structure for Error
  Covariance in Nonlinear Dynamical Systems.   [807]
Ignizio, J.P.:
  Stability Analysis: An Approach to the Evaluation of
  System Design.   [1318]
Ihara, J.:

A Structural Analysis of Criteria for Selecting Model
    Variables.   [808]
Iijima, J.:
  Serial Decomposition of General Systems.   [809]
Ikeshoji, H.:
  Characterization of the Causality of General Dynamic
    Systems.   [1655]
  A Methodology for Quality Goal-Seeking and
    Coordination, and the Practical Application.   [550]
Ingemarsson, I.:
  Toward a Theory of Unknown Functions.   [810]
Inmon, W.H.:
  Effective Data Base Design.   [811]
Innis, G.S.:
  New Directions in the Analysis of Ecological Systems. [812]
Innocent, P.R.:
  Toward Self-Adaptive Interface System.   [813]
Iri, M.:
  Use of Matroid Theory in Operations Research, Circuits
    and Systems Theory.   [814]
Iserman, R.:
  Identification and System Parameter Estimation.   [815]
  System Identification (Papers from an IFAC Symposium
    in Darmstadt, W.Germany , Sept. 1979).   [816]
Ishizuchi, H.:
  Precoordination in Quality Control.   [548]
  Refinement of Quality Control Systems.   [549]
Islam, S.:
  General Systems Methodology and Large-Scale Systems. [275]
Ivan, I.T.:
  System of Indicators for Estimating the Software
    Performances.   [817]
Ivey, E.S.:
  Principles of Mathematical Modeling.   [453]
Jaburek, W.J.:
  Teleconferencing via Videotex.   [818]
Jacak, W.:
  Events and Systems.   [819]
  On Goal-Oriented Systems.   [820]
  Modelling of Evenistic Systems.   [821]
Jachymczyk, W.:
  Karhunen-Loewe Transform Approximation for Certain
    Class of Random Processes.   [822]
Jack, W.:
  On the Eventistic Approach to the Notion of System. [823]
Jacob, F.:
  The Possible and the Actual.   [824]
Jacobs, B.E.:
  On Generalized Computational Complexity.   [825]
Jacoby, S.L.S.:
  Mathematical Modeling with Computers.   [826]
Jahoda, M.:
  Wholes and Parts, Meaning and Mechanism.   [827]
Jain, J.L.:
  Cybernetic Modeling of Queueing Complex.   [1086]
Jain, V.:
  A Systemic Assessment and Application of the Ekistic
    Typology.   [828]
Jakubowski, R.:
  General Algorithms of Complex Dynamic Systems
    Simulation.   [829]
  Syntactic Characterization of Machine Parts Shapes. [830]
James, D.J.G.:
  Case Studies in Mathematical Modelling.   [831]

Janiak, A.:
    Sequencing Problem with Resource Constraints.   [651]
Janko, J.:
    Systems Approach in Natural Science Reflecting
        Orientation Towards Practice: and Episode from the
        History of Czech Science in the Middle of the 19th
        Century.   [832]
Janssens, L.:
    The Application of Cybernetic Principles for the
        Training of Operator Crews of Nuclear Power Plants. [833]
Jantch, E.:
    The Self-Organizing Universe.   [834]
    The Evolutionary Vision: Toward a Unifying Paradigm of
        Physical, Biological and Sociocultural Evolution
        (AAAS Selected Symposium).   [835]
Jantzen, J.:
    A Generalized Fuzzy Relational Matrix.   [836]
Jarisch, W.:
    Stochastic Modeling of Fetal Heart Rate Variability
        with Applications to Clinical Risk-Assessment and
        Detection of Fetal Respiratory
        Movements-Preliminary Results.   [837]
    Signal Analysis of Visual Evoked Potentials.   [838]
Jaron, J.:
    Goal-Oriented Cybernetical Systems.   [839]
    Relations and Topologies in the Goal-Space of a System. [840]
Jean, R.V.:
    A Methodological Paradigm in Plant Biology.   [841]
Jeffers, J.N.R.:
    An Introduction to Systems Analysis: with Ecological
        Applications.   [842]
Jefferson, T.R.:
    On the Analysis of an Information Theoretic Model of
        Spacial Interaction.   [1518]
Jennergren, L.P.:
    The Multilevel Approach: a Systems Analysis
        Methodology.   [843]
    Systems Analysis by Multilevel Methods: with
        Applications for Economics and Management.   [424]
Jensen, H.L.:
    A Theoretical Framework for Quantifying Legal
        Decisions.   [844]
Jezek, J.:
    Universal Algebra and Theory of Models.   [845]
Joergensen, S.E.:
    State of the Art in Ecological Modelling.   [846]
Joerres, R.:
    Identification of the Temperature Masking of the
        Circadian System of Euglena Gracilis.   [847]
Johnk, L.:
    The Use of the QBE Relational Database System for a
        Gothic Ethymological Dictionary.   [496]
Johnson, D.S.:
    Computers and Intractability: a Guide to
        NP-Completeness.   [571]
Johnson, D.T.:
    A Planning Control Strategy that Allows for the Cost
        of Planning.   [848]
Johnson, J. H.:
    On the Theory of Lorrain-Categories.   [98]
Johnson, L.:
    A Theoretical Framework for Human Communication.   [849]
    Conversation Theory and Cognitive Coherence Theories.   [850]
    Conversation Theory.   [1257]

Design of Expert Systems from the Perspective of
  Conversation Theory Methodology.   [903]
Johnson, N.L.:
  Survival Models and Data Analysis.   [473]
Johnson, P.E.:
  Validating Computer Simulation Models of Expert
    Reasoning.   [851]
Johnson, R.W.:
  Properties of Cross-Entropy Minimization.   [1551]
Johnson, W.A.:
  Applications of Systems Theory to Ecology.   [852]
Jones, A.H.:
  Cybernetic Design Features for Computerised Business
    Information Systems.   [853]
Jones, B.:
  Determination of Reconstruction Families.   [854]
  General Systems Theory and Algorithm Theory.   [855]
Jones, D.D.:
  Catastrophe Theory Applied to Ecological Systems. [856]
  Lessons for Ecology Policy Design: a Case Study of
    Ecosystem Management.   [318]
Jones, L.M.:
  Defining Systems Boundaries in Practice: Some
    Proposals and Guidelines.   [857]
Jorgenson, S.E.:
  Hydrophysical and Ecological Modelling of Deep Lakes
    and Reservoirs.   [858]
  Hydrophysical and Ecological Models of Shallow Lakes
    (Summary of an IIASA Workshop).   [859]
  State of the Art in Ecological Modelling.   [860]
Jumarie, G.:
  Some Technical Applications of Relativistic
    Information Fuzzy Sets, Linguistics, Relativistic
    Sets and Communication.   [861]
  A Relativistic Approach to Modelling Dynamic Systems
    Involving Human Factors.   [862]
  A New Approach to the Transinformation of Random
    Experiments Involving Fuzzy Observations.   [863]
Jutila, S.T.:
  Social Gaming and Control by Dislocations.   [864]
Kaczmarek, L.K.:
  Self-Organization in Biological Systems with Multiple
    Cellular Contacts.   [78]
Kahne, S.:
  Introducing Systems Concepts to All University
    Students.   [865]
Kahneman, D.:
  Judgement Under Uncertainty: Heuristics and Biases. [866]
Kaindl, H.:
  Dynamic Control of the Quiescence Search in Computer
    Chess.   [867]
Kaiser, J.S.:
  Organizational Behavior: A Vector of Single-Valued
    Transformations within a Determinate Machine. [868]
Kak, S.:
  On Efficiency of Chemical Homeostasis - an
    Information-Theoretic Approach.   [869]
Kalaba, R.E.:
  Two Solution Techniques for Adaptive Reinvestment: A
    Small Sample Comparison.   [870]
Kalata, P.:
  On System Identification with and without Certainty. [871]
Kaliman, J.:
  Simulation and Analysis of Pathological Blood Pressure

Behaviour after Treadmill Test in Patients with
    Coarctation of the Aorta.   [215]
Kalman, R.E.:
  On Partial Realizations, Transfer Functions, and
    Canonical Forms.   [872]
Kandel, A.:
  Fuzzy Sets, Fuzzy Algebra, and Fuzzy Statistics.   [873]
  Fuzzy Switching and Automata: Theory and Applications. [874]
  Fuzzy Processes.   [875]
Kania, A.:
  On Stability of Formal Fuzziness Systems.   [876]
Kannellakis, P.C.:
  On the Computational Complexity of Cardinality
    Constraints in Relational Databases.   [877]
Kantor, J.R.:
  System Structure and Scientific Psychology.   [878]
Kaposi, A.:
  Research into the Assessment of Complexity.   [879]
Kapur, J.N.:
  On Maximum-Entropy Complexity Measures.   [880]
Karlin, S.:
  A Second Course in Stochastic Processes.   [881]
Kashyap, R. L.:
  A Bayesian Comparison of Different Classes of Dynamic
    Models Using Empirical Data.   [882]
Katz, M.J.:
  Questions of Uniqueness and Resolution in
    Reconstruction from Projections.   [883]
  Preformation of Autogenetic Patterns.   [884]
Kauffman, L.H.:
  Network Synthesis and Varela's Calculus.   [885]
Kaufmann, A.:
  Mathematical Models for the Study of the Reliability
    of Systems.   [886]
Kaufmann, G.:
  Visual Imagery and Its Relation to Problem Solving. [887]
Kaur, B.:
  Phased Error Correcting Perfect Codes.   [1540]
Kaushik, M.L.:
  Burst-Error-Correcting Codes with Weight-Constraints
    Under a New Metric.   [888]
  Single Error and Burst Error Correcting Codes Through
    a New Metric.   [889]
Kavakoglu, I.:
  Mathematical Modelling of Energy Systems.   [890]
Kawamura, K.:
  On Developing the Microelectronics Industry: A Systems
    Approach.   [891]
Kawato, M.:
  Mathematical and Phenomenological Studies of
    Biological Rhythms.   [1644]
Kaynak, E.:
  A Holistic Approach in Marketing Planning.   [892]
Kedem, B.:
  Binary Time Series.   [893]
Keen, G.P.W.:
  Decision Support Systems.   [894]
  Decision Support Systems: an Organizational
    Perspective.   [895]
Keeney, R.L.:
  Conflicting Objectives in Decisions.   [142]
Keeny, B.P.:
  Ecosystemic Epistemology: an Alternative Paradigm for
    Diagnosis.   [896]

Kefalas, A.G.St.:
  Management by Wisdom: On the Art and Science of
    Anticipatory Management.   [897]
Kekes, J.:
  The Centrality of Problem Solving.   [898]
Kellerman, W.:
  Computer-Aided Identification of System Structure: An
    Attempt at Implementation.   [899]
Kellermayr, K.H.:
  Performance Simulation of an Optical Bus Area Network. [900]
Kelly, J.R.:
  A Review of Decomposition and Design Methodologies. [1137]
Kennedy, M.H.:
  Factors Affecting the Comparative Advantages of
    Sequential and Batched Decision Making.   [766]
Kennedy, P.:
  A Guide to Econometrics.   [901]
Kent, W.:
  Data and Reality.   [902]
Keravnou, E.T.:
  Design of Expert Systems from the Perspective of
    Conversation Theory Methodology.   [903]
Kerefky, P.:
  Program Optimization and Manipulation on User Level. [904]
  Patient Registers on Microcomputers.   [905]
Kerin, T.:
  A Method for Performance Assessment of Medical
    Radioisotope Equipment.   [906]
Kerman, C.E.:
  Creative Tension: the Life and Thought of Kenneth
    Boulding.   [907]
Kerner, H.:
  Functional versus Imperative Programming - Complement
    or Contradiction?   [908]
Kesavan, H.K.:
  The Graph-Theoretic Field Model-I - Modelling and
    Formulations.   [1498]
Keyes, R.W.:
  Fundamental Limits in Digital Information Processing. [909]
Khatri, C.G.:
  An Introduction to Multivariate Statistics.   [1609]
Kickert, W.J.M.:
  An Example of Linguistic Modelling.   [910]
  Fuzzy Theories on Decision-Making. (Vol. 3 in
    Frontiers in Systems Research Book Series.)   [911]
  A Metasystem Approach to Organizational
    Decision-Making.   [912]
  A Formal Approach to Organization Design.   [913]
  Organization of Decision-Making: a Systems-Theoretical
    Approach.   [914]
Kieras, D.E.:
  Finite Automata and S-R Models.   [915]
Kijima, K.:
  A Unified Theory of Decision Principles.   [1656]
  A Structure of Rational Decision Principles.   [1658]
Kilmann, R.H.:
  Social Systems Design: Normative Theory and the Maps
    Technology.   [916]
  Methodological Approaches to Social Science.   [1172]
Kim, J.Y.:
  General Systems Theory and Human Communication.   [1455]
Kindler, E.:
  Classification of Simulation Programming Languages: I.
    Declaration of Necessary System Conceptions.   [917]

Classification of Simulation Programming Languages:
    II. Description of Types and Individual Typology.   [918]
Dynamic Systems and Theory of Simulation.   [919]
Simula 67 and Systems Analysis.   [920]
Simulation Programming Languages.   [921]
A Formalization of Some Simulation Language Concepts.   [922]
Systems of Material Flow.   [923]
Kindler, J.:
Proceedings of a Workshop on Modelling of Water
    Demands.   [924]
King, L.T.:
Problem Solving in a Project Environment.   [925]
Problem Solving in a Project Environment: A Consulting
    Process.   [926]
Kinston, W.:
Improving Health Care Institutions: An Action Research
    Approach to Organisation of Complex Systems.   [927]
Linking System Research and Delivery Systems. The
    Example of Health Care.   [928]
Kintala, C.M.R.:
Amounts of Nondeterminism in Finite Automata.   [929]
Kirby, J.R.:
Simultaneous and Successive Cognitive Processes.   [378]
Kirchmayer, L.K.:
Modelling of Large Scale Energy Systems. (IIASA
    Proceedings Series, Vol. 12).   [698]
Kirillov, A.B.:
Short-Term Memory as a Metastable State. II.
    Simulation Model.   [974]
Kirilova, F.:
The Qualitative Theory of Optimal Processes.
    (Translated From Russian by J. Casti).   [551]
Kittler, J.:
A Comment on Fast Monte Carlo Integration of P.D.F.
    Estimators.   [930]
Pattern Recognition Theory and Applications.   [931]
Kjurktchiev, T.:
A Method for Performance Assessment of Medical
    Radioisotope Equipment.   [906]
Kleijnen, J.P.C.:
Design and Analysis of Simualtion: Practical
    Statistical Techniques.   [932]
Minimizing Wafting Times Using Priority Classes: A
    Case Study in Response Surface Methodology.   [1793]
Klement, E.P.:
Proceedings of the International Seminar on Fuzzy Set
    Theory.   [933]
On the Cardinality of Fuzzy Sets.   [934]
Law of Large Numbers and Central Limit Theorem for
    Fuzzy Random Variables.   [935]
Kleyle, R.:
Switching Mechanisms in a Generalized Information
    System.   [936]
Klir, G.J.:
Basic and Applied General Systems Research: A
    Bibliography.   [937]
On the Problem of Computer-Aided Structure
    Identification: Some Experimental Observations and
    Resulting Guidelines.   [938]
Pattern Discovery in Activity Arrays.   [939]
A Conceptual Foundation for Systems Problem Solving.   [276]
Applied General Systems Research.   [940]
Frontiers in Systems Research: Implications for the
    Social Sciences (a New Book Series) .   [941]

Progress in Cybernetics and Systems Research, Vol.III. [1712]
The General Systems Research Movement.   [942]
The North-Holland Series in General Systems Research.   [943]
Architecture of Structure Systems: a Basis for the
   Reconstructability Analysis.   [944]
Computer-Aided Systems Modelling.   [945]
General Systems Problem Solving Methodology.   [946]
Procedures for Generating Reconstruction Hypotheses in
   the Reconstructability Analysis.   [947]
The Structure of Reconstructable Relations: a
   Comprehensive Study.   [282]
Reconstructability Analysis: Evaluation of
   Reconstruction Hypothesis.   [283]
Reconstructability Analysis: Overview and Bibliography. [284]
Decision Making in Reconstructability Analysis.   [285]
Measures of Uncertainty and Information Based on
   Possibility Distributions.   [754]
On Measures of Fuzziness and Fuzzy Completeness.   [755]
On Systems Methodology and Inductive Reasoning: The
   Issue of Parts and Wholes.   [948]
Progress in Cybernetics and Systems Research, Vol.VIII. [1722]
Reconstruction of Possibilistic Behavior Systems.   [286]
General Systems Concepts.   [949]
On the Notion of Distance Representing Information
   Closeness: Possibility and Probability
   Distributions.   [756]
Possibilistic Information Theory.   [950]
Basic and Applied General Systems Research: A
   Bibliography, 1977-1984.   [1729]
Kloeden, P.E.:
Fuzzy Dynamical Systems.   [951]
Knapp, R.W.:
Advanced System Development/Feasibility Techniques.   [351]
Knapp, S.:
Using Nursing Research: Discovery, Analysis, and
   Interpretation.   [1755]
Knappe, U.:
Microcomputer-Based System for Industrial Energy
   Control.   [952]
Knuth, D.E.:
Mathematics for the Analysis of Algorithms.   [662]
Kobayashi, H.:
Modelling and Analysis: an Introduction to System
   Performance Evaluation Methodology.   [953]
Kobayashi, I.:
Information and Information Processing Structure.   [954]
Kobsa, A.:
Knowledge Representation: The Representation of
   Knowledge in the Computer.   [955]
On Regarding AI Programs as Theories.   [956]
Knowledge Representation: A Survey of its Mechanisms,
   a Sketch of its Semantics.   [957]
Representing Belief Models in Semantic Networks.   [958]
Koczy, L.T.:
Interactive Sigma-Algebras and Fuzzy Objects of Type N. [959]
Kodratoff, Y.:
Non-Canonical Simplification.   [960]
Koeck, W.K.:
Autopoiesis, Communications, and Society: The Theory
   of Autopoietic System in the Social Sciences.   [157]
Koenderink, J.J.:
Perception of Solid Shape and Spatial Lay-Out through
   Photometric Invariants.   [961]
On Spotting Three-Dimensional Shape: Rigid Movement,

    Bending and Stretching by Eye: Implications for
    Robot Vision.   [962]
Koffman, E.B.:
    Problem Solving and Structured Programming in PASCAL.   [963]
Kohonen, T.:
    Associative Memory: a Systems-Theoretical Approach.   [964]
Kohout, L.J.:
    The Fuzzy Decade: a Bibliography of Fuzzy Systems and
    Closely Related Topics.   [555]
    Fuzzy Power Sets and Fuzzy Simplification Operators.   [99]
    Semantics of Implication Operators and Fuzzy
    Relational Products.   [100]
    The Algebraic Structure of the Spencer-Brown and
    Varela Calculi.   [965]
    Construction of an Expert Therapy Adviser as a Special
    Case of a General System Protection Design.   [966]
    The Four Modes of Inference in Fuzzy Expert Systems.   [102]
Kolodziej, W.J.:
    An Overview of Bilinear System Theory and Applications. [1180]
Komoda, N.:
    Structural Sensitivity Analysis and its Applications.   [732]
Kontur, I.F.:
    Runoff Simulation with a Random Walk Model.   [967]
Koppelaar, H.:
    Regular Career Systems II.   [968]
    Systematica.   [969]
Korhonen, P.:
    A Stepwise Procedure for Multivariate Clustering.   [970]
Kornai, J.:
    Descriptive-Explanatory Theoretical Models of the
    Socialist Economy: Review of a Research Direction.   [971]
Kornwachs, K.:
    Pragmatic Information and Nonclassical Systems.   [972]
    Some Notes on Information and Interaction.   [973]
Kortanek, K.O.:
    Extremal Methods and Systems Analysis.   [520]
Kovalenko, Y.I.:
    Short-Term Memory as a Metastable State. II.
    Simulation Model.   [974]
Kovalevsky, V.A.:
    Image Pattern Recognition.   [975]
Kowalaski, R.:
    Logic of Problem Solving.   [976]
Kowalik, J.S.:
    Mathematical Modeling with Computers.   [826]
Koyayashi, H.:
    Modeling and Analysis: an Introduction to System
    Performance Evaluation Methodology.   [977]
Kozen, D.:
    Positive First-Order Logic Is NP-Complete.   [978]
Kraft, R.W.:
    Symbols, Systems, Science & Survival.   [979]
Krallmann, H.:
    Enlarging the Systems Paradigm: Historical View from
    1973 until 1983.   [980]
Kramer, J.T.A.:
    Systems Thinking: Concepts and Notions.   [981]
Kramer, N.J.T.A.:
    A Taxonomy of Systems Science.   [1788]
Kriegel, U.:
    An Evolutionary Analysis of World Energy Consumption
    and World Population.   [982]
Krippendorff, K.:
    Communication and Control in Society.   [983]

On the Algorithm for Identifying Structures in
   Multi-Variate Data.   [984]
Krishana, V.S.:
   An Introduction to Category Theory.   [985]
Krol, J.:
   General Algorithms of Complex Dynamic Systems
   Simulation.   [829]
Kron-Morelli, R.:
   Random Ecological Systems with Structure:
   Stability-Complexity Relationship.   [655]
Krone, R.M.:
   Systems Analysis and Policy Sciences: Theory and
   Practice.   [986]
Kronsjo, L.I.:
   Algorithms: Their Complexity and Efficiency.   [987]
Kruijt, D.:
   Regular Career Systems II.   [968]
Kruskal, W.H.:
   Measures of Association for Cross Classification. [642]
Kryukov, V.I.:
   Short-Term Memory as a Metastable State. I. Master
   Equation Approach.   [988]
   Short-Term Memory as a Metastable State. II.
   Simulation Model.   [974]
Krzanik, L.:
   Evolutionary Projects: Dynamic Optimization of
   Delivery Step Structure.   [989]
Kubat, L.:
   Entropy and Information in Science and Philosophy. [990]
Kubinski, T.:
   An Outline of the Logical Theory of Questions.   [991]
Kubota, H.:
   A Stationarization Functor and its Concrete
   Realizations.   [1659]
Kucera, V.:
   Discrete Linear Control: the Polynomial Equation
   Approach.   [992]
Kuhlenbeck, H.:
   The Human Brain and its Universe. Vol.I: The World of
   Natural Sciences and its Phenomenology. Vol.II: The
   Brain and its Mind.   [993]
Kuhn, H.W.:
   Mathematical Systems Theory and Economics.   [994]
Kullback, S.:
   The Information in Contingency Tables.   [629]
Kurz, L.:
   Sequential Partition Detectors.   [451]
   Sequential Partition Detectors with Dependent Sampling. [452]
Kuyk, W.:
   Complementarity in Mathematics.   [995]
Kwakernaak, H.:
   Fuzzy Random Variables. Part 1: Definitions and
   Theorems. Part 2: Algorithms and Examples for the
   Discrete Case.   [996]
La Patra, J.W.:
   Applying the Systems Approach to Urban Development. [997]
La Porte, T.R.:
   Organized Complexity: Challenge to Politics and Policy. [998]
LaSalle, J.P.:
   The Stability of Dynamical Systems.   [999]
Labos, E.:
   Periodic and Non-Periodic Motions in Different Classes
   of Formal Neuronal Networks and Chaotic Spike
   Generators.   [1000]

Labudde, R.A.:
  Discrete Hamiltonian Mechanics.   [1001]
Ladanyi, O.G.:
  The Eighties - A World in Motion, A Challenge for
    Systems Engineering.   [1002]
  Modelling Economic Movements Using Published
    International Statistics.   [1003]
Laing, R.:
  Machines as Organisms: an Exploration of the Relevance
    of Recent Results.   [1004]
Lakshmivarahan, S.:
  Considerations for Fuzzifying Formal Languages and
    Synthesis of Fuzzy Grammars.   [1005]
Lamport, L.:
  Time, Clock and the Ordering of Events in Distributed
    Systems.   [1006]
Landau, Y.D.:
  Adaptive Control: the Model Reference Approach.   [1007]
Landweber, L.H.:
  Properties of Conflict-Free and Persistent Petri Nets. [1008]
Langdon, G.G.:
  Universal Modeling and Coding.   [1421]
Langefors, L.:
  Information and Data in Systems.   [1009]
Lasker, G.E.:
  Applied Systems and Cybernetics (6 Vols).   [1010]
  The Relation between Major World Problems and Systems
    Learning.   [1011]
Lasota, A.:
  The Extinction of Slowly Evolving Dynamical Systems. [1012]
Laszlo, C.A.:
  Champ - A System and Computer Program Package for
    Health Hazard Appraisal.   [1013]
  Cybernetics and Health Care.   [1170]
Laszlo, E.:
  GST: Prospects and Principles.   [1014]
  Some Reflections on Systems Theory's Critics.   [1015]
  Cybernetics and Global Planning.   [1016]
Lauenroth, H.-G.:
  Algorithmic Model Systems and Innovation Strategies
    for Automation of Industrial Production and
    Information Systems.   [1017]
Launer, R.L.:
  Modern Data Analysis.   [1018]
Lavorel, P.M.:
  How we Name or Misname Objects.   [1019]
Law, A.M.:
  The Art and Theory of Dynamic Programming.   [438]
Law, S.S.Y.:
  Adaptive Nuclear Reactor Control without Explicit
    Identification.   [1569]
Lazarevic, B.:
  Global and Large-Scale System Models. (Proc. of a
    Seminar in Dubrovnik, Yugoslavia, August 1978)   [1020]
Le Bonniec, G.P.:
  The Development of Model Reasoning.   [1021]
Le Moigne, J.-L.:
  Analyse de Systeme: Bibliographie et Commentaires
    (Systems Analysis: Bibliography and Comments).   [1022]
  Analyse de Systeme.   [1023]
LeMoigne, J.-L.:
  Modeling Complexity: A French History and its
    Prospects.   [1024]
Leal, A.:

An Interactive Program for Conversational Elicitation
    of Decision Structures.   [1025]
Lee, E.B.:
    Canonical Forms for Time Delay Systems.   [1026]
Lee, E.T.:
    Applications of Fuzzy Set Theory to Image Sciences.   [1027]
Lee, M.H.:
    System-Reliability Evaluation Techniques for
    Complex/Large Systems: A Review.   [804]
Lee, R.M.:
    Expert vs. Management Support Systems: Semantic Issues. [1028]
Lee, S.C.:
    Fuzzy Switching and Automata: Theory and Applications. [874]
Leeming, A.M.C.:
    A Systems Study of the Design of a User Interface.   [1029]
Lefebre, V.A.:
    Algebra of Conscience: A Comparative Analysis of
    Wester and Soviet Ethical Systems.   [1030]
    Second Order Cybernetics in the Soviet Union and the
    West.   [1031]
Legget, R.W.:
    A Reliability Index for Models.   [1032]
Lehmkule, S.W.:
    Visual Information Processing.   [1605]
Leigh, J.R.:
    Functional Analysis and Linear Control Theory. [1033]
Leinfellner, E.:
    Causality and Language: Phonetics / Phonology, Syntax,
    Semantics.   [1034]
    Disambiguation in German.   [1615]
    Ontology on the Computer.   [1719]
    VIE-LANG: A German Language Understandig System.   [238]
Leinhardt, S.:
    Special Issue of Mathematical Sociology on Social
    Network Research.   [1035]
Lenat, D.B.:
    Knowledge-Based Systems in Artificial Intelligence.   [383]
Lendaris, G.G.:
    Structural Modeling - A Tutorial Guide.   [1036]
Leon, J.:
    From Natural Language to a Canonical Representation of
    the Corresponding Semantic Relationships.   [504]
Lesmo, L.:
    An Interpreter of Fuzzy Production Rules.   [1037]
Lester, F.K.:
    Mathematical Problem Solving.   [1038]
Leung, Y.:
    A Fuzzy Set Analysis of Sociometric Structure. [1039]
Leung-yan-cheung, S.K.:
    Some Equivalences between Shannon Entropy and
    Kolmogorov Complexity.   [1040]
Levien, R.E.:
    Introduction to IIASA: Applying Systems Analysis in an
    International Setting.   [1041]
Levine, A.:
    Information Theoretic Analysis of Questionnaire Data.   [222]
Levine, R.D.:
    The Maximum Entropy Formalism.   [1042]
Levitan, G.L.:
    Use of Deductive Inference in Situation Control
    Systems.   [1043]
Levy, L.S.:
    Discrete Structures of Computer Science.   [1044]
Lewandovski, A.:

The Generation of Efficient Energy Supply Strategies
    using Multi-Criteria Optimization.   [658]
Lewicka-Strzalecka, A.:
    The Systems Approach as a Methodological Attitude. An
    Empirical Analysis.   [1045]
Lewin, D.:
    Theory and Design of Digital Computer Systems.   [1046]
Lewis, H.R.:
    The Efficiency of Algorithms.   [1047]
Lieberherr, K.J.:
    Complexity of Partial Satisfaction.   [1048]
Lilienfeld, R.:
    The Rise of Systems Theory: an Ideological Analysis. [1049]
Linger, R.C.:
    Structured Programming: Theory and Practice.   [1050]
Linkens, D.A.::
    Biological Systems Modelling and Control.   [1051]
Linstone, H.A.:
    The Multiple Perspective Concept with Applications to
    Technology and Other Decision Areas.   [1052]
    Cybernetics and the Future.   [1053]
Lions, J.I.:
    Some Methods in the Mathematical Analysis of Systems
    and Their Control.   [1054]
Lions, J.L.:
    New Trends in Systems Analysis: International
    Symposium, Versailles 1976.   [158]
    International Symposium on Systems Optimization and
    Analysis.   [159]
    Analysis and Optimization of Systems. (Proc. of a
    Conference in Versailles, France, Dec.1980).   [160]
    Some Methods in the Mathematical Analysis of Systems
    and Their Control.   [1055]
Lipset, D.:
    Gregory Bateson: The Legacy of a Scientist.   [1056]
Lipski, W.:
    Information Systems with Incomplete Information. [1057]
    On the Logic of Incomplete Information.   [1058]
Lipton, R.J.:
    Social Processes and Proofs of Theorems and Programs. [407]
Liu, P.S.:
    Statistical Pattern Classification with Binary
    Variables.   [1914]
Locker, A.:
    A New Look at the Description and Prescription of
    Systems.   [1059]
Loefgren, L.:
    Describability and Learnability of Complex Systems. [1060]
    Knowledge of Evolution and Evolution of Knowledge.   [1061]
    Excerpts from the Autology of Learning.   [1062]
Loo, S.G.:
    Measures of Fuzziness.   [1063]
    Fuzzy Relations in the Social and Behavioral Science. [1064]
Lopes, M.A.:
    Operations on Problems and Their Solution Spaces. [1065]
Lowen, W.:
    Creep-Creep: Out of Human Scale.   [1066]
    Dichotomies of the Mind: A Systems Science Model of
    the Mind and Personality.   [1067]
Lozinskii, E.L.:
    Construction of Relations in Relational Databases. [1068]
    Locality in Natural Language Processing.   [1069]
Lucadou, W.v.:
    Pragmatic Information and Nonclassical Systems.   [972]

Some Notes on Information and Interaction.   [973]
Lucas, R.G.:
  Political Determinants of Information System
    Configuration.   [1070]
Luenberger, D.G.:
  Introduction to Dynamic Systems: Theory, Models, and
    Applications.   [1071]
Luhman, N.:
  Trust and Power.   [1072]
  The World Society as a Social System.   [1073]
Lundeberg, M.:
  Information Systems Development: A Systematic Approach. [1074]
Lundquist, J.:
  Theory of Relatively Closed Systems and Applications. [675]
  The Axsater Integrated Production-Inventory System
    Interpreted in Terms of the Theory of Relatively
    Closed Systems.   [676]
Lusk, E.J.:
  A Discussion of External Validity in Systemic Research. [1075]
M'Pherson, P.K.:
  Systems Engineering: An Approach to Whole-System
    Design.   [1076]
  A Framework for Systems Engineering Design.   [1077]
MacDonald, B.A.:
  The Competence of a Multiple Context Learning System. [1078]
MacDonald, N.:
  Time Lags in Biological Models.   [1079]
MacElroy, R.D.:
  Complex Systems Tend to be Unstable. Is this Relevant
    to Ecology?   [334]
Machlup, F.:
  Knowledge: Its Creation, Distribution and Economic
    Significance. Vol.2: The Branches of Learning.   [1080]
Machtey, M.:
  An Introduction to the General Theory of Algorithms. [1081]
Maciejowski, J.M.:
  The Modelling of Systems with Small Observation Sets. [1082]
Mackenzie, K.D.:
  A Process Based Measure for the Degree of Hierarchy in
    a Group.   [1083]
  Leadership as a Task Process Uncertainty Control
    Process.   [1084]
Mackey, M.C.:
  The Extinction of Slowly Evolving Dynamical Systems. [1012]
Mackinnon, A.J.:
  Complexity and Decision-Making.   [1085]
Madan, S.:
  Cybernetic Modeling of Queueing Complex.   [1086]
  Basic Concepts on the Theoretical Treatment of
    Cybernetic Queueing Systems.   [1087]
Mager, P.P.:
  Biorhythms, System Organization and Bioactive
    Compounds.   [1088]
Mahmoud, M.S.:
  Large Scale Systems Modelling.   [1089]
Maione, U.:
  Mathematical Models for Surface Water Hydrology.
    (Proc. of the Workshop on Mathematical Models in
    Hydrology, Pisa, Italy, Dec. 9-12, 1974)   [316]
Majewska-Tronovicz, J.:
  Selected Praxiological Bibliography.   [1090]
Majone, G.:
  Pitfalls of Analysis.   [1091]
Majster, M.E.:

Extended Directed Graphs, a Formalism for Structured
    Data and Data Structures.   [1092]
Mak, K.L.:
  Dynamics and Control of Integrated
    Production-Inventory-Shipping Systems.   [1093]
Maker, P.P:
  Input-Output Relations in Multivariate Biosystems. [1094]
Makridakis, S.:
  Forecasting: Methods and Applications.   [1095]
  Automation or Autonomy in Organizational Design.   [509]
Malec, J.:
  Scenarios as a Tool for Dynamic Scene Representation. [1096]
Malik,, F.:
  Evolutionary Management.   [1097]
Mamdani, E.H.:
  Application of Fuzzy Logic to Approximate Reasoning
    Using Linguistic Synthesis.   [1098]
Mamdani, E.M.:
  Fuzzy Reasoning and Its Applications.   [1099]
Mandani, E.H.:
  A Fuzzy Logic Controller for a Traffic Junction.   [1280]
Mandriolo, D.:
  An Integrated Model of Problem Solver.   [679]
Manes, E.G.:
  Foundations of Systems Theory: Finitary and Infinitary
    Conditions.   [29]
  Foundations of System Theory: The Hankel Matrix.   [48]
Manescu, M.:
  Economic Cybernetics.   [1100]
Mann, K.H.:
  Ecology of Coastal Waters: A Systems Approach (Studies
    in Ecology, Volume 8).   [1101]
Manohar Raho, M.J.:
  Control Systems and Quantitative Economic Policy. [1102]
Marchuk, G.I.:
  Modelling and Optimization of Complex Systems. (Proc.
    of a Conference in Novosibirsk, USSR, July 1978) [1103]
Margolis, D.L.:
  Bond-graphs for Some Classic Dynamic Systems.   [1104]
Mark, J.W.:
  Application of Iterative Algorithms to Adaptive
    Predicitive Coding.   [1105]
Marko, H.:
  Methods of System Theory.   [1106]
Marlow, B.K.:
  Converting Run Code to Chain Code.   [122]
Marmarelis, P.Z.:
  Analysis of Physiological Systems.   [1107]
Marmarelis, V.Z.:
  Analysis of Physiological Systems.   [1107]
Marshall, J.E.:
  Control of Time-Delay Systems.   [1108]
Martin, C.F.:
  Applications of Algebraic Geometry to Systems Theory,
    Part 1   [752]
Martin, J.N.T.:
  On Mapping Real Systems.   [1109]
Martin, N.F.G.:
  Mathematical Theory of Entropy.   [1110]
Martin, P.E.:
  Simplification and Control of Information Patterns -
    Further Developments.   [1111]
Martin, T.B.:
  Automatic Speech and Speaker Recognition.   [426]

Martin, W.:
    Identification of the Temperature Masking of the
        Circadian System of Euglena Gracilis.   [847]
Martins, R.C.B.:
    On Reducibilities Among General Problems.   [1807]
Maruyama, M.:
    Heterogenistics: an Epistemological Restructuring of
        Biological and Social Sciences.   [1112]
Maryanski, F.I.:
    Digital Computer Simulation.   [1113]
Mason, R.O.:
    Measurement for Management Decision.   [1114]
Mastretta, M.:
    Development of Performance Adaptive Fuzzy Controllers
        with Application to Continuous Casting Plants.   [113]
Masud, A.S.:
    Multiple Objective Decision Making: Methods and
        Applications (a State-of-the-Art Survey).   [801]
Masuda, Y.:
    The Information Society as Past-Industrial Society.   [1115]
Matejko, A.:
    The Structural Criteria of Social System Maturity.   [1116]
Matevosyan, A.K.:
    The Invers Problem of Karhunen-Loeve.   [1117]
Matis, J.H.:
    Compartmental Analysis of Ecosystems Models.   [1118]
Matsuda, T.:
    Algebraic Properties of Satisficing Decision.   [1119]
Matsuno, K.:
    Compartmentalization of Self-Reproducing Machineries:
        Multiplication of Microsystems with
        Self-Instructing Polymerization of Amino Acids.   [1120]
    Disequilibrium Dynamics of Autonomous Systems and
        Their Structural Transformation.   [1121]
    Operational Description Microsystems Formation in
        Probiological Molecular Evolution.   [1122]
    Dynamics of Relations, Sequential Connectedness, and
        Feedback Loops in Open Systems.   [1123]
    Material Self-Assembly as a Physiochemical Process.   [1124]
    Self-Sustaining Multiplication and Reproduction of
        Microsystems in Protobiogenesis.   [1125]
Mattessich, R.:
    Instrumental Reasoning and Systems Methodology: an
        Epistemology of the Applied and Social Sciences. [1126]
    Axiomatic Representation of the Systems Framework:
        Similarities and Differences Between Mario Bunges's
        World of the Systems and My Own Systems Methodology. [1127]
Maturana, H.R.:
    Autopoiesis and Cognition: The Realization of the
        Living.   [1128]
Maurer, H.:
    The Austrian Approach to Videotex.   [1129]
Mauri, G.:
    Analysis and Compacting of Musical Texts.   [163]
May, R.M.:
    Will a Large Complex System be Stable?   [1130]
Maybeck, P.S.:
    Stochastic Models, Estimation, and Control.   [1131]
Maybee, J.S.:
    Computer-Assisted Analysis and Model Simplification. [660]
    Computer-Assisted Analysis and Model Simplification. [661]
McAleer, W.E.:
    Systems: A Concept for Business and Management.   [1132]
McClamroch, N.H.:

State Models of Dynamic Systems.   [1133]
State Models of Dynamic Systems: A Case Study Approach. [1134]
McCorkell, C.:
   A Hierarchical Adaptive Control System.   [1135]
McDonald, J.J.:
   Case Studies in Mathematical Modelling.   [831]
McDowall, D.:
   Interrupted Time Series Analysis.   [1136]
McGill, R.:
   Graphical Presentation of Results from Scientific
      Computer Models.   [652]
McGowan, C.L.:
   A Review of Decomposition and Design Methodologies.   [1137]
McIntosh, J.E.A.:
   Mathematical Modelling and Computers in Endocrinology. [1138]
McIntosh, R.P.:
   Mathematical Modelling and Computers in Endocrinology. [1138]
McIntyre, C.D.:
   A Hierarchical Model of Lotic Ecosystems.   [1139]
   Addendum to Internal Report 165: Mathematical
      Documentation for a Lotic Ecosystem Model.   [1140]
   Mathematical Documentation for a Lotic Ecosystem Model. [328]
McLean, J.M.:
   Identification and Analysis of the Subsystems
      Structure of Models.   [650]
   Methods of System Analysis and Model Building.   [1141]
McLone, R.R.:
   Mathematical Modelling.   [38]
McPherson, P.K.:
   On Understanding Modelling and Improving Human Systems. [1142]
Meadows, D.L.:
   Techniques for Implementing Computer-Based Policy
      Models.   [1143]
Mecca, S.J.:
   Problem Solving: a Systems Approach.   [1430]
Medgyessy, P.:
   Decomposition of Superpositions of Density Functions
      and Discrete Distributions.   [1144]
Medina, B.F.:
   Structured System Analysis: A New Technique.   [1145]
Mees, A.L.:
   Dynamics of Feedback Systems.   [1146]
Mehlhorm, K.:
   Nearly Optimal Binary Search Trees.   [1147]
Mehlmann, A.:
   Optimal Seducing Policies for Dynamic Continuous
      Lovers Under Risk of Being Killed a Rival.   [729]
Meissonnier, V.:
   From Natural Language to a Canonical Representation of
      the Corresponding Semantic Relationships.   [504]
Mel'cuk, I.A.:
   Cybernetics and Linguistics: Some Reasons for as Well
      as Some Consequences of Bringing them Together.   [1148]
   Cybernetics and Linguistics.   [1149]
Meldman, M.J.:
   Riss: a Relational Data Base Management System for
      Minicomputers.   [1150]
Melkumian, A.V.:
   Distortion Measures in Image Data Compression and
      Transmission Systems.   [1151]
Mello, J.M.C.:
   Decision Networks.   [734]
Memmi, D.:
   From Natural Language to a Canonical Representation of

the Corresponding Semantic Relationships.   [504]
Mende, W.:
   Application of Polyoptimization to Processes of
      Evolution.   [1320]
   Ecological Approach to Systems Analysis Based on
      Volterra Equations.   [1322]
   Ecological Approach to Systems Analysis Based on
      Volterra Equations.   [1323]
   An Evolutionary Analysis of World Energy Consumption
      and World Population.   [982]
Mergenthaler, W.:
   The Total Repair Cost in a Defective, Coherent Binary
      System.   [1152]
Mesarovic, M.D.:
   Theoretical Approaches to Non-Numerical Problem
      Solving.   [103]
   Multilevel Computer Model of World Development System. [1153]
Metz, H.A.J.:
   State Space Models for Animal Behaviour.   [1154]
Metzer, J.R.:
   Decision Table Languages and Systems.   [1155]
Meyer, A.R.:
   Weak Monadic Second Order Theory of Successor Is Not
      Elementary Recursive.   [1156]
Meyer, J.F.:
   Functional Dependence and its Application to System
      Evaluation.   [90]
   Closed-Form Solutions of Performability.   [1157]
Meyer, M.:
   A Statistical Scheme for Structural Analysis in Marine
      Ecosystems.   [185]
Meyer, W.:
   Data- and Behaviour Analysis with Classification
      Models.   [1321]
Michalski, R.S.:
   Pattern Recognition as Rule-Guided Inductive Inference. [1158]
Michel, A.N.:
   Qualitative Analysis of Large Scale Systems.   [1159]
Michie, D.:
   Expert Systems in the Microclectronic Age.   [1160]
   Machine Intelligence and Related Topics: An
      Information Scientist's Weekend Book.   [1161]
Michiels, H.:
   The Application of Cybernetic Principles for the
      Training of Operator Crews of Nuclear Power Plants. [833]
Migliarese, P.:
   A Methodological Inquiry of Organizational Noise in
      Sociotechnical Systems.   [315]
Mihram, G.A.:
   Simulation: Methodology for Decision Theorists. [1162]
Milgram, D.L.:
   Reduction Operation for Constraint Satisfaction. [725]
Milgram, M.:
   Barycentric Entropy of Mankov Systems.   [1163]
Millendorfer, J.:
   Growth Reducing Factors in Complex Systems.   [1164]
   Systemanalytical Approaches to Long Economic Waves. [1165]
Miller, D.:
   Design and System: Methodological Aspects, Vol.3. [577]
Miller, G.L.:
   Graph Isomorphism, General Remarks.   [1166]
Miller, J.G.:
   Living Systems.   [1167]
Miller, M.C.:

Mathematical Models in Medical Diagnosis.   [1168]
Miller, R.B.:
  Archetypes of Man-Computer Problem Solving.   [1169]
Miller, R.K.:
  Qualitative Analysis of Large Scale Systems.   [1159]
Mills, H.D.:
  Structured Programming: Theory and Practice.   [1050]
Milsum, J.H.:
  Cybernetics and Health Care.   [1170]
Minciardi, R.:
  Development of Performance Adaptive Fuzzy Controllers
    with Application to Continuous Casting Plants.   [113]
Mingers, J.C.:
  Towards an Appropriate Social Theory for Applied
    Systems Thinking: Critical Theory and Soft Systems
    Methodology.   [1171]
Mira, C.:
  Recurrences and Discrete Dynamic Systems.   [683]
Mitroff, I.I.:
  Methodological Approaches to Social Science.   [1172]
  On the Strength of Belief of Dialectical Belief
    Systems.   [1173]
Mitsch, W.J.:
  Energetics and Systems.   [1174]
Mittermeir, R.:
  CML-Graphs - A Notation for Systems Development.   [1175]
Mizumoto, M.:
  Note on the Arithmetic Rule by Zadeh for Fuzzy
    Conditional Inference.   [1176]
Mock, T.J.:
  Measurement, Accounting, and Organizational
    Information.   [1177]
Mohler, R.:
  Theory and Applications of Variable Structure Systems. [1178]
  Recent Developments in Variable Structure Systems,
    Economics and Biology.   [1179]
  An Overview of Bilinear System Theory and Applications. [1180]
Molina, F.W.:
  A Survey of Resource Directive Decomposition in
    Mathematical Programming.   [1181]
Molloy, K.J.:
  Insights into Complex Organisational Problem Solving. [1182]
Molloy, L.F.:
  The Unsteady State - Environmental Problems, Growth,
    and Culture.   [1849]
Monahan, G.E.:
  A Survey of Partially Observable Markov Decision
    Processes Theory, Models, and Algorithms.   [1183]
Montrucchio, L.:
  Bellamy's Equalitarian Utopia: A System-Theory
    Approach.   [364]
Moraga, C.:
  Introduction to Linear p-Adic Invariant Systems.   [1184]
Morein, R.:
  The Use of the Calculus of Variations and Functionals
    in the Analysis of Threshold Logic Unit Networks. [1185]
Morin, E.:
  Complexity.   [1186]
Morley, D.A.:
  Mathematical Modelling in Water and Wastewater
    Treatment.   [1187]
Morra, G.:
  Application of Loeve-Karhunen Expansion to Acoustic
    Diagnostic Problems for Internal Combustion Engines. [512]

Morris, C.N.:
    Introduction to Data Analysis and Statistical
        Inference.   [1188]
Morris, H.M.:
    Critical Stability in a Conserved Nutrient Ecosystem. [1189]
Morten, M.:
    Development of Performance Adaptive Fuzzy Controllers
        with Application to Continuous Casting Plants.   [113]
Morton, M.S.S.:
    Decision Support Systems.   [894]
    Decision Support Systems: an Organizational
        Perspective.   [895]
Moser, J.:
    Hidden Symmetrics in Dynamical Systems.   [1190]
Moss, Brian:
    Ecology of Fresh Waters.   [1191]
Moulin, Th.:
    Use of a Partially Self-Referent Formalism for
        Description of Natural Systems.   [1784]
Moylan, P.J.:
    Stability Criteria for Large-Scale Systems.   [1192]
Muhamad, M.A.:
    Chaotic Activity in Neural Systems.   [771]
Mulej, M.:
    Dialectical Systems Theory.   [1193]
    Dialectical Systems Theory and Work Simplification.   [1194]
    Method "Wholistic Creativity of Many (WCM)" Applied to
        Collect and Organize Associates' Ideas about New
        Products and Programs.   [1195]
Muller, D.J.:
    Algorithmic Aspects of Combinatorics.   [25]
Mumford, E.:
    Participative Systems Design: Structure and Method.   [1196]
Munro, I.:
    The Computational Complexity of Algebraic and Numeric
        Problems.   [190]
Munro, N.:
    Multivariate System Theory and Design.   [1289]
Murdick, R.G.:
    MIS: Concepts and Design.   [1197]
Murphy, R.E.:
    Adaptive Processes in Economic Systems.   [1198]
Murray, W.:
    Practical Optimization.   [613]
Nachane, D.M.:
    Optimization Methods in Multi-Level Systems.   [1199]
Nagel, E.:
    Teleology Revisited.   [1200]
Nahmias, S.:
    Fuzzy Variables.   [1201]
Nakano, B.:
    A Characterization of Interactions.   [1657]
    A Structure of Rational Decision Principles.   [1658]
    A Stationarization Functor and its Concrete
        Realizations.   [1659]
Narens, L.:
    A Critical Examination of the Analysis of Dichotomous
        Data.   [115]
Nash, P.:
    Systems Modelling and Optimization.   [1202]
Naughton, J.:
    Craft Knowledge in Systems Research: The Rules of the
        Game.   [1203]
Navarro, J.M.:

   Representation of Continuous Time State Equations by
      Discrete-Time State Equations.   [1546]
Naylor, A.W.:
   On Decomposition Theory: Generalized Dependence.   [1204]
Neal, M.:
   The Construction of Personality in the Novel.   [1205]
Neck, R.:
   Interactions between the Political and the Economic
      System in Austria.   [1206]
   A System-Theoretic Analysis of a Simple Macroeconomic
      Model.   [1207]
Nedeltchev, N.:
   A Method for Performance Assessment of Medical
      Radioisotope Equipment.   [906]
Neftci, S.:
   Canonical Forms for Time Delay Systems.   [1026]
Negoita, C.:
   Fuzzy Sets and Systems: a New International Journal.   [1208]
   Management Applications of Systems Theory.   [1209]
   Fuzzy Systems.   [1210]
Negoita, C.V.:
   On the Logic of Discrete Systems Dynamics.   [1211]
   Fuzzy Systems.   [1212]
Nelson, R.J.:
   Structure of Complex Systems.   [1213]
Nelson, S.H.:
   Problem Solving and Computer Programming.   [672]
Netchiporenko, V.I.:
   Structure Analysis of Systems.   [1214]
Neuwirth, E.:
   Fuzzy Similiarity Relations and Distance Coefficients
      in Thesauri.   [1215]
Nevitt, B.:
   ABC of Prophecy: Understanding the Environment.   [1216]
Newell, A.:
   Human Problem Solving.   [1217]
   The Knowledge Level.   [1218]
Newton-Smith, W.H.:
   The Structure of Time.   [1219]
Nguyen, H.T.:
   On Conditional Possibility Distributions.   [1220]
Nicholson, H.:
   Structure of Interconnected Systems.   [1221]
   Modelling of Dynamical Systems.   [1222]
   Modelling of Dynamical Systems.   [1223]
Nicolis, G.:
   Self-Organization in Non-Equilibrium Systems.   [1224]
Nicolis, J.S.:
   Inadequate Communication between Self-Organizing
      Systems and Desynchronization of Physiological
      Rhythms.   [1225]
   Bifurcation in Non-Negotiable Games: A Paradigm for
      Self-Organization in Cognitive Systems.   [1226]
Niehaus, R.J.:
   Manpower Planning and Organization Design.   [234]
Niemeyer, H.:
   Soziale Beziehungsgeflechte.   [1227]
Niezgodka, M.:
   Applications of Systems Theory to Economics,
      Management, and Technology.   [692]
Nigro, A.:
   Language Behavioral Feedback.   [1228]
Nijssen, G.M.:
   Modelling in Data Base Management Systems.   [1229]

Nikiforuk, P.N.:
  On the Design of an Adaptive Controller for Uncertain
    Plants via Liapunov Signal Synthesis Method.   [687]
Nilsson, A.:
  Information Systems Development: A Systematic Approach. [1074]
Nilsson, N.J.:
  Principles of Artificial Intelligence.   [1230]
  Readings in Artificial Intelligence.   [1850]
Ninneman, J.:
  The Survival of U.S.Business: A General Systems
    Approach.   [1231]
Nirenburg, S.:
  Locality in Natural Language Processing.   [1069]
Nishimura, H.:
  Realization of Dependency Structures.   [1232]
Nitecki, Z.:
  Global Theory of Dynamical Systems.   [1233]
Nobile, A.G.:
  On a Class of Discrete Models for Regulated Growth
    with Intrinsic Lower Bounds.   [1234]
Noguchi, S.:
  Properties of Analogue Neural Networks.   [1235]
Nokso-Koivisto, P.:
  Modeling and Simulating the Resource Allocation of the
    Regional Hospital System with Different Service
    Levels.   [1463]
Nolle, H.:
  Experimental Modelling in Engineering.   [379]
Norman, D.A.:
  Perspectives in Cognitive Science. (Papers from a
    Meeting in La Jolla, Cal., Aug. 1979).   [1236]
Norman, M.:
  Software Package for Economic Modelling.   [1237]
Nowak, S.:
  Methodology of Sociological Research.   [1238]
Nowakowska, M.:
  Methodological Problems of Measurement of Fuzzy
    Concepts in the Social Sciences.   [1239]
  The Logical Structure of the Development of a
    Scientific Discipline.   [1240]
  Theories of Research: Modelling Approaches.   [1241]
  Instability, Monopolies, and Alienation in Science.   [1242]
  Formal Semiotics and Multidimensional Semiotic Systems. [1243]
  New Theoretical, Methodological, and Empirical
    Possibilities in Decision Theory.   [1244]
  Cybernetics in the Social Sciences.   [1245]
Nowotny, H.:
  Not Quite Human: Science and Utopia.   [1246]
Nugyen, V.V.:
  Review and Unification of Linear Identifiability
    Concepts.   [1247]
Nurmi, H.:
  On Strategies of Cybernetic Model-Building.   [1248]
Nussbaumer, H.J.:
  Fast Fourier Transform and Convolution Algorithms.   [1249]
O'Donovan, T.M.:
  GSPS: Simulation Made Simple.   [1250]
O'Muircheartaigh, C.A.:
  The Analysis of Survey Data, Vol. 1: Exploring Data
    Structures. Vol. 2: Model Fitting.   [1251]
O'Neill, R.V.:
  Systems Ecology.   [1552]
Odell, P.L.:
  Cluster Analysis: a Survey.   [448]

Oden, A.S.:
   An Overview of Mathematical Modeling of the Behavior
      of Hydrocarbon Reservoirs.   [1252]
Odum, E.C.:
   Energy Basis for Man and Nature.   [1253]
Odum, E.P.:
   The Cybernetic Nature of Ecosystem.   [1295]
Odum, H.T.:
   Energy Basis for Man and Nature.   [1253]
Oeren, T.I.:
   Cybernetics and Modelling and Simulation of Large
      Scale Systems.   [1254]
   Rationale for Large Scale System Simulation Software
      Based on Cybernetic and General Systems Theories.   [1255]
   Simulation with Discrete Models.   [1256]
Ogborn, J.M.:
   Conversation Theory.   [1257]
Oguntade, O.O.:
   Implementing a Pragmatic Theory of Humanistic Systems. [1258]
Ohlsson, S.:
   On the Automated Learning of Problem Solving Rules.   [1259]
Ohmori, A.:
   A Methodology for Quality Goal-Seeking and
      Coordination, and the Practical Application.   [550]
Okeda, M.:
   Overlapping Decompositions, Expansions, and
      Contractions of Dynamic Systems.   [1260]
Okuda, T.H.:
   A Formulation of Fuzzy Decision Problems with Fuzzy
      Information Using Probability Measures of Fuzzy
      Events.   [1261]
Olbrot, A.:
   Canonical Forms for Time Delay Systems.   [1026]
Oldershaw, R.L.:
   Conceptual Foundations of the Self-Similar
      Hierarchical Cosmology.   [1262]
   On the Number of Levels in the Self-Similar
      Hierarchical Cosmology.   [1263]
   Empirical and Theoretical Support for Self-Similarity
      between Atomic and Stellar Systems.   [1264]
   New Evidence for the Principle of Self-Similarity.   [1265]
Ollero, A.:
   Fuzzy Methodologies for Interactive Multicriteria
      Optimization.   [104]
   The Structure of Relations in Personnel Management.   [1266]
Omatu, S.:
   Modeling, Estimation, and Their Applications for
      Distributed Parameter Systems.   [1499]
Optner, S.L.O.:
   Systems Analysis.   [1267]
Orava, P.J.:
   On the Concepts of Input-Output Model, Consolity, and
      State in the Theory of Dynamical Systems and
      Control.   [1268]
Ord, J.K.:
   Spacial Processes: Models and Applications.   [320]
Oren, T.I.:
   Simulation - As it Has Been, Is, and Should Be.   [1269]
   Concepts for Advanced Simulation Methodologies.   [1270]
Orlovsky, S.A.:
   Decision-Making with a Fuzzy Preference Relation.   [1271]
   Effective Alternatives for Multiple Fuzzy Preference
      Relations.   [1272]
Osgood, D.W.:

Cognitive Structure: Theory and Measurement of
   Individual Differences. [1519]
Otnes, R.K.:
  Applied Time Series Analysis. Vol. 1: Basic Techniques. [1273]
Ottendorfer:
  Decomposition of General Dynamic Systems. [1334]
Overton, W.S.:
  A Strategy of Model Construction. [1274]
  Evolution of a Hydrology Model - an Exercise in
   Modelling Strategy [1275]
  On Constructing a Hierarchical Model in the Flex
   Paradigm Illustrated by Structural Aspects of a
   Hydrology Model. [1276]
Owens, D.H.:
  Feedback and Multivariate Systems. [1277]
  Analysis and Control of Multipass Processes. [456]
Owsinski, J.W.:
  New Trends in Mathematical Modelling. [1626]
Pagliuca, A.:
  On Factor Analysis and Fisher's Linear Discriminant
   Analysis. [1492]
Palm, G.:
  Theoretical Approaches to Complex Systems. [746]
  Rules for Synaptic Changes and their Relevance for the
   Storage of Information in the Brain. [1278]
Palvoelgyi, L.:
  Network-Modelling of Learning. [1279]
Papadimitriou, C.H.:
  The Efficiency of Algorithms. [1047]
Pappis, C.P.:
  A Fuzzy Logic Controller for a Traffic Junction. [1280]
Pardo, L.:
  Information Energy of a Fuzzy Event and a Partition of
   Fuzzy Events. [1281]
Parisi, D.:
  A Lexicon Based Approach to Sentence Understanding. [269]
Paritsis, N.C.:
  Adaptional Problems in Natural Intelligent Systems
   with Changes in Environmental Variety. [1282]
Parks, M.S.:
  Dichotic Property and Teleogenesis. [1283]
Parsy, J.P.:
  A Method for Decomposing Interpreted Petri Nets and
   its Utilization. [1707]
Pask, G.:
  Progress in Cybernetics and Systems Research, Vol.IV. [1713]
  Consciousness. [1284]
  Development in Conversational Theory - Part 1 [1285]
  Progress in Cybernetics and Systems Research, Vol.IX. [1720]
  Progress in Cybernetics and Systems Research, Vol.VII. [1721]
  The Originality of Cybernetics and the Cybernetics of
   Originality. [1286]
  Cybernetics in Psychology and Education. [1287]
  The Architecture of Knowledge and Knowledge of
   Architecture. [1288]
Patel, R.V.:
  Multivariate System Theory and Design. [1289]
Patten, B.C.:
  Propagation of Cause in Ecosystems. [1290]
  Systems Analysis in Ecology. [1291]
  Systems Approach to the Concept of Environment. [1292]
  Compartmental Analysis of Ecosystems Models. [1118]
  Systems Approach to the Concept of Niche. [1293]
  System Theory and the Ecological Niche. [1294]

The Cybernetic Nature of Ecosystem.   [1295]
Pau, L.F.:
    Failure Diagnosis and Performance Monitoring.   [1296]
    Pattern Recognition Theory and Applications.   [931]
Paulre, B.E.:
    Systems Dynamics and the Analysis of Chance.   [1297]
Payne, C.:
    The Analysis of Survey Data, Vol. 1: Exploring Data
        Structures. Vol. 2: Model Fitting.   [1251]
Peacocke, C.:
    Holistic Explanation.   [1298]
Pearce, J.G.:
    Telecommunications Switching.   [1299]
Pearl, J.:
    On the Complexity of Inexact Computations.   [1300]
    On the Storage Economy of Inferential Question-Answer
        Systems.   [1301]
    A Framework for Processing Value Judgements.   [1302]
    An Interactive Program for Conversational Elicitation
        of Decision Structures.   [1025]
    On Summarizing Data Using Probabilistic Assertions. [1303]
    An Economic Basis for Certain Methods of Evaluating
        Probabilistic Forecasts.   [1304]
Pearlman, W.:
    Beyond Gestalt.   [1305]
Pearson, J.M.:
    The Use of Cybernetics in Organizational Design and
        Development: An Illustration within Air Force
        Logistics Command.   [1736]
Peay, E.R.:
    Structural Models with Qualitative Values.   [1306]
Pedretti, A.:
    Epistemology, Semantics, and Self-Reference.   [1307]
    The Cybernetics of Language.   [1308]
    Where Coincidence Coincides - Coining a Notion of
        Cybernetic Description?   [1309]
Pedrycz, W.:
    On Identification of Fuzzy Systems and Its Application
        in Control Problems.   [368]
    Control Problems in Fuzzy Systems.   [369]
    Fuzzy Relational Equations with Triangular Norms and
        their Resolutions.   [1310]
    Fuzzy Rules Generation for Fuzzy Control.   [370]
    Some Applicational Aspects of Fuzzy Relational
        Equations in Systems Analysis.   [1311]
    A Model of Decision-Making in a Fuzzy Environment. [1312]
    Construction of Fuzzy Relational Models.   [1313]
Peery, N.S.:
    General Systems Theory Approaches to Organizations:
        Some Problems in Application.   [1314]
Peixoto, M.M.:
    Dynamical Systems.   [1315]
Pelzmann, L.:
    Learned Helplessness - An Attribution-Model for
        Unemployed Workers?   [1316]
Perlis, A.J.:
    Social Processes and Proofs of Theorems and Programs. [407]
    Software Metrics: An Analysis and Evaluation.   [1317]
Perlis, J.H.:
    Stability Analysis: An Approach to the Evaluation of
        System Design.   [1318]
Pernici, B.:
    Problem Solving in Game Playing: Computer Chess.   [1319]
Peschel, M.:

Application of Polyoptimization to Processes of
    Evolution.   [1320]
Data- and Behaviour Analysis with Classification
    Models.   [1321]
Ecological Approach to Systems Analysis Based on
    Volterra Equations.   [1322]
Ecological Approach to Systems Analysis Based on
    Volterra Equations.   [1323]
Engineering Cybernetics.   [1324]
Socio-Economic Consequences of the Volterra Approach
    for Nonlinear Systems.   [1325]
Pestel, E.:
Multilevel Computer Model of World Development System. [1153]
Petersen, R.A.:
Applying Discrete Model Concepts to the Computation of
    Atmospheric Trajectories.   [1765]
Peterson, C.:
Cognitive Structure: Theory and Measurement of
    Individual Differences.   [1519]
Peterson, J.L.:
Petri Nets.   [1326]
Petri Net Theory and the Modeling of Systems.   [1327]
Petkoff, B.:
A Cybernetic Model of Scientific Research and
    Cognition.   [1328]
Petrov, B.N.:
Some Problems in the Theory of Searchless
    Self-Adaptive Systems, I.   [1329]
Petrovski, A.:
Medico-Demographic Models in Health Care Systems.   [1330]
Pfeilsticker, A.:
The Systems Approach and Fuzzy Set Theory Bridging the
    Gap between Mathematical and Language-Oriented
    Economics.   [1331]
Phillips, D.C.:
Holistic Thought in Social Science.   [1332]
Phillips, R.J.:
AESOP: an Architectural Relational Data Base.   [1333]
Pichet, J.:
Computer-Aided Identification of System Structure: An
    Attempt at Implementation.   [899]
Pichler, F.:
Decomposition of General Dynamic Systems.   [1334]
Progress in Cybernetics and Systems Research, Vol.VII. [1335]
Progress in Cybernetics and Systems Research, Vol.VI.   [1336]
Progress in Cybernetics and Systems Research, Vol.VIII. [1722]
Dynamical Systems Theory.   [1337]
Invariance-Relations in General Dynamic Systems.   [1553]
Robotics and Image Processing.   [1338]
General Systems Algorithms for Mathematical Systems
    Theory.   [1339]
Pickett, R.M.:
Evaluation of Diagnostic Systems: Methods from Signal
    Detection Theory.   [1646]
Pierce, J.R.:
Introduction to Communication Science and Systems.   [1340]
Pilsworth, B.W.:
Dynamic Programming for Fuzzy Systems with Fuzzy
    Environment.   [85]
Pinkava, V.:
The Algebraic Structure of the Spencer-Brown and
    Varela Calculi.   [965]
Pinske, M.S.:
Information and Information Stability of Random

Variables and Processes.   [1341]
Pipino, L.L.:
Potential Impact of Fuzzy Sets on the Social Sciences. [1342]
Pippenger, N.:
Information Theory and the Complexity of Boolean
Functions.   [1343]
Complexity Theory.   [1344]
Pirc, V.:
Method "Wholistic Creativity of Many (WCM)" Applied to
Collect and Organize Associates' Ideas about New
Products and Programs.   [1195]
Pitrik, R.:
Functional versus Imperative Programming - Complement
or Contradiction?   [908]
Pla Lopez, R.:
Systemic Transition from Ideological Learning to
Scientific Learning.   [1345]
Plotkin, A.A.:
Hierarchical Systems of Subsets.   [1346]
Poch, F.A.:
Some Applications of Fuzzy Sets to Statistics .   [1347]
Pogonowski, J.:
Set-Theoretical Approach to General Systems Theory. [1348]
Pokorny, D.:
Processing Sociological Data by the Guha Method: an
Example.   [737]
Popovici, A.:
Energy Management Planning Tools for the Transition
towards a Medium Developed Economy.   [605]
Popper, M.:
CODEX: Prototypes Driven Backward and Forward Chaining
Computer-Based Diagnostic Expert System.   [695]
Porati, A.:
Random Ecological Systems with Structure:
Stability-Complexity Relationship.   [655]
Porenta, G.:
A Mathematical Model of the Dopaminergic Synapse:
Stability and Sensitivity Analyses, and Simulation
of Parkinson's Disease and Aging Processes.   [1349]
A System Analysis of Information Transmission of Cat
Retinal Horizontal Cells.   [1822]
Porter, B.:
Synthesis of Dynamical Systems.   [1350]
Posner, E.C.:
Introduction to Communication Science and Systems. [1340]
Pospelov, D.A.:
Large Systems: Situational Control .   [1351]
Decision Making on Fuzzy Grounds.   [1913]
Logical-Linguistic Models in Control Systems.   [1352]
Poston, T.:
Catastrophe Theory and its Applications.   [1353]
Pot, John S.:
Scientific Relevance and the Rehabilitation of the
Goal Concept.   [1354]
Potter, J.M.:
Partial Prior Information and Decision-Making.   [1355]
Pozolov, R.V.:
On the Algorithm for Determining the Primary Structure
of Biopolymers.   [1585]
Prade, H.:
Fuzzy Sets and Systems: Theory and Applications. [443]
A Class of Fuzzy Measures Based on Triangular Norms. [444]
On the Best Way of Designating Objects in Sentence
Generation.   [508]

Pradhan, D.K.:
  A Graph-Structural Approach for the Generalization of
    Data Management Systems.   [293]
Pratt, W.K.:
  Digital Image Processing.   [1356]
Presern, S.:
  Plan Generating System for a Tactile Sensor.   [1357]
Pressman, R.S.:
  Software Engineering: A Practitioner's Approach.   [1358]
Prevost, P.:
  "Soft" Systems Methodology, Functionalism and the
    Social Sciences.   [1359]
Priemer, R.:
  On System Identification with and without Certainty.   [871]
Prigogine, I.:
  Self-Organization in Non-Equilibrium Systems.   [1224]
  From Being to Becoming - Time and Complexity in the
    Physical Sciences.   [1360]
Primas, H.:
  Theory Reduction and Non-Boolean Theories.   [1361]
Pritsker, A.A.B.:
  Modeling and Analysis Using Q-Gert Networks.   [1362]
Probst, B.:
  A Statistical Scheme for Structural Analysis in Marine
    Ecosystems.   [185]
Probst, G.J.B.:
  Centralization Versus Decentralization in Business
    Organisations: Cybernetic Rules for Effective
    Management.   [637]
  Evolutionary Management.   [1097]
  New ('second order') Cybernetics against Mismanagement.   [1363]
  The Necessity of Sensitivity Models for Managers,
    Planners and Politicians in Health Care Systems.   [1364]
Prokop, A.:
  Systems Analysis and Synthesis in Biology and
    Biotechnology.   [1365]
Prokop, C.:
  Systems of Material Flow.   [923]
Prosser F.:
  The Art of Digital Design: An Introduction to Top-Down
    Design.   [1867]
Protonotarios:
  Bifurcation in Non-Negotiable Games: A Paradigm for
    Self-Organization in Cognitive Systems.   [1226]
Protzen, J.-P.:
  Design Methods and Theories: Vol.10, No.2.   [1366]
Pszczolowski, T.:
  Praxiology - The Theory with Past and Future.   [1367]
Pugh, R.E.:
  Evaluation of Policy Simulation Models: a Conceptual
    Approach to Case Study.   [1368]
Puri, M.L.:
  Law of Large Numbers and Central Limit Theorem for
    Fuzzy Random Variables.   [935]
Putt, A.M.:
  General Systems Theory Applied to Nursing.   [1369]
Quade, E.:
  The State-of-the-Art Questionnaire on Applied Systems
    Analysis: a Report on the Response.   [1370]
  Pitfalls of Analysis.   [1091]
Qualitz, J.E.:
  Machines, Languages, and Computation.   [409]
Quine, W.V.:
  Theories and Things.   [1371]

Rabin, M.O.:
  Complexity of Computations.   [1372]
Radecki, T.:
  Level Fuzzy Sets.   [1373]
Ragade, R.K.:
  Advances in Fuzzy Set Theory and Applications.   [684]
  Fuzzy Interpretative Modeling.   [1374]
  Advances in Fuzzy Set Theory and Applications.   [686]
  General Systems Yearbook.   [1375]
Ragade, R.R.:
  General Systems: Yearbook of the Society for General
    Systems Research.   [1376]
Raghaven, V.V.:
  A Comparison of the Stability Characteristics of Some
    Graph Theoretic Clustering Methods.   [1377]
Raiffa, H.:
  Conflicting Objectives in Decisions.   [142]
Rainey, L.B.:
  The Use of Cybernetics in Organizational Design and
    Development: An Illustration within Air Force
    Logistics Command.   [1736]
Rajasethupathy, K.S.:
  Considerations for Fuzzifying Formal Languages and
    Synthesis of Fuzzy Grammars.   [1005]
Rajkovic, V.:
  Multiattribute Decision Making Using a Fuzzy Heuristic
    Approach.   [457]
Ralescu, D.:
  Toward a General Theory of Fuzzy Variables.   [1378]
  Law of Large Numbers and Central Limit Theorem for
    Fuzzy Random Variables.   [935]
Randers, J.:
  Elements of the System Dynamic Method.   [1379]
Rao, G.P.:
  Approximate Reconstruction of Mapping Functions from
    Linguistic Descriptions in Problems of Fuzzy Logic
    Applied to Systems Control.   [1380]
Rao, G.V.:
  Complex Digital Control Systems.   [1381]
Rapoport, A.:
  Response Models for Detection of Change.   [1382]
  Cybernetics as a Link between two Theories of
    Cognition.   [1383]
Rastogi, P.N.:
  Cybernetic Study of Societal Systems (Part I): Theory
    and Methodology.   [1384]
  The Behaviour of Societal Systems.   [1385]
  An Introduction to Social and Management Cybernetics. [1386]
Rasvan, V.:
  Some System Theory Ideas Connected with the Stability
    Problem.   [1387]
Ratko, I.:
  Computer Registration of Patients Waiting for Cardiac
    Operation.   [1388]
  On Evaluating of Logical Expressions in Programming
    Languages.   [1389]
  Patient Registers on Microcomputers.   [905]
Rauch, W.D.:
  Automatic Refereeing and Excerpting.   [1390]
  Automatic Extracting as an Interactive Process.   [1391]
Ray, W.H.:
  Advanced Process Control.   [1392]
Rayner, J.N.:
  Proximity and Preference: Problems in the

Multidimensional Analysis of Large Data Sets.   [632]
Read, R.C.:
  The Graph Isomorphism Disease.   [1393]
Reckmeyer, W.J.:
  General Systems Research and Design: Precursors and
    Futures.   [1394]
Reddy, N.P.:
  A Thermodynamic Theory of Knowledge.   [1395]
Redhead, M.:
  Models in Physics.   [1396]
Rees, R.K.D.:
  Software Reliability (2 Vols).   [1397]
Reggiani, M.G.:
  Adequacy of Ecosystem Models.   [711]
Reichle, D.E.:
  Dynamic Properties of Forest Ecosystems.   [1398]
Reineri, M.T.:
  First Passage Time Problems and Some Related
    Computational Methods.   [511]
  ECG Map Filtering by Means of Spherical Harmonics: A
    Simple Approximation and Results.   [93]
Reisig, G.H.R.:
  Statistical Data Analysis in the Information Domain. [1399]
Rerick, M.N.:
  Problem Solving: a Systems Approach.   [1430]
Rescher, N.:
  Methodological Pragmatism: a Systems-Theoretical
    Approach to the Theory of Knowledge.   [1400]
  Cognitive Systematization: a Systems-Theoretic
    Approach to a Coherent Theory of Knowledge.   [1401]
  Induction.   [1402]
  Empirical Inquiry.   [1403]
Resconi, G.:
  Logical Analysis of Dynamical Systems-Applications to
    Life Sciences.   [1404]
  Analysis of the Purine Biosynthesis by Mathematical
    Logic.   [1405]
  A New Formalization of the General Systems by Logical
    and Algebraic Structures.   [1406]
Retti, J.:
  SISSY - A Fast Interactive Simulation System.   [1407]
  Interactive Computer-Aided Planning: Estimating
    Health-Care Resource Requirements.   [782]
  Health Care in Austria: 1980-2000.   [783]
  A Framework for Interactive Engineering of Simulation
    Models of Time-Oriented Systems.   [1408]
  Frame-Oriented Knowledge Representation for Modelling
    and Simulation.   [1409]
  MOSES: A Schema-Based Knowledge Representation for
    Modelling and Simulation of Dynamical Systems.   [1410]
Reusch, B.:
  Continuous Systems with Digital Behavior.   [1411]
  Characterization of Continuous Systems that are
    Digital.   [1412]
Reuver, H.A.:
  Learning within the Context of General Systems Theory. [1413]
  Social Systems and Learning Systems.   [721]
Revell, N.:
  Systematic Methodologies for the Design of Small
    Database Systems.   [1414]
Reynolds, J.C.:
  Reasoning About Arrays.   [1415]
Rezenblit, J.:
  Modelling of Evenistic Systems.   [821]

Ribaric, M.:
  A General Systems Theory Approach to the Properties of
    a Periodic Discrete String.   [145]
  An Open System Composed of Non-Active Monotonous Parts
    in Passive Surroundings.   [188]
Ricci, F.J.:
  Computer Simulation of a Cybernetic Model of Man in
    his Cultural Environment.   [1416]
Ricciardi, L.:
  Progress in Cybernetics and Systems Research, Vol.III. [1712]
  A Cybernetic Approach to Population Dynamics Modeling. [260]
  Mean Variance and Skewness of the First Passage Time
    for the Ornstein-Uhlenbeck Process.   [290]
  Cybernetics in Biology and Medicine: An Introduction. [1417]
  First Passage Time Problems and Some Related
    Computational Methods.   [511]
  Progress in Cybernetics and Systems Research, Vol.IX. [1720]
  Progress in Cybernetics and Systems Research, Vol.VII. [1721]
  Cybernetics and Biology: Classic Attempts at Neuronal
    Modeling.   [1418]
  On a Class of Discrete Models for Regulated Growth
    with Intrinsic Lower Bounds.   [1234]
Richardson, D.:
  AESOP: an Architectural Relational Data Base.   [1333]
Richardt, J.:
  Data- and Behaviour Analysis with Classification
    Models.   [1321]
Riederer, P.:
  A Mathematical Model of the Dopaminergic Synapse:
    Stability and Sensitivity Analyses, and Simulation
    of Parkinson's Disease and Aging Processes.   [1349]
Rifkin, J.:
  Entropy: a New World View.   [1419]
Rijndorp, J.E.:
  Process Dynamics, Control and Protection.   [1435]
Ringle, M.:
  Philosophical Perspectives in Artificial Intelligence. [1420]
Rissanen, J.:
  Universal Modeling and Coding.   [1421]
Rivett, P.:
  Model Building for Decision Analysis.   [1422]
Robb, F.:
  Cybernetics in Management Thinking.   [1423]
Roberts, F.S.:
  Discrete Mathematical Models.   [1424]
  Measurement Theory with Applications to
    Decision-Making, Utility, and the Social Sciences. [1425]
Roberts, G.F.:
  An Input/Output Approach to the Structural Analysis of
    Digraphs.   [1426]
Roberts, P.C.:
  Modelling Large Systems.   [1427]
  Modelling Large Systems: Limits to Growth Revisited. [1428]
Roberts, S.L.:
  Systems Approach to Assessing Behavioral Problems of
    Critical Care Patients.   [1429]
Robertshaw, J.E.:
  Problem Solving: a Systems Approach.   [1430]
Robertson, A.:
  Perspectives on Lake Eco-System Modeling.   [1502]
Robertson, E.L.:
  Properties of Conflict-Free and Persistent Petri Nets. [1008]
Robinson, A.:
  Non-Standard Analysis.   [1431]

Robinson, C.:
  Global Theory of Dynamical Systems.   [1233]
Robinson, D.F.:
  Digraphs: Theory and Techniques.   [1432]
Robinson, E.A.:
  Digital Foundations of Time Series Analysis. Vol. 1:
    The Box-Jenkins Approach.   [1433]
Roedding, W.:
  Different Approaches to the Aggregation of Preferences. [1434]
Roffel, B.:
  Process Dynamics, Control and Protection.   [1435]
Rogers, G.:
  Basic and Applied General Systems Research: A
    Bibliography.   [937]
  Cybernetics and Artificial Intelligence.   [587]
Roider, B.:
  Input/Output Codings and Transition Functions in
    Effective Systems.   [237]
Rolph, J.E.:
  Introduction to Data Analysis and Statistical
    Inference.   [1188]
Roman, R.:
  On the Logic of Discrete Systems Dynamics.   [1211]
Romano, P.:
  A Methodological Inquiry of Organizational Noise in
    Sociotechnical Systems.   [315]
Rondon, R.J.:
  Statistical Pattern Classification with Binary
    Variables.   [1914]
Rose, J.:
  Modern Trends in Cybernetics and Systems (3 Vols.). [1436]
  Current Topics in Cybernetics and Systems (Proc.Fourth
    Mt.Congress on Cybernetics and Systems, Amsterdam,
    Aug. 21-25, 1978) .   [1437]
  Robotica (Int.Journal of Information, Education and
    Research in Robotics and Artificial Intelligence). [1438]
Rosen, R.:
  Dynamical Similarity and the Theory of Biological
    Transformations.   [1439]
  Feedforwards and Global Systems Failure.   [1440]
  Fundamentals of Measurement and Representation of
    Natural Systems.   [1441]
  On Anticipatory Systems.   [1442]
  Anticipatory Systems in Retrospect and Prospect.   [1443]
  Protein Folding: a Prototype for Control of Complex
    Systems.   [1444]
  Progress in Theoretical Biology, Vol.6.   [1445]
  The Physics of Complex Systems.   [1446]
Rosenberg, R.M.:
  Analytical Dynamics of Discrete Systems.   [1447]
Rosenfeld, A.:
  Reduction Operation for Constraint Satisfaction.   [725]
Rosenkrantz, R.D.:
  Inference, Method and Decision.   [1448]
Ross, S.D.:
  Learning and Discovery.   [1449]
Rossier E.:
  Condensed Forms of Large Scale Models.   [533]
Rossler, O.E.:
  Bifurcation Theory and Applications in Scientific
    Disciplines.   [691]
Rota, G.C.:
  Finite Operator Calculus.   [1450]
Roth, G.:

Self-Organizing Systems: An Interdisciplinary Approach. [1451]
Rousset, M.C.:
  Propagation: Another Way for Matching Patterns in KBS.   [346]
Rowe, R.E.:
  Selective Polymorphism of the Erythrocyte Acid
    Phosphate Locus in Humans.   [625]
Rowe, W.D.:
  An Anatomy of Risk.   [1452]
Royce, J.R.:
  The Role of General Systems and Information Theory in
    Multifactor Individuality Theory.   [1453]
Rozenberg, G.:
  The Mathematical Theory of LSystems.   [1454]
Ruben, B.D.:
  General Systems Theory and Human Communication.   [1455]
Ruberti, A.:
  Recent Developments in Variable Structure Systems,
    Economics and Biology.   [1179]
Ruberti, R.:
  Theory and Applications of Variable Structure Systems. [1178]
Rubinstein, R.Y.:
  Simulation and Monte Carlo Method.   [1456]
Ruble, T.L.:
  An Evaluation of the Effectiveness of Dialectical
    Inquiry Systems.   [349]
Ruda, M.:
  Program Optimization and Manipulation on User Level. [904]
  Patient Registers on Microcomputers.   [905]
Rudall, B.H.:
  Developing Automated Teaching Systems.   [1457]
Rudall, R.:
  Computers and Cybernetics.   [1458]
Rudawitz, L.M.:
  Client-Centered Design: Concepts and Experience.   [1459]
Rudwick, B.H.:
  Solving Management Problems: a Systems Approach to
    Planning and Control.   [1460]
Ruffner, M.A.H.:
  Human Factors in Systems Analysis.   [1461]
Rugh, W.J.:
  Nonlinear Systems Theory: The Volterra/Wiener Approach. [1462]
Ruotsalainen, P.:
  Modeling and Simulating the Resource Allocation of the
    Regional Hospital System with Different Service
    Levels.   [1463]
Rus, T.:
  Data Structures and Operating Systems.   [1464]
Rushinek, A.:
  An Analysis of Computer Software and Its Importance to
    the Accounting Information System.   [1465]
Rushinek, S.F.:
  An Analysis of Computer Software and Its Importance to
    the Accounting Information System.   [1465]
Rutherford, D.A.:
  Selection of Parameters for a Fuzzy Logic Controller. [209]
  Approximate Reconstruction of Mapping Functions from
    Linguistic Descriptions in Problems of Fuzzy Logic
    Applied to Systems Control.   [1380]
Ryabinin, I.:
  Reliability of Engineering Systems: Principles and
    Analysis.   [1466]
Ryan, R.M.:
  Conceptualizing the Integration of Social Welfare
    Delivery Systems: A Matrix Organization Approach. [1467]

Rzevski, G.:
  Research into the Assessment of Complexity.   [879]
  Some Philosophical Aspects of System Design.   [1468]
Sacerdote, L.:
  Mean Variance and Skewness of the First Passage Time
    for the Ornstein-Uhlenbeck Process.   [290]
  First Passage Time Problems and Some Related
    Computational Methods.   [511]
  On a Class of Discrete Models for Regulated Growth
    with Intrinsic Lower Bounds.   [1234]
Sachs, W.:
  Systemic Inactivism: A Modern Laissez-Faire Thinking. [1469]
Sadek, K.E.:
  Communication and Information: Their Relationship to
    Computer-Based Information Systems.   [796]
Sadovskiy, L.Y.:
  Questions Regarding Simulation of Hierarchical Systems. [1470]
Sadovsky, V.N.:
  Systems Theory: Philosophical and Methodological
    Problems.   [173]
  The System Principle and Leibnitz's Theoretical
    Heritage.   [1471]
  The Systemic Approach: Prerequisites, Problems and
    Difficulties.   [174]
Saeks, R.:
  Large-Scale Dynamical Systems.   [1472]
  System Theory: A Hilbert Space Approach.   [516]
Sagal, P.T.:
  Epistemology of Economics.   [1473]
Sage, A.P.:
  Methodology for Large Scale Systems.   [1474]
  A Multiple Objective Optimization-Based Approach to
    Choicemaking.   [1861]
  Behavioral and Organizational Considerations in the
    Design of Information Systems and Process of
    Planning and Decision Support.   [1475]
  Sensitivity Analysis in Systems for Planning and
    Decision Support.   [1476]
Sahal, D.:
  Principles of Regulation and Control.   [1477]
  Systemic Similitude.   [1478]
  A Theory of Evolution of Technology.   [1479]
  A Unified Theory of Self-Organization.   [1480]
  Patterns of Technological Innovations.   [1481]
Sahni, S.:
  Fundamentals of Data Structures.   [788]
Sain, M.K.:
  Introduction to Algebraic System Theory.   [1482]
Saitta, L.:
  An Interpreter of Fuzzy Production Rules.   [1037]
Sakawa, M.:
  An Interactive Goal Attainment Method for
    Multiobjective Nonconvex Problems.   [1483]
  Fuzzy Extension of Multiattribute Utility Analysis for
    Collective Choice.   [1527]
Salomaa, A.:
  The Mathematical Theory of LSystems.   [1454]
Salomon, J.:
  Random Boolean Networks.   [64]
Salzberg, P.M.:
  A Unified Theory of Attraction in General Systems.   [1484]
Sampson, J.R.:
  Adaptive Information Processing: An Introductory
    Survey.   [1485]

Samuelson, K.:
  Information and Data in Systems.   [1009]
  Information Systems and Networks.   [1486]
Sandblom, C.-L.:
  Optimization of Economic Policy Using Lagged Controls. [1487]
  Simulation Results on the Evaluation of Economic
    Policies.   [1488]
Sanders, C.W.:
  Decentralized Structures for State Estimation on Large
    Scale Systems.   [1653]
Sanderson, A.C.:
  Pattern Trajectory Analysis of Nonstationary
    Multivariate Data.   [1489]
Sanderson, J.G.:
  A Relational Theory of Computing.   [1490]
Sanderson, M.:
  Successful Problem Management.   [1491]
Santa, J.L.:
  Cognition.   [623]
Santis, F.de:
  On Factor Analysis and Fisher's Linear Discriminant
    Analysis.   [1492]
Santos, E.S.:
  Probabilistic Machines and Languages.   [1493]
Saridis, G.N.:
  Fuzzy Automata and Decision Processes.   [685]
  Self-Organizing Control of Stochastic Systems.   [1494]
  Linguistic Decision Structures for Hierarchical
    Systems.   [653]
Sarson, T.:
  Structural Systems Analysis: Tools and Techniques. [567]
Sastry, D.:
  A Two-Level Algorithm for the Simultaneous State and
    Parameter Identification.   [1495]
Sauer, C.H.:
  Computer Systems Performance Modeling.   [1496]
Saunders, P.T.:
  An Introduction to Catastrophe Theory.   [1497]
Savage, G.J.:
  The Graph-Theoretic Field Model-I - Modelling and
    Formulations.   [1498]
Sawaragi, Y.:
  Modeling, Estimation, and Their Applications for
    Distributed Parameter Systems.   [1499]
Sayward, F.:
  Software Metrics: An Analysis and Evaluation.   [1317]
Scandura, J.M.:
  Problem Solving: a Structural Process Approach with
    Instructional Implications.   [1500]
  Theoretical Foundations of Instruction: A Systems
    Alternative to Cognitive Psychology.   [1501]
Scarpetta, G.:
  A Systemic Study of Monetary Systems.   [255]
Scavia, D.:
  Perspectives on Lake Eco-System Modeling.   [1502]
  An Ecological Model of Lake Ontario.   [1503]
Schaefer, G.F.:
  The Application of General Systems Methodology to the
    Comparative Study of Public Policy.   [767]
Schal, R.K.:
  A Collection of Thoughts of General Systems Theorists
    Concerning the Process of Human Transformation.   [1504]
Schellenberg, E.:
  Project Management, Key to Successful Computer

Projects.   [1505]
Schellhorn, J.P.:
  Model Following in Multiple Control: Exact and
    Approximate.   [1506]
Schiaffino, A.:
  Bifurcation of Regular Nonnegative Stationary
    Solutions for a Quasilinear Parabolic System. [397]
Schieve, W.C.:
  Self-Organization and Dissipative Structures:
    Applications in the Physical and Social Sciences. [1507]
Schiffman, Y.M.:
  A Systems Approach for Estimating Energy Savings in
    Communities.   [1508]
Schilderinck, J.H.F.:
  Regression and Factor Analysis in Econometrics. [1509]
Schimanovich, W.:
  An Axiomatic Theory of Winning Mathematical Games and
    Economic Actions.   [1510]
Schkolnick, M.:
  A Clustering Algorithm for Hierarchical Structures. [1511]
Schmutzer, M.E.A.:
  Hi & Low - In & Out: Approaches to Social Status. [1512]
  The Valuton: An Automation Model of a Sociological
    Theory of Value Genesis.   [101]
Schnakenberg, J.:
  Thermodynamic Network Analysis of Biological Systems
    (Second Edition).   [1513]
Schoderbek, P.P.:
  A Cybernetic Model of the MBO Process.   [39]
Scholz, C.:
  The Architecture of Hierarchy.   [1514]
Schotter, A.:
  Applied Game Theory.   [211]
Schrattenholzer, L.:
  The Generation of Efficient Energy Supply Strategies
    using Multi-Criteria Optimization.   [658]
Schubert, L.K.:
  A Planning Control Strategy that Allows for the Cost
    of Planning.   [848]
Schuster, P.:
  The Hypercycle: a Principle of Natural
    Selforganization.   [462]
Schwartz, D.G.:
  Isomorphisms of Spencer-Brown's Laws of Form and
    Varela's Calculus for Self-Reference.   [1515]
Schwarzenback, J.:
  Systems Modelling and Control.   [1516]
Schwegler, H.:
  Self-Organizing Systems: An Interdisciplinary Approach. [1451]
Schwodiauer, G.:
  Applied Game Theory.   [211]
Scoltock, J.:
  A Survey of the Literature on Cluster Analysis. [1517]
Scott, C.H.:
  On the Analysis of an Information Theoretic Model of
    Spacial Interaction.   [1518]
Scott, W.A.:
  Cognitive Structure: Theory and Measurement of
    Individual Differences.   [1519]
Scott, W.R.:
  Organizations: Rational, Natural and Open Systems. [1520]
Sebus, J.:
  Biological Structures (translated from Dutch). [440]
Secker, J.A.:

Developing Automated Teaching Systems.   [1457]
Segel, L.A.:
  Mathematical Models in Molecular and Cellular Biology. [1521]
Seibert, P.:
  A Unified Theory of Attraction in General Systems.   [1484]
Sellers, P.H.:
  Combinatorial Complexes: A Mathematical Theory of
    Algorithms.   [1522]
Sembugamoorthy, V.:
  PLAS, a Paradigmatic Language Acquisition System.   [1523]
Sengupta, J.K.:
  Bilinear Models in Stochastic Programming.   [1524]
  Testing and Validation Problems in Stochastic Linear
    Programming.   [1525]
  Minimax Solutions in Stochastic Programming.   [1526]
Seo, F.:
  Fuzzy Extension of Multiattribute Utility Analysis for
    Collective Choice.   [1527]
Sernadas, A.:
  SYSTEMATICS: Its Syntax and Semantics as a Query
    Language.   [1528]
Sessa, S.:
  On the Fuzziness of Solutions of Alpha-Fuzzy Relation
    Equations on Finite Spaces.   [418]
Seton, F.:
  The Evalution of Economic Flows and Systemic
    Efficiency.   [1529]
Shaeffer, D.L.:
  A Model Evaluation Methodology Applicable to
    Environmental Assessment Models.   [1530]
Shahane, A.H.:
  Interdisciplinary Models of Water Systems.   [1531]
Shakun, M.F.:
  Formalizing Conflict Resolution in Policy Making.   [1532]
  Policy Making and Meaning as Design of Purposeful
    Systems.   [1533]
  Effectiveness, Productivity and Design of Purposeful
    Systems: The Profit-Making Case.   [1534]
Shankar, P.:
  Applications of Systems Theory to Ecology.   [852]
Shapiro, L.G.:
  Structural Descriptions and Inexact Matching.   [1535]
Sharif, N.:
  Proceedings of the Int. Conference on Systems
    Modelling in Developing Countries.   [1536]
  Systems Models for Decision Making.   [1537]
Sharma, B.D.:
  Adjacent-Error Correcting Binary Perfect Codes.   [1538]
  Block-Wise Error-Protecting Burst Codes with Weight
    Constraints.   [1539]
  Phased Error Correcting Perfect Codes.   [1540]
Shaw, C.:
  Dynamics: The Geometry of Behavior.   [3]
Shaw, M.:
  Software Metrics: An Analysis and Evaluation.   [1317]
Shaw, M.L.G.:
  On Becoming a Personal Scientist.   [1541]
  The Personal Scientist in the Community of Science.   [1542]
Shenhar, A.:
  Compound Bayes Predictors for Sequences with Apparent
    Markov Structure.   [353]
Shepard, R.N.:
  Multidimensional Scaling, Tree-fitting, and Clustering. [1543]
Shepherd, P.:

Identification and Analysis of the Subsystems
    Structure of Models.   [650]
Shepp, L.A.:
    On Prediction of Moving-Averaging Processes.   [1544]
Sherlock, R.A.:
    Analysis of the Behaviour of Kauffman Binary Networks:
        Part 1: State Space Description and Distribution of
        Limit Cycle Lengths; Part 2: the State Cycle
        Fraction for Networks of Different Connectivities. [1545]
Shieh, L.:
    Representation of Continuous Time State Equations by
        Discrete-Time State Equations.   [1546]
Shields, P.C.:
    Total Complexity and the Inference of Bent Programs. [517]
Shigan, E.N.:
    Systems Modeling in Health Care.   [1547]
Shmulyan, B.L.:
    Interaction of Entropy Systems.   [1548]
Shoesmith, D.J.:
    Multiple-Conclusion Logic.   [1549]
Shore, J.E.:
    Cross-Entropy Minimization Given Fully-Decomposable
        Subset and Aggregate Constraints.   [1550]
    Properties of Cross-Entropy Minimization.   [1551]
Shugart, H.H.:
    Systems Ecology.   [1552]
Siegel, A.F.:
    Modern Data Analysis.   [1018]
Sierocki, I.:
    Invariance-Relations in General Dynamic Systems.   [1553]
Siewiorek, D.P.:
    The Theory and Practice of Reliable System Design. [1554]
Siljak, D. D.:
    Overlapping Decompositions, Expansions, and
        Contractions of Dynamic Systems.   [1260]
Siljak, D.D.:
    Large-Scale Dynamic Systems.   [1555]
Silva, M.T.:
    Digital Foundations of Time Series Analysis. Vol. 1:
        The Box-Jenkins Approach.   [1433]
Silvern, L.C.:
    Systems Engineering Applied to Training.   [1556]
    Systems Engineering of Education I: the Evolution of
        Systems Thinking in Education.   [1557]
Silvert, W.:
    The Formulation and Evaluation of Predictions.   [1558]
Silviu, G.:
    Information Theory with Applications.   [1559]
Simberloff, D.S.:
    On the Intrinsic Structure of Differential Equation
        Models of Cosystems.   [715]
Simionescu, C.L.:
    On a Model Neurology.   [1560]
Simon, H.A.:
    Human Problem Solving.   [1217]
    How Complex Are Complex Systems?   [1561]
    Models of Discovery and Other Topics in the Methods of
        Science.   [1562]
    Models of Bounded Rationality (2 Volumes).   [1563]
Simon, J.C.:
    Computer Oriented Learning Processes.   [1564]
    Digital Image Processing.   [1565]
Simoncelli, G.:
    A Systemic Study of Monetary Systems.   [255]

Singh, I.:
    Economic Development as an Adaptive Process.   [384]
Singh, M.G.:
    Handbook of Large Scale Systems Engineering
        Applications.   [1566]
    Large Scale System Theory and Applications (Proc. of a
        Symposium in Toulouse, France, June 1980).   [1690]
    Large Scale Systems Modelling.   [1089]
Singh, R.:
    Systems Approaches to the Study of Industrial
        Relations.   [1567]
Singh, S.:
    Engineering Reliability: New Techniques and
        Applications.   [416]
Sinha, D.K.:
    Catastrophe Theory and Applications.   [1568]
Sinha, K.K.:
    Adaptive Nuclear Reactor Control without Explicit
        Identification.   [1569]
Sinha, N.K.:
    Applications of Projective Reduction Methods to
        Estimation and Control.   [753]
    Identification and Modeling for Linear Multivariable
        Discrete-Time Systems: A Survey.   [468]
    Identification of Multivariable Systems in the
        Transfer-Function Matrix Form.   [469]
Skala, H.J.:
    On Many-Valued Logics, Fuzzy Sets, Fuzzy Logics and
        Their Applications.   [1570]
Sklenar, J.:
    Apparatus for Formal Description of Discrete Systems. [1571]
Skorohod, A.V.:
    Controlled Stochastic Processes.   [611]
Skvoretz, J.:
    Information Diffusion in Formally Structured
        Populations: An Information Processing Approach.   [1572]
Slavtchev, A.:
    A Method for Performance Assessment of Medical
        Radioisotope Equipment.   [906]
Slawski, C.:
    A Systems Model of Negotiation: Arabs Versus Israeli. [1573]
    Developing the Learning Relationship: In Theory and
        Practice.   [1574]
    Identity Bargaining: A Policy Systems Research Model
        of Career Development.   [1575]
    Institutionalizing Innovation in an Organization: A
        Model and Case Study.   [1576]
    Love, Power and Conflict: A Systems Model of
        Interparty Negotiation.   [1577]
    Mapping Joint Action.   [1578]
    Social Actualization: A Theory of Community Living
        Arrangements and a Practical Utopia.   [1579]
    The Interpersonal Conflict Episode: A Systems Model.   [1580]
    Towards a Synthesis of Social Psychological "Theories". [1581]
Slepian, D.:
    On Prediction of Moving-Averaging Processes.   [1544]
Slovic, P.:
    Judgement Under Uncertainty: Heuristics and Biases.   [866]
Slyadz, N.N.:
    Analysis of Fuzzy Initial Information in
        Decision-Making Models.   [1582]
Smale, S.:
    The Mathematics of Time: Essays on Dynamical Systems,
        Economic Processes and Related Topics.   [1583]

Small, M.G.:
    Systemic and Global Learning.   [1584]
Smetanic, Y.S.:
    On the Algorithm for Determining the Primary Structure
        of Biopolymers.   [1585]
Smets, P.:
    The Degree of Belief in a Fuzzy Event.   [1586]
Smiley, T.J.:
    Multiple-Conclusion Logic.   [1549]
Smith, A.W.:
    Mangagement Systems Analyses and Applications.   [1587]
Smith, C.S.:
    A Search for Structure: Selected Essays on Science,
        Art and History.   [1588]
Smith, J.M.:
    Evolution and the Theory of Games.   [1589]
Smith, M.:
    Models in Ecology.   [1590]
Smith, P.M.:
    An Exploration of Shared Knowledge about Procedures
        Used in Fault Diagnosis Tasks.   [1591]
Smith, R.B.:
    How to Plan, Design & Implement a Bad System.   [1592]
Smith, R.G.:
    Frameworks for Cooperation in Distributed Problem
        Solving.   [1593]
Smithson, M.:
    Models for Fuzzy Nominal Data.   [1594]
Smyth, D.S.:
    Using a Systems Approach: the Structure of Root
        Definitions.   [1595]
Soeda, T.:
    Modeling, Estimation, and Their Applications for
        Distributed Parameter Systems.   [1499]
Soete, W.:
    Structural Methods for Decomposition of Large-Scale
        Systems.   [1596]
Soi, I.M.:
    An Overview of Computer-Orientied Learning Processes. [1597]
Somalvica, M.:
    An Integrated Model of Problem Solver.   [679]
Somalvico, M.:
    Deterministic and Nondeterministic Programming in
        Robot Systems.   [615]
    Interactivity and Incrementality in Natural Language
        Understanding Systems.   [680]
Sonde, B.S.:
    Introduction to System Design Using Integrated
        Circuits.   [1598]
Sonnier, I.L.:
    Holistic Education: Teaching of Science in the
        Affective Domain.   [1599]
Soong, T.T.:
    Probabilistic Modeling and Analysis in Science and
        Engineering.   [1600]
Sorenson, P.G.:
    An Introduction to Data Structures with Applications. [1735]
Sorrentino, N.:
    On the Model Order for Uncertain System Approximation. [590]
Southwood, T.R.E.:
    Ecological Methods (2nd Edition).   [1601]
Spagovich, V.G.:
    Theory of Adaptive Systems.   [1602]
Specker, E.:

  Complexity of Partial Satisfaction.   [1048]
Spedding, C.R.W.:
  Agricultural Systems.   [1603]
Sperb, R.P.:
  Maximum Principles and their Applications.   [1604]
Spoehr, K.T.:
  Visual Information Processing.   [1605]
Sprague, R.H.Jr.:
  Decision Support Systems: Issues and Challenges.   [521]
Spriet, J.A.:
  Computer-Aided Modelling and Simulation.   [1606]
Squire, E.:
  Introducing Systems Design.   [1607]
Srinivasan, S.K.:
  Probabilistic Analysis of Redundant Systems.   [1608]
Srivastava, M.S.:
  An Introduction to Multivariate Statistics.   [1609]
Stachowicz, M.:
  Karhunen-Loewe Transform Approximation for Certain
    Class of Random Processes.   [822]
Stantchev, I.:
  An Approach to Multiobjectives Decision Making with
    Ordinal Information.   [439]
Starr, M.K.:
  Multiple Criteria Decision Making.   [1610]
Starr, T.S.:
  Hierarchy: Perspectives for Ecological Complexity.   [24]
Steel, T.B.:
  The Logic of Questions and Answers.   [144]
Steele, G.L.:
  CONSTRAINTS - A Language for Expressing Almost
    Hierarchical Descriptions.   [1640]
Stefanski, J.:
  A Heuristic Search for the Best Hierarchical Structure. [1611]
Stegmuller, W.:
  The Structure and Dynamics of Theories.   [1612]
  The Structuralist View of Theories.   [1613]
Stein, W.E.:
  Response Models for Detection of Change.   [1382]
Steinacker, I.:
  Computer-Aided Research in Learning.   [1614]
  Disambiguation in German.   [1615]
  Ontology on the Computer.   [1719]
  VIE-LANG: A German Language Understandig System.   [238]
  Future, Impacts, and Future Impacts of Artificial
    Intelligence: A Bibliography.   [1616]
  VIE-LANG: A German Language Dialogue System.   [1753]
  Focussing as a Method for Disambiguation and Anaphora
    Resolution.   [1617]
Steinberg, E.:
  Dichotic Property and Teleogenesis.   [1283]
Steinberg, J.B.:
  An Informational Analysis of the Inter-Male Behavior
    of the Grasshopper Chortophaga Viridifaciata.   [1618]
Stemmer, N.:
  The Reliability of Inductive Inferences and Our Innate
    Capacities.   [1619]
Stephenson, R.C.:
  Computer Simulation of Lateral Inhibition in the
    Retina.   [389]
Stern, D.M.:
  Microprocessor System, Interfacing and Applications.   [168]
Steward, D.V.:
  Systems Analysis and Management: Structure, Strategy

and Design.   [1620]
Stewart, D.J.:
  Adaptional Problems in Natural Intelligent Systems
    with Changes in Environmental Variety.   [1282]
Stewart, I.:
  Catastrophe Theory and its Applications.   [1353]
Stock, O.:
  A Lexicon Based Approach to Sentence Understanding.   [269]
Stocker, G.:
  Biophysical Ecology and Ecosystem Research.   [1771]
Stockmeyer, L.J.:
  Intrinsically Difficult Problems.   [1621]
Stockwell, C.W.:
  Determination of Feedback for Human Posture Control
    without Physical Intervention.   [258]
Stoeva, S.:
  Intuitionstic L-Fuzzy Sets.   [57]
Stoica, M.:
  Energy Management Planning Tools for the Transition
    towards a Medium Developed Economy.   [605]
Stoklosa, J.:
  Transformations of Some Dynamical Systems-Networks of
    (a,k)-Machines Generating Cyclic Sequences.   [1622]
  On the Reachability Problem for Some Dynamical Systems. [1623]
Stone, L.D.:
  Theory of Optimal Search.   [1624]
Stoy, J.E.:
  Denotational Semantics: the Scott-Strachey Approach to
    Programming Languages.   [1625]
Straka, G.:
  Invariance-Relations in General Dynamic Systems.   [1553]
Straszak, A.:
  New Trends in Mathematical Modelling.   [1626]
Straube, P.:
  Data- and Behaviour Analysis with Classification
    Models.   [1321]
  Changes of Regional Passenger-Transport Demands by
    Society-Controlled Activities - A Fuzzy Simulation
    Approach.   [600]
Strauss, A.:
  An Introduction to Optimal Control Theory.   [1627]
Strejc, V.:
  State Space Theory of Discrete Linear Control.   [1628]
Strombach, W.:
  Wholeness, Gestalt, System: On the Meaning of these
    Concepts in German Language.   [1629]
Stucki, P.:
  Advances in Digital Image Processing.   [1630]
Stueckler, H.:
  Learned Helplessness - An Attribution-Model for
    Unemployed Workers?   [1316]
Subrahmaniam, K.:
  Multivariate Analysis: a Selected and Abstracted
    Bibliography, 1957-1972   [1631]
Subrahmanian, E.:
  A Generator for Models of Discrete-Event Systems.   [1632]
Subramanian, R.:
  Probabilistic Analysis of Redundant Systems.   [1608]
Sudit, E.F.:
  Effectiveness, Productivity and Design of Purposeful
    Systems: The Profit-Making Case.   [1534]
Sugawara, K.:
  Properties of Analogue Neural Networks.   [1235]
Sugeno, M.:

A Model of Learning Based on Fuzzy Information.   [1633]
Sugiyama, K.S.:
    Methods for Visual Understanding of Hierarchical
        System Structures.   [1634]
    Visual Q-Analysis (I): A Case of Future Computer
        Systems Development in Japan.   [1635]
    Visual Q-Analysis (II): Strategy Analysis for Future
        Computer Systems Development in Japan.   [1694]
Sukur, L.Y.:
    Practical Aspects of Alternatives Evaluating and
        Decision Making Under Uncertainty and Multiple
        Objectives.   [1817]
Sultan, L.H.:
    Some Considerations on Systems Organization for Fuzzy
        Information Processing.   [1636]
Summers, R.C.:
    Database Security and Integrity.   [518]
Suppe, F.:
    The Structure of Scientific Theories (Second Ed.)   [1637]
Suppes, P.:
    Some Remarks About Complexity.   [1638]
    PSA 1976.   [1639]
Sussman, G.J.:
    CONSTRAINTS - A Language for Expressing Almost
        Hierarchical Descriptions.   [1640]
Sussmann, H.J.:
    Catastrophe Theory as Applied to the Social and
        Biological Sciences: a Critique.   [1641]
Sutherland, J.W.:
    Societal Systems: Methodology, Modelling and
        Management.   [1642]
    Towards a System-Theoretical Decision Logic.   [1643]
Suzuki, R.:
    Mathematical and Phenomenological Studies of
        Biological Rhythms.   [1644]
Svoboda, A.:
    Advanced Logical Circuit Design Techniques.   [1645]
Swamy, M.N.S.:
    Graphs, Networks, and Algorithms.   [1688]
Swanick, B.H.:
    A Hierarchical Adaptive Control System.   [1135]
Swanson, E.B.:
    Measurement for Management Decision.   [1114]
Swarz, R.S.:
    The Theory and Practice of Reliable System Design. [1554]
Swets, J.A.:
    Evaluation of Diagnostic Systems: Methods from Signal
        Detection Theory.   [1646]
Sworder, D.D.:
    Influence of Sensor Failures on Log Regulators.   [1647]
Sydow, A.:
    Systems and Simulation 1980   [1648]
Syski, R.:
    Random Processes: a First Look.   [1649]
Systems and T:
    Some Methods in the Mathematical Analysis of Systems
        and Their Control.   [1055]
Szaniawksi, K.:
    Science as an Information-Seeking Process.   [1650]
Szanto, E.:
    Generation of Technological Information.   [1651]
Szego, G.P.:
    Mathematical Systems Theory and Economics.   [994]
Szuecs, E.:

    Similitude and Modelling.   [1652]
Szwillus, G.:
    Continuous Systems with Digital Behavior.   [1411]
    Characterization of Continuous Systems that are
        Digital.   [1412]
Tacker, E.C.:
    Decentralized Structures for State Estimation on Large
        Scale Systems.   [1653]
Tagawa, S.:
    Methods for Visual Understanding of Hierarchical
        System Structures.   [1634]
Takahara, Y.:
    A System Theoretic Consideration of the World Model. [1654]
    Characterization of the Causality of General Dynamic
        Systems.   [1655]
    A Unified Theory of Decision Principles.   [1656]
    A Characterization of Interactions.   [1657]
    A Structure of Rational Decision Principles.   [1658]
    A Stationarization Functor and its Concrete
        Realizations.   [1659]
Takatsu, S.:
    Algebraic Properties of Satisficing Decision.   [1119]
    Multiple-Objective Satisficing Decision Problems. [1660]
    Organizational Equilibrium under Uncertainty.   [1661]
Takatsuji, M.:
    An Information-Theoretical Approach to a System of
        Interacting Elements.   [1662]
Takeuchi, A.:
    Evolutionary Automata - Comparison of Automation
        Behavior and Restle's Learning Model.   [1663]
    The Foundations of Multivariate Analysis: A Unified
        Approach by Means of Projection onto Linear
        Subspeces.   [1664]
Talavage, J.:
    A Solution Procedure for the Hierarchical Coordination
        of Constrained Optimizing Systems.   [536]
Tamburrini, S.:
    Do Cybernetics, Systems Science and Fuzzy Sets Share
        Some Epistemological Problems?   [1665]
Tamura, H.:
    Identification of a Dynamic Lake Model by the Method
        of Data Handling: An Application to Lake Ontario. [1666]
Tanaka, A.:
    Properties of Analogue Neural Networks.   [1235]
Tancig, P.:
    SOVA - General Purpose Software Environment for
        NLU-Systems.   [1667]
Taneja, I.J.:
    A Note on Fano Inequality Involving Distortion.   [690]
Tani, S.T.:
    A Perspective on Modelling in Decision Analysis. [1668]
Tao, K.M.:
    On the Structure of Directed Graphs with Applications:
        A Rapprochement with Systems Theory (Part I). [1669]
    On the Structure of Directed Graphs with Applications:
        A Rapprochement with Systems Theory (Part II). [1670]
Taschdjian, E.:
    The Cybernetics of Stressed Systems.   [1671]
Tasso, C.:
    Natural Language Access to Online Data Bases.   [681]
    A Rule-Based Approach to Natural Language Text
        Representation and Comprehension.   [547]
Tatsumi, H.:
    Mathematical and Phenomenological Studies of

Biological Rhythms.   [1644]
Taylor, A.M.:
    A Systems Approach to the Political Organization of
    Space.   [1672]
Taylor, H.M.:
    A Second Course in Stochastic Processes.   [881]
Taylor, J.R.:
    An Introduction to Error Analysis.   [1673]
Taylor, W.R.:
    Using Systems Theory to Organize Confusion.   [1674]
Tazaki, E.:
    Structural Modeling in a Class of Systems Using Fuzzy
    Sets Theory.   [1675]
    Structural Modeling with Consensus for Societal
    Systems Using Fuzzy Set Theory.   [1676]
Tchon, K.:
    Events and Systems.   [819]
    On Goal-Oriented Systems.   [820]
    On the Eventistic Approach to the Notion of System. [823]
    A 'Transversal' Approach to Evenistic Systems.   [1677]
    Towards a Formal Theory of Description.   [1678]
    On Some Operations Preserving Generic Properties of
    Systems.   [1679]
    Towards a Global Analysis of Systems.   [1680]
Techo, R.:
    Data Communications: an Introduction to Concepts and
    Design.   [1681]
Teodorescu, D.:
    Cybernetical Actions and Markov Processes.   [1682]
Terano, T.:
    A Model of Learning Based on Fuzzy Information.   [1633]
Termini, S.:
    On the Convergence of Entropy Measures of a Fuzzy Set. [395]
    Do Cybernetics, Systems Science and Fuzzy Sets Share
    Some Epistemological Problems?   [1665]
Tesei, A.:
    Bifurcation of Regular Nonnegative Stationary
    Solutions for a Quasilinear Parabolic System.   [397]
Tesfatsion, L.:
    A New Approach to Filtering and Adaptive Control:
    Optimality Results   [1683]
    Two Solution Techniques for Adaptive Reinvestment: A
    Small Sample Comparison.   [870]
Thayse, A.:
    Discrete and Switching Functions.   [381]
Thissen, W.:
    Investigations into World 3 Model: Lessons for
    Understanding Complicated Models.   [1684]
Thomas, A.:
    Generating Tension for Constructive Change: The Use
    and Abuse of Systems Models.   [1685]
Thomas, H.:
    Formal Methods in Policy Analysis.   [242]
Thomason, M.G.:
    Syntactic Pattern Recognition.   [640]
Thompson, J.M.:
    The Instability of Evolving Systems.   [1686]
Thowsen, A.:
    Identifiability of Dynamic Systems.   [1687]
Thrall, R.M.:
    Decision Information.   [1759]
Thulasiraman, K.:
    Graphs, Networks, and Algorithms.   [1688]
Tighe, T.J.:

Modern Learning Theory: Foundations and Fundamental
  Issues. [1689]
Tillman, F.A.:
  System-Reliability Evaluation Techniques for
    Complex/Large Systems: A Review. [804]
Tilti, A.:
  Large Scale System Theory and Applications (Proc. of a
    Symposium in Toulouse, France, June 1980). [1690]
Titli, A.:
  Handbook of Large Scale Systems Engineering
    Applications. [1566]
Tjoa, A.M.:
  A System-Theoretical Approach to Database Systems. [1691]
  The Universal Relation as Logical View of Databases. [1692]
Toda, M.:
  Man, Robot, and Society: Models and Speculations. [1693]
  Visual Q-Analysis (I): A Case of Future Computer
    Systems Development in Japan. [1635]
  Visual Q-Analysis (II): Strategy Analysis for Future
    Computer Systems Development in Japan. [1694]
Toern, A.A.:
  Simulation Graphs: A General Tool for Modeling
    Simulation Designs. [1695]
Toffoli, T.:
  Computation and Construction Universality of
    Reversible Cellular Automata. [1696]
Tomlinson, R.:
  A Systems Approach to Planning in Organizations -
    Developing a Collaborative Research Study. [1697]
  Progress in Cybernetics and Systems Research, Vol.X. [1723]
Tompa, F.W.:
  A Practical Example of the Specification of Abstract
    Data Types. [1698]
Tonella, G.:
  Pluralistic Modelling for Development Planning. [1699]
Tong, C.H.:
  Stability of Dynamical Systems: a Constructive
    Approach. [214]
Tong, R.M.:
  Synthesis of Fuzzy Models for Industrial Processes -
    Some Recent Results. [1700]
  A Linguistic Approach to Decision Making with Fuzzy
    Sets. [1701]
  Some Properties of Fuzzy Feedback Systems. [1702]
  The Evaluation of Fuzzy Models Derived from
    Experimental Data. [1703]
Topolski, J.:
  Methodology of History. [1704]
Torasso, P.:
  An Interpreter of Fuzzy Production Rules. [1037]
Torelli, M.:
  Analysis and Compacting of Musical Texts. [163]
Tou, J.T.:
  Advances in Information Systems Science, Vol.6. [1705]
  Knowledge Engineering. [1706]
Toulotte, J.M.:
  A Method for Decomposing Interpreted Petri Nets and
    its Utilization. [1707]
Toutenburg, H.:
  Prior Information in Linear Models. [1708]
Tracy, L.:
  A Dynamic Living-Systems Model of Work Motivation. [1709]
Trappl, R.:
  A Second Order Model of the Optic Generator Potential

        and its Relation to Stevens' Power Function.   [1710]
    Non-Computerized Formal Methods with Computer
        Efficiency in Medicine.   [1711]
    Progress in Cybernetics and Systems Research, Vol.III. [1712]
    Progress in Cybernetics and Systems Research, Vol.IV.   [1713]
    Simple Forecasting Procedures in Health Care.   [1714]
    Computer Psychotherapy: Is it Acceptable, Feasible,
        Advisable?.   [1715]
    Health Care in Austria: 1980-2000.   [783]
    The Soft Side of Innovation: Medicine - a Hard Case.   [1716]
    A Health Hazard Appraisal Program for Austria.   [1717]
    A System Analysis of Information Transmission of Cat
        Retinal Horizontal Cells.   [1822]
    Cybernetics and Systems Research.   [1718]
    Fast Walsh vs. Fast Fourier Transform: A Comparison of
        Time-Efficiency.   [784]
    Ontology on the Computer.   [1719]
    Progress in Cybernetics and Systems Research, Vol.IX.   [1720]
    Progress in Cybernetics and Systems Research, Vol.VI.   [1336]
    Progress in Cybernetics and Systems Research, Vol.VII. [1721]
    Progress in Cybernetics and Systems Research, Vol.VIII. [1722]
    Progress in Cybernetics and Systems Research, Vol.X.   [1723]
    Progress in Cybernetics and Systems Research, Vol.XI.   [1724]
    The Structure of Manifestations in a Medical
        Consultation System.   [785]
    VIE-LANG: A German Language Understandig System.   [238]
    Cybernetics - Theory and Applications.   [1725]
    Future, Impacts, and Future Impacts of Artificial
        Intelligence: A Bibliography.   [1616]
    VIE-LANG: A German Language Dialogue System.   [1753]
    A Frame-Based Real-Time Graphic Interaction System.   [787]
    Cybernetics and Systems Research, Vol.II.   [1726]
    Impacts of Artificial Intelligence.   [1727]
    Artificial Intelligence: A One-Hour Course.   [1728]
    Basic and Applied General Systems Research: A
        Bibliography, 1977-1984.   [1729]
    Impacts of Artificial Intelligence: An Overview.   [1730]
    Power, Autonomy, Utopia: New Approaches Towards
        Complex Systems.   [1731]
    Reducing International Tension through Artificial
        Intelligence: A Proposal for 3 Projects.   [1732]
    The Impacts of Artificial Intelligence. Science,
        Technology, Military, Economy, Society, Culture,
        and Politics.   [1733]
Trattnig, W.:
    Denotational Semantics of a Software Specification
        Language Based on the Data Flow Model.   [1734]
Trayner, C.:
    Construction of an Expert Therapy Adviser as a Special
        Case of a General System Protection Design.   [966]
Tremblay, J.P.:
    An Introduction to Data Structures with Applications.   [1735]
Tribus,:
    The Maximum Entropy Formalism.   [1042]
Tripp, R.S.:
    The Use of Cybernetics in Organizational Design and
        Development: An Illustration within Air Force
        Logistics Command.   [1736]
Troncale, L.:
    Linkage Propositions between Fifty Principal Systems
        Concepts.   [1737]
    Origins of Hierarchical Levels: An Emergent
        Evolutionary Process Based on Systems Concepts.   [1738]
    Are Levels of Complexity in Bio-Systems Real?

Applications of Clustering Theory to Modeling
   Systems Emergence.   [1739]
On a Possible Discrimination between Bioevolution and
   a Theory of Systems Emergence.   [1740]
A General Survey of Systems Methodology.   [1741]
A General Survey of Systems Methodology. (Proc.of the
   Twenty-Sixth Annual Meeting of the SGSR).   [1742]
Linkage Propositions between Principal Systems
   Concepts.   [1743]
Some Key, Unanswered Questions about Hierarchies.   [1744]
Testing Hierarchy Models with Data Using Computerized,
   Empirical Data Bases.   [1745]
Towards a Formalization of Systems Linkage
   Propositions.   [1746]
A Hybrid Systems Method: Tests for Hierarchy & Links
   between Isomorphs.   [1747]
Future of General Systems Science: Obstaches,
   Potentials, Case Studies.   [1748]
Knowing Natural Systems Enables Better Design of
   Man-Made Systems: The Linkage Proposition Model.   [1749]
Systems Theory Applied to the Biological and Medical
   Sciences.   [1750]
Trost, H.:
Computer Simulation of Neurotic Processes.   [1751]
Outline of a Natural-Language Understanding System.   [1752]
Disambiguation in German.   [1615]
Ontology on the Computer.   [1719]
VIE-LANG: A German Language Understandig System.   [238]
VIE-LANG: A German Language Dialogue System.   [1753]
Declarative Knowledge Representation: An Overview and
   an Application in the Field of Natural Language
   Understanding.   [1754]
Focussing as a Method for Disambiguation and Anaphora
   Resolution.   [1617]
Representing Belief Models in Semantic Networks.   [958]
Trussel, P.:
Using Nursing Research: Discovery, Analysis, and
   Interpretation.   [1755]
Constructs from General Systems Theory to a Health
   History.   [1756]
Tsai, W.H.:
Attributed Grammar - A Tool for Combining Syntactic
   and Statistical Approaches to Pattern Recognition.   [1757]
Tse, E.:
Aggregation and Information Structuring in Large-Scale
   Dynamic Systems.   [46]
Tsetlin, M.L.:
Automata Theory and Modelling of Biological Systems.   [1758]
Tsokos, C.P.:
Decision Information.   [1759]
Tsypkin, Y.Z.:
The Theory of Adaptive and Learning Systems.   [1760]
Tuda, M.:
Methods for Visual Understanding of Hierarchical
   System Structures.   [1634]
Turksen, I.B.:
Bounds for Fuzzy Inference.   [1761]
Turney, P.:
On Laws of Form.   [1762]
Tussey, C.C.:
The Kentucky Longevity Information System.   [1763]
Tuteja, R.K.:
Two Generalized Measures of "Useful" Information.   [778]
Tversky, A.:

Judgement Under Uncertainty: Heuristics and Biases.   [866]
Tzafestas, S.G.:
   Optimization of System Reliability: a Survey of
      Problems and Techniques.   [1764]
Uccellini, L.W.:
   Applying Discrete Model Concepts to the Computation of
      Atmospheric Trajectories.   [1765]
Ulanowicz, R.E.:
   The Empirical Modelling of an Ecosystem.   [1766]
Ulgen, O.:
   A Sector Approach to the Formulation of Systems
      Dynamics.   [248]
   An Algorithm for Converting Signed Digraphs to
      Forrester Schematics.   [249]
Ullman, J.D.:
   The Theory of Joins in Relational Data Bases.   [20]
Ulman, J.D.:
   Introduction to Automata Theory, Languages, and
      Computation.   [780]
Ulrich, D.:
   A Frame-Based Real-Time Graphic Interaction System. [787]
Ulrich, W.:
   The Design of Problem Solving Systems.   [1767]
   A Critique of Pure Cybernetic Reason: The Chilean
      Experience with Cybernetics.   [1768]
Umnov, A.:
   An Approach to Distributed Modeling.   [1769]
Umpleby, S.:
   Methods for Making Social Organisation Adaptive.   [1770]
Unger, C.K.:
   Biophysical Ecology and Ecosystem Research.   [1771]
Unterberger, H.:
   Automatic Recognition of Voiced Signals using
      Articulation Models.   [1772]
Urban, H.B.:
   The Concept of Development for a Systems Perspective. [1773]
Utkin, V.I.:
   Variable Structure Systems with Sliding Modes.   [1774]
Uttenhove, H.J.:
   SAPS (Systems Approach Problem Solver): An
      Introduction and Guide.   [1775]
Uyeno, D.H.:
   PASSIM: A Discrete-Event Simulation Package for PASCAL. [1776]
Uyttenhove, H.J.J.:
   On the Problem of Computer-Aided Structure
      Identification: Some Experimental Observations and
      Resulting Guidelines.   [938]
   A Systems Approach to Grading of Flight Simulator
      Students.   [335]
   Procedures for Generating Reconstruction Hypotheses in
      the Reconstructability Analysis.   [947]
   On the Two-dimensionality in the Behavioral System
      Identification Problem (Part I).   [1777]
   The Systems Approach Problem Solver and a Stockmarket
      Example.   [1778]
Vaccari, E.:
   A System Theory Based Simulation Language.   [404]
   The Role of System Modelling in Natural Language
      Processing.   [1779]
Vagnucci, A.H.:
   Concatenation and Homogeneity of Biological Time
      Series.   [1780]
Valdes-Perez, R.E.:
   Information Loss Due to Data Quantization in

Reconstructability Analysis.   [1781]
Vallee, R.:
  Evolution of Dynamical Linear System with Random
    Initial Conditions.   [1782]
  "Eigen-Elements" for Observing and Interacting
    Subjects.   [1783]
Vallet, C.:
  Use of a Partially Self-Referent Formalism for
    Description of Natural Systems.   [1784]
Van Amerongen, C.:
  Model Analysis of Structures.   [791]
Van Amerongen, J.:
  An Autopilot for Ships Designed with Fuzzy Sets.   [1785]
Van Bendegem, J.P.:
  The Distribution Approach to Problem-Solving in
    Science: Prospects for GSM.   [257]
Van Campenhout, J.M.:
  Maximum Entropy and Conditional Probability.   [1786]
Van Doorn, A.J.:
  Perception of Solid Shape and Spatial Lay-Out through
    Photometric Invariants.   [961]
Van Esch, G.J.:
  Biological Structures (translated from Dutch).   [440]
Van Gigch, J.P.:
  A Metasystem Approach to Organizational
    Decision-Making.   [912]
  A Methodological Comparison of the Science, Systems
    and Metasystem Paradigms.   [1787]
  A Taxonomy of Systems Science.   [1788]
  Potential Impact of Fuzzy Sets on the Social Sciences. [1342]
Van Hezewijk, R.:
  Systems Theory and the Philosophy of Science.   [415]
Van Navta, L.:
  Digitial Computer Applications to Process Control.   [1789]
Van Rootselaar, B.:
  Annals of Systems Research, Vol. 7   [1790]
Van Steenkiste, G.C.:
  Modeling, Identification and Control in Environmental
    Systems.   [1791]
Van Straaten,:
  Proceedings of the Second Joint MTA/IIASA Task Force
    Meeting on Lake Balaton Modelling Veszprem
    (Hungary).   [1792]
Van Straten, G.:
  Uncertainty and Arbitrariness in Ecosystem Modelling:
    A Lake Modelling Example.   [515]
Van den Bogard, W.:
  Minimizing Wafting Times Using Priority Classes: A
    Case Study in Response Surface Methodology.   [1793]
Van der Ven, A.H.G.S.:
  Introduction to Scaling.   [1794]
Van der Zanden, J.:
  Measurement Theory and Signal Theory: Part I
    Investigations on the Foundations.   [1795]
Van der Zouwen, J.:
  Inventory of Systems Research in the Netherlands.   [1796]
  Testing Auxiliary Hypothesis Behind the Interview.   [421]
  Sociocybernetics, Vol.II.   [603]
  Inventory of Systems Research on the Netherlands.   [1797]
Vansteenkiste, G.C.:
  Computer-Aided Modelling and Simulation.   [1606]
Varela, F.J.:
  The Arithmetic of Closure.   [1798]
  Principles of Biological Autonomy. (Second Vol. in the

         North-Holland Series in General Systems Research,
         Edited by G.J. Klir.)   [1799]
      Autopoiesis and Cognition: The Realization of the
         Living.   [1128]
      Steps to a Cybernetics of Autonomy.   [1800]
Varshavskiy, V.I.:
      Asynchronous Processes I. Definition and
         Interpretation.   [1802]
      Asynchronous Processes II. Composition and Matching.   [1801]
Varshney, C.K.:
      The Unsteady State - Environmental Problems, Growth,
         and Culture.   [1849]
Vaschenko, N.D.:
      Adaptive Problem-Solving Systems.   [617]
Vaszary, P.:
      Computer Registration of Patients Waiting for Cardiac
         Operation.   [1388]
Vavrousek, J.:
      Multiple Structure of Systems.   [1803]
      An Approach to Modelling in Empirical Sciences.   [1804]
      Multiple Structure of Systems.   [1805]
Veelenturf, L.P.J.:
      An Automata-Theoretical Approach to Developing
         Learning Neural Networks.   [1806]
Veloso, P.A.S.:
      Operations on Problems and Their Solution Spaces.   [1065]
      On Reducibilities Among General Problems.   [1807]
Vemuri, V.:
      Modeling of Complex Systems: an Introduction.   [1808]
Venkatesan, M.:
      An Alternate Approach to Transitive Coupling in ISM.   [1809]
Ventriglia, F.:
      Kinetic Theory of Neural Systems: Memory Effects.   [1810]
      Kinetic Theory of Neural Systems: An Attempt Towards a
         Simulation of Cerebellar Cortex Activity.   [1811]
Verhagen, C.J.D.M.:
      Some General Remarks About Pattern Recognition; its
         Definitions; its Relation with Other Disciplines; a
         Literature Survey.   [1812]
Veroy, B.S.:
      Large Scale Network Architecture Synthesis:
         Interactive Strategy.   [1813]
      Overall Design of Reliable Centralized Voice/Data
         Communication Network.   [1814]
Vickers, G.:
      Education in Systems Thinking.   [1815]
Vidyasagar, M.:
      Nonlinear Systems: Stability Analysis.   [16]
Viliums, E.R.:
      Methods to Obtain Interlevel Consistency of Decisions
         under Uncertainty.   [1816]
      Practical Aspects of Alternatives Evaluating and
         Decision Making Under Uncertainty and Multiple
         Objectives.   [1817]
Villanueva, J.J.:
      Modularity Grade of the Neural Networks Synthesized
         with Z-Fivex Formal Neurones.   [19]
Vincent, T.L.:
      Optimality in Parametric Systems.   [1818]
Vitanyi, P.M.B.:
      Lindenmayer Systems: Structure, Languages, and Growth
         Functions.   [1819]
Vocarro, R.J.:
      Feature Extraction by System Identification.   [466]

Vogelsaenger, Th.:
  Recent Results in Modelling and Simulation of 3-D
     Tumor Growth.   [446]
Voigt, M.:
  Application of Polyoptimization to Processes of
     Evolution.   [1320]
Volta, G.:
  Synthesis and Analysis Methods for Safety and
     Reliability Studies (NATO Advanced Study Institute).   [40]
Von Bertalanffy, L.:
  Systems View of Man (edited by Paul La Violette).   [1820]
Von Glasersfeld, E.:
  Steps in the Construction of "Others" and "Reality": A
     Study of Self-Regulation.   [1821]
Von Luetzow, A.:
  A Second Order Model of the Optic Generator Potential
     and its Relation to Stevens' Power Function.   [1710]
  A System Analysis of Information Transmission of Cat
     Retinal Horizontal Cells.   [1822]
Von Plato, J.:
  Probability and Determinism.   [1823]
Von Wright, G.H.:
  The Logic of Preference.   [1824]
Voorhees, B.:
  Towards a Formalization of Systems Linkage
     Propositions.   [1746]
Voyat, G.E.:
  Piaget Systematized.   [1825]
Wade, J.W.:
  Architecture, Problems, and Purposes: Architectural
     Design as a Basic Problem-Solving Process.   [1826]
Wagner, G.P.:
  On the Properties of Randomly Connected
     McCulloch-Pitts Networks: Differences between
     Input-Constant and Input-Variant Networks.   [375]
Wagner, R.R.:
  A System-Theoretical Approach to Database Systems.   [1691]
  The Universal Relation as Logical View of Databases.   [1692]
Waldhauer, F.D.:
  Feedback.   [1827]
Walichiewicz, L.:
  Decomposition of Linguistic Rules in the Design of a
     Multi-Dimensional Fuzzy Control Algorithm.   [1828]
Walker, B.J.:
  Computer Security and Protection Structures.   [1829]
Walker, C.C.:
  A System Theoretic Approach to the Management of
     Complex Organizations: Management by Exception,
     Priority, and Input Span in a Class of
     Fixed-Structure Models.   [1830]
  A Systems-Theoretic Approach to the Management of
     Complex Organizations: Management by Consensus
     Level and its Interaction with Other Management
     Strategies.   [589]
Walker, J.A.:
  Dynamical Systems and Evolution Equations: Theory and
     Applications.   [1831]
Walker, V.R.:
  Criteria for Object Perception.   [1832]
Wallace, M.R.:
  Applications of Ultrastability.   [1833]
Wallenmaier, T.E.:
  Towards a Quantitative Concept of Teleological Systems. [1834]
Waller, R.J.:

Complexity and the Boundaries of Human Policy Making.   [1835]
Wallis, J.R.:
  Mathematical Models for Surface Water Hydrology.
     (Proc. of the Workshop on Mathematical Models in
     Hydrology, Pisa, Italy, Dec. 9-12, 1974)   [316]
Walter, J.R.:
  Systems Science in Health Care.   [323]
Wang, P.:
  Fuzzy Systems: Theory and Application to Policy
     Analysis and Information Systems.   [1836]
Wang, Z.:
  Application of Fuzzy Decision Making and Fuzzy Linear
     Programming in Production and Planning of Animal
     Husbandry System in Farming Region.   [1837]
  M-Fuzzy Numbers and Random Variables.   [1838]
Warfield, J.N.:
  Crossing Theory and Hierarchy Mapping.   [1839]
  Some Principles of Knowledge Organization.   [1840]
  Complementary Relations and Map Reading.   [1841]
Warner, W.K.:
  An Ashby Hierarchy for Human Action.   [483]
Warnier, J.-D.:
  Logical Construction of Systems.   [1842]
Warren, K.S.:
  Scientific Information Systems and the Principle of
     Selectivity.   [627]
Wasinowski, R.:
  An Interactive Model for Analysing Climate Change due
     to Increasing Carbon Dioxide.   [1872]
  Choosing among Alternative Strategies.   [1843]
Wasniowski, R.:
  The Use of Computer-Communication Systems in Futures
     Research.   [1844]
Watanabe, S.:
  A Generalized Fuzzy-Set Theory.   [1845]
  Pattern Recognition as a Quest for Minimum Entropy.   [1846]
Waterman, D.A.:
  Pattern-Directed Inference Systems.   [1847]
Watson, S.R.:
  Fuzzy Decision Analysis.   [1848]
Watt, J.:
  Management Information Systems: a System for Design.   [489]
Watt, K.E.F.:
  The Unsteady State - Environmental Problems, Growth,
     and Culture.   [1849]
Wearing, A.J.:
  Complexity and Decision-Making.   [1085]
Webb, O.J.:
  A Dual Model of the Brain.   [34]
Webber, B.L.:
  Readings in Artificial Intelligence.   [1850]
Wechler, W.:
  The Concept of Fuzziness in Automata and Language
     Theory.   [1851]
Wedde, H.F.:
  An Exercise in Flexible Formal Modeling under
     Realistic Assumptions.   [1852]
Weeks, D.:
  The Unsteady State - Environmental Problems, Growth,
     and Culture.   [1849]
Weinberg, D.:
  On the Design of Stable Systems.   [1853]
Weinberg, G.M.:
  On the Design of Stable Systems.   [1853]

Weingartner, P.:
    Analogy Among Systems.   [1854]
Weir, M.:
    Design for a Goal-Directed System.   [1855]
Weisberg, H.F.:
    An Introduction to Data Analysis.   [202]
Weisbuch, G.:
    Random Boolean Networks.   [64]
Weiss, J.J.:
    Fuzzy Decision Analysis.   [1848]
Wellstead, P.E.:
    Introduction to Physical System Modelling.   [1856]
Wendt, S.:
    The Programmed Action Module: an Element for System
        Modelling.   [1857]
Wenger, M.S::
    Thoughts on a Cybernetic View of Social Organization. [1858]
West, B.J.:
    Mathematical Models as a Tool for the Social Sciences. [1859]
Wheelwright, S.C.:
    Forecasting: Methods and Applications.   [1095]
Whinston, A.B.:
    Foundations of Decision Support Systems.   [189]
White, A.M.:
    Interdisciplinary Teaching.   [1860]
White, C.:
    Evolution of a Hydrology Model - an Exercise in
        Modelling Strategy   [1275]
    Addendum to Internal Report 165: Mathematical
        Documentation for a Lotic Ecosystem Model. [1140]
    A Multiple Objective Optimization-Based Approach to
        Choicemaking.   [1861]
    On Constructing a Hierarchical Model in the Flex
        Paradigm Illustrated by Structural Aspects of a
        Hydrology Model.   [1276]
White, D.E.:
    Advanced Logical Circuit Design Techniques.   [1645]
White, D.J.:
    Role and Effectiveness of Theories of Decision in
        Practice.   [1862]
White, G.C.:
    Compartmental Analysis of Ecosystems Models.   [1118]
Wilder, R.L.:
    Mathematics as a Cultural System.   [1863]
Wildhelm, J.:
    Objectives and Multiobjective Decision Making Under
        Uncertainty.   [1864]
Williams, J.G.:
    On the Strength of Belief of Dialectical Belief
        Systems.   [1173]
Williams, L.R.:
    A Reliability Index for Models.   [1032]
Williams, P.M.:
    Bayesian Conditionalisation and the Principle of
        Minimum Information.   [1865]
Willmer, M.A.P.:
    Communication and Information: Their Relationship to
        Computer-Based Information Systems.   [796]
Wilson, G.:
    Computer Simulation of a Cybernetic Model of Man in
        his Cultural Environment.   [1416]
Wilson, J.T.:
    Interactive Simulation on a Computer.   [475]
Windeknecht, T.G.:

System Theoretic Implications of Numerical Methods
    Applied to the Solution of Ordinary Differential
    Equations.   [1866]
Winkel, D.:
    The Art of Digital Design: An Introduction to Top-Down
    Design.   [1867]
Winkler, R.:
    Laws of the Game: How Principles of Nature Govern
    Chance.   [463]
Winlow, W.:
    Coherent and Incoherent Computation in Identified
    Molluscan Neurones.   [742]
Winograd, T.:
    Language as a Cognitive Process (Vol.1: Syntax).   [1868]
Winstead, W.H.:
    An Input/Output Approach to the Structural Analysis of
    Digraphs.   [250]
Winston, P.H.:
    Artificial Intelligence.   [1869]
    Learning and Reasoning by Analogy.   [1870]
    Artificial Intelligence: An MIT Perspective (2 Vols.).   [1871]
Wintz, P.:
    Digital Image Processing.   [639]
Wirosardjono, S.:
    The Unsteady State - Environmental Problems, Growth,
    and Culture.   [1849]
Wisniewski, T.:
    An Interactive Model for Analysing Climate Change due
    to Increasing Carbon Dioxide.   [1872]
Witt, B.I.:
    Structured Programming: Theory and Practice.   [1050]
Witten, I.H.:
    Probabilistic Behaviour-Structure Transformations
    Using Transitive Moore Models.   [1873]
    Some Recent Results on Nondeterministic Modelling of
    Behaviour Sequences.   [1874]
    Principles of Computer Speech.   [1875]
Wnuk, A.J.:
    The Block-Oriented Interactive Simulator BORIS.   [1876]
Wojciechowska, J.:
    Towards a Formal Theory of Description.   [1678]
Wolaver, T.G.:
    Effect of Systems Structure, Connectivity, and
    Recipient Control on the Sensitivity
    Characteristics of a Model Ecosystem.   [1877]
Wolf, D.J.:
    The Kentucky Longevity Information System.   [1763]
Wolkowski, Z.W.:
    Hierarchical Perception of Time and Periodicity.   [1878]
Wong, A.K.C.:
    Pattern Trajectory Analysis of Nonstationary
    Multivariate Data.   [1489]
Wood, A.D.:
    Mathematical Models in the Social, Management and Life
    Sciences.   [243]
Wood, C.:
    Database Security and Integrity.   [518]
Wood, E.F.:
    Review and Unification of Linear Identifiability
    Concepts.   [1247]
Wood-Harper, A.T.:
    A Taxonomy of Current Approaches to Systems Analysis.   [1879]
Wrathal, C.P.:
    Computer Acronyms, Abbreviations, etc.   [1880]

Wright, H.:
    A Discussion of External Validity in Systemic Research. [1075]
Wright, M.H.:
    Practical Optimization.   [613]
Wu, W.T.:
    Heuristic Approach to Chinese-Character Search.   [1881]
Wuendsch, L.:
    A Second Order Model of the Optic Generator Potential
        and its Relation to Stevens' Power Function.   [1710]
Wuensch, G.:
    System Theorie.   [1882]
Wulf, W.A.:
    Fundamental Structures of Computer Science.   [1883]
Wyner, A.D.:
    A Definition of Conditional Mutual Information for
        Arbitrary Ensembles.   [1884]
    On Prediction of Moving-Averaging Processes.   [1544]
    Fundamental Limits in Information Theory.   [1885]
Xuemou, W.:
    Pansystems Analysis and Fuzzy Sets.   [1886]
Yager, R.R.:
    Advances in Fuzzy Set Theory and Applications.   [684]
    Fuzzy Decision Making Including Unequal Objectives.   [1887]
    Linguistic Model and Fuzzy Truths.   [1888]
    Validation of Fuzzy-Linguistic Models.   [1889]
    Advances in Fuzzy Set Theory and Applications.   [686]
    On the Measure of Fuzziness and Negation. Part 1:
        Membership in the Unit Interval.   [1890]
    A Foundation for a Theory of Possibility.   [1891]
    A Linguistic Variable for Importance of Fuzzy Sets.   [1892]
    An Approach to Inference in Approximate Reasoning.   [1893]
    Aspects of Possibilistic Uncertainty.   [1894]
    Fuzzy Sets, Probabilities, and Decision.   [1895]
    Fuzzy Subsets of Type II in Decisions.   [1896]
    Fuzzy Thinking as Quick and Efficient.   [1897]
    On a General Class of Fuzzy Connectives.   [1898]
    Approximate Reasoning and Possibilistic Models in
        Classification.   [1899]
    Prototypical Values for Fuzzy Subsets.   [1900]
    Some Properties of Fuzzy Relationships.   [1901]
    Fuzzy Prediction Based on Regression Models.   [1902]
    Fuzzy Set and Possibility Theory.   [1903]
    Lingustic Hedges: Their Relation to Context and Their
        Experimental Realization.   [1904]
    Measuring Tranquility and Anxiety in Decision Making:
        An Application of Fuzzy Sets.   [1905]
    Some Procedures for Selecting Fuzzy Set-Theoretic
        Operators.   [1906]
    Entropy and Specificity in a Mathematical Theory of
        Evidence.   [1907]
    Membership in a Compound Fuzzy Subset.   [1908]
    On Different Classes of Linguistic Variables Defined
        via Fuzzy Subsets.   [1909]
Yakano, B.:
    A Unified Theory of Decision Principles.   [1656]
Yakowitz, S.J.:
    Computational Probability and Simulation.   [1910]
Yano, H.:
    An Interactive Goal Attainment Method for
        Multiobjective Nonconvex Problems.   [1483]
Yao, D.D.W.:
    Bounds for Fuzzy Inference.   [1761]
Yashin, A.:
    A New Proff and New Results in Conditional Gaussian

        Estimation Procedures.   [1911]
    Medico-Demographic Models in Health Care Systems.   [1330]
Yates, R.E.:
    Representation of Continuous Time State Equations by
        Discrete-Time State Equations.   [1546]
Yee, K.G.:
    Towards a System-Theoretical Decision Logic.   [1643]
Yeh, R.T.:
    Current Trends in Programming Methodology: Vol. IV -
        Data Structuring.   [1912]
Yezhkova, I.V.:
    Decision Making on Fuzzy Grounds.   [1913]
Ylinen, R.:
    Algebraic Theory for Multivariable Linear Systems.   [175]
Yoon, K.:
    Multiple Attribute Decision Making: Methods and
        Applications: A State-of-the-Art Survey.   [803]
Young, T.Y.:
    Statistical Pattern Classification with Binary
        Variables.   [1914]
Yu, C.T.:
    A Comparison of the Stability Characteristics of Some
        Graph Theoretic Clustering Methods.   [1377]
Yudin, E.G.:
    Systems Theory: Philosophical and Methodological
        Problems.   [173]
    The Systemic Approach: Prerequisites, Problems and
        Difficulties.   [174]
Zadeh, L.A.:
    Fuzzy Sets and Systems: a New International Journal. [1208]
Zahler, R.S.:
    Catastrophe Theory as Applied to the Social and
        Biological Sciences: a Critique.   [1641]
Zamzeer, M.J.:
    Computer Integrated Manufacturing (CIM) in Irish
        Industry.   [436]
Zarri, G.P.:
    The RESEDA Project: Using AI Techniques in Order to
        Solve the Problem of Information Incompleteness. [1915]
    From Natural Language to a Canonical Representation of
        the Corresponding Semantic Relationships.   [504]
Zedeck, S.:
    Measurement Theory for the Behavioral Sciences.   [606]
Zeeman, E.C.:
    Catastrophe Theory: Selected Papers 1972-1977   [1916]
Zeigler, B.P.:
    Canonical Realization of General Time Systems.   [1917]
    Structuring the Organization of Partial Models.   [1918]
    Concepts for Advanced Simulation Methodologies.   [1270]
    Methodology in Systems Modelling and Simulation.   [1919]
    Structuring Principles for Multifacet System Modelling. [1920]
Zeleny, M.:
    Multiple Criteria Decision Making.   [1610]
    Autopoiesis, Dissipative Structures, and Spontaneous
        Social Orders.   [1921]
    Autopoiesis: A Theory of Living Organization.   [1922]
    On the Squandering of Resources and Profits Via Linear
        Programming.   [1923]
    Multiple Criteria Decision Making.   [1924]
Zeman, J.:
    Entropy and Information in Science and Philosophy.   [990]
Zhukovin, V.E.:
    The Multicriteria Decision Making with Vector Fuzzy
        Preference Realtion.   [1925]

KEY WORD IN CONTEXT (KWIC-) INDEX

Wrathal, C.P: Computer Acronyms,   Abbreviations , etc.   [1880]
Nevitt, B.   :   ABC of Prophecy: Understanding the Env [1216]
Ferracin, A.:   Self-Organizing Ability in Living Systems.   [519]
Barfoot, C.B: tion of Conditional   Absorbing Markov Chains.   [107]
Bottaci, L. : onal Scheme for the   Abstract Specification of Production S [198]
Gallopin, G.:   The Abstract Concept of Environment.   [564]
Tompa, F.W. : he Specification of   Abstract Data Types.   [1698]
Subrahmaniam: sis: a Selected and   Abstracted Bibliography, 1957-1972   [1631]
Crimmins, E.:   The Art of   Abstracting .   [357]
Gruber, H. : Modularization and   Abstracting ; a Discussion on the Basis [677]
Brown, D.J.H: g Generalization an   Abstraction Techniques.   [229]
Thomas, A. : Change: The Use and   Abuse of Systems Models.   [1685]
Hellwig, Z.H: ome Problems of its   Acceptability and Interpretation.   [748]
Trappl, R. : sychotherapy: Is it   Acceptable , Feasible, Advisable?.   [1715]
Guida, G.   :   Natural Language Access to Online Data Bases.   [681]
Cavallo, R.E: t: Characteristics,   Accomplishments , and Current Developme [280]
Cavallo, R.E: t: Characteristics,   Accomplishments and Current Developmen [281]
Mock, T.J. :   Measurement,   Accounting , and Organizational Informa [1177]
Rushinek, A.: s Importance to the   Accounting Information System.   [1465]
Becker, P.W.:   Inherent Bounds on Achievable Reliability.   [130]
Becker, P.W.:   Inherent Bounds on Achievable Reliability.   [129]
Gloria-Botti: of the Erythrocyte Acid Phosphate Locus in Humans.   [625]
Matsuno, K. : merization of Amino Acids .   [1120]
Favella, L.F: rhunen Expansion to   Acoustic Diagnostic Problems for Inter [512]
Sembugamoort: radigmatic Language   Acquisition System.   [1523]
Wrathal, C.P: Computer   Acronyms , Abbreviations, etc.   [1880]
Bluth, B.J. : s General Theory of   Action . A Summary of the Basic Theory. [178]
England J.L.: Hierarchy for Human   Action .   [483]
Gomez, P.   : tems-Methodology in Action : Organic Problem Solving in a P [638]
Holden, A.V.: l and Local Mode of   Action of General Anestetics.   [770]
Kinston, W. : re Institutions: An   Action Research Approach to Organisati [927]
Slawski, C. :   Mapping Joint   Action .   [1578]
Wendt, S.   :   The Programmed   Action Module: an Element for System M [1857]
Schimanovich: Games and Economic   Actions .   [1510]
Teodorescu, :   Cybernetical Actions and Markov Processes.   [1682]
Gericke, M. : Society-Controlled   Activities - A Fuzzy Simulation Approa [600]
Fairhurst, M: Cyclic   Activity in Information Processing Sys [500]
Fararo, T.J.:   Social   Activity and Social Structure: A Contr [503]
Holden, A.V.:   Chaotic   Activity in Neural Systems.   [771]
Klir, G.J. : attern Discovery in   Activity Arrays.   [939]
Ventriglia, : f Cerebellar Cortex   Activity .   [1811]
Aulin-Ahmava: damental Theorem of   Actor-Systems .   [74]
Jacob, F.   : he Possible and the   Actual .   [824]
Slawski, C. :   Social   Actualization : A Theory of Community L [1579]
Conrad, M. :   Adaptability - The Significance of Var [343]
Paritsis, N.:   Adaptional Problems in Natural Intelli [1282]
Bartolini, G: ment of Performance   Adaptive Fuzzy Controllers with Applic [113]
Day, R.H.   : c Development as an   Adaptive Process.   [384]
De Greene, K: The   Adaptive Organization - Anticipation a [391]
De Jong, K. :   Adaptive System Design: A Genetic Appr [392]
De La Sen, M: Single Input-Output   Adaptive Control Systems to Avoid Dela [394]
Egardt, B. :   Stability of Adaptive Controllers.   [458]
Gladun, V.P.:   Adaptive Problem-Solving Systems.   [617]
Gupta, M.M. : On the Design of an   Adaptive Controller for Uncertain Plan [687]
Holling, C.S: Adaptive Environmental Assessment and [774]
Kalaba, R.E.: tion Techniques for   Adaptive Reinvestment: A Small Sample [870]
Landau, Y.D.:   Adaptive Control: the Model Reference [1007]
Mark, J.W. : ative Algorithms to   Adaptive Predicitive Coding.   [1105]
McCorkell, C: A Hierarchical   Adaptive Control System.   [1135]
Murphy, R.E.:   Adaptive Processes in Economic Systems [1198]
Sampson, J.R: Adaptive Information Processing: An In [1485]
Sinha, K.K. :   Adaptive Nuclear Reactor Control witho [1569]
Spagovich, V: Theory of Adaptive Systems.   [1602]
Tesfatsion, : ch to Filtering and   Adaptive Control: Optimality Results [1683]

```
Tsypkin, Y.Z:         The Theory of   Adaptive   and Learning Systems.         [1760]
Umpleby, S.  : Social Organisation    Adaptive .                               [1770]
            :   on Applications of    Adaptive   Systems Theory. Medicine.     [1940]
McIntyre, C.:                         Addendum   to Internal Report 165: Mathe [1140]
Halfon, E.  :                         Adequacy   of Ecosystem Models.          [711]
Sharma, B.D.:                         Adjacent-Error  Correcting Binary Perfe  [1538]
Couger, J.D.:                         Advanced   System Development/Feasibilit [351]
Hanson, O.J.: ng and Modifying an     Advanced   Business Systems Analysis Cou [723]
Hanson, O.J.: and Re-Design of an     Advanced   Systems Analysis Course.      [724]
Oren, T.I.  :         Concepts for    Advanced   Simulation Methodologies.     [1270]
Ray, W.H.   :                         Advanced   Process Control.              [1392]
Svoboda, A. :                         Advanced   Logical Circuit Design Techni [1645]
El-Sherief, :              Recent     Advances   in Multivariable System Model [471]
Gupta, M.M. :                         Advances   in Fuzzy Set Theory and Appli [684]
Gupta, M.M. :                         Advances   in Fuzzy Set Theory and Appli [686]
Halfon, E.  : al Systems Ecology:     Advances   and Case Studies.             [712]
Stucki, P.  :                         Advances   in Digital Image Processing.  [1630]
Tou, J.T.   :                         Advances   in Information Systems Scienc [1705]
Hoffer, J.A.: ing the Comparative     Advantages  of Sequential and Batched D  [766]
Hogeweg, P. : l Systems: Concrete     Advantages  of Discrete Event Formalism  [769]
Trappl, R.  : ceptable, Feasible,     Advisable ?.                             [1715]
Kohout, L.J.: f an Expert Therapy     Adviser    as a Special Case of a General [966]
Phillips, R.:                         AESOP : an Architectural Relational Dat  [1333]
Boyd, G.M.  :           Cybernetic    Aesthetics : Key Questions in the Desig  [205]
Hoffer, J.A.:             Factors     Affecting   the Comparative Advantages o [766]
Sonnier, I.L: g of Science in the     Affective   Domain.                      [1599]
Michie, D.  : the Microelectronic     Age .                                    [1160]
Shore, J.E. : mposable Subset and     Aggregate   Constraints.                 [1550]
Arbel, A.   :                         Aggregation  and Information Structurin  [46]
Barfoot, C.B:                         Aggregation  of Conditional Absorbing M  [107]
Roedding, W.: t Approaches to the     Aggregation  of Preferences.             [1434]
Porenta, G. : inson's Disease and     Aging   Processes.                       [1349]
Spedding, C.:                         Agricultural  Systems.                   [1603]
Dent, J.B.  : stems Simulation in     Agriculture .                            [410]
Kobsa, A.   :         On Regarding    AI  Programs as Theories.                [956]
Zarri, G.P. : SEDA Project: Using     AI  Techniques in Order to Solve the Pr  [1915]
Harris, L.R.: mantic Component to     Aid  in the Parsing of Natural Language  [727]
Tripp, R.S. : Illustration within     Air  Force Logistics Command.            [1736]
Glaser, F.B.: System Approach to      Alcohol   Treatment.                     [622]
Cliffe, M.J.:        Cybernetics of   Alcoholism .                             [321]
Gill, A.    :             Applied     Algebra  for the Computer Sciences.      [612]
Glushkov, V.:                         Algebra , Languages Programming.         [626]
Jezek, J.   :           Universal     Algebra  and Theory of Models.           [845]
Kandel, A.  :  Fuzzy Sets, Fuzzy      Algebra , and Fuzzy Statistics.          [873]
Lefebre, V.A:                         Algebra  of Conscience: A Comparative A  [1030]
Arigoni, A.C:                         Algebraic   Structure of Property Spaces [50]
Blomberg, H.:                         Algebraic   Theory for Multivariable Lin [175]
Borodin, A. : ional Complexity of     Algebraic   and Numeric Problems.        [190]
Csaki, P.   :                 An      Algebraic   Approach to Some General Pro [360]
Gottinger, H:           Toward an     Algebraic   Theory of Complexity in Dyna [647]
Hermann, R. :      Applications of    Algebraic   Geometry to Systems Theory,  [752]
Kohout, L.J.:                 The     Algebraic   Structure of the Spencer-Bro [965]
Matsuda, T. :                         Algebraic   Properties of Satisficing De [1119]
Resconi, G. : tems by Logical and     Algebraic   Structures.                  [1406]
Sain, M.K.  :        Introduction to  Algebraic   System Theory.               [1482]
Arakelian, A:                 On an   Algorithm   of Spectral Analysis.        [45]
Burns, J.R. :                 An      Algorithm   for Converting Signed Digrap [249]
Cheng, J.K. : ubgraph Isomorphism     Algorithm   Using Resolution.            [306]
Della Riccia: n a Factor Analysis     Algorithm .                             [405]
El-Shirbeeny:   Conjugate Gradient    Algorithm   for Solving Constrained Cont [472]
Gopal, K.   :    An Event Expansion   Algorithm   for Reliability Evaluation i [643]
Hakkala, L. : t-Type Coordination     Algorithm   for Dynamical Systems.       [709]
Jones, B.   :    Systems Theory and   Algorithm   Theory.                      [855]
Krippendorff:                 On the  Algorithm   for Identifying Structures i [984]
```

Sastry, D.   :   A Two-Level   Algorithm   for the Simultaneous State a [1495]
Schkolnick, :   A Clustering   Algorithm   for Hierarchical Structures. [1511]
Smetanic, Y.:   On the   Algorithm   for Determining the Primary [1585]
Walichiewicz: ional Fuzzy Control   Algorithm .   [1828]
Alspach, B. :   Algorithmic   Aspects of Combinatorics.   [25]
Chaitin, G.J: Algorithmic   Information Theory.   [291]
Lauenroth, H: Algorithmic   Model Systems and Innovati [1017]
Baase, S.   :   Computer   Algorithms : Introduction to Design and [77]
Baase, S.   :   Computer   Algorithms .   [76]
Bathe, K.J. : s and Computational   Algorithms   in Finite Element Analysis: [123]
Bezdek, J.C.:   Objective Function   Algorithms .   [167]
Cohen, J.   :   Non-Deterministic   Algorithms .   [326]
El-Sherief, : and Identification   Algorithms   and their Applications.   [471]
Greene, D.H.: for the Analysis of   Algorithms .   [662]
Jakubowski, :   General   Algorithms   of Complex Dynamic Systems [829]
Kronsjo, L.I: Algorithms : Their Complexity and Effic [987]
Kwakernaak, : d Theorems. Part 2:   Algorithms   and Examples for the Discre [996]
Lewis, H.R. :   The Efficiency of   Algorithms .   [1047]
Machtey, M. : e General Theory of   Algorithms .   [1081]
Mark, J.W.   : cation of Iterative   Algorithms   to Adaptive Prediciti ve Cod [1105]
Monahan, G.E: Theory, Models, and   Algorithms .   [1183]
Nussbaumer, : orm and Convolution   Algorithms .   [1249]
Pichler, F. :   General Systems   Algorithms   for Mathematical Systems Th [1339]
Sellers, P.H: hematical Theory of   Algorithms .   [1522]
Thulasiraman: aphs, Networks, and   Algorithms .   [1688]
Geyer, R.F. :   Alienation   Theories: a General Systems [604]
Nowakowska, : ty, Monopolies, and   Alienation   in Science.   [1242]
Ruotsalainen: lating the Resource   Allocation   of the Regional Hospital Sy [1463]
Johnson, D.T: ntrol Strategy that   Allows   for the Cost of Planning.   [848]
Dial, G.   :   Entropy of Order   Alpha   and Type Beta and Shannon's Ineq [420]
DiNola, A.   : ess of Solutions of   Alpha-Fuzzy   Relation Equations on Fini [418]
Venkatesan, :   An   Alternate   Approach to Transitive Coupl [1809]
Estep, M.L. :   Toward   Alternative   Methods of Systems Analysi [494]
Havranek, T.:   An   Alternative 'Approach to Missing Inform [738]
Keeny, B.P. : ic Epistemology: an   Alternative   Paradigm for Diagnosis.   [896]
Scandura, J.: truction: A Systems   Alternative   to Cognitive Psychology. [1501]
Wasinowski, :   Choosing among   Alternative   Strategies.   [1843]
Orlovsky, S.:   Effective   Alternatives   for Multiple Fuzzy Prefer [1272]
Viliums, E.R: ractical Aspects of   Alternatives   Evaluating and Decision M [1817]
Dolby, J.L. :   On the Notions of   Ambiguity   and Information Loss.   [427]
Matsuno, K. : g Polymerization of   Amino   Acids.   [1120]
Kintala, C.M: Amounts   of Nondeterminism in Finite Au [929]
Ben-Eli, M.U: Amplifying   Regulation and Variety Incr [149]
Chen, D.T.-W: Toward   Analogical   Reasoning in Problem Solvin [303]
Noguchi, S. :   Properties of   Analogue   Neural Networks.   [1235]
Bunge, M.   :   Analogy   between Systems.   [241]
Weingartner,:   Analogy   Among Systems.   [1854]
Winston, P.H: ng and Reasoning by   Analogy .   [1870]
Le Moigne, J: Analyse   de Systeme: Bibliographie et C [1022]
Le Moigne, J: Analyse   de Systeme.   [1023]
Porenta, G. : ity and Sensitivity   Analyses , and Simulation of Parkinson' [1349]
Smith, A.W. : Mangagement Systems   Analyses   and Applications.   [1587]
Wisniewski, : teractive Model for   Analysing   Climate Change due to Increa [1872]
Aarnes, E.   : tion in Compartment   Analysis .   [1]
Aggarwal, J.: Systems: Stability   Analysis .   [16]
Altmann, A. : disciplinary System   Analysis .   [27]
Anderson, T.: variate Statistical   Analysis .   [31]
Andersson, H: Analysis   of an Inventory Control Syste [32]
Apostolakis,:   Synthesis and   Analysis   Methods for Safety and Reliab [40]
Arakelian, A: gorithm of Spectral   Analysis .   [45]
Baase, S.   : ction to Design and   Analysis .   [77]
Barber, M.C.: for Ecosystem Flow   Analysis .   [105]
Basta, D.J. : ng Natural Systems: Analysis   for Regional Residuals-Enviro [114]
Batchelder, :   Examination of the   Analysis   of Dichotomous Data.   [115]

Bathe, K.J. : s in Finite Element   Analysis : U.S.-German Symposium.   [123]
Beltrami, E.: for Public Systems   Analysis .   [147]
Bennett, R.J: ystems: Philosophy,   Analysis and Control.   [152]
Bensoussan, : w Trends in Systems   Analysis : International Symposium, Ver   [158]
Bensoussan, : ms Optimization and   Analysis .   [159]
Bensoussan, :   Analysis and Optimization of Systems.   [160]
Bertoni, A. :   Analysis and Compacting of Musical Tex   [163]
Binder, D.A.:   Bayesian Cluster Analysis .   [169]
Bolter, M. : heme for Structural   Analysis in Marine Ecosystems.   [185]
Bowen, B.D. : ntroduction to Data   Analysis .   [202]
Branin, F.H.:   Problem   Analysis in Science and Engineering.   [212]
Breitenecker: Simulation and Analysis of Pathological Blood Pressur   [215]
Brillinger, :   Time Series: Data Analysis and Theory.   [221]
Brockett, P.: formation Theoretic   Analysis of Questionnaire Data.   [222]
Broekstra, G: Constraint Analysis and Structure Identification   [223]
Brown, D.B. :   Systems   Analysis and Design for Safety.   [228]
Bunn, D.W. : l Methods in Policy   Analysis .   [242]
Burns, D.W. : Models in Decision   Analysis .   [247]
Burns, J.R. : h to the Structural   Analysis of Digraphs.   [250]
Cavallo, R.E: Reconstructability Analysis : Evaluation of Reconstruction   [283]
Cavallo, R.E: Reconstructability Analysis : Overview and Bibliography.   [284]
Cavallo, R.E: Reconstructability Analysis .   [285]
Chang, Shi-K: Policy   Analysis and Information Systems, (A J   [294]
Chen, K.   :   System   Analysis in Different Social Settings.   [305]
Chikan, A.   :   Analysis of a Model System.   [307]
Chow, G.C.   :   Econometric   Analysis by Control Methods.   [309]
Cochin, I.   :   Analysis and Design of Dynamic Systems   [324]
Conant, R.C.:   Detection and   Analysis of Dependency Structures.   [339]
Cook, N.D.   : and Flexibility: an Analysis of Natural Systems.   [344]
Cotterman, W: Systems   Analysis and Design: A Foundation for   [350]
Cumani, A.   : tic Approach to the Analysis of Fuzzy Feedback Systems.   [366]
Dale, M.B.   :   Systems   Analysis and Ecology.   [372]
Della Riccia: Scores in a Factor   Analysis Algorithm.   [405]
Demarco, T. :   Structural Analysis and System Specification.   [406]
Desouza, G.R: nvironmental Impact   Analysis .   [411]
Dirickx, Y.M: Systems   Analysis by Multilevel Methods: with A   [424]
Drozen, V. : Approach to Pattern   Analysis .   [442]
Duda, R.O.   : ification and Scene Analysis .   [445]
Duran, B.S. :   Cluster   Analysis : a Survey.   [448]
Edwards, J.B: Analysis and Control of Multipass Proc   [456]
Elandt-Johns: val Models and Data   Analysis .   [473]
Elias, D.   : Reconstructability Analysis : APL Version.   [474]
Estep, M.L. : Methods of Systems   Analysis : The Case of Qualitative Know   [494]
Everitt, B.S: Problems in Cluster   Analysis .   [497]
Favella, L.F: cal Problems in ECG   Analysis .   [513]
Fiacco, A.V.: Methods and Systems   Analysis .   [520]
Freeling, A.: y Sets and Decision   Analysis .   [545]
Gams, M.   :   A Circuit   Analysis Program that Explains its Rea   [565]
Gane, C.   : Structural Systems   Analysis : Tools and Techniques.   [567]
Gehrlein, W.:   An   Analysis of Simple Counting Methods fo   [588]
Golledge, R.: he Multidimensional   Analysis of Large Data Sets.   [632]
Gourlay, A.R: Identification and Analysis of the Subsystems Structure o   [650]
Greenberg, H: Computer-Assisted   Analysis and Model Simplification.   [660]
Greenberg, H: Computer-Assisted   Analysis and Model Simplification.   [661]
Greene, D.H.: Mathematics for the   Analysis of Algorithms.   [662]
Grindley, K.: Approach to Systems   Analysis .   [670]
Hagg, C.   : nd Cost in Decision Analysis .   [699]
Haimes, Y.Y.:   Hierarchical Analysis of Water Resources Systems: M   [700]
Hanken, A.F.: ics and Society: An   Analysis of Social Systems.   [720]
Hanson, O.J.: The Role of Systems   Analysis in Designing Systems and Meth   [722]
Hanson, O.J.: ed Business Systems   Analysis Course Using Feedback from Pa   [723]
Hanson, O.J.: an Advanced Systems   Analysis Course.   [724]
Haruna, K. : uctural Sensitivity   Analysis and its Applications.   [732]
Hawkins, D.M: pplied Multivariate   Analysis (2 volumes).   [741]

Hossforf, H.:   Model Analysis of Structures.   [791]
Hostetter, G: amentals of Network Analysis .   [792]
Huber, O.   : rences: Theoretical Analysis of a Model.   [795]
Ihara, J.   :   A Structural Analysis of Criteria for Selecting Mod   [808]
Innis, G.S. : w Directions in the Analysis of Ecological Systems.   [812]
Jarisch, W. :   Signal Analysis of Visual Evoked Potentials.   [838]
Jeffers, J.N: oduction to Systems Analysis : with Ecological Applications   [842]
Jennergren, : Approach: a Systems Analysis Methodology.   [843]
Kindler, E. : mula 67 and Systems Analysis .   [920]
Kleijnen, J.:   Design and Analysis of Simualtion: Practical Stat   [932]
Klir, G.J.   : Reconstructability Analysis .   [944]
Klir, G.J.   : Reconstructability Analysis .   [947]
Kobayashi, H: Modelling and Analysis : an Introduction to System Pe   [953]
Koyayashi, H: Modeling and Analysis : an Introduction to System Pe   [977]
Kriegel, U. :   An Evolutionary Analysis of World Energy Consumption a   [982]
Krone, R.M. :   Systems Analysis and Policy Sciences: Theory a   [986]
Launer, R.L.:   Modern Data Analysis .   [1018]
Lefebre, V.A: ence: A Comparative Analysis of Wester and Soviet Ethical   [1030]
Leigh, J.R. :   Functional Analysis and Linear Control Theory.   [1033]
Leung, Y.   :   A Fuzzy Set Analysis of Sociometric Structure.   [1039]
Levien, R.E.: A: Applying Systems Analysis in an International Setting.   [1041]
Lewicka-Strz: itude. An Empirical Analysis .   [1045]
Lilienfeld, : ory: an Ideological Analysis .   [1049]
Lions, J.I. : in the Mathematical Analysis of Systems and Their Control.   [1054]
Lions, J.L. : in the Mathematical Analysis of Systems and Their Control.   [1055]
Majone, G. :   Pitfalls of Analysis .   [1091]
Marmarelis, :   Analysis of Physiological Systems.   [1107]
Matis, J.H. :   Compartmental Analysis of Ecosystems Models.   [1118]
McDowall, D.: rrupted Time Series Analysis .   [1136]
McLean, J.M. :   Methods of System Analysis and Model Building.   [1141]
Medina, B.F.:   Structured System Analysis : A New Technique.   [1145]
Michel, A.N.:   Qualitative Analysis of Large Scale Systems.   [1159]
Morein, R.   :   Functionals in the Analysis of Threshold Logic Unit Netwo   [1185]
Morris, C.N.: ntroduction to Data Analysis and Statistical Inference.   [1188]
Neck, R.   :   A System-Theoretic Analysis of a Simple Macroeconomic Mod   [1207]
Netchiporenk: Structure Analysis of Systems.   [1214]
O'Muircheart: The Analysis of Survey Data, Vol. 1: Explo   [1251]
Optner, S.L.:   Systems Analysis .   [1267]
Otnes, R.K. : Applied Time Series Analysis . Vol. 1: Basic Techniques.   [1273]
Patten, B.C.:   Systems Analysis in Ecology.   [1291]
Paulre, B.E.: ms Dynamics and the Analysis of Chance.   [1297]
Pedrycz, W. : quations in Systems Analysis .   [1311]
Perlis, A.J.: oftware Metrics: An Analysis and Evaluation.   [1317]
Perlis, J.H.:   Stability Analysis : An Approach to the Evaluatio   [1318]
Peschel, M. : Approach to Systems Analysis Based on Volterra Equations.   [1322]
Peschel, M. : Data- and Behaviour Analysis with Classification Models.   [1321]
Peschel, M. : Approach to Systems Analysis Based on Volterra Equations.   [1323]
Pritsker, A.:   Modeling and Analysis Using Q-Gert Networks.   [1362]
Prokop, A. :   Systems Analysis and Synthesis in Biology and   [1365]
Quade, E.   : on Applied Systems Analysis : a Report on the Response.   [1370]
Reisig, G.H.:   Statistical Data Analysis in the Information Domain.   [1399]
Resconi, G. :   Analysis of the Purine Biosynthesis by   [1405]
Resconi, G. :   Logical Analysis of Dynamical Systems-Applicat   [1404]
Rivett, P. : ilding for Decision Analysis .   [1422]
Roberts, G.F: h to the Structural Analysis of Digraphs.   [1426]
Robinson, A.:   Non-Standard Analysis .   [1431]
Robinson, A.: ions of Time Series Analysis . Vol. 1: The Box-Jenkins Appr   [1433]
Ruffner, M.A: Factors in Systems Analysis .   [1461]
Rushinek, A.:   An Analysis of Computer Software and Its   [1465]
Ryabinin, I.: ems: Principles and Analysis .   [1466]
Sage, A.P. :   Sensitivity Analysis in Systems for Planning and D   [1476]
Sanderson, A: Pattern Trajectory Analysis of Nonstationary Multivariate   [1489]
Santis, F.de: On Factor Analysis and Fisher's Linear Discrimin   [1492]
Santis, F.de: Linear Discriminant Analysis .   [1492]

```
Schilderinck: gression and Factor   Analysis  in Econometrics.              [1509]
Schnakenberg: ermodynamic Network   Analysis  of Biological Systems (Second [1513]
Scoltock, J.: terature on Cluster   Analysis .                              [1517]
Scott, C.H. :              On the   Analysis  of an Information Theoretic M  [1518]
Seo, F.     : tiattribute Utility   Analysis  for Collective Choice.        [1527]
Sherlock, R.:                       Analysis  of the Behaviour of Kauffman  [1545]
Slyadz, N.N.:                       Analysis  of Fuzzy Initial Information   [1582]
Soong, T.T. : listic Modeling and   Analysis  in Science and Engineering.   [1600]
Srinivasan, :       Probabilistic   Analysis  of Redundant Systems.         [1608]
Steinberg, J:    An Informational   Analysis  of the Inter-Male Behavior of [1618]
Steward, D.V:             Systems   Analysis  and Management: Structure, St  [1620]
Subrahmaniam:        Multivariate   Analysis  : a Selected and Abstracted Bi [1631]
Takeuchi, A.: ons of Multivariate   Analysis  : A Unified Approach by Means  [1664]
Tani, S.T.  : delling in Decision   Analysis .                              [1668]
Taylor, J.R.: troduction to Error   Analysis .                              [1673]
Tchon, K.   :     Towards a Global  Analysis  of Systems.                   [1680]
Toda, M.    : ysis (II): Strategy   Analysis  for Future Computer Systems D  [1694]
Trussel, P. : esearch: Discovery,   Analysis , and Interpretation.          [1755]
Valdes-Perez: Reconstructability    Analysis .                              [1781]
Von Luetzow,:           A System    Analysis  of Information Transmission o  [1822]
Wang, P.    : plication to Policy   Analysis  and Information Systems.       [1836]
Watson, S.R.:     Fuzzy Decision    Analysis .                              [1848]
Wood-Harper,: proaches to Systems   Analysis .                              [1879]
Xuemou, W.  :          Pansystems   Analysis  and Fuzzy Sets.               [1886]
            : Symposium on Policy    Analysis  and Information Systems.       [1934]
Groen, J.C.F:         A Constraint  Analytic  Measure for the Distance betw  [671]
Rosenberg, R:                       Analytical Dynamics of Discrete System  [1447]
Basta, D.J. :                       Analyzing Natural Systems: Analysis fo   [114]
Goodman, L.A:                       Analyzing Qualitative / Categorical Da   [641]
Horacek, H. :     On Generation of  Anaphora  and Gapping in German.         [781]
Steinacker, : Disambiguation and    Anaphora  Resolution.                    [1617]
Rowe, W.D.  :                  An   Anatomy   of Risk.                       [1452]
Holden, A.V.: f Action of General   Anestetics .                            [770]
Hazlett, B.A: ods in the Study of   Animal    Behavior.                      [744]
Metz, H.A.J.: te Space Models for   Animal    Behaviour.                    [1154]
Wang, Z.    : ion and Planning of   Animal    Husbandry System in Farming Reg [1837]
Van Rootsela:                       Annals    of Systems Research, Vol. 7   [1790]
Gati, G.    :            Further    Annotated Bibliography on the Isomorph   [585]
Banathy, B.H: eedings of the 29th   Annual    Meeting.                       [97]
Beer, S.    : s Equifinal: Eighth   Annual    Ludwig Von Bertalanffy Memorial [135]
Boulding, K.: eral System: Fourth   Annual    Ludwig von Bertalanffy Memorial [199]
Cordier, M.O:        Propagation:   Another   Way for Matching Patterns in K  [346]
Belnap, N.P.: ic of Questions and   Answers .                               [144]
Cornelis, A.:                       Anthropological Transformations in the   [347]
De Greene, K: tive Organization -   Anticipation and Management of Crisis.   [391]
Kefalas, A.G:    Art and Science of Anticipatory Management.                 [897]
Rosen, R.   :                       Anticipatory Systems in Retrospect and  [1443]
Rosen, R.   :                  On   Anticipatory Systems.                   [1442]
Yager, R.R. : ing Tranquility and   Anxiety   in Decision Making: An Applica [1905]
Breitenecker: Coarctation of the    Aorta .                                 [215]
De La Sen, M: ng of Multivariable   Aperiodic Sampling Systems.              [393]
Elias, D.   : ctability Analysis:   APL       Version.                       [474]
Franksen, O.: Design Rules Using    APL .                                   [541]
Sklenar, J. :                       Apparatus for Formal Description of Di  [1571]
Cover, T.M. : for Sequences with    Apparent  Markov Structure.             [353]
Bowler, T.D.: king: Its Scope and   Applicability .                         [203]
Busch, J.A. : s IV: A System-Type   Applicable to Human Groups.             [252]
Shaeffer, D.: luation Methodology   Applicable to Environmental Assessment  [1530]
Ballance, R.: Dependence and its    Application to System Evaluation.         [90]
Bartolini, G: zy Controllers with   Application to Continuous Casting Plan   [113]
Best, D.P.  : tic Approach to the   Application of Information Technology.    [165]
Czogala, E. : zzy Systems and Its   Application in Control Problems.          [368]
De Raadt, J.:                  An   Application of Ashby's Law to the Skew    [398]
Fandel, G.  : n Making Theory and   Application . (Proc. of a conf. in Hage  [502]
```

Favella, L.F: Application   of Loeve-Karhunen Expansio [512]
Furukawa, O.: , and the Practical   Application .   [550]
Green, M.A. : hnology, and System   Application .   [659]
Hellerman, L: Initiative and its   Application   to Computers.   [747]
Hofferbert, :   The   Application   of General Systems Methodo   [767]
Jain, V.   : emic Assessment and   Application   of the Ekistic Typology.   [828]
Janssens, L.:   The   Application   of Cybernetic Principles f   [833]
Mamdani, E.H: Application   of Fuzzy Logic to Approxim [1098]
Mark, J.W. :   Application   of Iterative Algorithms to [1105]
Peery, N.S. : s: Some Problems in   Application .   [1314]
Peschel, M. :   Application   of Polyoptimization to Pro [1320]
Tamura, H. : f Data Handling: An   Application   to Lake Ontario.   [1666]
Trost, H.   :   An Overview and an   Application   in the Field of Natural La [1754]
Wang, P.   : Systems: Theory and   Application   to Policy Analysis and Inf [1836]
Wang, Z.   :   Application   of Fuzzy Decision Making a [1837]
Yager, R.R. : Decision Making: An   Application   of Fuzzy Sets.   [1905]
Pedrycz, W. :   Some   Applicational Aspects of Fuzzy Relatio   [1311]
Andersen, E.: with Social Science   Applications .   [28]
Bibbero, R.J: em, Interfacing and   Applications .   [168]
Bose, N.K. : Systems: Theory and   Applications .   [192]
Brauer, W. :   Net Theory and   Applications . (Proc. of the Advanced C   [213]
Casti, J.   : l Systems and Their   Applications : Linear Theory.   [271]
Christensen,: lementation; Vol.4:   Applications .   [311]
Cliff, A.D. : ocesses: Models and   Applications .   [320]
Courtois, P.: and Computer System   Applications .   [352]
Dhillon, B.S: New Techniques and   Applications .   [416]
Dirickx, Y.M: level Methods: with   Applications   for Economics and Managem [424]
Dubois, D. : Systems: Theory and   Applications .   [443]
El-Sherief, : lgorithms and their   Applications .   [471]
Fosko, P.D. :   Applications   of General Systems Theory   [537]
Golubitsky, : ophe Theory and its   Applications .   [635]
Gottinger, H: ty and Catastrophe: Applications   of Dynamic System Theory. [646]
Grubbstrom, :   Closed Systems and   Applications .   [675]
Gupta, M.M. : uzzy Set Theory and   Applications .   [684]
Gupta, M.M. : uzzy Set Theory and   Applications .   [686]
Gupta, M.M. :   Feedback Control   Applications   of Fuzzy Set Theory: A Su [689]
Gurel, O.   : urcation Theory and   Applications   in Scientific Disciplines [691]
Gutenbaum, J: Applications   of Systems Theory to Econ   [692]
Hartnett, W.: proaches, Theories,   Applications .   [731]
Haruna, K. : ty Analysis and its   Applications .   [732]
Herman, G.T.: Implementation and   Applications .   [751]
Hermann, R. :   Applications   of Algebraic Geometry to   [752]
Hickin, J. :   Applications   of Projective Reduction M   [753]
Hwang, C.L. : Making: Methods and   Applications   (a State-of-the-Art Surve [801]
Hwang, C.L. : Making: Methods and   Applications : A State-of-the-Art Surve   [803]
Jarisch, W. : te Variability with   Applications   to Clinical Risk-Assessme [837]
Jeffers, J.N: is: with Ecological   Applications .   [842]
Johnson, W.A: Applications   of Systems Theory to Ecol   [852]
Jumarie, G. :   Some Technical   Applications   of Relativistic Informati   [861]
Kandel, A. : utomata: Theory and   Applications .   [874]
Kittler, J. : ognition Theory and   Applications .   [931]
Lee, E.T.   :   Applications   of Fuzzy Set Theory to Im [1027]
Linstone, H.: ective Concept with   Applications   to Technology and Other D [1052]
Luenberger, : Theory, Models, and   Applications .   [1071]
Makridakis, : asting: Methods and   Applications .   [1095]
Mamdani, E.M: y Reasoning and Its   Applications .   [1099]
Mohler, R. :   Theory and   Applications   of Variable Structure Sys [1178]
Mohler, R. : r System Theory and   Applications .   [1180]
Negoita, C. :   Management   Applications   of Systems Theory.   [1209]
Poch, F.A. :   Some   Applications   of Fuzzy Sets to Statisti [1347]
Poston, T. : ophe Theory and its   Applications .   [1353]
Roberts, F.S: urement Theory with   Applications   to Decision-Making, Utili [1425]
Sawaragi, Y.: timation, and Their   Applications   for Distributed Parameter [1499]
Schieve, W.C: ipative Structures: Applications   in the Physical and Socia [1507]

Silviu, G.   : rmation Theory with Applications .   [1559]
Singh, M.G.   : Systems Engineering   Applications .   [1566]
Sinha, D.K.   : astrophe Theory and   Applications .   [1568]
Skala, H.J.   : zy Logics and Their   Applications .   [1570]
Smith, A.W.   : ystems Analyses and   Applications .   [1587]
Sperb, R.P.   : rinciples and their   Applications .   [1604]
Tao, K.M.   : irected Graphs with   Applications : A Rapprochement with Sys [1669]
Tao, K.M.   : irected Graphs with   Applications : A Rapprochement with Sys [1670]
Tilti, A.   : e System Theory and   Applications (Proc. of a Symposium in [1690]
Trappl, R.   : netics - Theory and   Applications .   [1725]
Tremblay, J.: ata Structures with   Applications .   [1735]
Troncale, L.: n Bio-Systems Real?   Applications of Clustering Theory to M [1739]
Van Navta, L: Digital Computer   Applications to Process Control.   [1789]
Walker, J.A.: uations: Theory and   Applications .   [1831]
Wallace, M.R: Applications of Ultrastability.   [1833]
              : of the Workshop on   Applications of Adaptive Systems Theor [1940]
Bose, N.K.   :   Applied Multidimensional Systems Theor [193]
Brams, S.J.   :   Applied Game Theory.   [211]
Ghosal, A.   :   Applied Cybernetics: its Relevance to   [607]
Gill, A.   :   Applied Algebra for the Computer Scien [612]
Greenspan, D: Arithmetic   Applied Mathematics.   [664]
Haley, K.B.   :   Applied Operatons Research in Fishing. [710]
Hawkins, D.M: Topics in   Applied Multivariate Analysis (2 volum [741]
Jones, D.D.   : Catastrophe Theory   Applied to Ecological Systems.   [856]
Klir, G.J.   :   Applied General Systems Research.   [940]
Klir, G.J.   :   Basic and   Applied General Systems Research: A Bi [937]
Lasker, G.E.:   Applied Systems and Cybernetics (6 Vol [1010]
Mattessich, : Epistemology of the   Applied and Social Sciences.   [1126]
Mingers, J.C: e Social Theory for   Applied Systems Thinking: Critical The [1171]
Mulej, M.   : vity of Many (WCM)"   Applied to Collect and Organize Associ [1195]
Otnes, R.K.   :   Applied Time Series Analysis. Vol. 1:   [1273]
Putt, A.M.   : eral Systems Theory   Applied to Nursing.   [1369]
Quade, E.   : rt Questionnaire on   Applied Systems Analysis: a Report on [1370]
Rao, G.P.   : lems of Fuzzy Logic   Applied to Systems Control.   [1380]
Silvern, L.C: Systems Engineering   Applied to Training.   [1556]
Sussmann, H.: tastrophe Theory as   Applied to the Social and Biological S [1641]
Trappl, R.   :   Basic and   Applied General Systems Research: A Bi [1729]
Troncale, L.:   Systems Theory   Applied to the Biological and Medical [1750]
Windeknecht,: f Numerical Methods   Applied to the Solution of Ordinary Di [1866]
La Patra, J.:   Applying the Systems Approach to Urban [997]
Levien, R.E.: roduction to IIASA: Applying Systems Analysis in an Intern [1041]
Uccellini, L: Applying Discrete Model Concepts to th [1765]
Laszlo, C.A.: e for Health Hazard   Appraisal .   [1013]
Trappl, R.   :   A Health Hazard   Appraisal Program for Austria.   [1717]
Ackerman, E.: ciences: a Computer   Approach .   [7]
Agrawal, V.D: formation Theoretic   Approach to Digital Fault Testing.   [18]
Beloglavec, : eral Systems Theory   Approach to the Properties of a Period [145]
Best, D.P.   :   A Cybernetic   Approach to the Application of Informa   [165]
Blauberg, I.:   The Systemic   Approach : Prerequisites, Problems and   [174]
Brayton, R.K: ems: a Constructive   Approach .   [214]
Brightman, H: ogical and Creative   Approach .   [219]
Brightman, H: ogical and Creative   Approach .   [220]
Brown, J.H.U: rocesses: a Systems   Approach to Endocrinology.   [230]
Burns, J.R.   :   A Sector   Approach to the Formulation of Systems [248]
Burns, J.R.   :   An Input/Output   Approach to the Structural Analysis of [250]
Callebaut, W: The Distribution Approach to Problem-Solving in Science [257]
Capocelli, R: A Cybernetic   Approach to Population Dynamics Modeli   [260]
Carlsson, C.:   An   Approach to Handling Fuzzy Problem Str   [264]
Castelfranch: A Lexicon Based   Approach to Sentence Understanding.   [269]
Chang, L.C.   : A Graph-Structural   Approach for the Generalization of Dat [293]
Checkland, P: ethinking a Systems   Approach .   [300]
Churchman, C: The Systems   Approach and its Enemies.   [312]
Ciupa, M.   :   A Cybernetic   Approach to Packet Switched Data Commu [317]
Cliff, A.D.   : ure: a Quantitative   Approach .   [319]

Comstock, F.:   A Systems   Approach to Grading of Flight Simulato [335]
Csaki, P.   :   An Algebraic   Approach to Some General Problems of M [360]
Cugno, F.   : ia: A System-Theory   Approach .   [364]
Cumani, A.   : On a possibilistic   Approach to the Analysis of Fuzzy Feed [366]
Davies, A.J.: ent Method: A First   Approach .   [380]
De Jong, K.   : m Design: A Genetic   Approach .   [392]
De Malherbe,: ogy: a Hierarchical   Approach .   [396]
Driankow, D.:   An   Approach to Multiobjectives Decision M [439]
Drozen, V.   : inatory Spaces - An   Approach to Pattern Analysis.   [442]
Efstathiou, : g a Fuzzy Heuristic   Approach .   [457]
Ehrenkrantz,:   The Systems Approach to Design.   [460]
Erdi, P.   :   System-Theoretical   Approach to the Neural Organization: F [484]
Feintuch, A.: ry: A Hilbert Space   Approach .   [516]
Fleddermann,: eaching the Systems   Approach with a Group Thesis.   [529]
Fum, D.   :   A Rule-Based Approach to Natural Language Text Repr [547]
Gelfand, A.E: A Systems-Theoretic   Approach to the Management of Complex   [589]
Gericke, M.   : A Fuzzy Simulation   Approach .   [600]
Geyer, R.F.   : : a General Systems   Approach   [604]
Gigley, H.   : eory. An Integrated   Approach to Simulation Modelling of Na [610]
Glaser, F.B.:   A System   Approach to Alcohol Treatment.   [622]
Greenspan, D: A New Computer   Approach to the Modeling of Atoms, Ion [666]
Grindley, K.: Systematics: a New   Approach to Systems Analysis.   [670]
Havranek, T.:   An Alternative   Approach to Missing Information in the [738]
Holden, A.V.:   Cybernetic   Approach to Global and Local Mode of A [770]
Jack, W.   :   On the Eventistic   Approach to the Notion of System.   [823]
Janko, J.   :   Systems   Approach in Natural Science Reflecting [832]
Jennergren, :   The Multilevel   Approach : a Systems Analysis Methodolo [843]
Jumarie, G.   :   A Relativistic   Approach to Modelling Dynamic Systems   [862]
Jumarie, G.   :   A New   Approach to the Transinformation of Ra [863]
Kak, S.   :   formation-Theoretic   Approach .   [869]
Kawamura, K.: Industry: A Systems   Approach .   [891]
Kaynak, E.   :   A Holistic   Approach in Marketing Planning.   [892]
Kickert, W.J: A Metasystem   Approach to Organizational Decision-Ma [912]
Kickert, W.J: Systems-Theoretical   Approach .   [914]
Kickert, W.J: A Formal   Approach to Organization Design.   [913]
Kinston, W.   : An Action Research   Approach to Organisation of Complex Sy [927]
Kohonen, T.   : Systems-Theoretical   Approach .   [964]
Kryukov, V.I: I. Master Equation   Approach .   [988]
Kucera, V.   : Polynomial Equation   Approach .   [992]
La Patra, J.: pplying the Systems   Approach to Urban Development.   [997]
Landau, Y.D.: the Model Reference   Approach .   [1007]
Lewicka-Strz: The Systems   Approach as a Methodological Attitude. [1045]
Lundeberg, M: pment: A Systematic   Approach .   [1074]
M'Pherson, P: ems Engineering: An   Approach to Whole-System Design.   [1076]
Mann, K.H.   : l Waters: A Systems   Approach (Studies in Ecology, Volume 8 [1101]
Maurer, H.   :   The Austrian   Approach to Videotex.   [1129]
McClamroch, : stems: A Case Study   Approach .   [1134]
Ninneman, J.: : A General Systems   Approach .   [1231]
Patten, B.C.:   Systems   Approach to the Concept of Environment [1292]
Patten, B.C.:   Systems   Approach to the Concept of Niche.   [1293]
Perlis, J.H.: bility Analysis: An   Approach to the Evaluation of System D [1318]
Peschel, M.   :   Ecological   Approach to Systems Analysis Based on   [1322]
Peschel, M.   :   Ecological   Approach to Systems Analysis Based on   [1323]
Peschel, M.   : ces of the Volterra   Approach for Nonlinear Systems.   [1325]
Pfeilsticker: The Systems   Approach and Fuzzy Set Theory Bridging [1331]
Pogonowski, :   Set-Theoretical   Approach to General Systems Theory.   [1348]
Pressman, R.: g: A Practitioner's   Approach .   [1358]
Pugh, R.E.   :   odels: a Conceptual   Approach to Case Study.   [1368]
Rescher, N.   : Systems-Theoretical   Approach to the Theory of Knowledge.   [1400]
Rescher, N.   : a Systems-Theoretic   Approach to a Coherent Theory of Knowl [1401]
Roberts, G.F: An Input/Output   Approach to the Structural Analysis of [1426]
Roberts, S.L: Systems   Approach to Assessing Behavioral Probl [1429]
Robertshaw, : Solving: a Systems   Approach .   [1430]
Robinson, E.: 1: The Box-Jenkins   Approach .   [1433]

Roth, G.   : n Interdisciplinary Approach .   [1451]
Rudwick, B.H: Problems: a Systems Approach to Planning and Control.   [1460]
Rugh, W.J.   : The Volterra/Wiener Approach .   [1462]
Ryan, R.M.   : Matrix Organization Approach .   [1467]
Scandura, J.:   Structural Process Approach with Instructional Implicatio   [1500]
Schiffman, Y: A Systems Approach for Estimating Energy Savings   [1508]
Skvoretz, J.: ormation Processing Approach .   [1572]
Smyth, D.S. :   Using a Systems Approach : the Structure of Root Defini   [1595]
Stoy, J.E.   :   the Scott-Strachey Approach to Programming Languages.   [1625]
Takatsuji, M: rmation-Theoretical Approach to a System of Interacting El   [1662]
Takeuchi, A.: Analysis: A Unified Approach by Means of Projection onto L   [1664]
Taylor, A.M.:   A Systems Approach to the Political Organization   [1672]
Tchon, K.   :   A 'Transversal' Approach to Evenistic Systems.   [1677]
Tesfatsion, :   A New Approach to Filtering and Adaptive Con   [1683]
Tjoa, A.M.   : System-Theoretical Approach to Database Systems.   [1691]
Tomlinson, R: A Systems Approach to Planning in Organizations   [1697]
Tong, R.M.   :   A Linguistic Approach to Decision Making with Fuzzy   [1701]
Umnov, A.   :   An Approach to Distributed Modeling.   [1769]
Uyttenhove, :   The Systems Approach Problem Solver and a Stockmar   [1778]
Vavrousek, J: An Approach to Modelling in Empirical Sci   [1804]
Veelenturf, : utomata-Theoretical Approach to Developing Learning Neural   [1806]
Venkatesan, :   An Alternate Approach to Transitive Coupling in ISM   [1809]
Walker, C.C.: A System Theoretic Approach to the Management of Complex   [1830]
White, C.   : Optimization-Based Approach to Choicemaking.   [1861]
Wu, W.T.   :   Heuristic Approach to Chinese-Character Search.   [1881]
Yager, R.R. :   An Approach to Inference in Approximate R   [1893]
Banerji, R.B: Theoretical Approaches to Non-Numerical Problem So   [103]
Batchelor, B: al and Programmatic Approaches to Pattern Recognition.   [116]
Bruckmann, G: Input-Output Approaches in Global Modelling. (Proc.   [231]
Coffman, C.V: Constructive Approaches to Mathematical Models.   [325]
Gottinger, H: Systems Approaches and Environmental Problems.   [645]
Hartnett, W.:   Systems: Approaches , Theories, Applications.   [731]
Heim, R.   :   Theoretical Approaches to Complex Systems.   [746]
Millendorfer: Systemanalytical Approaches to Long Economic Waves.   [1165]
Mitroff, I.I: Methodological Approaches to Social Science.   [1172]
Nowakowska, : Research: Modelling Approaches .   [1241]
Peery, N.S. : eral Systems Theory Approaches to Organizations: Some Prob   [1314]
Roedding, W.:   Different Approaches to the Aggregation of Prefe   [1434]
Schmutzer, M: i & Low - In & Out: Approaches to Social Status.   [1512]
Singh, R.   :   Systems Approaches to the Study of Industrial   [1567]
Trappl, R.   : tonomy, Utopia: New Approaches Towards Complex Systems.   [1731]
Tsai, W.H.   : tic and Statistical Approaches to Pattern Recognition.   [1757]
Wood-Harper,: Taxonomy of Current Approaches to Systems Analysis.   [1879]
Mingers, J.C: Towards an Appropriate Social Theory for Applied   [1171]
Ibidapo-Obe,:   An Approximate Differential Structure for   [807]
Mamdani, E.H: n of Fuzzy Logic to Approximate Reasoning Using Linguistic   [1098]
Rao, G.P.   :   Approximate Reconstruction of Mapping   [1380]
Schellhorn, : Control: Exact and Approximate .   [1506]
Yager, R.R. :   Approximate Reasoning and Possibilisti   [1899]
Yager, R.R. : ach to Inference in Approximate Reasoning.   [1893]
Balossino, N: Harmonics: A Simple Approximation and Results.   [93]
Della Riccia: The Approximation of Factor Scores in a Fa   [405]
Genesio, R. : or Uncertain System Approximation .   [590]
Grollmann, J: eedback Systems via Approximation by Delayed Systems.   [673]
Jachymczyk, : nen-Loewe Transform Approximation for Certain Class of Ran   [822]
Slawski, C. : del of Negotiation: Arabs Versus Israeli.   [1573]
Fedra, K.   :   Uncertainty and Arbitrariness in Ecosystem Modelling: [515]
Wyner, A.D. : ual Information for Arbitrary Ensembles.   [1884]
Chatterjee, :   Comparison of Arbitration Procedures: Models with Co   [295]
Miller, R.B.:   Archetypes of Man-Computer Problem Sol   [1169]
Phillips, R.:   AESOP: an Architectural Relational Data Base.   [1333]
Wade, J.W.   : lems, and Purposes: Architectural Design as a Basic Proble   [1826]
Klir, G.J.   :   Architecture of Structure Systems: a B   [944]
Pask, G.   :   The Architecture of Knowledge and Knowledg   [1288]

Pask, G.   : ge and Knowledge of Architecture .   [1288]
Scholz, C.   :   The Architecture of Hierarchy.   [1514]
Veroy, B.S. : Large Scale Network Architecture Synthesis: Interactive St [1813]
Wade, J.W.   :   Architecture , Problems, and Purposes: [1826]
Kellermayr, : n of an Optical Bus Area Network.   [900]
Linstone, H.: and Other Decision   Areas .   [1052]
Glanville, R: ensions: A Form for   Argument .   [619]
Greenspan, D: A Completely   Arithmetic Formulation of Classical an [663]
Greenspan, D: Arithmetic Applied Mathematics.   [664]
Mizumoto, M.:   Note on the Arithmetic Rule by Zadeh for Fuzzy Con [1176]
Varela, F.J.:   The   Arithmetic of Closure.   [1798]
Slawski, C. : of Community Living Arrangements and a Practical Utopia. [1579]
Klir, G.J.   : scovery in Activity Arrays .   [939]
Reynolds, J.:   Reasoning About Arrays .   [1415]
Ackoff, R.L.:   The Art and Science of Mess Management.   [10]
Bekey, G.A. : Reflections on the   Art and Science of Simulation.   [140]
Crimmins, E.:   The Art of Abstracting.   [357]
Dreyfus, S.E: The Art and Theory of Dynamic Programming. [438]
Joergensen, :   State of the   Art in Ecological Modelling.   [846]
Jorgenson, S: State of the   Art in Ecological Modelling.   [860]
Kefalas, A.G: t by Wisdom: On the   Art and Science of Anticipatory Manage [897]
Smith, C.S. :   Essays on Science, Art and History.   [1588]
Winkel, D.   :   The Art of Digital Design: An Introduction [1867]
Hassell, M.P: The Dynamics of   Arthropod Predator-Prey Systems.   [733]
Unterberger,: oiced Signals using Articulation Models.   [1772]
Barr, A.   :   The Handbook of   Artificial Intelligence (Vol.1).   [109]
Barr, A.   :   The Handbook of   Artificial Intelligence (Vol.2).   [110]
Boden, M.   :   Artificial Intelligence and Natural Ma [180]
Chen, C.N.   : ern Recognition and Artificial Intelligence.   [302]
Cohen, P.R. :   The Handbook of   Artificial Intelligence (Vol.3).   [327]
Davis, R.   : ge-Based Systems in Artificial Intelligence.   [383]
Gause, D.C. :   Cybernetics and Artificial Intelligence.   [587]
Gigley, H.   :   Artificial Intelligence Meets Brain Th [610]
Hayes, P.   :   Trends in Artificial Intelligence.   [743]
Horn, W.   :   An Artificial Intelligence System for Med [786]
Hunt, E.B.   :   Artificial Intelligence.   [797]
Hyde, M.O.   : nk?: The Search for Artificial Intelligence.   [805]
Nilsson, N.J: Principles of   Artificial Intelligence.   [1230]
Ringle, M.   : cal Perspectives in Artificial Intelligence.   [1420]
Steinacker, : d Future Impacts of   Artificial Intelligence: A Bibliograph [1616]
Trappl, R.   : nal Tension through Artificial Intelligence: A Proposal fo [1732]
Trappl, R.   :   The Impacts of   Artificial Intelligence. Science, Tech [1733]
Trappl, R.   :   Impacts of Artificial Intelligence: An Overview. [1730]
Trappl, R.   :   Artificial Intelligence: A One-Hour Co [1728]
Trappl, R.   :   Impacts of Artificial Intelligence.   [1727]
Webber, B.L.:   Readings in   Artificial Intelligence.   [1850]
Winston, P.H: Artificial Intelligence.   [1869]
Winston, P.H: Artificial Intelligence: An MIT Perspe [1871]
Ashby, R.   : Intelligence: Ross   Ashby 's Writings on Cybernetics.   [55]
De Raadt, J.:   An Application of   Ashby 's Law to the Skewness of Income [398]
England J.L.:   An Ashby Hierarchy for Human Action.   [483]
Airaksinen, :   Normative Aspects of Some Stabilizing Social Sys [22]
Alspach, B. :   Algorithmic Aspects of Combinatorics.   [25]
De La Sen, M: Some Aspects of the Modeling of Multivariab [393]
Eugene, J.   :   Aspects of the General Theory of Syste [495]
Findler, N.V: Aspects of Computer Learning.   [525]
Franksen, O.:   Qualitative Aspects of Large Scale Systems: Develo [541]
Gasparski, W: tem: Methodological   Aspects , Vol.3.   [577]
Hoernig, K.M: Aspects of Automatic Program Construct [765]
Overton, W.S: rated by Structural   Aspects of a Hydrology Model.   [1276]
Pedrycz, W. :   Some Applicational Aspects of Fuzzy Relational Equations [1311]
Rzevski, G. :   Some Philosophical Aspects of System Design.   [1468]
Viliums, E.R: Practical Aspects of Alternatives Evaluating and [1817]
Yager, R.R. :   Aspects of Possibilistic Uncertainty. [1894]

Pearl, J.   : Using Probabilistic Assertions .   [1303]
Hanson, O.J.:   Assessing   and Modifying an Advanced Bu [723]
Roberts, S.L: Systems Approach to Assessing   Behavioral Problems of Criti [1429]
Boroush, M.A: Technology Assessment : Creative Futures.   [191]
Broekstra, G: for Organizational Assessment   and Change.   [227]
Craine, R.   : ument: An Empirical Assessment .   [354]
Holling, C.S: ptive Environmental Assessment   and Management.   [774]
Jain, V.   :   A Systemic Assessment   and Application of the Ekis [828]
Kaposi, A.   :   Research into the Assessment   of Complexity.   [879]
Kerin, T.   : hod for Performance Assessment   of Medical Radioisotope Equ [906]
Shaeffer, D.: le to Environmental Assessment   Models.   [1530]
Mulej, M.   : ollect and Organize Associates ' Ideas about New Products a [1195]
Goodman, L.A: Measures of Association   for Cross Classification. [642]
Findler, N.V: Associative   Networks: Representation a [524]
Kohonen, T. :   Associative   Memory: a Systems-Theoreti [964]
Wedde, H.F. : ing under Realistic Assumptions .   [1852]
Barenblatt, : ty and Intermediate Asymptotics .   [106]
Varshavskiy,:   Asynchronous   Processes. I. Definition [1802]
Varshavskiy,:   Asynchronous   Processes II. Composition [1801]
Uccellini, L: the Computation of Atmospheric   Trajectories.   [1765]
Oldershaw, R: -Similarity between Atomic   and Stellar Systems.   [1264]
Greenspan, D: to the Modeling of Atoms , Ions and Molecules. (Part I: Hy [666]
Casti, J.   : the Mathematics of Attainable   Goals and Irreducible Uncer [270]
Sakawa, M.   : An Interactive Goal Attainment   Method for Multiobjective N [1483]
Kellerman, W: ystem Structure: An Attempt   at Implementation.   [899]
Ventriglia, :   Neural Systems: An Attempt   Towards a Simulation of Cerebe [1811]
Ricciardi, L: nd Biology: Classic Attempts   at Neuronal Modeling.   [1418]
Lewicka-Strz: as a Methodological Attitude . An Empirical Analysis.   [1045]
Salzberg, P.: A Unified Theory of Attraction   in General Systems.   [1484]
Hwang, C.L. :   Multiple Attribute   Decision Making: Methods and [803]
Tsai, W.H.   :   Attributed   Grammar - A Tool for Combin [1757]
Pelzmann, L.: d Helplessness - An Attribution-Model   for Unemployed Worke [1316]
Gage, W.L.   :   Auditorium   Management Information Syst [552]
El-Shirbeeny: A Sensitivity Augmented   Conjugate Gradient Algorithm [472]
Augustine, N: Augustine 's Laws and Major System Deve [71]
Horn, W.   :   Health Care in Austria : 1980-2000.   [783]
Neck, R.   :   Economic System in Austria .   [1206]
Trappl, R.   : praisal Program for Austria .   [1717]
Maurer, H.   :   The Austrian   Approach to Videotex.   [1129]
Katz, M.J.   :   Preformation of Autogenetic   Patterns.   [884]
Loefgren, L.: Excerpts from the Autology   of Learning.   [1062]
Ehrig, H.   : Universal Theory of Automata .   [461]
Gollmann, D.: -Linear Networks of Automata .   [633]
Gupta, M.M. :   Fuzzy Automata   and Decision Processes.   [685]
Hopcroft, J.:   Introduction to Automata   Theory, Languages, and Comput [780]
Kandel, A.   : Fuzzy Switching and Automata : Theory and Applications. [874]
Kieras, D.E.:   Finite Automata   and S-R Models.   [915]
Kintala, C.M: terminism in Finite Automata .   [929]
Takeuchi, A.:   Evolutionary Automata   - Comparison of Automation Be [1663]
Toffoli, T. : Reversible Cellular Automata .   [1696]
Tsetlin, M.L: Automata   Theory and Modelling of Biolo [1758]
Wechler, W. : ept of Fuzziness in Automata   and Language Theory.   [1851]
Veelenturf, :   An Automata-Theoretical   Approach to Devel [1806]
Ohlsson, S. :   On the Automated   Learning of Problem Solving [1259]
Rudall, B.H.:   Developing Automated   Teaching Systems.   [1457]
Dixon, N.R. :   Automatic   Speech and Speaker Recogniti [426]
Hoernig, K.M: Aspects of Automatic   Program Construction.   [765]
Rauch, W.D. :   Automatic   Extracting as an Interactive [1391]
Rauch, W.D. :   Automatic   Refereeing and Excerpting. [1390]
Unterberger,:   Automatic   Recognition of Voiced Signal [1772]
Bandler, W. :   The Valuton: An Automation   Model of a Sociological The [101]
Binder, Z. : n in the Integrated Automation   of Complex Systems.   [170]
Faucheuz, C.:   Automation   or Autonomy in Organization [509]
Lauenroth, H: tion Strategies for Automation   of Industrial Production an [1017]

Takeuchi, A.: ata - Comparison of Automation Behavior and Restle's Learn [1663]
Matsuno, K. : librium Dynamics of Autonomous Systems and Their Structura [1121]
Faucheuz, C.:   Automation or Autonomy in Organizational Design.   [509]
Trappl, R. :   Power, Autonomy , Utopia: New Approaches Towar [1731]
Varela, F.J.: iples of Biological Autonomy . (Second Vol. in the North-Ho [1799]
Varela, F.J.: to a Cybernetics of Autonomy .   [1800]
Van Ameronge: An Autopilot for Ships Designed with Fuzz [1785]
Andrew, A.M.:   Autopoiesis and Self-Organization.   [35]
Benseler, F.:   Autopoiesis , Communications, and Socie [157]
Maturana, H.:   Autopoiesis and Cognition: The Realiza [1128]
Zeleny, M. :   Autopoiesis : A Theory of Living Organi [1922]
Zeleny, M. :   Autopoiesis , Dissipative Structures, a [1921]
Benseler, F.: iety: The Theory of Autopoietic System in the Social Scien [157]
Dijkstra, W.:   Testing Auxiliary Hypothesis Behind the Interv [421]
De La Sen, M: Control Systems to Avoid Delays in the Input Generation.   [394]
Bezdek, J.C.: s and Relations: an Axiomatic Basis for Clustering.   [166]
Mattessich, :   Axiomatic Representation of the System [1127]
Schimanovich: An Axiomatic Theory of Winning Mathematic [1510]
Grubbstrom, :   The Axsater Integrated Production-Inventor [676]
Gyarfas, F. : : Prototypes Driven Backward and Forward Chaining Computer [695]
Smith, R.B. : esign & Implement a Bad System.   [1592]
Van Straaten: rce Meeting on Lake Balaton Modelling Veszprem (Hungary). [1792]
Slawski, C. :   Identity Bargaining : A Policy Systems Research [1575]
Milgram, M. :   Barycentric Entropy of Mankov Systems. [1163]
Atre, S.   :   Data Base : Structured Techniques for Design   [66]
Harris, L.R.:   Using the Data Base as a Semantic Component to Aid in [727]
Harris, L.R.: tural Language Data Base Queries.   [727]
Hutt, A.T.F.:   A Relational Data Base Management System.   [799]
Inmon, W.H. :   Effective Data Base Design.   [811]
Meldman, M.J: : a Relational Data Base Management System for Minicompute [1150]
Nijssen, G.M: Modelling in Data Base Management Systems.   [1229]
Phillips, R.: ral Relational Data Base .   [1333]
Batchelor, B: l Shape Description Based Upon Convex Hulls of Concavities [119]
Castelfranch: A Lexicon Based Approach to Sentence Understandi [269]
Delaney, W. :   A System Theory Based Simulation Language.   [404]
Dubois, D. : s of Fuzzy Measures Based on Triangular Norms.   [444]
Hanassab, S.:   An Interpolator Based Controller for Cybernetic System [719]
Higashi, M. : nty and Information Based on Possibility Distributions.   [754]
Mackenzie, K: A Process Based Measure for the Degree of Hierar [1083]
Oeren, T.I. : Simulation Software Based on Cybernetic and General System [1255]
Peschel, M. : to Systems Analysis Based on Volterra Equations.   [1322]
Peschel, M. : to Systems Analysis Based on Volterra Equations.   [1323]
Sugeno, M. : A Model of Learning Based on Fuzzy Information.   [1633]
Trattnig, W.: cification Language Based on the Data Flow Model.   [1734]
Troncale, L.: volutionary Process Based on Systems Concepts.   [1738]
Yager, R.R. :   Fuzzy Prediction Based on Regression Models.   [1902]
Aho, A.V.   :   in Relational Data Bases .   [20]
Guida, G.   : cess to Online Data Bases .   [681]
Troncale, L.: zed, Empirical Data Bases .   [1745]
Bluth, B.J. : n. A Summary of the Basic Theory.   [178]
Bogdanski, C: Basic Elements of Cybernetic Physics. [182]
Destouches, :   Basic Concepts of System Theory.   [413]
Hellwig, Z.H: Basic Problems of Modern Econometrics. [748]
Klir, G.J. :   Basic and Applied General Systems Rese [937]
Madan, V.   :   Basic Concepts on the Theoretical Trea [1087]
Otnes, R.K. : s Analysis. Vol. 1: Basic Techniques.   [1273]
Trappl, R. :   Basic and Applied General Systems Rese [1729]
Wade, J.W. : ectural Design as a Basic Problem-Solving Process.   [1826]
Balkus, K. :   Necessities as the Basis of the Social Self-Organization. [88]
Bezdek, J.C.: tions: an Axiomatic Basis for Clustering.   [166]
Gruber, H. : a Discussion on the Basis of Sequential Systems.   [677]
Klir, G.J. : tructure Systems: a Basis for the Reconstructability Analy [944]
Odum, H.T. :   Energy Basis for Man and Nature.   [1253]
Pearl, J.   :   An Economic Basis for Certain Methods of Evaluatin [1304]

Hoffer, J.A.: s of Sequential and   Batched   Decision Making.   [766]
Lipset, D. :   Gregory   Bateson : The Legacy of a Scientist.   [1056]
Cover, T.M. :   Compound   Bayes   Predictors for Sequences with Ap   [353]
Dial, G.   :   A Note on   Bayes   Risk with Utilities.   [419]
Binder, D.A.:   Bayesian   Cluster Analysis.   [169]
Kashyap, R. :   A   Bayesian   Comparison of Different Class   [882]
Williams, P.:   Bayesian   Conditionalisation and the Pr   [1865]
Prigogine, I: From Being to   Becoming   - Time and Complexity in the   [1360]
Shaw, M.L.G.:   On   Becoming   a Personal Scientist.   [1541]
Oren, T.I. : ulation - As it Has   Been , Is, and Should Be.   [1269]
Henize, J. : mic Systems: Can We   Begin   to Do Better.   [750]
Bridge, J. :   Beginning   Model Theory.   [217]
Abraham, R. : cs: The Geometry of   Behavior .   [3]
Baltes, P.B.: pan Development and   Behavior , Vol.I.   [94]
Cavallo, R.E: on of Possibilistic   Behavior   Systems.   [286]
Dalenoort, G: Modularity of Behavior   of Networks.   [373]
Hazlett, B.A: the Study of Animal   Behavior .   [744]
Kaiser, J.S.:   Organizational Behavior : A Vector of Single-Valued Tr [868]
Oden, A.S. : cal Modeling of the   Behavior   of Hydrocarbon Reservoirs.   [1252]
Reusch, B. : ystems with Digital   Behavior .   [1411]
Steinberg, J: s of the Inter-Male   Behavior   of the Grasshopper Chortophag [1618]
Takeuchi, A.: rison of Automation   Behavior   and Restle's Learning Model.   [1663]
Bennett, S. : ques for Social and   Behavioral   Sciences.   [153]
Casti, J.   : hods for Social and   Behavioral   Systems.   [274]
Ghiselli, E.: ment Theory for the   Behavioral   Sciences.   [606]
Loo, S.G.   : s in the Social and   Behavioral   Science.   [1064]
Nigro, A.   :   Language   Behavioral   Feedback.   [1228]
Roberts, S.L: proach to Assessing   Behavioral   Problems of Critical Care P [1429]
Sage, A.P. :   Behavioral   and Organizational Consider   [1475]
Uyttenhove, : mensionality in the   Behavioral   System Identification Probl [1777]
Breitenecker: ical Blood Pressure   Behaviour   after Treadmill Test in Pati   [215]
Metz, H.A.J.: e Models for Animal   Behaviour .   [1154]
Peschel, M. :   Data- and   Behaviour   Analysis with Classification [1321]
Rastogi, P.N: The   Behaviour   of Societal Systems.   [1385]
Sherlock, R.:   Analysis of the   Behaviour   of Kauffman Binary Networks: [1545]
Witten, I.H.: nistic Modelling of   Behaviour   Sequences.   [1874]
Witten, I.H.:   Probabilistic   Behaviour-Structure   Transformations Us [1873]
Dijkstra, W.: uxiliary Hypothesis   Behind   the Interview.   [421]
Hartl, R.F. : overs Under Risk of   Being   Killed a Rival.   [729]
Prigogine, I: From   Being   to Becoming - Time and Complexit   [1360]
Airenti, G. : el of Knowledge and   Belief .   [23]
Kobsa, A.   :   Representing   Belief   Models in Semantic Networks.   [958]
Mitroff, I.I: On the Strength of   Belief   of Dialectical Belief Systems.   [1173]
Mitroff, I.I: lief of Dialectical   Belief   Systems.   [1173]
Smets, P.   :   The Degree of   Belief   in a Fuzzy Event.   [1586]
Cugno, F.   :   Bellamy 's Equalitarian Utopia: A Syste   [364]
Koenderink, : pe: Rigid Movement,   Bending   and Stretching by Eye: Implica   [962]
Feldman, J.A: nd the Inference of   Bent   Programs.   [517]
Beer, S.   : h Annual Ludwig Von   Bertalanffy   Memorial Lecture.   [135]
Boulding, K.: h Annual Ludwig von   Bertalanffy   Memorial Lecture.   [199]
Farreny, H. :   On the   Best   Way of Designating Objects in Sen   [508]
Stefanski, J: stic Search for the   Best   Hierarchical Structure.   [1611]
Dial, G.   : rder Alpha and Type   Beta   and Shannon's Inequality.   [420]
Franova, M. :   Proofs Obtained by Beth 's Tableaux.   [542]
Henize, J. : Can We Begin to Do   Better .   [750]
Troncale, L.: ral Systems Enables   Better   Design of Man-Made Systems: The [1749]
Bunge, M.   : Concept of Society: Beyond   Individualism and Holism.   [240]
Cramer, F. : logical Science and   Beyond .   [355]
Pearlman, W.:   Beyond   Gestalt.   [1305]
Niemeyer, H.:   Soziale   Beziehungsgeflechte .   [1227]
Kahneman, D.: nty: Heuristics and   Biases .   [866]
Le Moigne, J: Analyse de Systeme: Bibliographie   et Commentaires (Systems [1022]
Anderson, T.:   A   Bibliography   of Multivariate Statistic   [31]
Cavallo, R.E: lysis: Overview and   Bibliography .   [284]

Gaines, B.R.: The Fuzzy Decade: a   Bibliography   of Fuzzy Systems and Clos [555]
Gati, G.   :   Further Annotated   Bibliography   on the Isomorphism Diseas [585]
Klir, G.J.   : Systems Research: A   Bibliography .   [937]
Majewska-Tro: ected Praxiological   Bibliography .   [1090]
Steinacker, : ial Intelligence: A   Bibliography .   [1616]
Subrahmaniam: cted and Abstracted   Bibliography , 1957-1972   [1631]
Trappl, R.   : Systems Research: A   Bibliography , 1977-1984.   [1729]
De Mottoni, :   Bifurcation   of Regular Nonnegative Sta [397]
Gurel, O.   :   Bifurcation   Theory and Applications in [691]
Nicolis, J.S: Bifurcation   in Non-Negotiable Games: A [1226]
Gollmann, D.:   On the Theory of Bilinear   Shift Registers.   [634]
Mohler, R.   :   An Overview of   Bilinear   System Theory and Application [1180]
Sengupta, J.:   Bilinear   Models in Stochastic Programm [1524]
Garbade, K. : ructure of Treasury   Bills .   [568]
Conant, R.C.: ion of Entropy of a   Binary   Variable: Satisfying a Reliabil [336]
Haralik, R.M: Characterization of   Binary   Relation Homomorphisms.   [726]
Hwang, C.L. :   Decomposition of Binary   Sequences of Finite Periods and [802]
Kedem, B.   :   Binary   Time Series.   [893]
Mehlhorn, K.:   Nearly Optimal   Binary   Search Trees.   [1147]
Mergenthaler: Defective, Coherent   Binary   System.   [1152]
Sharma, B.D.: nt-Error Correcting   Binary   Perfect Codes.   [1538]
Sherlock, R.: haviour of Kauffman   Binary   Networks: Part 1: State Space D [1545]
Young, T.Y. : Classification with   Binary   Variables.   [1914]
Troncale, L.: ls of Complexity in   Bio-Systems   Real? Applications of Clus [1739]
Mager, P.P. : em Organization and   Bioactive   Compounds.   [1088]
Troncale, L.: crimination between   Bioevolution   and a Theory of Systems E [1740]
Babloyantz, : elf-Organization in   Biological   Systems with Multiple Cellu [78]
Cramer, F.   : exity: a Concept in   Biological   Science and Beyond.   [355]
Drommenhoek,:   Biological   Structures (translated from [440]
Gold, H.J.   : matical Modeling of   Biological   Systems: an Introductory Gu [630]
Jantch, E.   : radigm of Physical,   Biological   and Sociocultural Evolution [835]
Linkens, D.A: Biological   Systems Modelling and Contr [1051]
MacDonald, N: Time Lags in   Biological   Models.   [1079]
Maruyama, M.: al Restructuring of   Biological   and Social Sciences.   [1112]
Rosen, R.   : y and the Theory of   Biological   Transformations.   [1439]
Schnakenberg: Network Analysis of   Biological   Systems (Second Edition). [1513]
Sussmann, H.: d to the Social and   Biological   Sciences: a Critique.   [1641]
Suzuki, R.   : ological Studies of   Biological   Rhythms.   [1644]
Troncale, L.: eory Applied to the   Biological   and Medical Sciences.   [1750]
Tsetlin, M.L: ry and Modelling of   Biological   Systems.   [1758]
Vagnucci, A.:   and Homogeneity of   Biological   Time Series.   [1780]
Varela, F.J.:   Principles of Biological   Autonomy. (Second Vol. in t [1799]
Eisen, M.   : ical Models in Cell   Biology   and Cancer Chemotherapy.   [465]
Getz, W.M.   : atical Modelling in   Biology   and Ecology.   [602]
Jean, R.V.   : l Paradigm in Plant   Biology .   [841]
Mohler, R.   : tems, Economics and   Biology .   [1179]
Prokop, A.   : is and Synthesis in   Biology   and Biotechnology.   [1365]
Ricciardi, L: Cybernetics in   Biology   and Medicine: An Introduction. [1417]
Ricciardi, L: Cybernetics and Biology : Classic Attempts at Neuronal [1418]
Rosen, R.   : ress in Theoretical   Biology , Vol.6.   [1445]
Segel, L.A. : ecular and Cellular   Biology .   [1521]
Apostolico, : in the Modeling of   Biomolecular   Systems.   [41]
Unger, C.K. :   Biophysical   Ecology and Ecosystem Rese [1771]
Smetanic, Y.: rimary Structure of   Biopolymers .   [1585]
Mager, P.P. :   Biorhythms , System Organization and Bi [1088]
Resconi, G. : lysis of the Purine   Biosynthesis   by Mathematical Logic.   [1405]
Maker, P.P   : ons in Multivariate   Biosystems .   [1094]
Prokop, A.   : esis in Biology and   Biotechnology .   [1365]
Brugger, K. : thematical Model of   Biphasic   Insulin Secretion.   [232]
Apter, M.J. :   On the Concept of Bistability .   [43]
Glanville, R: e Box there are Two   Black   Boxes Trying to Get Out.   [620]
Wnuk, A.J.   :   The   Block-Oriented   Interactive Simulator B [1876]
Sharma, B.D.:   Block-Wise   Error-Protecting Burst Code [1539]
Halmes, A.   : Honour of Prof.Hans   Blomberg .   [717]

Breitenecker: sis of Pathological Blood Pressure Behaviour after Treadmi [215]
Margolis, D.:   Bond-graphs for Some Classic Dynamic S [1104]
Bradley, R. : Modelling: A Course Book for Engineers and Scientists.   [210]
Michie, D.   : Scientist's Weekend Book .   [1161]
Atlan, H.   :   Random   Boolean Networks.   [64]
Pippenger, N: d the Complexity of   Boolean Functions.   [1343]
Wnuk, A.J.   : teractive Simulator BORIS .   [1876]
Gomez, P.   :   Top-Down Versus Bottom-Up Organizational Design: a Cyb [636]
Kerman, C.E.: Thought of Kenneth   Boulding .   [907]
Jones, L.M. :   Defining Systems Boundaries in Practice: Some Proposals [857]
Waller, R.J.: Complexity and the   Boundaries of Human Policy Making.   [1835]
Eilbert, R.F: Contrivedness: The   Boundary Between Pattern Recognition a [464]
Fogel, E.   : tem Identification: Bounded Noise Case.   [532]
Simon, H.A. :   Models of Bounded Rationality (2 Volumes).   [1563]
Astola, J.   : The Lee-Scheme and Bounds for Lee-Codes.   [56]
Becker, P.W.: and Their Inherent   Bounds on Achievable Reliability.   [130]
Becker, P.W.: and their Inherent   Bounds on Achievable Reliability.   [129]
Nobile, A.G.: ith Intrinsic Lower   Bounds .   [1234]
Turksen, I.B: Bounds for Fuzzy Inference.   [1761]
Glanville, R: Inside Every White Box there are Two Black Boxes Trying t [620]
Robinson, E.: alysis. Vol. 1: The   Box-Jenkins Approach.   [1433]
Glanville, R: there are Two Black   Boxes Trying to Get Out.   [620]
Andreae, J.H: A Dual Model of the   Brain .   [34]
Beer, S.   :   Brain of the Firm (second edition).   [134]
Gigley, H.   : Intelligence Meets Brain Theory. An Integrated Approach t [610]
Kuhlenbeck, :   The Human   Brain and its Universe. Vol.I: The Wor [993]
Kuhlenbeck, : nology. Vol.II: The Brain and its Mind.   [993]
Palm, G.   : Information in the   Brain .   [1278]
Machlup, F. : ficance. Vol.2: The Branches of Learning.   [1080]
Pfeilsticker: nd Fuzzy Set Theory   Bridging the Gap between Mathematical   [1331]
Fowler, T.B.:   Brillouin and the Concept of Informati [538]
Mel'cuk, I.A: ome Consequences of   Bringing them Together.   [1148]
Checkland, P: Guidelines. Part 2:   Building Conceptual Models.   [297]
De Malherbe,:   Model   Building in Ecology: a Hierarchical Ap [396]
McLean, J.M.: Analysis and Model   Building .   [1141]
Rivett, P.   :   Model   Building for Decision Analysis.   [1422]
Agassi, J.   : in Honour of Mario Bunge .   [14]
Mattessich, : ences Between Mario Bunges 's World of the Systems and My O [1127]
Acar, W.   :   Bureaucracy as Organizational Patholog [5]
Kaushik, M.L: Single Error and Burst Error Correcting Codes Through a [889]
Sharma, B.D.: se Error-Protecting   Burst Codes with Weight Constraints.   [1539]
Kaushik, M.L: Burst-Error-Correcting Codes with Weig [888]
Kellermayr, : ation of an Optical   Bus Area Network.   [900]
Gomez, P.   : Decentralization in Business Organisations: Cybernetic Rul [637]
Hanson, O.J.: difying an Advanced   Business Systems Analysis Course Using [723]
Jones, A.H. : es for Computerised Business Information Systems.   [853]
McAleer, W.E: tems: A Concept for   Business and Management.   [1132]
Ninneman, J.: The Survival of U.S. Business : A General Systems Approach.   [1231]
Broekstra, G: C-Analysis of C-Structures: Representa [226]
Broekstra, G: C-Analysis of C-Structures : Representation and Evalu [226]
Kohout, L.J.: er-Brown and Varela   Calculi .   [965]
Dreyfus, S.E: Programming and the   Calculus of Variations.   [437]
Kauffman, L.: thesis and Varela's   Calculus .   [885]
Morein, R.   :   The Use of the Calculus of Variations and Functionals [1185]
Rota, G.C.   :   Finite Operator Calculus .   [1450]
Schwartz, D.: f Form and Varela's   Calculus for Self-Reference.   [1515]
Eisen, M.   : in Cell Biology and Cancer Chemotherapy.   [465]
Faribault, M: tural Language to a   Canonical Representation of the Corres [504]
Kalman, R.E.: sfer Functions, and   Canonical Forms.   [872]
Lee, E.B.   :   Canonical Forms for Time Delay Systems [1026]
Zeigler, B.P: Canonical Realization of General Time   [1917]
Stemmer, N. : nces and Our Innate Capacities .   [1619]
Wisniewski, : e due to Increasing Carbon Dioxide.   [1872]
Ratko, I.   : atients Waiting for Cardiac Operation.   [1388]

Klement, E.P: f Large Numbers and Central Limit Theorem for Fuzzy Random [935]
Kekes, J.   :   The Centrality of Problem Solving.   [898]
Gomez, P.   :   Centralization Versus Decentralization [637]
Veroy, B.S. : Design of Reliable Centralized Voice/Data Communication N [1814]
Janko, J.   : Middle of the 19th Century .   [832]
Ventriglia, : rds a Simulation of Cerebellar Cortex Activity.   [1811]
De Cleris, M: Certain System Concepts in Law and Pol [388]
Gollmann, D.: e Identification of Certain Non-Linear Networks of Automat [633]
Jachymczyk, : m Approximation for Certain Class of Random Processes.   [822]
Pearl, J.   : Economic Basis for Certain Methods of Evaluating Probabil [1304]
Kalata, P. : on with and without Certainty .   [871]
Batchelor, B: verting Run Code to Chain Code.   [122]
Gyarfas, F. : ackward and Forward Chaining Computer-Based Diagnostic Exp [695]
Barfoot, C.B: al Absorbing Markov Chains .   [107]
Ericson, R.F: The General Systems Challenge .   [486]
La Porte, T.: ganized Complexity: Challenge to Politics and Policy.   [998]
Ladanyi, O.G: World in Motion, A Challenge for Systems Engineering.   [1002]
Alter, S.L. : tice and Continuing Challenges .   [26]
Fick, G.   : Systems: Issues and Challenges .   [521]
Laszlo, C.A.:   Champ - A System and Computer Program [1013]
Benoit, B.M.:   Fractals: Form, Chance , and Dimension.   [156]
Carvallo, M.:   Chance and the Necessity of the Third [267]
Eigen, M.   : es of Nature Govern Chance .   [463]
Paulre, B.E.: and the Analysis of Chance .   [1297]
Broekstra, G: onal Assessment and Change .   [227]
Rapoport, A.: ls for Detection of Change .   [1382]
Thomas, A. : on for Constructive Change : The Use and Abuse of Systems M [1685]
Wisniewski, : r Analysing Climate Change due to Increasing Carbon Dioxid [1872]
Gericke, M. :   Changes of Regional Passenger-Transpor [600]
Palm, G.   : Rules for Synaptic Changes and their Relevance for the St [1278]
Paritsis, N.: ligent Systems with Changes in Environmental Variety.   [1282]
Holden, A.V.:   Chaotic Activity in Neural Systems.   [771]
Labos, E.   : uronal Networks and Chaotic Spike Generators.   [1000]
Cavallo, R.E: Research Movement: Characteristics , Accomplishments, and   [280]
Cavallo, R.E: Research Movement: Characteristics , Accomplishments and C   [281]
Gottinger, H: Structural Characteristics Economical Models: a S [648]
Raghaven, V.: on of the Stability Characteristics of Some Graph Theoreti [1377]
Wolaver, T.G: on the Sensitivity Characteristics of a Model Ecosystem.   [1877]
Haralik, R.M: The Characterization of Binary Relation Ho [726]
Jakubowski, :   Syntactic Characterization of Machine Parts Shap [830]
Reusch, B. :   Characterization of Continuous Systems [1412]
Takahara, Y.:   Characterization of the Causality of G [1655]
Takahara, Y.:   A Characterization of Interactions.   [1657]
Kak, S.   : On Efficiency of Chemical Homeostasis - an Information-   [869]
Eisen, M.   : Biology and Cancer Chemotherapy .   [465]
Csendes, T. : lation Study on the Chemoton .   [363]
Kaindl, H. : Search in Computer Chess .   [867]
Pernici, B. : e Playing: Computer Chess .   [1319]
Espejo, R. : ment of Industry in Chile 1970-1973.   [490]
Ulrich, W. : ernetic Reason: The Chilean Experience with Cybernetics.   [1768]
Wu, W.T.   : uristic Approach to Chinese-Character Search.   [1881]
Seo, F.   : ysis for Collective Choice .   [1527]
White, C.   : n-Based Approach to Choicemaking .   [1861]
Wasinowski, :   Choosing among Alternative Strategies. [1843]
Steinberg, J: of the Grasshopper Chortophaga Viridifaciata.   [1618]
Churchman, C: Churchman 's Conversations.   [313]
Joerres, R. : ture Masking of the Circadian System of Euglena Gracilis.   [847]
Benningfield: Circuit and Systems Theory.   [155]
Gams, M.   :   A Circuit Analysis Program that Explains [565]
Svoboda, A. : Advanced Logical Circuit Design Techniques.   [1645]
Iri, M.   : perations Research, Circuits and Systems Theory.   [814]
Sonde, B.S. : gn Using Integrated Circuits .   [1598]
Gengoux, K.G: Citizen Participation: A New Factor in [591]
Butzer, K.W.:   Civilizations : Organisms or Systems? [253]

Dubois, D. :   A Class of Fuzzy Measures Based on Trian [444]
Hwang, C.L. : inite Periods and a Class of Error Correcting Codes.   [802]
Jachymczyk, : imation for Certain Class of Random Processes.   [822]
Nobile, A.G.:   On a Class of Discrete Models for Regulated [1234]
Tazaki, E. : tural Modeling in a Class of Systems Using Fuzzy Sets Theo   [1675]
Walker, C.C.: and Input Span in a Class of Fixed-Structure Models.   [1830]
Yager, R.R. :   On a General Class of Fuzzy Connectives.   [1898]
Kashyap, R. : arison of Different Classes of Dynamic Models Using Empiri   [882]
Labos, E.   : otions in Different Classes of Formal Neuronal Networks an   [1000]
Van den Boga: imes Using Priority Classes : A Case Study in Response Surf   [1793]
Andersson, H: Analysis of a Classes tory Control System Using the T [32]
Margolis, D.: ond-graphs for Some Classic Dynamic Systems.   [1104]
Ricciardi, L: netics and Biology: Classic Attempts at Neuronal Modeling.   [1418]
Carvallo, M.: essity of the Third Classical Paradigm.   [267]
Greenspan, D: etic Formulation of Classical and Special Relativistic Phy   [663]
Batchelor, B: The Classification of Pole Maps in the Com   [118]
Bozinovski, : entation of Pattern Classification Teaching.   [208]
Duda, R.O. :   Pattern Classification and Scene Analysis.   [445]
Goodman, L.A: sociation for Cross Classification .   [642]
Kindler, E. :   Classification of Simulation Programmi   [917]
Kindler, E. :   Classification of Simulation Programmi   [918]
Peschel, M. : viour Analysis with Classification Models.   [1321]
Yager, R.R. : ibilistic Models in Classification .   [1899]
Young, T.Y. : Statistical Pattern Classification with Binary Variables.   [1914]
Rudawitz, L.:   Client-Centered Design: Concepts and E   [1459]
Wisniewski, : Model for Analysing Climate Change due to Increasing Carbo   [1872]
Jarisch, W. : ith Applications to Clinical Risk-Assessment and Detection   [837]
Lamport, L. :   Time, Clock and the Ordering of Events in Di [1006]
Andersson, H: heory of Relatively Closed Systems.   [32]
Grubbstrom, : heory of Relatively Closed Systems and Applications.   [675]
Grubbstrom, : heory of Relatively Closed Systems.   [676]
Meyer, J.F. :   Closed-Form Solutions of Performabilit   [1157]
Gaines, B.R.: f Fuzzy Systems and Closely Related Topics.   [555]
Higashi, M. : senting Information Closeness : Possibility and Probability   [756]
Foo, N.Y.   :   Closure Properties and Homomorphisms o   [535]
Varela, F.J.:   The Arithmetic of Closure .   [1798]
Binder, D.A.:   Bayesian Cluster Analysis.   [169]
Duran, B.S. :   Cluster Analysis: a Survey.   [448]
Everitt, B.S: esolved Problems in Cluster Analysis.   [497]
Scoltock, J.: f the Literature on Cluster Analysis.   [1517]
Bell, D.A. :   Physical Record Clustering in Databases.   [141]
Bezdek, J.C.: Axiomatic Basis for Clustering .   [166]
Korhonen, P.: re for Multivariate Clustering .   [970]
Raghaven, V.: ome Graph Theoretic Clustering Methods.   [1377]
Schkolnick, :   A Clustering Algorithm for Hierarchical [1511]
Shepard, R.N: , Tree-fitting, and Clustering .   [1543]
Troncale, L.: al? Applications of Clustering Theory to Modeling Systems   [1739]
Mittermeir, :   CML-Graphs - A Notation for Systems De   [1175]
Binder, Z. :   Stabilization, Co-ordination and Structuralization in [170]
Breitenecker: st in Patients with Coarctation of the Aorta.   [215]
Mann, K.H. :   Ecology of Coastal Waters: A Systems Approach (St [1101]
Batchelor, B: Converting Run Code to Chain Code.   [122]
Batchelor, B: g Run Code to Chain Code .   [122]
Hwang, C.L. : of Error Correcting Codes .   [802]
Kaushik, M.L.: st-Error-Correcting Codes with Weight-Constraints Under a   [888]
Kaushik, M.L.: st Error Correcting Codes Through a New Metric.   [889]
Sharma, B.D.: or-Protecting Burst Codes with Weight Constraints.   [1539]
Sharma, B.D.: ting Binary Perfect Codes .   [1538]
Sharma, B.D.: Correcting Perfect Codes .   [1540]
Gyarfas, F. :   CODEX : Prototypes Driven Backward and   [695]
Mark, J.W. : daptive Predicitive Coding .   [1105]
Rissanen, J.: versal Modeling and Coding .   [1421]
Buchberger, :   Input/Output Codings and Transition Functions in Ef [237]
Neuwirth, E.: ations and Distance Coefficients in Thesauri.   [1215]

Glass, A.L. :   Cognition .   [623]
Maturana, H.:   Autopoiesis and Cognition : The Realization of the Livi [1128]
Petkoff, B. : ntific Research and   Cognition .   [1328]
Rapoport, A.: een two Theories of   Cognition .   [1383]
Bobrow, D.G.: tanding: Studies in   Cognitive Science.   [179]
Bogner, S. : Cybernetic Model of   Cognitive Processes.   [183]
Chytil, M.K.: ematical Methods as   Cognitive Problem-Solvers.   [314]
Das, J.P.   : eous and Successive   Cognitive Processes.   [378]
Johnson, L. : ersation Theory and   Cognitive Coherence Theories.   [850]
Nicolis, J.S: elf-Organization in   Cognitive Systems.   [1226]
Norman, D.A.:   Perspectives in Cognitive Science. (Papers from a Meet [1236]
Rescher, N. :   Cognitive Systematization: a Systems-T [1401]
Scandura, J.: tems Alternative to   Cognitive Psychology.   [1501]
Scott, W.A. :   Cognitive Structure: Theory and Measur [1519]
Winograd, T.:   Language as a   Cognitive Process (Vol.1: Syntax).   [1868]
Caws, P.   :   Coherence , System, and Structure.   [288]
Johnson, L. : heory and Cognitive   Coherence Theories.   [850]
Ben-Dov, Y. : ecial Structures of   Coherent   Systems.   [148]
Haydon, P.G.:   Coherent   and Incoherent Computation in [742]
Mergenthaler: ost in a Defective,   Coherent   Binary System.   [1152]
Rescher, N. : retic Approach to a   Coherent   Theory of Knowledge.   [1401]
Pedretti, A.:   Where Coincidence Coincides - Coining a Noti [1309]
Pedretti, A.:   Where Coincidence   Coincides - Coining a Notion of Cybern [1309]
Pedretti, A.: cidence Coincides -   Coining   a Notion of Cybernetic Descrip [1309]
Tomlinson, R: ions - Developing a   Collaborative Research Study.   [1697]
Mulej, M.   : y (WCM)" Applied to   Collect   and Organize Associates' Ideas [1195]
Schal, R.K. :   A Collection of Thoughts of General Syst [1504]
Auger, P.   :   nteractions Between Collective and Individual Levels of Or   [68]
Seo, F.   : tility Analysis for   Collective Choice.   [1527]
Sellers, P.H: Combinatorial Complexes: A Mathematica [1522]
Alspach, B. : orithmic Aspects of   Combinatorics .   [25]
Drozen, V. :   Combinatory Spaces - An Approach to Pa   [442]
Tsai, W.H. : rammar - A Tool for   Combining Syntactic and Statistical Ap [1757]
Favella, L.F: oblems for Internal   Combustion Engines.   [512]
Tripp, R.S. : Air Force Logistics   Command .   [1736]
Kittler, J. :   A Comment   on Fast Monte Carlo Integratio [930]
Le Moigne, J: e: Bibliographie et   Commentaires (Systems Analysis: Biblio [1022]
Dombi, J.   :   Comments   on the Gamma-Model.   [429]
Gyula, P.   :   Common   Laws of Sciences and Systems.   [696]
Caianiello, :   Energetics Versus Communication in the Nervous System.   [256]
Ciupa, M.   : acket Switched Data   Communication Systems.   [317]
Hull, R.   :   Communication and Information: Their R   [796]
Johnson, L. : Framework for Human   Communication .   [849]
Jumarie, G. : lativistic Sets and   Communication .   [861]
Krippendorff: Communication and Control in Society.   [983]
Nicolis, J.S: Inadequate Communication between Self-Organizing [1225]
Pierce, J.R.:   Introduction to Communication Science and Systems. [1340]
Ruben, B.D. : ms Theory and Human   Communication .   [1455]
Veroy, B.S. : tralized Voice/Data   Communication Network.   [1814]
Benseler, F.:   Autopoiesis,   Communications , and Society: The Theor   [157]
Techo, R.   :   Data Communications : an Introduction to Con [1681]
Schiffman, Y: g Energy Savings in   Communities .   [1508]
Galitsky, V.:   Modeling the Plant   Community   Dynamics.   [562]
Shaw, M.L.G.: al Scientist in the   Community   of Science.   [1542]
Slawski, C. : zation: A Theory of   Community   Living Arrangements and a Pr [1579]
Bertoni, A. :   Analysis and   Compacting of Musical Texts.   [163]
Bavel, Z.   :   Math Companion   for Computer Science.   [126]
Best, D.P. : n a Medium Sized UK   Company .   [164]
Gomez, P.   : ing in a Publishing   Company .   [638]
Hoffer, J.A.: ctors Affecting the   Comparative Advantages of Sequential a [766]
Hofferbert, : Methodology to the   Comparative Study of Public Policy.   [767]
Lefebre, V.A: ra of Conscience: A   Comparative Analysis of Wester and Sov [1030]
Caselles, A.:   A Method to   Compare   Theories in the Light of Gener [268]
Chatterjee, :   Comparison of Arbitration Procedures: [295]

Halldin, S. :   Comparison of Forest Water and Energy   [716]
Horn, W.   : ourier Transform: A   Comparison of Time-Efficiency.   [784]
Kalaba, R.E.: ent: A Small Sample   Comparison .   [870]
Kashyap, R. :   A Bayesian   Comparison of Different Classes of Dyn   [882]
Raghaven, V.:   A   Comparison of the Stability Characteri   [1377]
Takeuchi, A.: utionary Automata -   Comparison of Automation Behavior and   [1663]
Van Gigch, J: A Methodological   Comparison of the Science, Systems and   [1787]
Aarnes, E.   : f Identification in   Compartment Analysis.   [1]
Matis, J.H. :   Compartmental Analysis of Ecosystems M   [1118]
Matsuno, K. :   Compartmentalization of Self-Reproduci   [1120]
MacDonald, B: The   Competence of a Multiple Context Learn   [1078]
Cushing, J.M: Two Species   Competition in a Periodic Environment.   [367]
Kerner, H.   : ative Programming -   Complement or Contradiction?   [908]
Espejo, R.   : nd Information: The   Complementarity Control-Autonomy.   [492]
Kuyk, W.   :   Complementarity in Mathematics.   [995]
Warfield, J.:   Complementary Relations and Map Readin   [1841]
Chatterjee, : edures: Models with   Complete and Incomplete Information .   [295]
Greenspan, D: A   Completely Arithmetic Formulation of C   [663]
Higashi, M. : Fuzziness and Fuzzy   Completeness .   [755]
Batchelor, B: of Pole Maps in the   Complex Plane.   [118]
Binder, Z.   : rated Automation of   Complex Systems.   [170]
Bubnicki, Z.: l Identification of   Complex Systems with Cascade Structure   [235]
Colombano, S: Complex Systems Tend to be Unstable. I   [334]
Gelfand, A.E: o the Management of   Complex Organizations: Management by C   [589]
Goh, B.S.   : and Persistence of   Complex Ecosystems.   [628]
Gopal, K.   : ility Evaluation in   Complex Systems.   [643]
Heim, R.   : tical Approaches to   Complex Systems.   [746]
Jakubowski, : neral Algorithms of   Complex Dynamic Systems Simulation.   [829]
Kinston, W. : to Organisation of   Complex Systems.   [927]
Loefgren, L.: and Learnability of   Complex Systems.   [1060]
Madan, S.   : odeling of Queueing   Complex .   [1086]
Marchuk, G.I: and Optimization of   Complex Systems. (Proc. of a Conferenc   [1103]
May, R.M.   :   Will a Large   Complex System be Stable?   [1130]
Millendorfer: Reducing Factors in   Complex Systems.   [1164]
Molloy, K.J.:   Insights into   Complex Organisational Problem Solving   [1182]
Nelson, R.J.:   Structure of   Complex Systems.   [1213]
Rao, G.V.   :   Complex Digital Control Systems.   [1381]
Rosen, R.   : type for Control of   Complex Systems.   [1444]
Rosen, R.   :   The Physics of   Complex Systems.   [1446]
Simon, H.A. :   How   Complex Are Complex Systems?   [1561]
Simon, H.A. :   How Complex Are   Complex Systems?   [1561]
Trappl, R.   : Approaches Towards   Complex Systems.   [1731]
Vemuri, V.   :   Modeling of   Complex Systems: an Introduction.   [1808]
Walker, C.C.: o the Management of   Complex Organizations: Management by E   [1830]
Hwang, C.L. : tion Techniques for   Complex/Large Systems: A Review.   [804]
Sellers, P.H: Combinatorial   Complexes : A Mathematical Theory of Al   [1522]
Allen, T.F.H: ives for Ecological   Complexity .   [24]
Ball, M.O.   :   Complexity of Network Reliability Comp   [89]
Borodin, A. :   The Computational   Complexity of Algebraic and Numeric Pr   [190]
Casti, J.   :   Connectivity,   Complexity and Catastrophe in Large-Sc   [272]
Cramer, F.   :   Fundamental   Complexity : a Concept in Biological Sc   [355]
Deutsch, K.W: ation: Reduction of   Complexity in Data Processing for Worl   [414]
Feldman, J.A: Total   Complexity and the Inference of Bent P   [517]
Gottinger, H: Complexity and Catastrophe: Applicatio   [646]
Gottinger, H: Models: a Study of   Complexity .   [648]
Gottinger, H: Algebraic Theory of   Complexity in Dynamic Systems.   [647]
Jacobs, B.E.: lized Computational   Complexity .   [825]
Kannellakis,: n the Computational   Complexity of Cardinality Constraints   [877]
Kaposi, A.   : o the Assessment of   Complexity .   [879]
Kapur, J.N. : On Maximum-Entropy   Complexity Measures.   [880]
Kronsjo, L.I: Algorithms: Their   Complexity and Efficiency.   [987]
La Porte, T.:   Organized   Complexity : Challenge to Politics and   [998]
LeMoigne, J.:   Modeling   Complexity : A French History and its P   [1024]
Leung-yan-ch: ropy and Kolmogorov   Complexity .   [1040]

Lieberherr, :   Complexity   of Partial Satisfaction.   [1048]
Mackinnon, A: Complexity   and Decision-Making.   [1085]
Morin, E.   :   Complexity .   [1186]
Pearl, J.   :   On the   Complexity   of Inexact Computations.   [1300]
Pippenger, N: tion Theory and the   Complexity   of Boolean Functions.   [1343]
Pippenger, N: Complexity   Theory.   [1344]
Prigogine, I: Becoming - Time and   Complexity   in the Physical Sciences. [1360]
Rabin, M.O. :   Complexity   of Computations.   [1372]
Suppes, P.   :   Some Remarks About   Complexity .   [1638]
Troncale, L.:   Are Levels of Complexity   in Bio-Systems Real? Applic [1739]
Waller, R.J.:   Complexity   and the Boundaries of Human [1835]
Thissen, W. : s for Understanding   Complicated Models.   [1684]
Harris, L.R.: Base as a Semantic   Component   to Aid in the Parsing of Nat [727]
Boncina, R. :   An Open System Composed   of Non-Active Monotonous Part [188]
Varshavskiy,: onous Processes II.   Composition and Matching.   [1801]
Cover, T.M. :   Compound   Bayes Predictors for Sequence [353]
Yager, R.R. :   Membership in a Compound   Fuzzy Subset.   [1908]
Mager, P.P. : ation and Bioactive   Compounds .   [1088]
Fum, D.   :   Representation and Comprehension .   [547]
Cavallo, R.E: ctable Relations: a   Comprehensive Study.   [282]
Melkumian, A: sures in Image Data   Compression and Transmission Systems. [1151]
Arbib, M.A. :   Cooperative   Computation and the Cybernetic Society   [49]
Denning, P.J: nes, Languages, and   Computation .   [409]
Haydon, P.G.: rent and Incoherent   Computation in Identified Molluscan Ne [742]
Hopcroft, J.: ory, Languages, and   Computation .   [780]
Hu, N.C.   : lti-Dimensional FFT   Computation .   [794]
Toffoli, T. :   Computation and Construction Universal [1696]
Uccellini, L: del Concepts to the   Computation of Atmospheric Trajectorie [1765]
Bathe, K.J. :   Formulations and   Computational Algorithms in Finite Ele [123]
Borodin, A. :   The   Computational Complexity of Algebraic   [190]
Favella, L.F: ms and Some Related   Computational Methods.   [511]
Jacobs, B.E.:   On Generalized   Computational Complexity.   [825]
Kannellakis,:   On the   Computational Complexity of Cardinalit   [877]
Yakowitz, S.:   Computational Probability and Simulati [1910]
Ball, M.O. : Network Reliability   Computations .   [89]
Pearl, J.   : mplexity of Inexact   Computations .   [1300]
Rabin, M.O. :   Complexity of   Computations .   [1372]
Ackerman, E.: Health Sciences: a   Computer   Approach.   [7]
Adam, N.R. :   Current Topics in   Computer   Simulation.   [12]
Baase, S.   :   Computer   Algorithms: Introduction to D [77]
Baase, S.   :   Computer   Algorithms.   [76]
Ballard, D.H: Computer   Vision.   [91]
Bavel, Z.   : Math Companion for   Computer   Science.   [126]
Christensen,: cal Origins; Vol.3:   Computer   Implementation; Vol.4: Applic [311]
Courtois, P.: ility: Queueing and   Computer   System Applications.   [352]
De Delara, G: Computer   Simulation of Lateral Inhibit [389]
Drechsler, F: Computer   Integrated Manufacturing (CIM [436]
Elison, D.I.: ive Simulation on a   Computer .   [475]
Findler, N.V: Aspects of   Computer   Learning.   [525]
Gill, A.   :   ied Algebra for the Computer   Sciences.   [612]
Graedel, T.E: lts from Scientific   Computer   Models.   [652]
Greenspan, D: A New   Computer   Approach to the Modeling of A [666]
Grogono, P. : Problem Solving and   Computer   Programming.   [672]
Gruska, J. : ical Foundations of   Computer   Science.   [678]
Hall, E.L. :   Computer   Image Processing and Recognit [714]
Holst, P.   :   Computer   Simulation 1951-1976: An Inde [775]
Johnson, P.E: Validating   Computer   Simulation Models of Expert R [851]
Kaindl, H. : uiescence Search in   Computer   Chess.   [867]
Kobsa, A.   : of Knowledge in the   Computer .   [955]
Laszlo, C.A.: hamp - A System and   Computer   Program Package for Health Ha [1013]
Levy, L.S. : crete Structures of   Computer   Science.   [1044]
Lewin, D.   : d Design of Digital   Computer   Systems.   [1046]
Maryanski, F: Digital   Computer   Simulation.   [1113]
Mesarovic, M: Multilevel   Computer   Model of World Development Sy [1153]

Pernici, B. : ng in Game Playing: Computer Chess.   [1319]
Ratko, I.   :   Computer Registration of Patients Wait [1388]
Ricci, F.J. :   Computer Simulation of a Cybernetic Mo [1416]
Rushinek, A.:   An Analysis of   Computer Software and Its Importance t [1465]
Sauer, C.H. :   Computer Systems Performance Modeling. [1496]
Schellenberg: , Key to Successful   Computer Projects.   [1505]
Simon, J.C. :   Computer Oriented Learning Processes. [1564]
Sugiyama, K.:): A Case of Future   Computer Systems Development in Japan. [1635]
Toda, M.   : Analysis for Future   Computer Systems Development in Japan. [1694]
Trappl, R. :   Computer Psychotherapy: Is it Acceptab [1715]
Trappl, R. : Formal Methods with   Computer Efficiency in Medicine.   [1711]
Trappl, R. :   Ontology on the Computer .   [1719]
Trost, H.   :   Computer Simulation of Neurotic Proces [1751]
Van Navta, L: Digitial   Computer Applications to Process Contr [1789]
Walker, B.J.:   Computer Security and Protection Struc [1829]
Witten, I.H.:   Principles of Computer Speech.   [1875]
Wrathal, C.P: Computer Acronyms, Abbreviations, etc. [1880]
Wulf, W.A. : ental Structures of   Computer Science.   [1883]
Horn, W.   :   Interactive Computer-Aided Planning: Estimating He [782]
Kellerman, W: Computer-Aided Identification of Syste [899]
Klir, G.J. :   On the Problem of   Computer-Aided Structure Identificatio [938]
Klir, G.J. :   Computer-Aided Systems Modelling.   [945]
Spriet, J.A.:   Computer-Aided Modelling and Simulatio [1606]
Steinacker, :   Computer-Aided Research in Learning.   [1614]
Greenberg, H: Computer-Assisted Analysis and Model S [660]
Greenberg, H: Computer-Assisted Analysis and Model S [661]
Gyarfas, F. : nd Forward Chaining   Computer-Based Diagnostic Expert Syste [695]
Hull, R.   : eir Relationship to   Computer-Based Information Systems.   [796]
Meadows, D.L: es for Implementing   Computer-Based Policy Models.   [1143]
Wasniowski, :   The Use of   Computer-Communication Systems in Futu [1844]
Soi, I.M.   :   An Overview of   Computer-Orientied Learning Processes. [1597]
Jones, A.H. : Design Features for   Computerised Business Information Syst [853]
Troncale, L.: els with Data Using   Computerized , Empirical Data Bases.   [1745]
Arbib, M.A. :   Computers and the Cybernetic Society.   [47]
Chen, D.T.-W: Problem Solving by   Computers .   [303]
Findler, N.V: Use of Knowledge by   Computers .   [524]
Garey, M.R. :   Computers and Intractability: a Guide [571]
Hellerman, L: its Application to   Computers .   [747]
Hyde, M.O. :   Computers that Think?: The Search for [805]
Jacoby, S.L.: tical Modeling with   Computers .   [826]
McIntosh, J.: tical Modelling and   Computers in Endocrinology.   [1138]
Rudall, R. :   Computers and Cybernetics.   [1458]
Sanderson, J: elational Theory of   Computing .   [1490]
             : ing and Statistical   Computing .   [1943]
Vagnucci, A.:   Concatenation and Homogeneity of Biolo [1780]
Batchelor, B: pon Convex Hulls of   Concavities .   [119]
Batchelor, B: iption for Labeling   Concavity Trees.   [120]
Apter, M.J. :   On the   Concept of Bistability.   [43]
Brown, D.J.H: A Task-Free   Concept Learning Model Employing Gener [229]
Bunge, M.   :   A Systems   Concept of Society: Beyond Individuali [240]
Collot, F. : A New Mathematical   Concept for General Systems Theory.   [333]
Cramer, F. : ental Complexity: a   Concept in Biological Science and Beyo [355]
Fowler, T.B.:   Brillouin and the   Concept of Information.   [538]
Gallopin, G.:   The Abstract   Concept of Environment.   [564]
Linstone, H.: ultiple Perspective   Concept with Applications to Technolog [1052]
McAleer, W.E: Systems: A   Concept for Business and Management.   [1132]
Patten, B.C.: ems Approach to the   Concept of Environment.   [1292]
Patten, B.C.: ems Approach to the   Concept of Niche.   [1293]
Pot, John S.: itation of the Goal   Concept .   [1354]
Urban, H.B. :   The   Concept of Development for a Systems P [1773]
Wallenmaier,: ards a Quantitative   Concept of Teleological Systems.   [1834]
Wechler, W. :   The   Concept of Fuzziness in Automata and L [1851]
Bahm, A.J. :   Holons: Three   Conceptions .   [81]
Kindler, E. : of Necessary System   Conceptions .   [917]

De Cleris, M: Certain System Concepts in Law and Politics.   [388]
Destouches, :   Basic Concepts of System Theory.   [413]
Flavin, M.   :   Fundamental Concepts of Information Modelling.   [528]
Kahne, S.   : Introducing Systems   Concepts to All University Students.   [865]
Kindler, E. : Simulation Language   Concepts .   [922]
Klir, G.J.   :   General Systems Concepts .   [949]
Kramer, J.T.:   Systems Thinking: Concepts and Notions.   [981]
Madan, S.   :   Basic Concepts on the Theoretical Treatment   [1087]
Murdick, R.G: MIS: Concepts and Design.   [1197]
Nowakowska, : easurement of Fuzzy   Concepts in the Social Sciences.   [1239]
Nugyen, V.V.: ear Identifiability   Concepts .   [1247]
Orava, P.J. :   On the Concepts of Input-Output Model, Consol   [1268]
Oren, T.I.   :   Concepts for Advanced Simulation Metho   [1270]
Rudawitz, L.: nt-Centered Design: Concepts and Experience.   [1459]
Strombach, W: he Meaning of these   Concepts in German Language.   [1629]
Techo, R.   :   an Introduction to Concepts and Design.   [1681]
Troncale, L.: y Principal Systems   Concepts .   [1737]
Troncale, L.: ss Based on Systems   Concepts .   [1738]
Troncale, L.: n Principal Systems   Concepts .   [1743]
Uccellini, L: ying Discrete Model   Concepts to the Computation of Atmosph   [1765]
Cavallo, R.E: A Conceptual Foundation for Systems Prob   [276]
Checkland, P: s. Part 2: Building   Conceptual Models.   [297]
Oldershaw, R: Conceptual Foundations of the Self-Sim   [1262]
Pugh, R.E.   : imulation Models: a   Conceptual Approach to Case Study.   [1368]
Ryan, R.M.   :   Conceptualizing the Integration of Soc   [1467]
Schal, R.K. : l Systems Theorists   Concerning the Process of Human Transf   [1504]
Hogeweg, P. : elopmental Systems: Concrete Advantages of Discrete Event   [769]
Takahara, Y.: ion Functor and its   Concrete Realizations.   [1659]
Fontela, E. :   Condensed Forms of Large Scale Models.   [533]
Barfoot, C.B: Aggregation of Conditional Absorbing Markov Chains.   [107]
Hisdal, E.   :   Conditional Possibilities, Independenc   [761]
Mizumoto, M.: by Zadeh for Fuzzy   Conditional Inference.   [1176]
Nguyen, H.T.:   On   Conditional Possibility Distributions.   [1220]
Van Campenho: Maximum Entropy and   Conditional Probability.   [1786]
Wyner, A.D. :   A Definition of Conditional Mutual Information for Arb   [1884]
Yashin, A.   :   and New Results in Conditional Gaussian Estimation Proced   [1911]
Williams, P.:   Bayesian   Conditionalisation and the Principle o   [1865]
Anderson, B.: tary and Infinitary   Conditions .   [29]
Vallee, R.   : with Random Initial   Conditions .   [1782]
Gasparski, W: Papers from the Conference in Design Methods.   [579]
Sharif, N.   : eedings of the Int.   Conference on Systems Modelling in Dev   [1536]
Lucas, R.G. :   Information System   Configuration .   [1070]
Shakun, M.F.:   Formalizing Conflict Resolution in Policy Making.   [1532]
Slawski, C. :   Love, Power and Conflict : A Systems Model of Interpart   [1577]
Slawski, C. :   The Interpersonal   Conflict Episode: A Systems Model.   [1580]
Landweber, L: Properties of   Conflict-Free and Persistent Petri Net   [1008]
Bell, D.E.   :   Conflicting Objectives in Decisions.   [142]
Taylor, W.R.:   Theory to Organize   Confusion .   [1674]
El-Shirbeeny: nsitivity Augmented   Conjugate Gradient Algorithm for Solvi   [472]
Dammasch, I.: perties of Randomly   Connected McCulloch-Pitts Networks: Di   [375]
Rasvan, V.   : System Theory Ideas   Connected with the Stability Problem.   [1387]
Matsuno, K. : lations, Sequential   Connectedness , and Feedback Loops in O   [1123]
Yager, R.R. : eral Class of Fuzzy   Connectives .   [1898]
Zimmermann, :   Latent   Connectives in Human Decision Making.   [1926]
Sherlock, R.: tworks of Different   Connectivities .   [1545]
Casti, J.   :   Connectivity , Complexity and Catastrop   [272]
Wolaver, T.G: Systems Structure,   Connectivity , and Recipient Control on   [1877]
Lefebre, V.A: Algebra of   Conscience : A Comparative Analysis of   [1030]
Glanville, R: Consciousness : And so on.   [618]
Pask, G.   :   Consciousness .   [1284]
Gelfand, A.E: ions: Management by   Consensus Level and its Interaction wi   [589]
Tazaki, E.   : tural Modeling with   Consensus for Societal Systems Using F   [1676]
Mel'cuk, I.A: for as Well as Some   Consequences of Bringing them Together   [1148]
Peschel, M. :   Socio-Economic Consequences of the Volterra Approach   [1325]

Morris, H.M.: ical Stability in a Conserved   Nutrient Ecosystem.   [1189]
Takahara, Y.:   A System Theoretic Consideration of the World Model.   [1654]
Csanyi, V. :   Considerations for the Scientific Form [361]
Lakshmivarah: Considerations for Fuzzifying Formal L [1005]
Sage, A.P. :   and Organizational Considerations in the Design of Inform [1475]
Sultan, L.H.:   Some   Considerations on Systems Organization [1636]
Broekstra, G: ment by Matching. A Consistency   Model for Organizational A [227]
Viliums, E.R: o Obtain Interlevel Consistency   of Decisions under Uncerta [1816]
Orava, P.J. :   Input-Output Model, Consolity , and State in the Theory of [1268]
El-Shirbeeny: gorithm for Solving   Constrained   Control Design Problems.   [472]
Fletcher, R.: ptimization; Vol.2: Constrained   Optimization.   [530]
Fortuny-Amat: cal Coordination of Constrained   Optimizing Systems.   [536]
Broekstra, G: Constraint   Analysis and Structure Iden [223]
Freuder, E.C: Synthesizing Constraint Expressions.   [546]
Groen, J.C.F: A Constraint   Analytic Measure for the Di [671]
Haralik, R.M: ction Operation for Constraint   Satisfaction.   [725]
Grabowski, J: oblem with Resource Constraints .   [651]
Kannellakis,: xity of Cardinality Constraints   in Relational Databases.   [877]
Sharma, B.D.: t Codes with Weight Constraints .   [1539]
Shore, J.E. : ubset and Aggregate Constraints .   [1550]
Sussman, G.J: CONSTRAINTS   - A Language for Expressin [1640]
Overton, W.S: On   Constructing a Hierarchical Model in t [1276]
Czogala, E. : Contribution to the Construction of the Gamma-Operator.   [371]
Dominique, J: Logical   Construction of Systems.   [430]
Hoernig, K.M: f Automatic Program Construction .   [765]
Kohout, L.J.:   Construction of an Expert Therapy Advi [966]
Lozinskii, E: Construction of Relations in Relationa [1068]
Neal, M.   :   The   Construction of Personality in the Nov [1205]
Overton, W.S: A Strategy of Model Construction .   [1274]
Pedrycz, W. :   Construction of Fuzzy Relational Model [1313]
Toffoli, T. :   Computation and   Construction Universality of Reversibl [1696]
Von Glasersf: Steps in the Construction of "Others" and "Reality" [1821]
Warnier, J.-:   Logical   Construction of Systems.   [1842]
Brayton, R.K: ynamical Systems: a Constructive Approach.   [214]
Coffman, C.V: Constructive Approaches to Mathematica [325]
Franova, M. : ogram Synthesis and Constructive Proofs Obtained by Beth's [542]
Thomas, A. : erating Tension for Constructive Change: The Use and Abuse [1685]
Trussel, P. :   Constructs   from General Systems Theory [1756]
Horn, W.   :   ations in a Medical Consultation System.   [785]
King, L.T. :   ject Environment: A Consulting   Process.   [926]
Kriegel, U. :   sis of World Energy Consumption and World Population.   [982]
Babloyantz, :   h Multiple Cellular Contacts .   [78]
Adams, W.M. :   elative Information Content .   [13]
MacDonald, B: tence of a Multiple Context   Learning System.   [1078]
Reuver, H.A.: Learning within the Context   of General Systems Theory.   [1413]
Yager, R.R. :   : Their Relation to Context   and Their Experimental Realiza [1904]
Gokhale, D.V: The Information in   Contingency Tables.   [629]
Alter, S.L. : urrent Practice and Continuing   Challenges.   [26]
Andrew, A.M.:   Logic and Continuity - a Systems Dichotomy.   [36]
Barto, A.G. :   Discrete and Continuous Models.   [112]
Bartolini, G: with Application to Continuous Casting Plants.   [113]
Hartl, R.F. : olicies for Dynamic Continuous Lovers Under Risk of Being   [729]
Reusch, B. :   Continuous Systems with Digital Behavi [1411]
Reusch, B. : Characterization of Continuous Systems that are Digital.   [1412]
Shieh, L.   :   Representation of   Continuous Time State Equations by Dis [1546]
Okeda, M.   : ns, Expansions, and Contractions of Dynamic Systems.   [1260]
Kerner, H. : ing - Complement or Contradiction ?   [908]
Bernard, N. :   Multiensembles, a Contribution to the Search of Fuzzy Se [162]
Cliffe, M.J.:   The   Contribution of Cybernetics to the Sci   [322]
Czogala, E. :   A   Contribution to the Construction of th   [371]
Fararo, T.J.: Social Structure: A Contribution to the Theory of Social S   [503]
Zouwen, J.v.: es in Interviews; a Contribution of Systems Methodology.   [1930]
Eilbert, R.F: Contrivedness : The Boundary Between Pa   [464]
Andersson, H: sis of an Inventory Control   System Using the Theory of Rel [32]

Atherton, D.:   Nonlinear Control Engineering.   [60]
Bailey, N.T.: ernetics of Disease Control .   [83]
Bellman, R. : hematical Theory of Control Processes: Nonlinear Processes [143]
Belova, N. :   Valuation and Control .   [146]
Bennett, R.J: sophy, Analysis and Control .   [152]
Boltyanskii,:   Optimal Control of Discrete Systems. (Translat [187]
Boltyanskii,:   Optimal Control of Discrete Systems.   [186]
Boyd, G.M. : he Design of Mutual Control in Education.   [205]
Camana, P.C.: k for Human Posture Control without Physical Intervention. [258]
Chow, G.C. : ometric Analysis by Control Methods.   [309]
Cook, N.D. :   An Isomorphism of Control in Natural, Social, and Cybern [345]
Czogala, E. : Its Application in Control Problems.   [368]
Czogala, E. : eneration for Fuzzy Control .   [370]
Czogala, E. :   Control Problems in Fuzzy Systems.   [369]
Dalenoort, G: e Representation of Control in Systems in General, and in   [374]
De La Sen, M: put-Output Adaptive Control Systems to Avoid Delays in the [394]
Dimirivski, : ries in Systems and Control Theory.   [422]
Edwards, J.B: Analysis and Control of Multipass Processes.   [456]
El-Fattah, Y: on, Simulation, and Control .   [467]
El-Sherief, :   Identification and Control of Linear Discrete Multivariab [470]
El-Shirbeeny: Solving Constrained Control Design Problems.   [472]
Erdi, P.   : ation: Feed-Forward Control of the Ontogenetic Development [484]
Findeisen, W: Control and Coordination in Hierarchic [523]
Furukawa, O.: finement of Quality Control Systems.   [549]
Furukawa, O.: dination in Quality Control .   [548]
Gupta, M.M. :   Feedback Control Applications of Fuzzy Set Theo [689]
Hatze, H.   : nt in Neuromuscular Control .   [736]
Hickin, J. : s to Estimation and Control .   [753]
Johnson, D.T: A Planning Control Strategy that Allows for the C [848]
Jutila, S.T.: Social Gaming and Control by Dislocations.   [864]
Kaindl, H. :   Dynamic Control of the Quiescence Search in Co [867]
Knappe, U. : r Industrial Energy Control .   [952]
Krippendorff: Communication and Control in Society.   [983]
Kucera, V. :   Discrete Linear Control : the Polynomial Equation Appro [992]
Landau, Y.D.:   Adaptive Control : the Model Reference Approach. [1007]
Leigh, J.R. : Analysis and Linear Control Theory.   [1033]
Levitan, G.L: erence in Situation Control Systems.   [1043]
Linkens, D.A: stems Modelling and Control .   [1051]
Lions, J.I. : f Systems and Their Control .   [1054]
Lions, J.L. : f Systems and Their Control .   [1055]
Mackenzie, K: Process Uncertainty Control Process.   [1084]
Mak, K.L.   :   Dynamics and Control of Integrated Production-Inven [1093]
Manohar Raho: Control Systems and Quantitative Econo [1102]
Marshall, J.:   Control of Time-Delay Systems.   [1108]
Martin, P.E.: Simplification and Control of Information Patterns - Furt [1111]
Maybeck, P.S: ls, Estimation, and Control .   [1131]
McCorkell, C: erarchical Adaptive Control System.   [1135]
Orava, P.J. : namical Systems and Control .   [1268]
Pospelov, D.: ystems: Situational Control   .   [1351]
Pospelov, D.: inguistic Models in Control Systems.   [1352]
Rao, G.P.   :   Applied to Systems Control .   [1380]
Rao, G.V.   :   Complex Digital Control Systems.   [1381]
Ray, W.H.   :   Advanced Process Control .   [1392]
Roffel, B. :   Process Dynamics, Control and Protection.   [1435]
Rosen, R.   : ng: a Prototype for Control of Complex Systems.   [1444]
Rudwick, B.H: ach to Planning and Control .   [1460]
Sahal, D.   : s of Regulation and Control .   [1477]
Saridis, G.N: Self-Organizing Control of Stochastic Systems.   [1494]
Schellhorn, : llowing in Multiple Control : Exact and Approximate.   [1506]
Schwarzenbac: stems Modelling and Control .   [1516]
Sinha, K.K. : ive Nuclear Reactor Control without Explicit Identificatio [1569]
Strauss, A. : oduction to Optimal Control Theory.   [1627]
Strejc, V. : of Discrete Linear Control .   [1628]
Tesfatsion, : tering and Adaptive Control : Optimality Results   [1683]

Van Navta, L: ications to Process Control .   [1789]
Van Steenkis: Identification and Control   in Environmental Systems.   [1791]
Walichiewicz: i-Dimensional Fuzzy Control   Algorithm.   [1828]
Wolaver, T.G: vity, and Recipient Control   on the Sensitivity Characteris   [1877]
Espejo, P.   : The Complementarity Control-Autonomy .   [492]
Hunt, L.R.   :   Controllability of General Nonlinear S   [798]
Gihman, I.I.:   Controlled Stochastic Processes.   [611]
Braae, M.   : s for a Fuzzy Logic Controller .   [209]
Gupta, M.M. : sign of an Adaptive Controller for Uncertain Plants via Li   [687]
Hanassab, S.: Interpolator Based Controller for Cybernetic Systems.   [719]
Pappis, C.P.:   A Fuzzy Logic Controller for a Traffic Junction.   [1280]
Baldwin, J.F: Modelling Controllers Using Fuzzy Relations.   [84]
Bartolini, G: ance Adaptive Fuzzy Controllers with Application to Contin   [113]
Egardt, B.   : ability of Adaptive Controllers .   [458]
Sandblom, C.: Policy Using Lagged Controls .   [1487]
De Luca, A. :   On the Convergence of Entropy Measures of a F   [395]
Johnson, L. :   Conversation Theory and Cognitive Cohe   [850]
Keravnou, E.: the Perspective of Conversation Theory Methodology.   [903]
Ogborn, J.M.:   Conversation Theory.   [1257]
Leal, A.   : ractive Program for Conversational Elicitation of Decision   [1025]
Pask, G.   :   Development in Conversational   Theory - Part 1   [1285]
Churchman, C: Churchman's Conversations .   [313]
Batchelor, B: Converting Run Code to Chain Code.   [122]
Burns, J.R. :   An Algorithm for Converting Signed Digraphs to Forreste   [249]
Batchelor, B: Methods for Finding Convex Hulls of Planar Figures.   [121]
Batchelor, B: cription Based Upon Convex Hulls of Concavities.   [119]
Nussbaumer, : urier Transform and Convolution Algorithms.   [1249]
Smith, R.G. :   Frameworks for Cooperation in Distributed Problem Sol   [1593]
Arbib, M.A. :   Cooperative Computation and the Cybern   [49]
Golemanov, L: On the Coordinability of Hierarchical Systems   [631]
Brown, J.H.U: Integration and Coordination of Metabolic Processes: a   [230]
Findeisen, W: Control and Coordination in Hierarchical Systems ([523]
Fortuny-Amat: or the Hierarchical Coordination of Constrained Optimizing   [536]
Furukawa, O.: ty Goal-Seeking and Coordination , and the Practical Applic   [550]
Hakkala, L. : sible Gradient-Type Coordination Algorithm for Dynamical S   [709]
Anderson, R.:   Proving Programs Correct .   [30]
Hwang, C.L. : nd a Class of Error Correcting Codes.   [802]
Kaushik, M.L: ror and Burst Error Correcting Codes Through a New Metric.   [889]
Sharma, B.D.:   Adjacent-Error Correcting Binary Perfect Codes.   [1538]
Sharma, B.D.:   Phased Error Correcting Perfect Codes.   [1540]
Arigoni, A.C: Correctness of the Semiotic Transforma   [51]
Faribault, M: presentation of the Corresponding Semantic Relationships.   [504]
Ventriglia, : ation of Cerebellar Cortex Activity.   [1811]
Zorbas, Y.G.: Reliability during Cosmic Flights.   [1929]
Oldershaw, R: imilar Hierarchical Cosmology .   [1262]
Oldershaw, R: imilar Hierarchical Cosmology .   [1263]
Hagg, C.   :   Possibility and Cost in Decision Analysis.   [699]
Johnson, D.T: that Allows for the Cost of Planning.   [848]
Mergenthaler: The Total Repair Cost in a Defective, Coherent Binary S   [1152]
Hallam, T.G.: Equation Models of Cosystems .   [715]
Gehrlein, W.: Analysis of Simple Counting Methods for Ordering Incomple   [588]
Chen, K.   : ology in Developing Countries .   [304]
Ghosal, A.   : Study of Developing Countries .   [608]
Sharif, N.   : lling in Developing Countries .   [1536]
Gasparski, W: r a German Speaking Country .   [582]
Gasparski, W: nor German Speaking Country .   [584]
Auger, P.   :   Coupling between N Levels of Observati   [67]
Venkatesan, : roach to Transitive Coupling in ISM.   [1809]
Bradley, R. : atical Modelling: A Course Book for Engineers and Scientis   [210]
Hanson, O.J.: ss Systems Analysis Course Using Feedback from Past Studen   [723]
Hanson, O.J.: ed Systems Analysis Course .   [724]
Karlin, S.   :   A Second Course in Stochastic Processes.   [881]
Trappl, R.   : ligence: A One-Hour Course .   [1728]
Ibidapo-Obe,: Structure for Error Covariance in Nonlinear Dynamical Syst   [807]

	: dies. A New Journal	Covering the Study, Research and Techn	[1938]
Naughton, J.:		Craft Knowledge in Systems Research: T	[1203]
Auger, P. :	sical) Resulting in	Creation of Structures.	[67]
Machlup, F. :	Knowledge: Its	Creation , Distribution and Economic Si	[1080]
Boroush, M.A:	hnology Assessment:	Creative Futures.	[191]
Brightman, H:	ving: A Logical and	Creative Approach.	[219]
Brightman, H:	ving: A Logical and	Creative Approach.	[220]
Kerman, C.E.:		Creative Tension: the Life and Thought	[907]
Mulej, M. :	Method "Wholistic	Creativity of Many (WCM)" Applied to C	[1195]
Lowen, W. :		Creep-Creep : Out of Human Scale.	[1066]
Janssens, L.:	raining of Operator	Crews of Nuclear Power Plants.	[833]
De Greene, K:	n and Management of	Crisis .	[391]
Fandel, G. :	Multiple	Criteria Decision Making Theory and Ap	[502]
Ihara, J. :	uctural Analysis of	Criteria for Selecting Model Variables	[808]
Matejko, A. :	The Structural	Criteria of Social System Maturity.	[1116]
Moylan, P.J.:	Stability	Criteria for Large-Scale Systems.	[1192]
Starr, M.K. :	Multiple	Criteria Decision Making.	[1610]
Walker, V.R.:		Criteria for Object Perception.	[1832]
Zeleny, M. :	Multiple	Criteria Decision Making.	[1924]
Zionts, S. :	Multiple	Criteria Problem Solving.	[1928]
Conant, R.C.:	fying a Reliability	Criterion .	[336]
Aulin, A. :	Progress: Towards a	Critical Social Philosophy and a Criti	[73]
Batchelder, :	A	Critical Examination of the Analysis o	[115]
Mingers, J.C:	d Systems Thinking:	Critical Theory and Soft Systems Metho	[1171]
Morris, H.M.:		Critical Stability in a Conserved Nutr	[1189]
Roberts, S.L:	avioral Problems of	Critical Care Patients.	[1429]
Aulin, A. :	al Philosophy and a	Criticism of Marixsm.	[73]
Laszlo, E. :	on Systems Theory's	Critics .	[1015]
Sussmann, H.:	logical Sciences: a	Critique .	[1641]
Ulrich, W. :	A	Critique of Pure Cybernetic Reason: Th	[1768]
Goodman, L.A:	of Association for	Cross Classification.	[642]
Shore, J.E. :		Cross-Entropy Minimization Given Fully	[1550]
Shore, J.E. :	Properties of	Cross-Entropy Minimization.	[1551]
Warfield, J.:		Crossing Theory and Hierarchy Mapping.	[1839]
Ricci, F.J. :	Model of Man in his	Cultural Environment.	[1416]
Wilder, R.L.:	Mathematics as a	Cultural System.	[1863]
Trappl, R. :	, Economy, Society,	Culture , and Politics.	[1733]
Watt, K.E.F.:	oblems, Growth, and	Culture .	[1849]
Adam, N.R. :		Current Topics in Computer Simulation.	[12]
Alter, S.L. :	on Support Systems:	Current Practice and Continuing Challe	[26]
Cavallo, R.E:	ccomplishments, and	Current Developments. (a Report Sponso	[280]
Cavallo, R.E:	Accomplishments and	Current Developments.	[281]
Hoye, R.E. :		Current Status of Hospital Management	[793]
Rose, J. :		Current Topics in Cybernetics and Syst	[1437]
Wood-Harper,:	A Taxonomy of	Current Approaches to Systems Analysis	[1879]
Yeh, R.T. :		Current Trends in Programming Methodol	[1912]
Estep, M.L. :	iplinary Inquiry in	Curriculum Design.	[493]
Georgiev, A.:	vidual Human Growth	Curve .	[597]
Zwick, M. :	The	Cusp Catastrophe and the Laws of Diale	[1932]
Aplin, J.C. :	A	Cybernetic Model of the MBO Process.	[39]
Arbib, M.A. :	Computers and the	Cybernetic Society.	[47]
Arbib, M.A. :	Computation and the	Cybernetic Society.	[49]
Aulin, A. :	The	Cybernetic Laws of Social Progress: To	[73]
Best, D.P. :	Implementation of a	Cybernetic Organisation Structure in a	[164]
Best, D.P. :	A	Cybernetic Approach to the Application	[165]
Bogdanski, C:	Basic Elements of	Cybernetic Physics.	[182]
Bogner, S. :	A	Cybernetic Model of Cognitive Processe	[183]
Boyd, G.M. :		Cybernetic Aesthetics: Key Questions i	[205]
Capocelli, R:	A	Cybernetic Approach to Population Dyna	[260]
Ciupa, M. :	A	Cybernetic Approach to Packet Switched	[317]
Cook, N.D. :	atural, Social, and	Cybernetic Systems.	[345]
Espejo, R. :		Cybernetic Praxis in Government: The M	[490]
Gomez, P. :	ization Design: a	Cybernetic Perspective.	[636]
Gomez, P. :	ness Organisations:	Cybernetic Rules for Effective Managem	[637]

Hanassab, S.: llel Processors for   Cybernetic   Systems.   [718]
Hanassab, S.: ased Controller for   Cybernetic   Systems.   [719]
Holden, A.V.:   Cybernetic   Approach to Global and Loca   [770]
Janssens, L.:   The Application of   Cybernetic   Principles for the Training   [833]
Jones, A.H. :   Cybernetic   Design Features for Compute   [853]
Madan, S.   :   Cybernetic   Modeling of Queueing Comple   [1086]
Madan, S.   : etical Treatment of   Cybernetic   Queueing Systems.   [1087]
Nurmi, H.   :   On Strategies of   Cybernetic   Model-Building.   [1248]
Oeren, T.I. : n Software Based on   Cybernetic   and General Systems Theorie   [1255]
Patten, B.C.:   The   Cybernetic   Nature of Ecosystem.   [1295]
Pedretti, A.: Coining a Notion of   Cybernetic   Description?   [1309]
Petkoff, B. :   A   Cybernetic   Model of Scientific Researc   [1328]
Rastogi, P.N: Cybernetic   Study of Societal Systems ([1384]
Ricci, F.J. : ter Simulation of a   Cybernetic   Model of Man in his Cultura   [1416]
Ulrich, W.   : A Critique of Pure   Cybernetic   Reason: The Chilean Experie   [1768]
Wenger, M.S::   Thoughts on a Cybernetic   View of Social Organization   [1858]
Jaron, J.   :   Goal-Oriented   Cybernetical Systems.   [839]
Teodorescu, :   Cybernetical Actions and Markov Proces   [1682]
Ashby, R.   : Ashby's Writings on   Cybernetics . [55]
Bailey, N.T.:   Some Issues in the   Cybernetics   of Disease Control.   [83]
Burton, D.J.: gy for Second-Order   Cybernetics .   [251]
Busch, J.A. :   Cybernetics   IV: A System-Type Applicab   [252]
Cliffe, M.J.: The Contribution of   Cybernetics   to the Science of Psychopa   [322]
Cliffe, M.J.:   Cybernetics   of Alcoholism.   [321]
Dimitrova, L: Cybernetics   and the Organization of Ma   [423]
Espejo, R.   :   Cybernetics   in Management and Organiza   [491]
Fosko, P.D. :   Systems Theory and   Cybernetics   to the Management of/ Rese   [537]
Gause, D.C. :   Cybernetics   and Artificial Intelligenc   [587]
George, F.H.:   Cybernetics   and the Environment.   [593]
George, F.H.:   The Foundations of   Cybernetics .   [594]
George, F.H.: ical Foundations of   Cybernetics .   [595]
Ghosal, A.   :   Applied   Cybernetics : its Relevance to Operatio   [607]
Ghosal, A.   :   Relevance of   Cybernetics   in the Study of Developing   [608]
Hanken, A.F.:   Cybernetics   and Society: An Analysis o   [720]
Lasker, G.E.: Applied Systems and   Cybernetics   (6 Vols).   [1010]
Laszlo, E.   :   Cybernetics   and Global Planning.   [1016]
Lefebre, V.A: Second Order   Cybernetics   in the Soviet Union and th   [1031]
Linstone, H.:   Cybernetics   and the Future.   [1053]
Manescu, M. :   Economic   Cybernetics .   [1100]
Mel'cuk, I.A: Cybernetics   and Linguistics: Some Reas   [1148]
Mel'cuk, I.A: Cybernetics   and Linguistics.   [1149]
Milsum, J.H.:   Cybernetics   and Health Care.   [1170]
Nowakowska, :   Cybernetics   in the Social Sciences.   [1245]
Oeren, T.I. :   Cybernetics   and Modelling and Simulati   [1254]
Pask, G.   :   The Originality of   Cybernetics   and the Cybernetics of Ori   [1286]
Pask, G.   : Cybernetics and the   Cybernetics   of Originality.   [1286]
Pask, G.   :   Cybernetics   in Psychology and Educatio   [1287]
Pedretti, A.:   The   Cybernetics   of Language.   [1308]
Peschel, M. :   Engineering   Cybernetics .   [1324]
Pichler, F. :   Progress in   Cybernetics   and Systems Research, Vol.   [1336]
Pichler, F. :   Progress in   Cybernetics   and Systems Research, Vol.   [1335]
Probst, G.J.: ew ('second order')   Cybernetics   against Mismanagement.   [1363]
Rapoport, A.:   Cybernetics   as a Link between two Theo   [1383]
Rastogi, P.N: cial and Management   Cybernetics .   [1386]
Ricciardi, L: Cybernetics   in Biology and Medicine: A   [1417]
Ricciardi, L: Cybernetics   and Biology: Classic Attem   [1418]
Robb, F.   :   Cybernetics   in Management Thinking.   [1423]
Rose, J.   :   Modern Trends in Cybernetics   and Systems (3 Vols.).   [1436]
Rose, J.   :   Current Topics in Cybernetics   and Systems (Proc.Fourth M   [1437]
Rudall, R.   :   Computers and   Cybernetics .   [1458]
Tamburrini, :   Do   Cybernetics , Systems Science and Fuzzy   [1665]
Taschdjian, :   The   Cybernetics   of Stressed Systems.   [1671]
Trappl, R.   :   Progress in   Cybernetics   and Systems Research, Vol.   [1712]
Trappl, R.   :   Progress in   Cybernetics   and Systems Research, Vol.   [1713]

Trappl, R.   :   Progress in Cybernetics and Systems Research, Vol. [1721]
Trappl, R.   :   Progress in Cybernetics and Systems Research, Vol. [1720]
Trappl, R.   :   Progress in Cybernetics and Systems Research, Vol. [1723]
Trappl, R.   :   Progress in Cybernetics and Systems Research, Vol. [1724]
Trappl, R.   :   Cybernetics and Systems Research.   [1718]
Trappl, R.   :   Cybernetics and Systems Research, Vol. [1726]
Trappl, R.   :   Cybernetics - Theory and Applications. [1725]
Trappl, R.   :   Progress in Cybernetics and Systems Research, Vol. [1722]
Tripp, R.S. :   The Use of Cybernetics in Organizational Design a [1736]
Ulrich, W.   : ean Experience with Cybernetics .   [1768]
Varela, F.J.:   Steps to a Cybernetics of Autonomy.   [1800]
             : ournal "Progress in Cybernetics " Devoted to Papers Present [1942]
Sherlock, R.: stribution of Limit   Cycle Lengths; Part 2: the State Cycle [1545]
Sherlock, R.: ; Part 2: the State   Cycle Fraction for Networks of Differe [1545]
Arora, P.N. : cted Divergence and Cyclic Symmetry.   [54]
Fairhurst, M: Cyclic Activity in Information Process [500]
Stoklosa, J.: Machines Generating   Cyclic Sequences.   [1622]
Arora, P.N. : annon's Entropy and Cyclicity .   [53]
Janko, J.   : from the History of Czech Science in the Middle of the 19t [832]
Bergeron, R.: aginary, Fluid, and   Dancing Systems.   [161]
Aho, A.V.   : Joins in Relational   Data Bases.   [20]
Atre, S.   :   Data Base: Structured Techniques for D [66]
Batchelder, : ysis of Dichotomous   Data .   [115]
Bowen, B.D. : An Introduction to   Data Analysis.   [202]
Brillinger, :   Time Series: Data Analysis and Theory.   [221]
Brockett, P.: is of Questionnaire   Data .   [222]
Chang, L.C. : e Generalization of   Data Management Systems.   [293]
Ciupa, M.   : to Packet Switched   Data Communication Systems.   [317]
Collins, D.M.: ages from Imperfect   Data by Iterative Entropy Maximization [331]
Deutsch, K.W: on of Complexity in   Data Processing for World Models.   [414]
Elandt-Johns: Survival Models and   Data Analysis.   [473]
Gehrlein, W.:   Incomplete Ordinal   Data .   [588]
Golledge, R.: l Analysis of Large   Data Sets.   [632]
Goodman, L.A: ative / Categorical   Data .   [641]
Guida, G.   : ge Access to Online   Data Bases.   [681]
Harris, L.R.:   Using the   Data Base as a Semantic Component to A [727]
Harris, L.R.: of Natural Language   Data Base Queries.   [727]
Havranek, T.: essing Sociological   Data by the Guha Method: an Example.   [737]
Hoaglin, D.C: Data for Decisions: Information Strate   [763]
Horowitz, E.:   Fundamentals of Data Structures.   [788]
Hutt, A.T.F.:   A Relational Data Base Management System.   [799]
Inmon, W.H. :   Effective   Data Base Design.   [811]
Kashyap, R. : els Using Empirical   Data .   [882]
Kent, W.   :   Data and Reality.   [902]
Krippendorff: es in Multi-Variate   Data .   [984]
Langefors, L: Information and   Data in Systems.   [1009]
Launer, R.L.:   Modern   Data Analysis.   [1018]
Majster, M.E: lism for Structured   Data and Data Structures.   [1092]
Majster, M.E: Structured Data and   Data Structures.   [1092]
Meldman, M.J: Riss: a Relational   Data Base Management System for Minico [1150]
Melkumian, E: n Measures in Image   Data Compression and Transmission Syst [1151]
Morris, C.N.:   Introduction to   Data Analysis and Statistical Inferenc [1188]
Nijssen, G.M: Modelling in   Data Base Management Systems.   [1229]
O'Muircheart: Analysis of Survey   Data , Vol. 1: Exploring Data Structure [1251]
O'Muircheart: , Vol. 1: Exploring   Data Structures. Vol. 2: Model Fitting [1251]
Pearl, J.   :   On Summarizing   Data Using Probabilistic Assertions.   [1303]
Phillips, R.: tectural Relational   Data Base.   [1333]
Reisig, G.H.:   Statistical   Data Analysis in the Information Domai [1399]
Rus, T.   :   Data Structures and Operating Systems. [1464]
Sanderson, A: ionary Multivariate   Data .   [1489]
Smithson, M.: s for Fuzzy Nominal   Data .   [1594]
Tamura, H.   : el by the Method of   Data Handling: An Application to Lake [1666]
Techo, R.   :   Data Communications: an Introduction t [1681]
Tompa, F.W. : ication of Abstract   Data Types.   [1698]

Takatsu, S. : jective Satisficing Decision   Problems.   [1660]
Tani, S.T.   : ive on Modelling in Decision   Analysis.   [1668]
Tong, R.M.   : guistic Approach to Decision   Making with Fuzzy Sets.   [1701]
Tsokos, C.P.:   Decision   Information.   [1759]
Viliums, E.R: ives Evaluating and Decision   Making Under Uncertainty and   [1817]
Wang, Z.   : pplication of Fuzzy Decision   Making and Fuzzy Linear Progr   [1837]
Watson, S.R.:   Fuzzy   Decision   Analysis.   [1848]
White, D.J. : ness of Theories of Decision   in Practice.   [1862]
Wildhelm, J.: and Multiobjective   Decision   Making Under Uncertainty.   [1864]
Yager, R.R. :   Fuzzy   Decision   Making Including Unequal Obje   [1887]
Yager, R.R. : Probabilities, and   Decision .   [1895]
Yager, R.R. : lity and Anxiety in Decision   Making: An Application of Fuz   [1905]
Yezhkova, I.:   Decision   Making on Fuzzy Grounds.   [1913]
Zeleny, M.   :   Multiple Criteria Decision   Making.   [1924]
Zhukovin, V.:   The Multicriteria Decision   Making with Vector Fuzzy Pref   [1925]
Zimmermann, : onnectives in Human   Decision   Making.   [1926]
Dompere, K.K: On Epistemology and   Decision-Choice Rationality.   [433]
Biswas, A.K.: g and Environmental   Decision-Making .   [171]
Kickert, W.J: Fuzzy Theories on Decision-Making . (Vol. 3 in Frontiers   [911]
Kickert, W.J: h to Organizational   Decision-Making .   [912]
Kickert, W.J: Organization of Decision-Making : a Systems-Theoretical   [914]
Mackinnon, A: Complexity and Decision-Making .   [1085]
Orlovsky, S.:   Decision-Making with a Fuzzy Preferenc   [1271]
Pedrycz, W. :   A Model of Decision-Making in a Fuzzy Environment   [1312]
Potter, J.M.: ior Information and   Decision-Making .   [1355]
Roberts, F.S: ith Applications to   Decision-Making , Utility, and the Soci   [1425]
Slyadz, N.N.: tial Information in   Decision-Making Models.   [1582]
Bell, D.E.   : cting Objectives in Decisions .   [142]
Gage, W.L.   : dence for Strategic Decisions .   [554]
Hoaglin, D.C: Data for   Decisions : Information Strategies for   [763]
Jensen, H.L.: r Quantifying Legal   Decisions .   [844]
Viliums, E.R: evel Consistency of   Decisions under Uncertainty.   [1816]
Yager, R.R. : bsets of Type II in Decisions .   [1896]
Kindler, E. : mming Languages: I. Declaration of Necessary System Concep   [917]
Trost, H.   :   Declarative Knowledge Representation: [1754]
Courtois, P.:   Decomposability : Queueing and Computer   [352]
Dessauchoy, : ormation Theory and Decomposability of Systems.   [412]
Toulotte, J.:   A Method for Decomposing Interpreted Petri Nets and   [1707]
Hwang, C.L. :   Decomposition of Binary Sequences of F   [802]
Iijima, J.   :   Serial   Decomposition of General Systems.   [809]
McGowan, C.L: A Review of Decomposition and Design Methodologies   [1137]
Medgyessy, P: Decomposition of Superpositions of Den   [1144]
Molina, F.W.: Resource Directive   Decomposition in Mathematical Programm   [1181]
Naylor, A.W.:   On   Decomposition Theory: Generalized Depe   [1204]
Pichler, F. :   Decomposition of General Dynamic Syste   [1334]
Soete, W.   : uctural Methods for Decomposition of Large-Scale Systems.   [1596]
Walichiewicz: Decomposition of Linguistic Rules in t   [1828]
Okeda, M.   :   Overlapping Decompositions , Expansions, and Contra   [1260]
Levitan, G.L: Use of   Deductive Inference in Situation Contr   [1043]
Jorgenson, S: ogical Modelling of   Deep Lakes and Reservoirs.   [858]
Mergenthaler: al Repair Cost in a   Defective , Coherent Binary System.   [1152]
Andersson, H: Control System Using Defined eory of Relatively Closed Syste   [32]
Bauer, M.A. :   Defining Structural Descriptions.   [125]
Jones, L.M. :   Defining Systems Boundaries in Practic   [857]
Varshavskiy,: onous Processes. I.   Definition and Interpretation.   [1802]
Wyner, A.D. :   A   Definition of Conditional Mutual Infor   [1884]
Kwakernaak, : Variables. Part 1:   Definitions and Theorems. Part 2: Algo   [996]
Smyth, D.S. : e Structure of Root Definitions .   [1595]
Verhagen, C.: rn Recognition; its   Definitions ; its Relation with Other D   [1812]
Mackenzie, K: sed Measure for the   Degree of Hierarchy in a Group.   [1083]
Smets, P.   :   The   Degree of Belief in a Fuzzy Event.   [1586]
Lee, E.B.   : ical Forms for Time Delay Systems.   [1026]
Grollmann, J: ia Approximation by   Delayed Systems.   [673]
De La Sen, M: ol Systems to Avoid   Delays in the Input Generation.   [394]

Kinston, W.   : System Research and   Delivery   Systems. The Example of Healt   [928]
Krzanik, L.   : mic Optimization of   Delivery   Step Structure.   [989]
Ryan, R.M.   : n of Social Welfare   Delivery   Systems: A Matrix Organizatio [1467]
Gericke, M.   : Passenger-Transport   Demands   by Society-Controlled Activiti [600]
Kindler, J.   : Modelling of Water   Demands .   [924]
Ballonoff, P: Mathematical   Demography of Social Systems, II.   [92]
Stoy, J.E.   :   Denotational   Semantics: the Scott-Stra   [1625]
Trattnig, W.:   Denotational   Semantics of a Software S   [1734]
Collins, D.M: Electron   Density   Images from Imperfect Data by   [331]
Medgyessy, P: f Superpositions of   Density   Functions and Discrete Distrib   [1144]
Ballance, R.:   Functional   Dependence   and its Application to Syst   [90]
Domotor, Z.   : bilistic and Causal   Dependence   Structures.   [432]
Naylor, A.W.: Theory: Generalized   Dependence .   [1204]
Conant, R.C.: ion and Analysis of   Dependency   Structures.   [339]
Nishimura, H: Realization of   Dependency   Structures.   [1232]
Dwyer, R.F.   : tion Detectors with   Dependent   Sampling.   [452]
Tong, R.M.   : ion of Fuzzy Models   Derived   from Experimental Data.   [1703]
Loefgren, L.:   Describability   and Learnability of Com   [1060]
Batchelor, B: Hierarchical Shape   Description   Based Upon Convex Hulls of   [119]
Batchelor, B: Shape   Description   for Labeling Concavity Tre   [120]
Christensen,: ook. Vol.1: General   Description ; Vol.2: Philosophical Orig   [311]
Csaki, P.   : l Problems of Model   Description .   [360]
Grollmann, J: Description   of Feedback Systems via Ap   [673]
Kindler, E.   : ming Languages: II.   Description   of Types and Individual Ty   [918]
Locker, A.   : A New Look at the   Description   and Prescription of System   [1059]
Matsuno, K.   :   Operational   Description   Microsystems Formation in   [1122]
Pedretti, A.: otion of Cybernetic   Description ?   [1309]
Sherlock, R.: Part 1: State Space   Description   and Distribution of Limit   [1545]
Sklenar, J.   : pparatus for Formal   Description   of Discrete Systems.   [1571]
Tchon, K.   : a Formal Theory of   Description .   [1678]
Vallet, C.   : erent Formalism for   Description   of Natural Systems.   [1784]
Bauer, M.A.   : Defining Structural   Descriptions .   [125]
Rao, G.P.   : ons from Linguistic   Descriptions   in Problems of Fuzzy Logi [1380]
Shapiro, L.G: Structural   Descriptions   and Inexact Matching.   [1535]
Sussman, G.J: Almost Hierarchical   Descriptions .   [1640]
Kornai, J.   :   Descriptive-Explanatory   Theoretical Mo   [971]
Atre, S.   : ured Techniques for   Design , Performance and Management.   [66]
Baase, S.   : ms: Introduction to   Design   and Analysis.   [77]
Boyd, G.M.   : ey Questions in the   Design   of Mutual Control in Education.   [205]
Brown, D.B.   : ystems Analysis and   Design   for Safety.   [228]
Bryant, D.T.: ng and Organization   Design .   [234]
Clark, W.C.   : for Ecology Policy   Design : a Case Study of Ecosystem Mana   [318]
Cochin, I.   :   Analysis and   Design   of Dynamic Systems.   [324]
Cotterman, W: ystems Analysis and   Design : A Foundation for the 1980's.   [350]
De Jong, K.   :   Adaptive System   Design : A Genetic Approach.   [392]
Ehrenkrantz,: Systems Approach to   Design .   [460]
El-Shirbeeny: Constrained Control   Design   Problems.   [472]
Espejo, R.   : stems: a System for   Design .   [489]
Estep, M.L.   : quiry in Curriculum   Design .   [493]
Faucheuz, C.: y in Organizational   Design .   [509]
Franksen, O.: Systems: Developing   Design   Rules Using APL.   [541]
Gasparski, W: Problems of   Design   Methodology.   [574]
Gasparski, W: e Logical Theory of   Design .   [576]
Gasparski, W: Design   and System: Methodological Aspe   [577]
Gasparski, W: Studies in   Design   Methodology in Poland (Special   [575]
Gasparski, W: m the Conference in   Design   Methods.   [579]
Gasparski, W: Non Est Idem: Or on   Design   Methodology as Seen from Neithe   [582]
Gasparski, W: non est idem: Or on   Design   Methodology as Seen from neithe   [584]
Gomez, P.   : m-Up Organizational   Design : a Cybernetic Perspective.   [636]
Grant, D,P.   :   Design   Methods and Theories: Vol.15, N   [656]
Grant, D,P.   :   Design   Methods and Theories; Vol.15, N   [657]
Gupta, M.M.   :   On the   Design   of an Adaptive Controller for U   [687]
Honkasalo, A: Design   of Work Tasks and Education of   [776]
Inmon, W.H.   : Effective Data Base   Design .   [811]

Jones, A.H.   :   Cybernetic Design   Features for Computerised Busin [853]
Keravnou, E.:   Design   of Expert Systems from the Pers [903]
Kickert, W.J: ach to Organization Design . [913]
Kilmann, R.H: Social Systems Design   : Normative Theory and the Maps [916]
Kleijnen, J.:   Design   and Analysis of Simualtion: Pra [932]
Kohout, L.J.: l System Protection Design . [966]
Leeming, A.M: ystems Study of the Design   of a User Interface. [1029]
Lewin, D.   :   Theory and Design   of Digital Computer Systems. [1046]
M'Pherson, P: ach to Whole-System Design . [1076]
M'Pherson, P: Systems Engineering Design . [1077]
McGowan, C.L: f Decomposition and Design   Methodologies. [1137]
Mumford, E. : rticipative Systems Design   : Structure and Method. [1196]
Murdick, R.G: MIS: Concepts and Design . [1197]
Patel, R.V. : e System Theory and Design . [1289]
Perlis, J.H.: valuation of System Design . [1318]
Protzen, J.-:   Design   Methods and Theories: Vol.10, N [1366]
Reckmeyer, W: ystems Research and Design   : Precursors and Futures. [1394]
Revell, N.   : thodologies for the Design   of Small Database Systems. [1414]
Rudawitz, L.:   Client-Centered Design   : Concepts and Experience. [1459]
Rzevski, G. : l Aspects of System Design . [1468]
Sage, A.P.   : nsiderations in the Design   of Information Systems and Proc [1475]
Shakun, M.F.: king and Meaning as Design   of Purposeful Systems. [1533]
Shakun, M.F.: s, Productivity and Design   of Purposeful Systems: The Prof [1534]
Siewiorek, D: of Reliable System Design . [1554]
Smith, R.B. :   How to Plan, Design   & Implement a Bad System. [1592]
Sonde, B.S. : roduction to System Design   Using Integrated Circuits. [1598]
Squire, E.   : Introducing Systems Design . [1607]
Steward, D.V: cture, Strategy and Design . [1620]
Svoboda, A. : ced Logical Circuit Design   Techniques. [1645]
Techo, R.   : ion to Concepts and Design . [1681]
Tripp, R.S. : s in Organizational Design   and Development: An Illustratio [1736]
Troncale, L.: tems Enables Better Design   of Man-Made Systems: The Linkag [1749]
Ulrich, W.   :   The Design   of Problem Solving Systems. [1767]
Veroy, B.S. :   Overall Design   of Reliable Centralized Voice/D [1814]
Wade, J.W.   : oses: Architectural Design   as a Basic Problem-Solving Proc [1826]
Walichiewicz: uistic Rules in the Design   of a Multi-Dimensional Fuzzy Co [1828]
Weinberg, G.:   On the Design   of Stable Systems. [1853]
Weir, M.   :   Design   for a Goal-Directed System. [1855]
Winkel, D.   :   The Art of Digital Design   : An Introduction to Top-Down De [1867]
Winkel, D.   : duction to Top-Down Design . [1867]
             :   Design   Studies. A New Journal Covering [1938]
             : d Techniques of All Design   (Published Quarterly in Co-oper [1938]
Farreny, H. :   On the Best Way of Designating   Objects in Sentence Genera [508]
Van Ameronge: Autopilot for Ships Designed   with Fuzzy Sets. [1785]
Hanson, O.J.: Systems Analysis in Designing   Systems and Methods of Train [722]
Toern, A.A. : Modeling Simulation Designs . [1695]
Nicolis, J.S: anizing Systems and Desynchronization   of Physiological Rhy [1225]
Conant, R.C.:   Detection   and Analysis of Dependency S [339]
Jarisch, W. : Risk-Assessment and Detection   of Fetal Respiratory Movemen [837]
Rapoport, A.: Response Models for Detection   of Change. [1382]
Swets, J.A. : Methods from Signal Detection   Theory. [1646]
Dwyer, R.F. : equential Partition Detectors . [451]
Dwyer, R.F. : equential Partition Detectors   with Dependent Sampling. [452]
Balkus, K.   :   Organizational Determinants   of National Development. [87]
Lucas, R.G. :   Political Determinants   of Information System Con [1070]
Kaiser, J.S.: formations within a Determinate   Machine. [868]
Camana, P.C.:   Determination   of Feedback for Human Po [258]
Jones, B.   :   Determination   of Reconstruction Famili [854]
Smetanic, Y.: n the Algorithm for Determining   the Primary Structure of B [1585]
Von Plato, J: Probability and Determinism . [1823]
Aggarwal, S.:   Deterministic   Representation of Probab [17]
Burks, A.W. :   Models of Deterministic   Systems. [246]
Dempster, A.:   Deterministic   and Stochastic Schedulin [408]
Gini, G.   :   Deterministic   and Nondeterministic Pro [615]

Gheorghe, A.: on towards a Medium   Developed   Economy.   [605]
Banathy, B.H: Developing   a Systems View of Education   [95]
Chen, K.   : ation Technology in   Developing   Countries.   [304]
Franksen, O.: arge Scale Systems: Developing   Design Rules Using APL.   [541]
Ghosal, A. : ics in the Study of   Developing   Countries.   [608]
Kawamura, K.:   On   Developing   the Microelectronics Indust   [891]
Rudall, B.H.:   Developing   Automated Teaching Systems.   [1457]
Sharif, N. : ystems Modelling in   Developing   Countries.   [1536]
Slawski, C. :   Developing   the Learning Relationship: [1574]
Tomlinson, R: in Organizations -   Developing   a Collaborative Research St   [1697]
Veelenturf, : retical Approach to   Developing   Learning Neural Networks.   [1806]
Augustine, N: ws and Major System   Development   Programs.   [71]
Balkus, K. : ms in Sociocultural   Development .   [86]
Balkus, K. : minants of National   Development .   [87]
Baltes, P.B.:   Life Span   Development   and Behavior, Vol.I.   [94]
Bartolini, G: Development   of Performance Adaptive Fu   [113]
Cole, I.D. : Naturalized System   Development .   [330]
Day, R.H.   :   Economic   Development   as an Adaptive Process.   [384]
Erdi, P.   : of the Ontogenetic   Development .   [484]
Fosko, P.D. : nt of/ Research and   Development .   [537]
Hatze, H.   : cybernetics - A New   Development   in Neuromuscular Control.   [736]
La Patra, J.: s Approach to Urban   Development .   [997]
Le Bonniec, :   The   Development   of Model Reasoning.   [1021]
Lundeberg, M: Information Systems   Development : A Systematic Approach.   [1074]
Mesarovic, M: uter Model of World   Development   System.   [1153]
Mittermeir, : otation for Systems   Development .   [1175]
Nowakowska, : al Structure of the   Development   of a Scientific Discipline   [1240]
Pask, G.   :   Development   in Conversational Theory -   [1285]
Slawski, C. : rch Model of Career   Development .   [1575]
Sugiyama, K.: re Computer Systems   Development   in Japan.   [1635]
Toda, M.   : re Computer Systems   Development   in Japan.   [1694]
Tonella, G. : istic Modelling for   Development   Planning.   [1699]
Tripp, R.S. : zational Design and   Development : An Illustration within Ai   [1736]
Urban, H.B. :   The Concept of Development   for a Systems Perspective.   [1773]
Couger, J.D.:   Advanced System Development/Feasibility Techniques.   [351]
Hogeweg, P. : ocally Synchronised   Developmental Systems: Concrete Advant   [769]
Cavallo, R.E: Recent Developments in Systems Methodology fo   [278]
Cavallo, R.E: hments, and Current   Developments . (a Report Sponsored by t   [280]
Cavallo, R.E: shments and Current   Developments .   [281]
Cavallo, R.E: ce Research: Recent   Developments .   [287]
Hartley, R. :   Recent Developments in Markov Decision Proces   [730]
Martin, P.E.: Patterns - Further   Developments .   [1111]
Mohler, R. :   Recent Developments in Variable Structure Sys   [1179]
            : ess in Cybernetics"   Devoted   to Papers Presented at the Sec   [1942]
Ganascia, J.: tion Facilities for   Diagnosis   Systems.   [566]
Keeny, B.P. : native Paradigm for   Diagnosis .   [896]
Miller, M.C.: l Models in Medical   Diagnosis .   [1168]
Pau, L.F.   :   Failure   Diagnosis   and Performance Monitoring.   [1296]
Smith, P.M. : dures Used in Fault   Diagnosis   Tasks.   [1591]
Favella, L.F: pansion to Acoustic   Diagnostic   Problems for Internal Combu   [512]
Gyarfas, F. : ning Computer-Based   Diagnostic   Expert System.   [695]
Swets, J.A. :   Evaluation of Diagnostic   Systems: Methods from Signa   [1646]
Checkland, P: ce. Part 1: Systems   Diagram   - Some Tentative Guidelines. P   [297]
Cosier, R.A.: he Effectiveness of   Dialectical   Inquiry Systems.   [349]
Mitroff, I.I: rength of Belief of   Dialectical   Belief Systems.   [1173]
Mulej, M.   :   Dialectical   Systems Theory.   [1193]
Mulej, M.   :   Dialectical   Systems Theory and Work Si   [1194]
Zwick, M.   : phe and the Laws of   Dialectics .   [1932]
Trost, H.   : : A German Language   Dialogue   System.   [1753]
Garfolo, B.T: The Renal   Dialysis   Medical Information System.   [573]
Parks, M.S. :   Dichotic   Property and Teleogenesis.   [1283]
Lowen, W.   :   Dichotomies   of the Mind: A Systems Sci   [1067]
Batchelder, : of the Analysis of   Dichotomous   Data.   [115]
Andrew, A.M.: tinuity - a Systems   Dichotomy .   [36]

Eveland, S.   : othic Ethymological   Dictionary .   [496]
Gasparski, W: Si Duo   Dicunt   Idem Non Est Idem: Or on Design   [582]
Dammasch, I.: och-Pitts Networks: Differences between Input-Constant and   [375]
Mattessich, : k: Similarities and   Differences Between Mario Bunges's Wor   [1127]
Scott, W.A.   : ement of Individual   Differences .   [1519]
Chen, K.   : System Analysis in   Different   Social Settings.   [305]
Kashyap, R.   : esian Comparison of   Different   Classes of Dynamic Models Us   [882]
Labos, E.   : Periodic Motions in   Different   Classes of Formal Neuronal N   [1000]
Roedding, W.:   Different   Approaches to the Aggregatio   [1434]
Ruotsalainen: ospital System with   Different   Service Levels.   [1463]
Sherlock, R.: ion for Networks of   Different   Connectivities.   [1545]
Andersson, H: Ana   Different an Inventory Control System U   [32]
Hallam, T.G.: rinsic Structure of   Differential Equation Models of Cosyst   [715]
Ibidapo-Obe,:   An Approximate   Differential Structure for Error Covar   [807]
Windeknecht,: olution of Ordinary   Differential Equations.   [1866]
Stockmeyer, :   Intrinsically   Difficult   Problems.   [1621]
Blauberg, I.: sites, Problems and   Difficulties .   [174]
Skvoretz, J.:   Information   Diffusion   in Formally Structured Popul   [1572]
Agrawal, V.D: eoretic Approach to   Digital   Fault Testing.   [18]
Bozic, S.M. :   Digital   and Kalman Filtering.   [206]
Gonzales, R.:   Digital   Image Processing.   [639]
Keyes, R.W. : ndamental Limits in   Digital   Information Processing.   [909]
Lewin, D.   : heory and Design of   Digital   Computer Systems.   [1046]
Maryanski, F: Digital   Computer Simulation.   [1113]
Pratt, W.K. :   Digital   Image Processing.   [1356]
Rao, G.V.   :   Complex   Digital   Control Systems.   [1381]
Reusch, B.   : inuous Systems with   Digital   Behavior.   [1411]
Reusch, B.   : us Systems that are   Digital .   [1412]
Robinson, E.:   Digital   Foundations of Time Series Ana   [1433]
Simon, J.C. :   Digital   Image Processing.   [1565]
Stucki, P.   :   Advances in   Digital   Image Processing.   [1630]
Winkel, D.   :   The Art of   Digital   Design: An Introduction to Top   [1867]
Van Navta, L: Digitial   Computer Applications to Proc   [1789]
Burns, J.R. : r Converting Signed   Digraphs   to Forrester Schematics.   [249]
Burns, J.R. : uctural Analysis of   Digraphs .   [250]
Roberts, G.F: uctural Analysis of   Digraphs .   [1426]
Robinson, D.:   Digraphs : Theory and Techniques.   [1432]
Benoit, B.M.: : Form, Chance, and   Dimension .   [156]
Blunden, M. : ent - The Political   Dimension .   [177]
Glanville, R: The Model's   Dimensions : A Form for Argument.   [619]
Wisniewski, : o Increasing Carbon   Dioxide .   [1872]
Arora, P.N. :   Directed   Divergence and Cyclic Symmetr   [54]
Majster, M.E: Extended   Directed   Graphs, a Formalism for Struc   [1092]
Tao, K.M.   : On the Structure of   Directed   Graphs with Applications: A R   [1669]
Tao, K.M.   : On the Structure of   Directed   Graphs with Applications: A R   [1670]
Kornai, J.   : eview of a Research   Direction .   [971]
Innis, G.S. :   New   Directions   in the Analysis of Ecologic   [812]
Molina, F.W.: Survey of Resource   Directive   Decomposition in Mathematica   [1181]
Steinacker, :   Disambiguation   in German.   [1615]
Steinacker, : ing as a Method for   Disambiguation   and Anaphora Resolution   [1617]
Beckett, J.A: nt: The Case of the   Disappearing   Model.   [132]
Anderton, R.: enties: an Emerging   Discipline ?   [33]
Nowakowska, : ent of a Scientific   Discipline .   [1240]
Gurel, O.   : tions in Scientific   Disciplines .   [691]
Verhagen, C.: Relation with Other   Disciplines ; a Literature Survey.   [1812]
De Beaufrand: Text,   Discourse , and Process: Towards a Mult   [386]
De Beaugrand: Text,   Discourse , and Process.   [387]
Klir, G.J.   :   Pattern   Discovery   in Activity Arrays.   [939]
Ross, S.D.   :   Learning and   Discovery .   [1449]
Simon, H.A. :   Models of   Discovery   and Other Topics in the Meth   [1562]
Trussel, P. : g Nursing Research: Discovery , Analysis, and Interpretatio   [1755]
Adam, A.   :   Discrete   Systems Methodology.   [11]
Andersen, E.:   Discrete   Statistical Models with Socia   [28]
Barto, A.G. :   Discrete   and Continuous Models.   [112]

Beloglavec, : rties of a Periodic Discrete String.   [145]
Boltyanskii,: Optimal Control of   Discrete Systems. (Translated From Rus   [187]
Boltyanskii,: Optimal Control of   Discrete Systems.   [186]
Davio, M.   :   Discrete and Switching Functions.   [381]
El-Sherief, : d Control of Linear Discrete Multivariable Systems with Pr   [470]
Gentile, S. :   A Discrete Model for the Study of a Lake   [592]
Greenspan, D: Discrete Modeling in the Microcosm and   [665]
Gumovski, I.:   Recurrences and Discrete Dynamic Systems.   [683]
Gupta, M.M. : mization for Linear Discrete Systems.   [688]
Hogeweg, P. : crete Advantages of Discrete Event Formalism.   [769]
Kucera, V. :   Discrete Linear Control: the Polynomia   [992]
Kwakernaak, : nd Examples for the Discrete Case.   [996]
Labudde, R.A: Discrete Hamiltonian Mechanics.   [1001]
Levy, L.S. :   Discrete Structures of Computer Scienc   [1044]
Medgyessy, P: nsity Functions and Discrete Distributions.   [1144]
Negoita, C.V: On the Logic of Discrete Systems Dynamics.   [1211]
Nobile, A.G.:   On a Class of Discrete Models for Regulated Growth w [1234]
Oeren, T.I. :   Simulation with Discrete Models.   [1256]
Roberts, F.S: Discrete Mathematical Models.   [1424]
Rosenberg, R: lytical Dynamics of Discrete Systems.   [1447]
Sklenar, J. : rmal Description of   Discrete Systems.   [1571]
Strejc, V. : ate Space Theory of Discrete Linear Control.   [1628]
Uccellini, L: Applying Discrete Model Concepts to the Computa   [1765]
Subrahmanian: rator for Models of Discrete-Event Systems.   [1632]
Uyeno, D.H. :   PASSIM: A Discrete-Event Simulation Package for   [1776]
El-Sherief, : inear Multivariable Discrete-Time Systems: A Survey.   [468]
Shieh, L.   :   State Equations by Discrete-Time State Equations.   [1546]
Santis, F.de: and Fisher's Linear Discriminant Analysis.   [1492]
Troncale, L.:   On a Possible Discrimination between Bioevolution an [1740]
Gruber, H. :   and Abstracting; a Discussion on the Basis of Sequential   [677]
Lusk, E.J. :   A Discussion of External Validity in Sys [1075]
Bailey, N.T.: the Cybernetics of   Disease Control.   [83]
Gati, G.   : on the Isomorphism   Disease .   [585]
Porenta, G. : tion of Parkinson's Disease and Aging Processes.   [1349]
Read, R.C. : e Graph Isomorphism Disease .   [1393]
Matsuno, K. :   Disequilibrium Dynamics of Autonomous   [1121]
Jutila, S.T.: ming and Control by Dislocations .   [864]
Auger, P.   :   Order, Disorder in Hierarchically Organized S   [69]
Schieve, W.C: lf-Organization and Dissipative Structures: Applications i [1507]
Zeleny, M. :   Autopoiesis, Dissipative Structures, and Spontaneou [1921]
Groen, J.C.F: tic Measure for the Distance between Neighbouring Structur   [671]
Higashi, M. :   On the Notion of Distance Representing Information Clos [756]
Neuwirth, E.: arity Relations and Distance Coefficients in Thesauri.   [1215]
Glanville, R: Distinguished and Exact Lies.   [621]
Gurdial,   : nequality Involving Distortion .   [690]
Melkumian, A: Distortion Measures in Image Data Comp   [1151]
Chandrasekar: ystem Metaphors for Distributed Problem Solving.   [292]
Fox, M.S.   : anizational View of Distributed Systems.   [539]
Lamport, L. : dering of Events in Distributed Systems.   [1006]
Sawaragi, Y.: ir Applications for Distributed Parameter Systems.   [1499]
Smith, R.G. :   for Cooperation in Distributed Problem Solving.   [1593]
Umnov, A.   :   An Approach to Distributed Modeling.   [1769]
Callebaut, W: The Distribution Approach to Problem-Solvi   [257]
Machlup, F. : edge: Its Creation, Distribution and Economic Significance [1080]
Sherlock, R.: ace Description and Distribution of Limit Cycle Lengths; P [1545]
De Raadt, J.: f Income and Output Distributions .   [398]
Higashi, M. : ased on Possibility Distributions .   [754]
Higashi, M. : ity and Probability Distributions .   [756]
Medgyessy, P: ctions and Discrete Distributions .   [1144]
Nguyen, H.T.: itional Possibility Distributions .   [1220]
Arora, P.N. :   Directed Divergence and Cyclic Symmetry.   [54]
Colby, J.A. :   Mathematical Documentation for a Lotic Ecosystem Mo   [328]
McIntyre, C.: t 165: Mathematical Documentation for a Lotic Ecosystem Mo [1140]
Reisig, G.H.: in the Information   Domain .   [1399]

Sonnier, I.L: ce in the Affective   Domain .   [1599]
Porenta, G. : atical Model of the   Dopaminergic Synapse: Stability and Se [1349]
Gyarfas, F. :   CODEX: Prototypes Driven Backward and Forward Chaining C [695]
Andreae, J.H: A Dual   Model of the Brain.   [34]
Valdes-Perez: Information Loss   Due to Data Quantization in Reconstruc [1781]
Gasparski, W: Si Duo   Dicunt Idem Non Est Idem: Or on De [582]
Arbel, A.   : ring in Large-Scale   Dynamic Systems.   [46]
Baldwin, J.F: Dynamic Programming for Fuzzy Systems   [85]
Casti, J.   : stems: a Polyhedral   Dynamic Representation.   [273]
Cochin, I.   : lysis and Design of   Dynamic Systems.   [324]
Dreyfus, S.E: Dynamic Programming and the Calculus o [437]
Dreyfus, S.E: e Art and Theory of   Dynamic Programming.   [438]
Gottinger, H: he: Applications of   Dynamic System Theory.   [646]
Gottinger, H: ry of Complexity in   Dynamic Systems.   [647]
Gumovski, I.: rences and Discrete   Dynamic Systems.   [683]
Gupta, M.M. :   Hierarchical Dynamic Optimization for Linear Discre [688]
Haken, H.   : attern Formation by   Dynamic Systems and Pattern Recognitio [707]
Hartl, R.F. : ducing Policies for   Dynamic Continuous Lovers Under Risk o [729]
Jakubowski, : gorithms of Complex   Dynamic Systems Simulation.   [829]
Jumarie, G. : proach to Modelling   Dynamic Systems Involving Human Factor [862]
Kaindl, H.   :   Dynamic Control of the Quiescence Sear [867]
Kashyap, R. : ifferent Classes of   Dynamic Models Using Empirical Data.   [882]
Kindler, E. :   Dynamic Systems and Theory of Simulati [919]
Krzanik, L. : lutionary Projects: Dynamic Optimization of Delivery Step   [989]
Luenberger, :   Introduction to Dynamic Systems: Theory, Models, and A [1071]
Malec, J.   : arios as a Tool for   Dynamic Scene Representation.   [1096]
Margolis, D.: hs for Some Classic   Dynamic Systems.   [1104]
McClamroch, :   State Models of Dynamic Systems: A Case Study Approach [1134]
McClamroch, :   State Models of Dynamic Systems.   [1133]
Okeda, M.   : and Contractions of   Dynamic Systems.   [1260]
Pichler, F. : position of General   Dynamic Systems.   [1334]
Randers, J. : ments of the System   Dynamic Method.   [1379]
Reichle, D.E: Dynamic Properties of Forest Ecosystem [1398]
Sierocki, I.: elations in General   Dynamic Systems.   [1553]
Siljak, D.D.:   Large-Scale Dynamic Systems.   [1555]
Takahara, Y.: ausality of General   Dynamic Systems.   [1655]
Tamura, H.   : Identification of a   Dynamic Lake Model by the Method of Da [1666]
Thowsen, A. : Identifiability of   Dynamic Systems.   [1687]
Tracy, L.   :   A Dynamic Living-Systems Model of Work M [1709]
Bednarek, A.:   Dynamical Systems II.   [133]
Blaquire, A.:   Dynamical Systems and Microphysics. (S   [172]
Brayton, R.K: Stability of Dynamical Systems: a Constructive Appr [214]
Casti, J.   :   Dynamical Systems and Their Applicatio   [271]
Hakkala, L. : ation Algorithm for   Dynamical Systems.   [709]
Ibidapo-Obe,: riance in Nonlinear   Dynamical Systems.   [807]
Kloeden, P.E: Fuzzy Dynamical Systems.   [951]
LaSalle, J.P: The Stability of Dynamical Systems.   [999]
Lasota, A.   : of Slowly Evolving   Dynamical Systems.   [1012]
Moser, J.   : idden Symmetrics in   Dynamical Systems.   [1190]
Nicholson, H: Modelling of   Dynamical Systems.   [1222]
Nicholson, H: Modelling of   Dynamical Systems.   [1223]
Nitecki, Z. : Global Theory of   Dynamical Systems.   [1233]
Orava, P.J. : te in the Theory of   Dynamical Systems and Control.   [1268]
Peixoto, M.M: Dynamical Systems.   [1315]
Pichler, F. :   Dynamical Systems Theory.   [1337]
Porter, B.   :   Synthesis of Dynamical Systems.   [1350]
Resconi, G. : Logical Analysis of   Dynamical Systems-Applications to Life [1404]
Retti, J.   : g and Simulation of   Dynamical Systems.   [1410]
Rosen, R.   :   Dynamical Similarity and the Theory of [1439]
Saeks, R.   :   Large-Scale Dynamical Systems.   [1472]
Smale, S.   : of Time: Essays on   Dynamical Systems, Economic Processes   [1583]
Stoklosa, J.: sformations of Some   Dynamical Systems-Networks of (a,k)-Ma [1622]
Stoklosa, J.: ty Problem for Some   Dynamical Systems.   [1623]
Vallee, R.   :   Evolution of Dynamical Linear System with Random In [1782]

Millendorfer: Approaches to Long Economic Waves.   [1165]
Murphy, R.E.: aptive Processes in Economic Systems.   [1198]
Neck, R.   : e Political and the Economic System in Austria.   [1206]
Norman, M.   : oftware Package for Economic Modelling.   [1237]
Pearl, J.   :   An   Economic Basis for Certain Methods of   [1304]
Sandblom, C.:   Optimization of Economic Policy Using Lagged Controls.   [1487]
Sandblom, C.: n the Evaluation of   Economic Policies.   [1488]
Schimanovich: hematical Games and   Economic Actions.   [1510]
Seton, F.   :   The Evalution of Economic Flows and Systemic Efficiency   [1529]
Smale, S.   : Dynamical Systems, Economic Processes and Related Topics.   [1583]
Gottinger, H: ral Characteristics   Economical Models: a Study of Complexi   [648]
Carter, C.F.: and Expectations in   Economics : Essays in Honour of G.L.S.   [265]
Dirickx, Y.M: th Applications for   Economics and Management.   [424]
Gutenbaum, J: f Systems Theory to   Economics , Management, and Technology.   [692]
Kuhn, H.W.   : Systems Theory and Economics .   [994]
Mohler, R.   :   Structure Systems, Economics and Biology.   [1179]
Pfeilsticker: d Language-Oriented   Economics .   [1331]
Sagal, P.T.   :   Epistemology of Economics .   [1473]
Athanassov,   : ts for the National Economy .   [58]
Gheorghe, A.: a Medium Developed   Economy .   [605]
Kornai, J.   : ls of the Socialist Economy : Review of a Research Directio   [971]
Pearl, J.   :   On the Storage Economy of Inferential Question-Answer   [1301]
Trappl, R.   : chnology, Military, Economy , Society, Culture, and Politic   [1733]
Barber, M.C.: Markovian Model for   Ecosystem Flow Analysis.   [105]
Clark, W.C.   : gn: a Case Study of Ecosystem Management.   [318]
Colby, J.A.   : ntation for a Lotic Ecosystem Model.   [328]
Conrad, M.   : ty from Molecule to Ecosystem .   [343]
Fedra, K.   : nd Arbitrariness in Ecosystem Modelling: A Lake Modelling   [515]
Gutierrez, L: Ecosystem Succession.   [693]
Halfon, E.   :   Adequacy of Ecosystem Models.   [711]
Hall, C.A.S.:   Ecosystem Modeling in Theory and Pract   [713]
McIntyre, C.: ntation for a Lotic   Ecosystem Model.   [1140]
Morris, H.M.: Conserved Nutrient   Ecosystem .   [1189]
Patten, B.C.: ybernetic Nature of   Ecosystem .   [1295]
Ulanowicz, R: cal Modelling of an   Ecosystem .   [1766]
Unger, C.K.   : hysical Ecology and Ecosystem Research.   [1771]
Wolaver, T.G: eristics of a Model   Ecosystem .   [1877]
Keeny, B.P.   :   Ecosystemic Epistemology: an Alternati   [896]
Bolter, M.   : Analysis in Marine Ecosystems .   [185]
Casti, J.   :   Lake   Ecosystems : a Polyhedral Dynamic Repre   [273]
Goh, B.S.   : sistence of Complex Ecosystems .   [628]
Hirata, H.   : del of Hierarchical Ecosystems with Utility Efficiency Mas   [759]
Hirata, H.   : del of Hierarchical Ecosystems with Migration.   [758]
Hirata, H.   : and Energy Flow in Ecosystems .   [757]
Matis, J.H.   : tmental Analysis of Ecosystems Models.   [1118]
McIntyre, C.: ical Model of Lotic   Ecosystems .   [1139]
Patten, B.C.: agation of Cause in   Ecosystems .   [1290]
Reichle, D.E: roperties of Forest   Ecosystems .   [1398]
Athanassov,   : tem for Training of EDP-Specialists for the National Econo   [58]
Banathy, B.H: g a Systems View of   Education .   [95]
Bayraktar, B: Education in Systems Science.   [127]
Boyd, G.M.   : f Mutual Control in Education .   [205]
Erickson, S.: the Foundations of   Education .   [485]
Honkasalo, A: n of Work Tasks and   Education of Labor Force.   [776]
Pask, G.   : s in Psychology and Education .   [1287]
Silvern, L.C: tems Engineering in   Education I: the Evolution of Systems   [1557]
Silvern, L.C: Systems Thinking in   Education .   [1557]
Sonnier, I.L: Holistic   Education : Teaching of Science in the   [1599]
Vickers, G.   :   Education in Systems Thinking.   [1815]
Elohim, J.L.: earning to Organize   Educational Systems.   [477]
Brewer, J.N.: cture and Cause and   Effect Relations in Social System Simu   [216]
Wolaver, T.G: Effect of Systems Structure, Connectiv   [1877]
Buchberger,   : sition Functions in Effective Systems.   [237]
Gomez, P.   : ybernetic Rules for Effective Management.   [637]

Inmon, W.H. :   Effective   Data Base Design.   [811]
Orlovsky, S.:   Effective   Alternatives for Multiple Fu [1272]
Cosier, R.A.: n Evaluation of the Effectiveness of Dialectical Inquiry S   [349]
Shakun, M.F.:   Effectiveness , Productivity and Design   [1534]
White, D.J. :   Role and Effectiveness of Theories of Decision   [1862]
Ventriglia, : ral Systems: Memory Effects .   [1810]
Hirata, H. : ystems with Utility Efficiency   Mass and its Stability.   [759]
Kak, S.   :   On Efficiency   of Chemical Homeostasis - a   [869]
Kronsjo, L.I: heir Complexity and Efficiency .   [987]
Lewis, H.R. :   The Efficiency   of Algorithms.   [1047]
Seton, F.   : Flows and Systemic Efficiency .   [1529]
Trappl, R. : thods with Computer Efficiency   in Medicine.   [1711]
Conant, R.C.:   Efficient   Proofs of Identities in N-Di   [340]
Grauer, M. : The Generation of   Efficient   Energy Supply Strategies usi   [658]
Yager, R.R. : inking as Quick and Efficient .   [1897]
Vallee, R. :   " Eigen-Elements " for Observing and Inte [1783]
Beer, S.   : Death is Equifinal: Eighth   Annual Ludwig Von Bertalanffy M [135]
Ladanyi, O.G: The Eighties - A World in Motion, A Challe   [1002]
Jain, V.   : Application of the   Ekistic   Typology.   [828]
Collins, D.M: Electron   Density Images from Imperfect   [331]
Bathe, K.J. : lgorithms in Finite Element   Analysis: U.S.-German Symposiu [123]
Davies, A.J.:   The Finite   Element   Method: A First Approach.   [380]
Wendt, S.   : d Action Module: an Element   for System Modelling.   [1857]
Meyer, A.R. : of Successor Is Not Elementary   Recursive.   [1156]
Bogdanski, C: Basic Elements   of Cybernetic Physics.   [182]
Cliff, A.D. :   Elements   of Spatial Structure: a Quant   [319]
Randers, J. :   Elements   of the System Dynamic Method. [1379]
Takatsuji, M: stem of Interacting Elements .   [1662]
Leal, A.   : for Conversational Elicitation of Decision Structures.   [1025]
Troncale, L.: to Modeling Systems Emergence .   [1739]
Troncale, L.: a Theory of Systems Emergence .   [1740]
De Greene, K: Force Fields and Emergent   Phenomena in Sociotechnical M [390]
Troncale, L.: archical Levels: An Emergent   Evolutionary Process Based on [1738]
Elstob, C.M.:   Emergentism and Mind.   [479]
Anderton, R.: n the Seventies: an Emerging   Discipline?   [33]
Dedons, A. : ional: Survey of an Emerging   Field.   [401]
Craine, R. : tary Instrument: An Empirical   Assessment.   [354]
Kashyap, R. : ynamic Models Using Empirical   Data.   [882]
Lewicka-Strz: ogical Attitude. An Empirical   Analysis.   [1045]
Nowakowska, : Methodological, and Empirical   Possibilities in Decision Th [1244]
Oldershaw, R: Empirical   and Theoretical Support for [1264]
Rescher, N. :   Empirical   Inquiry.   [1403]
Troncale, L.: Using Computerized, Empirical   Data Bases.   [1745]
Ulanowicz, R: The Empirical   Modelling of an Ecosystem.   [1766]
Vavrousek, J: ach to Modelling in Empirical   Sciences.   [1804]
Brown, D.J.H: cept Learning Model Employing   Generalization an Abstractio [229]
Troncale, L.: ing Natural Systems Enables   Better Design of Man-Made Syst [1749]
Brown, J.H.U: Systems Approach to Endocrinology .   [230]
McIntosh, J.: ng and Computers in Endocrinology .   [1138]
Churchman, C: ms Approach and its Enemies .   [312]
Caianiello, :   Energetics   Versus Communication in the [256]
Mitsch, W.J.:   Energetics   and Systems.   [1174]
Caputo, R.S.: blishing a Rational Energy   Policy for Western Europe.   [261]
Gheorghe, A.:   Energy   Management Planning Tools for t   [605]
Grauer, M. : ration of Efficient Energy   Supply Strategies using Multi-C [658]
Hafele, W. : ling of Large Scale Energy   Systems. (IIASA Proceedings Ser [698]
Halldin, S. : of Forest Water and Energy   Exchange Models.(Proc. of an IU [716]
Hirata, H. : A Model of Mass and Energy   Flow in Ecosystems.   [757]
Kavakoglu, I: atical Modelling of Energy   Systems.   [890]
Knappe, U. : stem for Industrial Energy   Control.   [952]
Kriegel, U. : y Analysis of World Energy   Consumption and World Populatio [982]
Odum, H.T. :   Energy   Basis for Man and Nature.   [1253]
Pardo, L.   :   Information Energy   of a Fuzzy Event and a Partitio [1281]
Schiffman, Y: oach for Estimating Energy   Savings in Communities.   [1508]

Holler, M.J.: lf-Policing Quality Enforcement Systems.   [772]
Atherton, D.:   Nonlinear Control Engineering .   [60]
Branin, F.H.: ysis in Science and Engineering .   [212]
David, F.W. : mental Modelling in Engineering .   [379]
Dhillon, B.S: Engineering Reliability: New Technique   [416]
Ladanyi, O.G: allenge for Systems Engineering .   [1002]
M'Pherson, P: Systems Engineering : An Approach to Whole-Syst   [1076]
M'Pherson, P: amework for Systems Engineering Design.   [1077]
Peschel, M. :   Engineering Cybernetics.   [1324]
Pressman, R.:   Software Engineering : A Practitioner's Approach   [1358]
Retti, J.   : ork for Interactive Engineering of Simulation Models of Ti   [1408]
Ryabinin, I.:   Reliability of Engineering Systems: Principles and An   [1466]
Silvern, L.C: Systems Engineering Applied to Training.   [1556]
Silvern, L.C: Systems Engineering of Education I: the Evolut   [1557]
Singh, M.G. : Large Scale Systems Engineering Applications.   [1566]
Soong, T.T. : ysis in Science and Engineering .   [1600]
Tou, J.T.   :   Knowledge Engineering .   [1706]
Bradley, R. : : A Course Book for Engineers and Scientists.   [210]
Gilmore, R. :   for Scientists and Engineers .   [614]
Hyvarinen, L: Theory for Systems Engineers .   [806]
Favella, L.F: Internal Combustion Engines .   [512]
Gasparski, W: een from Neither an English nor a German Speaking Country.   [582]
Gasparski, W: s Seen from neither English nor German Speaking Country.   [584]
          :   Special English Issue of the Polish Journal "P   [1942]
Krallmann, H: Enlarging the Systems Paradigm: Histor   [980]
Wyner, A.D. : ation for Arbitrary Ensembles .   [1884]
Beer, S.   :   The Heart of Enterprise .   [136]
Dimitrova, L: w in the Industrial Enterprise .   [423]
Honkasalo, A: Entropic Processes and Economic System   [777]
Hoehle, U. :   A Remark on Entropies with Respect to Plausibility   [764]
Arora, P.N. :   Shannon's Entropy and Cyclicity.   [53]
Bailey, K.D.:   Equilibrium, Entropy and Homeostasis: A Multidiscip   [82]
Campbell, J.: l Man: Information, Entropy , Language, and Life.   [259]
Christensen,:   Entropy Minimax Sourcebook. Vol.1: Gen   [311]
Collins, D.M: t Data by Iterative Entropy Maximization.   [331]
Conant, R.C.:   Estimation of Entropy of a Binary Variable: Satisfyi   [336]
De Luca, A. :   the Convergence of Entropy Measures of a Fuzzy Set.   [395]
Dial, G.   :   Entropy of Order Alpha and Type Beta a   [420]
Kubat, L.   :   Entropy and Information in Science and   [990]
Leung-yan-ch: ces between Shannon Entropy and Kolmogorov Complexity.   [1040]
Levine, R.D.:   The Maximum Entropy Formalism.   [1042]
Martin, N.F.: hematical Theory of Entropy .   [1110]
Milgram, M. :   Barycentric Entropy of Mankov Systems.   [1163]
Rifkin, J. :   Entropy : a New World View.   [1419]
Shmulyan, B.:   Interaction of Entropy Systems.   [1548]
Van Campenho: Maximum Entropy and Conditional Probability.   [1786]
Watanabe, S.: a Quest for Minimum Entropy .   [1846]
Yager, R.R. :   Entropy and Specificity in a Mathemati   [1907]
Baldwin, J.F: Systems with Fuzzy Environment .   [85]
Cushing, J.M: ition in a Periodic Environment .   [367]
Gallopin, G.: Abstract Concept of Environment .   [564]
George, F.H.: Cybernetics and the Environment .   [593]
King, L.T. : olving in a Project Environment : A Consulting Process.   [926]
King, L.T. : olving in a Project Environment .   [925]
Nevitt, B. : : Understanding the Environment .   [1216]
Patten, B.C.: h to the Concept of Environment .   [1292]
Pedrycz, W. : n-Making in a Fuzzy Environment .   [1312]
Ricci, F.J. : Man in his Cultural Environment .   [1416]
Tancig, P. : al Purpose Software Environment for NLU-Systems.   [1667]
Beck, M.B. :   Hard or Soft Environmental Systems?   [128]
Bennett, R.J: Environmental Systems: Philosophy, Ana   [152]
Biswas, A.K.: tical Modelling and Environmental Decision-Making.   [171]
Desouza, G.R: r Socioeconomic and Environmental Impact Analysis.   [411]
Gottinger, H: tems Approaches and Environmental Problems.   [645]

Holling, C.S: Adaptive Environmental Assessment and Managemen [774]
Paritsis, N.: ems with Changes in   Environmental Variety.   [1282]
Shaeffer, D.: ology Applicable to   Environmental Assessment Models.   [1530]
Van Steenkis: tion and Control in   Environmental Systems.   [1791]
Watt, K.E.F.: he Unsteady State -   Environmental Problems, Growth, and Cu [1849]
Frauenthal, : atical Modelling in   Epidemology .   [544]
Janko, J.   : wards Practice: and   Episode from the History of Czech Scie [832]
Slawski, C. : erpersonal Conflict   Episode : A Systems Model.   [1580]
Cornelis, A.:   Epistemological Indicators of Scientif [348]
Maruyama, M.: Heterogenistics: an   Epistemological Restructuring of Biolo [1112]
Tamburrini, : zzy Sets Share Some   Epistemological Problems?   [1665]
Burton, D.J.:   Methodology and   Epistemology for Second-Order Cybernet [251]
Csanyi, V. : ific Formulation of   Epistemology .   [361]
Dompere, K.K: On   Epistemology and Decision-Choice Ratio [433]
Keeny, B.P. :   Ecosystemic Epistemology : an Alternative Paradigm [896]
Mattessich, : ems Methodology: an   Epistemology of the Applied and Social [1126]
Pedretti, A.:   Epistemology , Semantics, and Self-Refe [1307]
Sagal, P.T. :   Epistemology of Economics.   [1473]
Cugno, F.   :   Bellamy's   Equalitarian Utopia: A System-Theory A [364]
Hallam, T.G.: ure of Differential   Equation Models of Cosystems.   [715]
Kryukov, V.I: le State. I. Master   Equation Approach.   [988]
Kucera, V. : rol: the Polynomial   Equation Approach.   [992]
DiNola, A. : lpha-Fuzzy Relation   Equations on Finite Spaces.   [418]
Pedrycz, W. :   Fuzzy Relational   Equations with Triangular Norms and th [1310]
Pedrycz, W. : of Fuzzy Relational   Equations in Systems Analysis.   [1311]
Peschel, M. : s Based on Volterra   Equations .   [1322]
Peschel, M. : s Based on Volterra   Equations .   [1323]
Shieh, L.   : ntinuous Time State   Equations by Discrete-Time State Equat [1546]
Shieh, L.   : Discrete-Time State   Equations .   [1546]
Walker, J.A.: stems and Evolution   Equations : Theory and Applications.   [1831]
Windeknecht,: dinary Differential   Equations .   [1866]
Beer, S.   :   Death is   Equifinal : Eighth Annual Ludwig Von Be [135]
Bailey, K.D.:   Equilibrium , Entropy and Homeostasis: [82]
Takatsu, S. :   Organizational Equilibrium under Uncertainty.   [1661]
Kerin, T.   : edical Radioisotope   Equipment .   [906]
Leung-yan-ch: Some   Equivalences between Shannon Entropy a [1040]
Aggarwal, S.: bilistic Systems by   Ergodic Machines.   [17]
            :   Logik,   Erkenntnistheorie , Wissenschaftstheori [1935]
Hwang, C.L. : iods and a Class of   Error Correcting Codes.   [802]
Ibidapo-Obe,: ntial Structure for   Error Covariance in Nonlinear Dynamica [807]
Kaushik, M.L: Single   Error and Burst Error Correcting Codes [889]
Kaushik, M.L: gle Error and Burst   Error Correcting Codes Through a New M [889]
Sharma, B.D.:   Phased   Error Correcting Perfect Codes.   [1540]
Taylor, J.R.: An Introduction to   Error Analysis.   [1673]
Sharma, B.D.:   Block-Wise   Error-Protecting Burst Codes with Weig [1539]
Gloria-Botti: Polymorphism of the   Erythrocyte Acid Phosphate Locus in Hu [625]
Gall, J.   : ow Systems Work and   Especially How They Fail.   [563]
Agassi, J. : c Philosophy Today: Essays in Honour of Mario Bunge.   [14]
Carter, C.F.: tions in Economics: Essays in Honour of G.L.S. Shackle.   [265]
Smale, S.   : athematics of Time: Essays on Dynamical Systems, Economic [1583]
Smith, C.S. : Structure: Selected   Essays on Science, Art and History.   [1588]
Caputo, R.S.:   Establishing a Rational Energy Policy [261]
Horn, W.   : ter-Aided Planning: Estimating Health-Care Resource Requir [782]
Ivan, I.T. : m of Indicators for   Estimating the Software Performances. [817]
Schiffman, Y: ystems Approach for   Estimating Energy Savings in Communiti [1508]
Conant, R.C.:   Estimation of Entropy of a Binary Vari [336]
Hickin, J. : eduction Methods to   Estimation and Control.   [753]
Iserman, R. : nd System Parameter   Estimation .   [815]
Maybeck, P.S: Stochastic Models,   Estimation , and Control.   [1131]
Sawaragi, Y.:   Modeling,   Estimation , and Their Applications for [1499]
Tacker, E.C.: tructures for State   Estimation on Large Scale Systems.   [1653]
Yashin, A. : onditional Gaussian   Estimation Procedures.   [1911]
Kittler, J. : tegration of P.D.F.   Estimators .   [930]
Lefebre, V.A: f Wester and Soviet   Ethical Systems.   [1030]

Eveland, S. : System for a Gothic Ethymological Dictionary.   [496]
Joerres, R. : Circadian System of Euglena Gracilis.   [847]
Caputo, R.S.: Policy for Western   Europe .   [261]
Elton, M.C.J: Evaluating   New Telecommunication Servi [480]
Pearl, J.   : Certain Methods of   Evaluating   Probabilistic Forecasts.   [1304]
Ratko, I.   :   On   Evaluating   of Logical Expressions in P [1389]
Viliums, E.R: cts of Alternatives   Evaluating   and Decision Making Under U [1817]
Ballance, R.: plication to System   Evaluation .   [90]
Broekstra, G: Representation and   Evaluation   of Reconstruction Hypothese [226]
Cavallo, R.E: ctability Analysis: Evaluation   of Reconstruction Hypothesi [283]
Cosier, R.A.:   An   Evaluation   of the Effectiveness of Dia [349]
Gopal, K.   : thm for Reliability Evaluation   in Complex Systems.   [643]
Hwang, C.L. : System-Reliability Evaluation   Techniques for Complex/Larg [804]
Kobayashi, H: System Performance   Evaluation   Methodology.   [953]
Koyayashi, H: System Performance   Evaluation   Methodology.   [977]
Perlis, A.J.: cs: An Analysis and   Evaluation .   [1317]
Perlis, J.H.: An Approach to the   Evaluation   of System Design.   [1318]
Pugh, R.E.   :   Evaluation   of Policy Simulation Models [1368]
Sandblom, C.: tion Results on the   Evaluation   of Economic Policies.   [1488]
Shaeffer, D.:   A Model   Evaluation   Methodology Applicable to E [1530]
Silvert, W. : The Formulation and Evaluation   of Predictions.   [1558]
Swets, J.A. :   Evaluation   of Diagnostic Systems: Meth [1646]
Tong, R.M.   :   The   Evaluation   of Fuzzy Models Derived fro [1703]
Seton, F.   :   The   Evalution   of Economic Flows and System [1529]
Jacak, W.   :   Modelling of Evenistic   Systems.   [821]
Tchon, K.   : versal' Approach to Evenistic   Systems.   [1677]
Gopal, K.   :   An   Event   Expansion Algorithm for Reliabil [643]
Hogeweg, P. : antages of Discrete Event   Formalism.   [769]
Pardo, L.   : n Energy of a Fuzzy Event   and a Partition of Fuzzy Events. [1281]
Smets, P.   : f Belief in a Fuzzy Event .   [1586]
Jack, W.   :   On the Eventistic Approach to the Notion of S [823]
Jacak, W.   :   Events   and Systems.   [819]
Lamport, L. : and the Ordering of Events   in Distributed Systems.   [1006]
Okuda, T.H. : y Measures of Fuzzy Events .   [1261]
Pardo, L.   : Partition of Fuzzy Events .   [1281]
Glanville, R: Inside   Every   White Box there are Two Black Bo [620]
Gage, W.L.   :   Evidence   for Strategic Decisions.   [554]
Oldershaw, R: New   Evidence   for the Principle of Self-Sim [1265]
Yager, R.R. : hematical Theory of Evidence .   [1907]
Jarisch, W. : Analysis of Visual Evoked   Potentials.   [838]
Boulding, K.: Theory of Societal   Evolution .   [200]
Csanyi, V.   :   General Theory of Evolution .   [362]
Jantch, E.   : l and Sociocultural Evolution   (AAAS Selected Symposium). [835]
Loefgren, L.:   Knowledge of   Evolution   and Evolution of Knowledge. [1061]
Loefgren, L.: ge of Evolution and   Evolution   of Knowledge.   [1061]
Matsuno, K. : iological Molecular Evolution .   [1122]
Overton, W.S: Evolution   of a Hydrology Model - an Ex [1275]
Peschel, M. : ion to Processes of Evolution .   [1320]
Sahal, D.   :   A Theory of Evolution   of Technology.   [1479]
Silvern, L.C: of Education I: the   Evolution   of Systems Thinking in Educa [1557]
Smith, J.M. :   Evolution   and the Theory of Games.   [1589]
Vallee, R.   :   Evolution   of Dynamical Linear System w [1782]
Walker, J.A.: namical Systems and   Evolution   Equations: Theory and Applic [1831]
Byerly, H.   :   Teleology and Evolutionary Theory Mechanisms and Mea [254]
Jantch, E.   :   The   Evolutionary Vision: Toward a Unifying [835]
Kriegel, U. :   An   Evolutionary Analysis of World Energy   [982]
Krzanik, L. :   Evolutionary Projects: Dynamic Optimiz [989]
Malik,, F.   :   Evolutionary Management.   [1097]
Takeuchi, A.:   Evolutionary Automata - Comparison of   [1663]
Troncale, L.: Levels: An Emergent   Evolutionary Process Based on Systems   [1738]
Acalugaritei: Hierarchies of Evolutions and Stabilities.   [4]
Fairtlough, : tems Thinking Might   Evolve .   [501]
Ben-Eli, M.U: Variety Increase in   Evolving   Systems.   [149]
Lasota, A.   : xtinction of Slowly Evolving   Dynamical Systems.   [1012]

Thompson, J.:   The Instability of   Evolving Systems.   [1686]
Glanville, R: Distinguished and   Exact Lies.   [621]
Schellhorn, : n Multiple Control: Exact   and Approximate.   [1506]
Batchelder, :   A Critical   Examination of the Analysis of Dichoto [115]
Fedra, K.   : g: A Lake Modelling   Example .   [515]
Havranek, T.: the Guha Method: an   Example .   [737]
Kickert, W.J: An   Example of Linguistic Modelling.   [910]
Kinston, W. : livery Systems. The   Example of Health Care.   [928]
Tompa, F.W. :   A Practical   Example of the Specification of Abstra [1698]
Uyttenhove, : r and a Stockmarket   Example .   [1778]
Kwakernaak, : t 2: Algorithms and   Examples for the Discrete Case.   [996]
Walker, C.C.: ions: Management by   Exception , Priority, and Input Span in [1830]
Rauch, W.D. : atic Refereeing and   Excerpting .   [1390]
Loefgren, L.:   Excerpts from the Autology of Learning [1062]
Halldin, S. : st Water and Energy   Exchange Models.(Proc. of an IUFRO Wor [716]
Overton, W.S: ydrology Model - an   Exercise in Modelling Strategy   [1275]
Wedde, H.F. :   An   Exercise in Flexible Formal Modeling u [1852]
Favella, L.F: n of Loeve-Karhunen   Expansion to Acoustic Diagnostic Probl [512]
Gopal, K.   :   An Event   Expansion Algorithm for Reliability Ev [643]
Okeda, M.   : ing Decompositions,   Expansions , and Contractions of Dynami [1260]
Carter, C.F.:   Uncertainty and   Expectations in Economics: Essays in H [265]
Rudawitz, L.: esign: Concepts and   Experience .   [1459]
Ulrich, W. : Reason: The Chilean   Experience with Cybernetics.   [1768]
Batchelor, B: Experimental and Programmatic Approach [116]
David, F.W. :   Experimental Modelling in Engineering. [379]
Klir, G.J. : dentification: Some   Experimental Observations and Resultin [938]
Tong, R.M. : Models Derived from   Experimental Data.   [1703]
Yager, R.R. : o Context and Their   Experimental Realization.   [1904]
Gerardy, R. :   Experiments with Some Methods for the [599]
Jumarie, G. : formation of Random   Experiments Involving Fuzzy Observatio [863]
Bandler, W. :   Inference in Fuzzy   Expert Systems.   [102]
Bottaci, L. : em Programs Used in   Expert Systems.   [198]
Gyarfas, F. : er-Based Diagnostic   Expert System.   [695]
Johnson, P.E: imulation Models of   Expert Reasoning.   [851]
Keravnou, E.:   Design of   Expert Systems from the Perspective of [903]
Kohout, L.J.: Construction of an   Expert Therapy Adviser as a Special Ca [966]
Lee, R.M.   :   Expert vs. Management Support Systems: [1028]
Michie, D. :   Expert Systems in the Microelectronic [1160]
Gams, M.   : alysis Program that   Explains its Reasoning.   [565]
Achinstein, : There Be a Model of   Explanation ?   [6]
Ganascia, J.:   Explanation Facilities for Diagnosis S [566]
Garfinkel, A: Forms and   Explanation .   [572]
Peacocke, C. :   Holistic   Explanation .   [1298]
Sinha, K.K. : tor Control without   Explicit Identification.   [1569]
Laing, R.   : es as Organisms: an   Exploration of the Relevance of Recent [1004]
Smith, P.M. :   An   Exploration of Shared Knowledge about [1591]
O'Muircheart: urvey Data, Vol. 1:   Exploring Data Structures. Vol. 2: Mod [1251]
Sussman, G.J: TS - A Language for   Expressing Almost Hierarchical Descrip [1640]
Freuder, E.C: hesizing Constraint   Expressions .   [546]
Ratko, I.   : aluating of Logical   Expressions in Programming Languages. [1389]
Majster, M.E: Extended Directed Graphs, a Formalism [1092]
Seo, F.   :   Fuzzy   Extension of Multiattribute Utility An [1527]
Lusk, E.J. :   A Discussion of   External Validity in Systemic Research [1075]
Lasota, A. :   The   Extinction of Slowly Evolving Dynamica [1012]
Rauch, W.D. :   Automatic   Extracting as an Interactive Process.   [1391]
Eisenstein, :   Feature   Extraction by System Identification.   [466]
Fiacco, A.V.:   Extremal Methods and Systems Analysis. [520]
Koenderink, : g and Stretching by   Eye : Implications for Robot Vision.   [962]
Ganascia, J.:   Explanation   Facilities for Diagnosis Systems.   [566]
Della Riccia: he Approximation of   Factor Scores in a Factor Analysis Alg [405]
Della Riccia: Factor Scores in a   Factor Analysis Algorithm.   [405]
Gengoux, K.G: articipation: A New   Factor in Social Systems.   [591]
Santis, F.de: On   Factor Analysis and Fisher's Linear Di [1492]
Schilderinck: Regression and   Factor Analysis in Econometrics.   [1509]

Hoffer, J.A.:   Factors   Affecting the Comparative Adva   [766]
Jumarie, G. : ems Involving Human Factors .   [862]
Millendorfer: Growth Reducing Factors   in Complex Systems.   [1164]
Ruffner, M.A: Human Factors   in Systems Analysis.   [1461]
Gardner, M.R: Predicting Novel Facts .   [570]
Conant, R.C.:   Faculty   Observations of Variables and   [337]
Gall, J.   : Especially How They Fail .   [563]
Checkland, P: s Movement and the " Failure " of Management Science.   [299]
Pau, L.F.   :   Failure   Diagnosis and Performance Moni   [1296]
Rosen, R.   : and Global Systems Failure .   [1440]
Sworder, D.D: Influence of Sensor Failures   on Log Regulators.   [1647]
Jones, B.   : n of Reconstruction Families .   [854]
Gurdial,   :   A Note on Fano   Inequality Involving Distortion.   [690]
Wang, Z.   : Husbandry System in Farming   Region.   [1837]
Brigham, E.O: The Fast   Fourier Transform.   [218]
Horn, W.   :   Fast   Walsh vs. Fast Fourier Transform: [784]
Horn, W.   :   Fast Walsh vs. Fast   Fourier Transform: A Comparison o   [784]
Kittler, J. :   A Comment on Fast   Monte Carlo Integration of P.D.F.   [930]
Nussbaumer, :   Fast   Fourier Transform and Convolution   [1249]
Retti, J.   :   SISSY - A Fast   Interactive Simulation System.   [1407]
Agrawal, V.D: Approach to Digital Fault   Testing.   [18]
Smith, P.M. :   Procedures Used in Fault   Diagnosis Tasks.   [1591]
Trappl, R. : : Is it Acceptable, Feasible , Advisable?.   [1715]
Eisenstein, :   Feature   Extraction by System Identific   [466]
Jones, A.H. :   Cybernetic Design Features   for Computerised Business Inf   [853]
Erdi, P.   : eural Organization: Feed-Forward   Control of the Ontogeneti   [484]
Bogart, D.H.:   Feedback , Feedforward, and Feedwithin: [181]
Camana, P.C. :   Determination of Feedback   for Human Posture Control wit   [258]
Cumani, A. : e Analysis of Fuzzy Feedback   Systems.   [366]
Engell, S. : ty, Information and Feedback .   [481]
Grollmann, J: Description of Feedback   Systems via Approximation by   [673]
Gupta, M.M. :   Feedback   Control Applications of Fuzzy   [689]
Hanson, O.J.: alysis Course Using Feedback   from Past Students.   [723]
Matsuno, K. :   Connectedness, and Feedback   Loops in Open Systems.   [1123]
Mees, A.L. :   Dynamics of Feedback   Systems.   [1146]
Nigro, A.   : Language Behavioral Feedback .   [1228]
Owens, D.H. :   Feedback   and Multivariate Systems.   [1277]
Tong, R.M. : Properties of Fuzzy Feedback   Systems.   [1702]
Waldhauer, F: Feedback .   [1827]
Bogart, D.H.:   Feedback, Feedforward , and Feedwithin: Strategic   [181]
Rosen, R.   :   Feedforwards   and Global Systems Failur   [1440]
Bogart, D.H.: k, Feedforward, and Feedwithin : Strategic Information in S   [181]
Jarisch, W. : chastic Modeling of Fetal   Heart Rate Variability with Appl   [837]
Jarisch, W. : nt and Detection of Fetal   Respiratory Movements-Preliminar   [837]
Hu, N.C.   :   Multi-Dimensional FFT   Computation.   [794]
Dedons, A. : rvey of an Emerging Field .   [401]
Savage, G.J.: The Graph-Theoretic Field   Model-I - Modelling and Formulat   [1498]
Trost, H.   :   Application in the Field   of Natural Language Understandin   [1754]
De Greene, K: Force Fields   and Emergent Phenomena in Socio   [390]
Troncale, L.: ropositions between Fifty   Principal Systems Concepts.   [1737]
Batchelor, B: vex Hulls of Planar Figures .   [121]
Balossino, N: ECG Map Filtering   by Means of Spherical Harmon   [93]
Bozic, S.M. :   Digital and Kalman Filtering .   [206]
Tesfatsion, :   A New Approach to Filtering   and Adaptive Control: Optima   [1683]
Egiazarian, : ansforms and Linear Filters .   [459]
Batchelor, B: Two Methods for Finding   Convex Hulls of Planar Figures   [121]
Anderson, B.: of Systems Theory: Finitary   and Infinitary Conditions.   [29]
Bathe, K.J. : ional Algorithms in Finite   Element Analysis: U.S.-German S   [123]
Davies, A.J.:   The Finite   Element Method: A First Approac   [380]
DiNola, A. : lation Equations on Finite   Spaces.   [418]
Gerardy, R. :   Probabilistic Finite   State System Identification.   [598]
Hwang, C.L. : Binary Sequences of Finite   Periods and a Class of Error Co   [802]
Kieras, D.E.:   Finite   Automata and S-R Models.   [915]
Kintala, C.M: f Nondeterminism in Finite   Automata.   [929]

Rota, G.C.	:	Finite Operator Calculus.	[1450]
Gerardy, R.	: e Identification of	Finite-State Systems.	[599]
Beer, S.	: Brain of the	Firm (second edition).	[134]
Cerbone, G.	: and Skewness of the	First Passage Time for the Ornstein-Uh	[290]
Davies, A.J.:	e Element Method: A	First Approach.	[380]
Favella, L.F:		First Passage Time Problems and Some R	[511]
Syski, R.	: Random Processes: a	First Look.	[1649]
	: Proceedings of the	First International Symposium on Polic	[1934]
Kozen, D.	: Positive	First-Order Logic Is NP-Complete.	[978]
Santis, F.de:	Factor Analysis and	Fisher 's Linear Discriminant Analysis.	[1492]
Haley, K.B.	: erations Research in	Fishing . (Vol. 10 in NATO Conference S	[710]
O'Muircheart:	ures. Vol. 2: Model	Fitting .	[1251]
Bahm, A.J.	: Methodologies of	Five Types of Philosophy.	[80]
Bahm, A.J.	:	Five Types of Systems Philosophy.	[79]
Walker, C.C.:	Span in a Class of	Fixed-Structure Models.	[1830]
Cull, P.	:	Flaws of Form.	[365]
Overton, W.S:	chical Model in the	Flex Paradigm Illustrated by Structura	[1276]
Cook, N.D.	: Stability and	Flexibility : an Analysis of Natural Sy	[344]
Gage, W.L.	:	Flexibility in Network Planning.	[553]
Wedde, H.F.	: An Exercise in	Flexible Formal Modeling under Realist	[1852]
Comstock, F.:	roach to Grading of	Flight Simulator Students.	[335]
Zorbas, Y.G.:	ility during Cosmic	Flights .	[1929]
Barber, M.C.:	Model for Ecosystem	Flow Analysis.	[105]
Dimitrova, L:	ization of Material	Flow in the Industrial Enterprise.	[423]
Hirata, H.	: of Mass and Energy	Flow in Ecosystems.	[757]
Kindler, E.	: Systems of Material	Flow .	[923]
Trattnig, W.:	e Based on the Data	Flow Model.	[1734]
Seton, F.	: alution of Economic	Flows and Systemic Efficiency.	[1529]
Bergeron, R.:	Imaginary,	Fluid , and Dancing Systems.	[161]
Steinacker, :		Focussing as a Method for Disambiguati	[1617]
Rosen, R.	: Protein	Folding : a Prototype for Control of Co	[1444]
Schellhorn, :	Model	Following in Multiple Control: Exact a	[1506]
De Greene, K:		Force Fields and Emergent Phenomena in	[390]
Honkasalo, A:	Education of Labor	Force .	[776]
Tripp, R.S.	: stration within Air	Force Logistics Command.	[1736]
Van Straaten:	oint MTA/IIASA Task	Force Meeting on Lake Balaton Modellin	[1792]
Hirzinger, G:	Robot-Teaching via	Force-Torque-Sensors .	[760]
Burns, D.W.	: The Synthesis of	Forecasting Models in Decision Analysi	[247]
Makridakis, :		Forecasting : Methods and Applications.	[1095]
Trappl, R.	: Simple	Forecasting Procedures in Health Care.	[1714]
Pearl, J.	: ating Probabilistic	Forecasts .	[1304]
Halldin, S.	: Comparison of	Forest Water and Energy Exchange Model	[716]
Reichle, D.E:	namic Properties of	Forest Ecosystems.	[1398]
Benoit, B.M.:	Fractals:	Form , Chance, and Dimension.	[156]
Cull, P.	: Flaws of	Form .	[365]
El-Sherief, :	fer-Function Matrix	Form .	[469]
Glanville, R:	del's Dimensions: A	Form for Argument.	[619]
Schwartz, D.:	cer-Brown's Laws of	Form and Varela's Calculus for Self-Re	[1515]
Turney, P.	: On Laws of	Form .	[1762]
Agusti, J.	: esized with Z-Fivex	Formal Neurones.	[19]
Arigoni, A.C:	On the Synthesis of	Formal Properties.	[50]
Bunn, D.W.	:	Formal Methods in Policy Analysis.	[242]
Elstob, C.M.:	The	Formal and the Physical.	[478]
Harrison, M.:	Introduction to	Formal Language Theory.	[728]
Kania, A.	: On Stability of	Formal Fuzziness Systems.	[876]
Kickert, W.J:	A	Formal Approach to Organization Design	[913]
Labos, E.	: ifferent Classes of	Formal Neuronal Networks and Chaotic S	[1000]
Lakshmivarah:	ions for Fuzzifying	Formal Languages and Synthesis of Fuzz	[1005]
Nowakowska, :		Formal Semiotics and Multidimensional	[1243]
Sklenar, J.	: Apparatus for	Formal Description of Discrete Systems	[1571]
Tchon, K.	: Towards a	Formal Theory of Description.	[1678]
Trappl, R.	: Non-Computerized	Formal Methods with Computer Efficienc	[1711]
Wedde, H.F.	: xercise in Flexible	Formal Modeling under Realistic Assump	[1852]
Hogeweg, P.	: s of Discrete Event	Formalism .	[769]

Levine, R.D.: The Maximum Entropy Formalism .   [1042]
Majster, M.E: Directed Graphs, a Formalism for Structured Data and Data [1092]
Vallet, C. : ially Self-Referent Formalism for Description of Natural S [1784]
Kindler, E. :   A Formalization of Some Simulation Langu [922]
Resconi, G. :   A New Formalization of the General Systems b [1406]
Troncale, L.:   Towards a Formalization of Systems Linkage Propo [1746]
Shakun, M.F.:   Formalizing Conflict Resolution in Pol [1532]
Skvoretz, J.: mation Diffusion in Formally Structured Populations: An In [1572]
Hajek, P.   : chanized Hypothesis Formation .   [704]
Haken, H.   :   Pattern Formation by Dynamic Systems and Patte [707]
Matsuno, K. : iption Microsystems Formation in Probiological Molecular E [1122]
            : chanized Hypothesis Formation .   [1941]
Hajek, P.   : hanizing Hypothesis Formationmathematical Foundations of a [703]
Fontela, E. :   Condensed Forms of Large Scale Models.   [533]
Garfinkel, A: Forms and Explanation.   [572]
Kalman, R.E.: ions, and Canonical Forms .   [872]
Lee, E.B.   :   Canonical Forms for Time Delay Systems.   [1026]
Burns, J.R. : tor Approach to the Formulation of Systems Dynamics.   [248]
Csanyi, V. : for the Scientific Formulation of Epistemology.   [361]
Greenspan, D: mpletely Arithmetic Formulation of Classical and Special R [663]
Okuda, T.H. :   A Formulation of Fuzzy Decision Problems [1261]
Silvert, W. :   The Formulation and Evaluation of Predicti [1558]
Bathe, K.J. :   Formulations and Computational Algorit [123]
Savage, G.J.: l-I - Modelling and Formulations .   [1498]
Burns, J.R. : Signed Digraphs to Forrester Schematics.   [249]
Gyarfas, F. : Driven Backward and Forward Chaining Computer-Based Diagno [695]
Cavallo, R.E: A Conceptual Foundation for Systems Problem Solving [276]
Cotterman, W: lysis and Design: A Foundation for the 1980's.   [350]
Yager, R.R. :   A Foundation for a Theory of Possibility [1891]
Anderson, B.:   Foundations of Systems Theory: Finitar [29]
Arbib, M.A. :   Foundations of System Theory: The Hank [48]
Aulin, A.   :   On the Foundations of Systems Theory.   [72]
Bonczek, R.H: Foundations of Decision Support System [189]
Broekstra, G: On the Foundations of GIT (General Informatio [225]
Christensen,:   Foundations of Inductive Reasoning.   [310]
Erickson, S.: Systems View of the Foundations of Education.   [485]
Favella, L.F: Mathematical Foundations of ECG Map Reconstruction. [510]
Gaines, B.R.:   Logical Foundations for Database Systems.   [559]
George, F.H.:   The Foundations of Cybernetics.   [594]
George, F.H.:   Philosophical Foundations of Cybernetics.   [595]
Gruska, J. :   Mathematical Foundations of Computer Science.   [678]
Hajek, P.   : rmationmathematical Foundations of a General Theory.   [703]
Oldershaw, R: Conceptual Foundations of the Self-Similar Hierar [1262]
Robinson, E.:   Digital Foundations of Time Series Analysis. V [1433]
Scandura, J.:   Theoretical Foundations of Instruction: A Systems [1501]
Takeuchi, A.:   The Foundations of Multivariate Analysis: [1664]
Tighe, T.J. : rn Learning Theory: Foundations and Fundamental Issues.   [1689]
Van der Zand: vestigations on the Foundations .   [1795]
Bandler, W. :   The Four Modes of Inference in Fuzzy Exper [102]
Brigham, E.O: The Fast Fourier Transform.   [218]
Horn, W.   : Fast Walsh vs. Fast Fourier Transform: A Comparison of Tim [784]
Nussbaumer, :   Fast Fourier Transform and Convolution Algo [1249]
Boulding, K.: s a General System: Fourth Annual Ludwig von Bertalanffy M [199]
Benoit, B.M.:   Fractals : Form, Chance, and Dimension. [156]
Sherlock, R.: 2: the State Cycle Fraction for Networks of Different Con [1545]
Horn, W.   :   A Frame-Based Real-Time Graphic Interact [787]
Retti, J.   :   Frame-Oriented Knowledge Representatio [1409]
Jensen, H.L.:   A Theoretical Framework for Quantifying Legal Decisi [844]
Johnson, L. :   A Theoretical Framework for Human Communication.   [849]
M'Pherson, P: A Framework for Systems Engineering Desi [1077]
Mattessich, : tion of the Systems Framework : Similarities and Difference [1127]
Pearl, J.   :   A Framework for Processing Value Judgeme [1302]
Retti, J.   :   A Framework for Interactive Engineering [1408]
Smith, R.G. :   Frameworks for Cooperation in Distribu [1593]

LeMoigne, J.: eling Complexity: A   French   History and its Prospects.   [1024]
Moss, Brian :   Ecology of Fresh   Waters.   [1191]
Klir, G.J.   :   Frontiers in Systems Research: Implica   [941]
Shore, J.E. : Minimization Given   Fully-Decomposable Subset and Aggregat [1550]
Bezdek, J.C.: ith Fuzzy Objective   Function Algorithms.   [167]
Trappl, R.   : n to Stevens' Power   Function .   [1710]
Ballance, R.:   Functional   Dependence and its Applicat   [90]
Kerner, H.   :   Functional   versus Imperative Programmi   [908]
Leigh, J.R. :   Functional   Analysis and Linear Control [1033]
Prevost, P. : ystems Methodology,   Functionalism and the Social Sciences. [1359]
Morein, R.   : s of Variations and   Functionals in the Analysis of Thresho [1185]
Buchberger, : ings and Transition   Functions in Effective Systems.   [237]
Davio, M.   : crete and Switching   Functions .   [381]
Hoskins, R.F: Generalised Functions .   [790]
Ingemarsson,: a Theory of Unknown   Functions .   [810]
Kalman, R.E.: lizations, Transfer   Functions , and Canonical Forms.   [872]
Medgyessy, P: ositions of Density   Functions and Discrete Distributions.   [1144]
Pippenger, N: mplexity of Boolean   Functions .   [1343]
Rao, G.P.   : truction of Mapping   Functions from Linguistic Descriptions [1380]
Vitanyi, P.M: nguages, and Growth   Functions .   [1819]
Takahara, Y.: A Stationarization   Functor and its Concrete Realizations.   [1659]
Aulin-Ahmava: teering Machines: A   Fundamental   Theorem of Actor-Systems.   [74]
Cramer, F.   :   Fundamental   Complexity: a Concept in B   [355]
Flavin, M.   :   Fundamental   Concepts of Information Mo   [528]
Keyes, R.W. :   Fundamental   Limits in Digital Informat   [909]
Tighe, T.J. : ry: Foundations and   Fundamental   Issues.   [1689]
Wulf, W.A.   :   Fundamental   Structures of Computer Sci [1883]
Wyner, A.D. :   Fundamental   Limits in Information Theo [1885]
Gross, D.   :   Fundamentals of Queueing Theory.   [674]
Horowitz, E.:   Fundamentals of Data Structures.   [788]
Hostetter, G: Fundamentals of Network Analysis.   [792]
Rosen, R.   :   Fundamentals of Measurement and Repres [1441]
Ackoff, R.L.:   The Future of Operations Research is Past.   [9]
Ackoff, R.L.:   Resurrecting the   Future of Operational Research.   [8]
Linstone, H.: Cybernetics and the   Future .   [1053]
Pszczolowski: heory with Past and   Future .   [1367]
Steinacker, :   Future , Impacts, and Future Impacts of [1616]
Steinacker, : uture, Impacts, and   Future Impacts of Artificial Intellige [1616]
Sugiyama, K.: ysis (I): A Case of   Future Computer Systems Development in   [1635]
Toda, M.   : rategy Analysis for   Future Computer Systems Development in   [1694]
Troncale, L.:   Future of General Systems Science: Obs   [1748]
Boroush, M.A: ssessment: Creative   Futures .   [191]
Reckmeyer, W: ign: Precursors and   Futures .   [1394]
Wasniowski, : nication Systems in   Futures Research.   [1844]
Lakshmivarah: Considerations for   Fuzzifying Formal Languages and Synthe [1005]
DiNola, A.   :   On the Fuzziness of Solutions of Alpha-Fuzzy   [418]
Higashi, M. :   On Measures of   Fuzziness and Fuzzy Completeness.   [755]
Kania, A.   : Stability of Formal   Fuzziness Systems.   [876]
Loo, S.G.   :   Measures of   Fuzziness .   [1063]
Wechler, W. :   The Concept of   Fuzziness in Automata and Language The [1851]
Yager, R.R. :   On the Measure of   Fuzziness and Negation. Part 1: Member [1890]
Baldwin, J.F: g Controllers Using   Fuzzy Relations.   [84]
Baldwin, J.F: mic Programming for   Fuzzy Systems with Fuzzy Environment.   [85]
Baldwin, J.F: Fuzzy Systems with   Fuzzy Environment.   [85]
Bandler, W. : ation Operators and   Fuzzy Relational Products.   [100]
Bandler, W. :   Fuzzy Power Sets and Fuzzy Simplificat   [99]
Bandler, W. : uzzy Power Sets and   Fuzzy Simplification Operators.   [99]
Bandler, W. : des of Inference in   Fuzzy Expert Systems.   [102]
Baptistella,:   Fuzzy Methodologies for Interactive Mu   [104]
Bartolini, G: erformance Adaptive   Fuzzy Controllers with Application to   [113]
Bernard, N. : on to the Search of   Fuzzy Sets Theory.   [162]
Bezdek, J.C.:   Fuzzy Partitions and Relations: an Axi   [166]
Bezdek, J.C.: ns Recognition with   Fuzzy Objective Function Algorithms.   [167]
Braae, M.   : of Parameters for a   Fuzzy Logic Controller.   [209]

Carlsson, C.: pproach to Handling Fuzzy Problem Structures.   [264]
Cumani, A.   : to the Analysis of Fuzzy Feedback Systems.   [366]
Czogala, E. : n Identification of Fuzzy Systems and Its Application in C   [368]
Czogala, E. :   Fuzzy Rules Generation for Fuzzy Contr [370]
Czogala, E. : ules Generation for Fuzzy Control.   [370]
Czogala, E. : Control Problems in Fuzzy Systems.   [369]
De Luca, A. : tropy Measures of a Fuzzy Set.   [395]
Dubois, D. :   Fuzzy Sets and Systems: Theory and App [443]
Dubois, D. :   A Class of Fuzzy Measures Based on Triangular Nor   [444]
Edmons, E.A.:   Lattice Fuzzy Logics.   [455]
Efstathiou, : sion Making Using a Fuzzy Heuristic Approach.   [457]
Freeling, A.:   Fuzzy Sets and Decision Analysis.   [545]
Gaines, B.R.:   The Fuzzy Decade: a Bibliography of Fuzzy   [555]
Gaines, B.R.: : a Bibliography of Fuzzy Systems and Closely Related Topi   [555]
Gaines, B.R.:   Fuzzy and Probability Uncertainty Logi [556]
Gaines, B.R.:   Sequential Fuzzy System Identification.   [560]
Gericke, M. : lled Activities - A Fuzzy Simulation Approach.   [600]
Gupta, M.M. :   Advances in Fuzzy Set Theory and Applications.   [684]
Gupta, M.M. :   Fuzzy Automata and Decision Processes. [685]
Gupta, M.M. :   Advances in Fuzzy Set Theory and Applications.   [686]
Gupta, M.M. : rol Applications of Fuzzy Set Theory: A Survey.   [689]
Haack, S.   :   Do We Need ' Fuzzy Logic'?   [697]
Higashi, M. : es of Fuzziness and Fuzzy Completeness.   [755]
Hisdal, E. :   Generalized Fuzzy Set Systems and Particularizatio [762]
Jantzen, J. :   A Generalized Fuzzy Relational Matrix.   [836]
Jumarie, G. : ivistic Information Fuzzy Sets, Linguistics, Relativistic   [861]
Jumarie, G. : periments Involving Fuzzy Observations.   [863]
Kandel, A. :   Fuzzy Switching and Automata: Theory a [874]
Kandel, A. :   Fuzzy Processes.   [875]
Kandel, A. :   Fuzzy Sets, Fuzzy Algebra, and Fuzzy S [873]
Kandel, A. :   Fuzzy Sets, Fuzzy Algebra, and Fuzzy Statistics. [873]
Kandel, A. :   Fuzzy Algebra, and Fuzzy Statistics.   [873]
Kickert, W.J: Fuzzy Theories on Decision-Making. (Vo [911]
Klement, E.P: national Seminar on Fuzzy Set Theory.   [933]
Klement, E.P: the Cardinality of Fuzzy Sets.   [934]
Klement, E.P: l Limit Theorem for Fuzzy Random Variables.   [935]
Kloeden, P.E: Fuzzy Dynamical Systems.   [951]
Koczy, L.T. : Sigma-Algebras and Fuzzy Objects of Type N.   [959]
Kwakernaak, :   Fuzzy Random Variables. Part 1: Defini [996]
Lakshmivarah: es and Synthesis of Fuzzy Grammars.   [1005]
Lee, E.T.   :   Applications of Fuzzy Set Theory to Image Sciences.   [1027]
Lesmo, L.   :   An Interpreter of Fuzzy Production Rules.   [1037]
Leung, Y.   :   A Fuzzy Set Analysis of Sociometric Stru [1039]
Loo, S.G.   :   Fuzzy Relations in the Social and Beha [1064]
Mamdani, E.H: Application of Fuzzy Logic to Approximate Reasoning U [1098]
Mamdani, E.M: Fuzzy Reasoning and Its Applications.   [1099]
Mizumoto, M.: c Rule by Zadeh for Fuzzy Conditional Inference.   [1176]
Nahmias, S. :   Fuzzy Variables.   [1201]
Negoita, C. :   Fuzzy Sets and Systems: a New Internat [1208]
Negoita, C. :   Fuzzy Systems.   [1210]
Negoita, C.V: Fuzzy Systems.   [1212]
Neuwirth, E.:   Fuzzy Similiarity Relations and Distan [1215]
Nowakowska, : s of Measurement of Fuzzy Concepts in the Social Sciences. [1239]
Okuda, T.H. :   A Formulation of Fuzzy Decision Problems with Fuzzy Inf [1261]
Okuda, T.H. : ision Problems with Fuzzy Information Using Probability Me   [1261]
Okuda, T.H. : ability Measures of Fuzzy Events.   [1261]
Orlovsky, S.: ision-Making with a Fuzzy Preference Relation.   [1271]
Orlovsky, S.: atives for Multiple Fuzzy Preference Relations.   [1272]
Pappis, C.P.:   A Fuzzy Logic Controller for a Traffic J [1280]
Pardo, L.   : rmation Energy of a Fuzzy Event and a Partition of Fuzzy E [1281]
Pardo, L.   :   and a Partition of Fuzzy Events.   [1281]
Pedrycz, W. :   Fuzzy Relational Equations with Triang [1310]
Pedrycz, W. : ecision-Making in a Fuzzy Environment.   [1312]
Pedrycz, W. : cational Aspects of Fuzzy Relational Equations in Systems   [1311]

Pedrycz, W. :   Construction of Fuzzy Relational Models.   [1313]
Pfeilsticker: ystems Approach and Fuzzy Set Theory Bridging the Gap betw [1331]
Pipino, L.L.: Potential Impact of Fuzzy Sets on the Social Sciences.   [1342]
Poch, F.A. : ome Applications of Fuzzy Sets to Statistics .   [1347]
Radecki, T. :   Level Fuzzy Sets.   [1373]
Ragade, R.K.:   Fuzzy Interpretative Modeling.   [1374]
Ralescu, D. : a General Theory of Fuzzy Variables.   [1378]
Rao, G.P.   : ions in Problems of Fuzzy Logic Applied to Systems Control [1380]
Seo, F.   :   Fuzzy Extension of Multiattribute Util [1527]
Skala, H.J. : Many-Valued Logics, Fuzzy Sets, Fuzzy Logics and Their App [1570]
Skala, H.J. : Logics, Fuzzy Sets, Fuzzy Logics and Their Applications.   [1570]
Slyadz, N.N.:   Analysis of Fuzzy Initial Information in Decision- [1582]
Smets, P.   : gree of Belief in a Fuzzy Event.   [1586]
Smithson, M.:   Models for Fuzzy Nominal Data.   [1594]
Sugeno, M. : f Learning Based on Fuzzy Information.   [1633]
Sultan, L.H.: ms Organization for Fuzzy Information Processing.   [1636]
Tamburrini, : Systems Science and Fuzzy Sets Share Some Epistemological [1665]
Tazaki, E. : ss of Systems Using Fuzzy Sets Theory.   [1675]
Tazaki, E. : ietal Systems Using Fuzzy Set Theory.   [1676]
Tong, R.M. :   Synthesis of Fuzzy Models for Industrial Processes [1700]
Tong, R.M. : Some Properties of Fuzzy Feedback Systems.   [1702]
Tong, R.M. : The Evaluation of Fuzzy Models Derived from Experimental [1703]
Tong, R.M. : ecision Making with Fuzzy Sets.   [1701]
Turksen, I.B: Bounds for Fuzzy Inference.   [1761]
Van Ameronge: Ships Designed with Fuzzy Sets.   [1785]
Walichiewicz: a Multi-Dimensional Fuzzy Control Algorithm.   [1828]
Wang, P.   :   Fuzzy Systems: Theory and Application [1836]
Wang, Z.   :   Application of Fuzzy Decision Making and Fuzzy Linear [1837]
Wang, Z.   : Decision Making and Fuzzy Linear Programming in Production [1837]
Watson, S.R.:   Fuzzy Decision Analysis.   [1848]
Xuemou, W. : ystems Analysis and Fuzzy Sets.   [1886]
Yager, R.R. :   Fuzzy Decision Making Including Unequa [1887]
Yager, R.R. : inguistic Model and Fuzzy Truths.   [1888]
Yager, R.R. : a General Class of Fuzzy Connectives.   [1898]
Yager, R.R. : Some Properties of Fuzzy Relationships.   [1901]
Yager, R.R. : otypical Values for Fuzzy Subsets.   [1900]
Yager, R.R. :   Fuzzy Set and Possibility Theory.   [1903]
Yager, R.R. :   Fuzzy Sets, Probabilities, and Decisio [1895]
Yager, R.R. :   Fuzzy Subsets of Type II in Decisions. [1896]
Yager, R.R. : e for Importance of Fuzzy Sets.   [1892]
Yager, R.R. :   Fuzzy Thinking as Quick and Efficient. [1897]
Yager, R.R. :   Fuzzy Prediction Based on Regression M [1902]
Andersson, H: em Using the Theory Fuzzy latively Closed Systems.   [32]
Yager, R.R. : dures for Selecting Fuzzy Set-Theoretic Operators.   [1906]
Yager, R.R. : : An Application of Fuzzy Sets.   [1905]
Yager, R.R. : rship in a Compound Fuzzy Subset.   [1908]
Yezhkova, I.: Decision Making on Fuzzy Grounds.   [1913]
Zhukovin, V.: Making with Vector Fuzzy Preference Realtion.   [1925]
Yager, R.R. :   Validation of Fuzzy-Linguistic Models.   [1889]
Watanabe, S.:   A Generalized Fuzzy-Set Theory.   [1845]
Brams, S.J. :   Applied Game Theory.   [211]
Eigen, M.   :   Laws of the Game : How Principles of Nature Govern [463]
Naughton, J.: h: The Rules of the Game .   [1203]
Pernici, B. : Problem Solving in Game Playing: Computer Chess.   [1319]
Nicolis, J.S: n in Non-Negotiable Games : A Paradigm for Self-Organizatio [1226]
Schimanovich: inning Mathematical Games and Economic Actions.   [1510]
Smith, J.M. : n and the Theory of Games .   [1589]
Jutila, S.T.:   Social Gaming and Control by Dislocations.   [864]
Dombi, J.   :   Comments on the Gamma-Model .   [429]
Czogala, E. : Construction of the Gamma-Operator .   [371]
Pfeilsticker: Theory Bridging the Gap between Mathematical and Language- [1331]
Horacek, H. : ion of Anaphora and Gapping in German.   [781]
Yashin, A. : ults in Conditional Gaussian Estimation Procedures.   [1911]
Battista, J.: listic Paradigm and General Systems Theory.   [124]

Beloglavec, :   A General Systems Theory Approach to the   [145]
Bluth, B.J. :   Parson's General Theory of Action. A Summary of   [178]
Bosserman, R: curity Processes in   General Living Systems.   [196]
Botez, M.C. : c Representation of   General Systems: Open Multimodelling.   [197]
Boulding, K.:   The Universe as a   General System: Fourth Annual Ludwig v   [199]
Bowler, T.D.:   General Systems Thinking: Its Scope an   [203]
Bunge, M.   :   General Systems and Holism.   [239]
Caselles, A.: ies in the Light of   General Systems Theory.   [268]
Cavallo, R.E: General Systems Methodology and Large-   [275]
Christensen,:   Sourcebook. Vol.1:   General Description; Vol.2: Philosophi   [311]
Collot, F. : matical Concept for   General Systems Theory.   [333]
Csaki, P.   : ic Approach to Some   General Problems of Model Description.   [360]
Csanyi, V. :   General Theory of Evolution.   [362]
Dalenoort, G: ntrol in Systems in   General , and in Neural Networks in Par   [374]
Durkin, J.E.: p Psychotherapy and   General System Theory.   [449]
Ericson, R.F: The   General Systems Challenge.   [486]
Eugene, J. :   Aspects of the   General Theory of Systems.   [495]
Fosko, P.D. :   Applications of   General Systems Theory and Cybernetics   [537]
Gaines, B.R.:   General Systems. Yearbook of the Socie   [558]
Gaines, B.R.: of the Society for   General Systems Research.   [558]
Gaines, B.R.:   General Systems Research: Quo Vadis?   [557]
Geyer, R.F. : enation Theories: a   General Systems Approach   [604]
Hajek, P.   : al Foundations of a   General Theory.   [703]
Hofferbert, :   The Application of   General Systems Methodology to the Com   [767]
Holden, A.V.: l Mode of Action of   General Anestetics.   [770]
Hunt, L.R. :   Controllability of   General Nonlinear Systems.   [798]
Iijima, J. : al Decomposition of   General Systems.   [809]
Jakubowski, :   General Algorithms of Complex Dynamic   [829]
Jones, B.   :   General Systems Theory and Algorithm T   [855]
Klir, G.J. :   The   General Systems Research Movement.   [942]
Klir, G.J. : h-Holland Series in   General Systems Research.   [943]
Klir, G.J. :   General Systems Problem Solving Method   [946]
Klir, G.J. :   Applied   General Systems Research.   [940]
Klir, G.J. :   Basic and Applied   General Systems Research: A Bibliograp   [937]
Klir, G.J. :   General Systems Concepts.   [949]
Kohout, L.J.: a Special Case of a   General System Protection Design.   [966]
Machtey, M. : Introduction to the   General Theory of Algorithms.   [1081]
Miller, G.L.:   Graph Isomorphism,   General Remarks.   [1166]
Ninneman, J.: of U.S.Business: A   General Systems Approach.   [1231]
Oeren, T.I. : d on Cybernetic and   General Systems Theories.   [1255]
Peery, N.S. :   General Systems Theory Approaches to O   [1314]
Pichler, F. :   Decomposition of   General Dynamic Systems.   [1334]
Pichler, F. :   General Systems Algorithms for Mathema   [1339]
Pogonowski, : retical Approach to   General Systems Theory.   [1348]
Putt, A.M. :   General Systems Theory Applied to Nurs   [1369]
Ragade, R.K.:   General Systems Yearbook.   [1375]
Ragade, R.R.:   General Systems: Yearbook of the Socie   [1376]
Ragade, R.R.: of the Society for   General Systems Research.   [1376]
Ralescu, D. :   Toward a   General Theory of Fuzzy Variables.   [1378]
Reckmeyer, W: General Systems Research and Design: P   [1394]
Resconi, G. : ormalization of the   General Systems by Logical and Algebra   [1406]
Reuver, H.A.: thin the Context of   General Systems Theory.   [1413]
Royce, J.R. :   The Role of   General Systems and Information Theory   [1453]
Ruben, B.D. :   General Systems Theory and Human Commu   [1455]
Salzberg, P.: ry of Attraction in   General Systems.   [1484]
Schal, R.K. : tion of Thoughts of   General Systems Theorists Concerning t   [1504]
Sierocki, I.: riance-Relations in   General Dynamic Systems.   [1553]
Takahara, Y.: of the Causality of   General Dynamic Systems.   [1655]
Tancig, P. :   SOVA -   General Purpose Software Environment f   [1667]
Toern, A.A. : imulation Graphs: A   General Tool for Modeling Simulation D   [1695]
Trappl, R. :   Basic and Applied   General Systems Research: A Bibliograp   [1729]
Troncale, L.:   A   General Survey of Systems Methodology.   [1742]
Troncale, L.:   Future of   General Systems Science: Obstaches, Po   [1748]
Troncale, L.:   A   General Survey of Systems Methodology.   [1741]

Trussel, P. :   Constructs from General Systems Theory to a Health His [1756]
Veloso, P.A.: educibilities Among   General Problems.   [1807]
Verhagen, C.:   Some   General Remarks About Pattern Recognit [1812]
Yager, R.R. :   On a   General Class of Fuzzy Connectives.   [1898]
Zeigler, B.P: ical Realization of   General Time Systems.   [1917]
          :   General Systems Research: A Science, a [1933]
Hoskins, R.F: Generalised Functions.   [790]
Brown, D.J.H: ing Model Employing   Generalization an Abstraction Techniqu [229]
Chang, L.C. : al Approach for the   Generalization of Data Management Syst [293]
Averkin, A.N: Generalized Strategies in Problem Solv [75]
Dessauchoy, :   Generalized Information Theory and Dec [412]
Egiazarian, :   The   Generalized Hadamard Transforms and Li [459]
Hisdal, E. :   Generalized Fuzzy Set Systems and Part [762]
Hooda, D.S. :   Two   Generalized Measures of "Useful" Infor [778]
Jacobs, B.E.:   On   Generalized Computational Complexity.   [825]
Jantzen, J. :   A   Generalized Fuzzy Relational Matrix.   [836]
Kleyle, R. : ing Mechanisms in a   Generalized Information System.   [936]
Naylor, A.W.: composition Theory: Generalized Dependence.   [1204]
Watanabe, S.:   A   Generalized Fuzzy-Set Theory.   [1845]
Elias, D.   :   Procedures for   Generating Reconstruction Hypotheses i [474]
Klir, G.J. :   Procedures for   Generating Reconstruction Hypotheses i [947]
Presern, S. :   Plan   Generating System for a Tactile Sensor [1357]
Stoklosa, J.: s of (a,k)-Machines   Generating Cyclic Sequences.   [1622]
Thomas, A. :   Generating Tension for Constructive Ch [1685]
Czogala, E. :   Fuzzy Rules   Generation for Fuzzy Control.   [370]
De La Sen, M: Delays in the Input   Generation .   [394]
Farreny, H. : Objects in Sentence   Generation .   [508]
Grauer, M. :   The   Generation of Efficient Energy Supply   [658]
Hajek, P.   :   On   Generation of Inductive Hypotheses.   [702]
Horacek, H. :   On   Generation of Anaphora and Gapping in   [781]
Szanto, E. :   Generation of Technological Informatio [1651]
Subrahmanian: A   Generator for Models of Discrete-Event [1632]
Trappl, R. : Model of the Optic   Generator Potential and its Relation t [1710]
Labos, E.   : s and Chaotic Spike   Generators .   [1000]
Tchon, K.   : erations Preserving   Generic Properties of Systems.   [1679]
Bandler, W. : cal Theory of Value   Genesis .   [101]
De Jong, K. : ve System Design: A   Genetic Approach.   [392]
Aulin-Ahmava: he Impossibility of   Genuinly Self-Steering Machines: A Fun [74]
Gould, P.   : Social Scientists,   Geographers and Planners.   [649]
Dutton, G. : Harvard Papers on   Geographic Information Systems.   [450]
Abraham, R. :   Dynamics: The   Geometry of Behavior.   [3]
Atkin, R.   : al Dynamics and the   Geometry of Systems.   [61]
Hermann, R. : ations of Algebraic   Geometry to Systems Theory, Part 1   [752]
Bathe, K.J. : ment Analysis: U.S.-   German Symposium.   [123]
Buchberger, :   VIE-LANG: A   German Language Understandig System.   [238]
Gasparski, W: er an English nor a   German Speaking Country.   [582]
Gasparski, W: neither English nor   German Speaking Country.   [584]
Horacek, H. : hora and Gapping in   German .   [781]
Steinacker, :   Disambiguation in German .   [1615]
Strombach, W: f these Concepts in   German Language.   [1629]
Trost, H.   :   VIE-LANG: A   German Language Dialogue System.   [1753]
Pearlman, W.:   Beyond   Gestalt .   [1305]
Strombach, W: Wholeness, Gestalt , System: On the Meaning of the [1629]
Glanville, R: ack Boxes Trying to   Get Out.   [620]
Globus, G.G::   Is There a Ghost in the Machine after all?   [624]
Broekstra, G: the Foundations of   GIT (General Information Theory).   [225]
Shore, J.E. : ntropy Minimization   Given Fully-Decomposable Subset and Ag [1550]
Bruckmann, G: utput Approaches in   Global Modelling. (Proc.of the Fifth I [231]
Bubnicki, Z.:   Global and Local Identification of Com [235]
Holden, A.V.: ernetic Approach to   Global and Local Mode of Action of Gen [770]
Laszlo, E. :   Cybernetics and Global Planning.   [1016]
Lazarevic, B: Global and Large-Scale System Models.   [1020]
Nitecki, Z. :   Global Theory of Dynamical Systems.   [1233]
Rosen, R.   :   Feedforwards and Global Systems Failure.   [1440]

Small, M.G. :   Systemic and Global   Learning.   [1584]
Tchon, K.   :   Towards a Global   Analysis of Systems.   [1680]
Pot, John S.: habilitation of the Goal   Concept.   [1354]
Sakawa, M.   :   An Interactive Goal   Attainment Method for Multiobject [1483]
Weir, M.   :   Design for a Goal-Directed System.   [1855]
Jacak, W.   :   On Goal-Oriented Systems.   [820]
Jaron, J.   :   Goal-Oriented Cybernetical Systems.   [839]
Furukawa, O.: odology for Quality Goal-Seeking and Coordination, and the [550]
Jaron, J.   : d Topologies in the Goal-Space of a System.   [840]
Casti, J.   : atics of Attainable Goals   and Irreducible Uncertainties. [270]
Eveland, S. : tabase System for a Gothic   Ethymological Dictionary.   [496]
Eigen, M.   : rinciples of Nature Govern   Chance.   [463]
Espejo, R.   : ybernetic Praxis in Government : The Management of Industry [490]
Joerres, R. : n System of Euglena Gracilis .   [847]
Agusti, J.   :   Modularity Grade   of the Neural Networks Synthesiz [19]
El-Shirbeeny: Augmented Conjugate Gradient Algorithm for Solving Constra [472]
Hakkala, L. :   An Infeasible Gradient-Type Coordination Algorithm f [709]
Comstock, F.: Systems Approach to Grading of Flight Simulator Students.   [335]
Tsai, W.H.   :   Attributed Grammar - A Tool for Combining Syntact [1757]
Lakshmivarah: Synthesis of Fuzzy Grammars .   [1005]
Campbell, J.:   Grammatical Man: Information, Entropy, [259]
Miller, G.L.:   Graph   Isomorphism, General Remarks.   [1166]
Raghaven, V.: acteristics of Some Graph   Theoretic Clustering Methods.   [1377]
Read, R.C.   :   The Graph   Isomorphism Disease.   [1393]
Chang, L.C. :   A Graph-Structural Approach for the Gene [293]
Savage, G.J.:   The Graph-Theoretic Field Model-I - Modell   [1498]
Horn, W.   : ame-Based Real-Time Graphic Interaction System.   [787]
Graedel, T.E: Graphical Presentation of Results from [652]
Hazony, Y.   : ng with Interactive Graphics .   [745]
Majster, M.E: Extended Directed Graphs , a Formalism for Structured Dat [1092]
Tao, K.M.   : ructure of Directed Graphs with Applications: A Rapprochem [1669]
Tao, K.M.   : ructure of Directed Graphs with Applications: A Rapprochem [1670]
Thulasiraman: Graphs , Networks, and Algorithms.   [1688]
Toern, A.A. :   Simulation Graphs : A General Tool for Modeling Si [1695]
Steinberg, J: ale Behavior of the Grasshopper Chortophaga Viridifaciata. [1618]
Lipset, D.   :   Gregory Bateson: The Legacy of a Scien [1056]
Yezhkova, I.: ion Making on Fuzzy Grounds .   [1913]
Durkin, J.E.:   Living Groups: Group   Psychotherapy and General System [449]
Fleddermann,: ems Approach with a Group   Thesis.   [529]
Mackenzie, K: e of Hierarchy in a Group .   [1083]
Busch, J.A. : Applicable to Human Groups .   [252]
Durkin, J.E.:   Living Groups : Group Psychotherapy and Genera [449]
Duechting, W: lation of 3-D Tumor Growth .   [446]
Georgiev, A.: om Individual Human Growth   Curve.   [597]
Millendorfer: Growth   Reducing Factors in Complex Sys [1164]
Nobile, A.G.: odels for Regulated Growth   with Intrinsic Lower Bounds.   [1234]
Roberts, P.C: Systems: Limits to Growth   Revisited.   [1428]
Vitanyi, P.M: ure, Languages, and Growth   Functions.   [1819]
Watt, K.E.F.: ronmental Problems, Growth , and Culture.   [1849]
Hendel, R.J.:   The Grue   Paradox: An Information-Theoretic [749]
Callebaut, W: ence: Prospects for GSM .   [257]
O'Donovan, T: GSPS : Simulation Made Simple.   [1250]
Laszlo, E.   :   GST : Prospects and Principles.   [1014]
Hajek, P.   : hine Studies on the Guha   Method of Mechanized Hypothesis F [704]
Havranek, T.: logical Data by the Guha   Method: an Example.   [737]
Havranek, T.: Information on the Guha   Method.   [738]
            : pecial Issue on the Guha   Method of Mechanized Hypothesis F [1941]
Garey, M.R. : d Intractability: a Guide   to NP-Completeness.   [571]
Kennedy, P. :   A Guide   to Econometrics.   [901]
Lendaris, G.: deling - A Tutorial Guide .   [1036]
Uttenhove, H: An Introduction and Guide .   [1775]
Gold, H.J.   : ms: an Introductory Guidebook .   [630]
Checkland, P: am - Some Tentative Guidelines . Part 2: Building Conceptua [297]
Jones, L.M. :   Some Proposals and Guidelines .   [857]

Sugiyama, K.: al Understanding of Hierarchical System Structures.   [1634]
Sussman, G.J: r Expressing Almost Hierarchical Descriptions.   [1640]
Troncale, L.:   Origins of Hierarchical Levels: An Emergent Evolu [1738]
Wolkowski, Z: Hierarchical Perception of Time and Pe [1878]
Auger, P.   : f Organization in a Hierarchically Organized System.   [68]
Auger, P.   : Order, Disorder in Hierarchically Organized Systems.   [69]
Acalugaritei: Hierarchies of Evolutions and Stabilit [4]
Troncale, L.: red Questions about Hierarchies .   [1744]
Allen, T.F.H: Hierarchy : Perspectives for Ecological [24]
Apostolico, : of Information and Hierarchy in the Modeling of Biomolecu [41]
England J.L.:   Hierarchy Theory and Sociology.   [482]
England J.L.:   An Ashby Hierarchy for Human Action.   [483]
Mackenzie, K: e for the Degree of Hierarchy in a Group.   [1083]
Scholz, C. : The Architecture of Hierarchy .   [1514]
Troncale, L.: s Method: Tests for Hierarchy & Links between Isomorphs. [1747]
Troncale, L.:   Testing   Hierarchy Models with Data Using Compu [1745]
Warfield, J.: Crossing Theory and Hierarchy Mapping.   [1839]
Feintuch, A.:   System Theory: A Hilbert Space Approach.   [516]
Adams, W.M. :   Hiragana and Katakana Syllabaries: Rel [13]
Krallmann, H: e Systems Paradigm: Historical View from 1973 until 1983. [980]
Janko, J.   : nd Episode from the History of Czech Science in the Middle [832]
LeMoigne, J.: omplexity: A French History and its Prospects.   [1024]
Smith, C.S. : on Science, Art and History .   [1588]
Topolski, J.:   Methodology of History .   [1704]
Trussel, P. : Theory to a Health History .   [1756]
Bunge, M.   : General Systems and Holism .   [239]
Bunge, M.   : d Individualism and Holism .   [240]
Battista, J.:   The   Holistic Paradigm and General Systems [124]
Kaynak, E. :   A   Holistic Approach in Marketing Plannin [892]
Peacocke, C.:   Holistic Explanation.   [1298]
Phillips, D.:   Holistic Thought in Social Science. [1332]
Sonnier, I.L: Holistic Education: Teaching of Scienc [1599]
Haimes, Y.Y.:   Hierarchical Holographic Modeling.   [701]
Bahm, A.J. :   Holons : Three Conceptions.   [81]
Bailey, K.D.: ibrium, Entropy and Homeostasis : A Multidisciplinary Legac [82]
Kak, S.   : iciency of Chemical Homeostasis - an Information-Theoretic [869]
Vagnucci, A.:   Concatenation and Homogeneity of Biological Time Series. [1780]
Foo, N.Y.   :   Preservation under Homomorphisis .   [534]
Foo, N.Y.   : sure Properties and Homomorphisms of Timevarying Systems. [535]
Haralik, R.M: of Binary Relation Homomorphisms .   [726]
Agassi, J. : hy Today: Essays in Honour of Mario Bunge.   [14]
Carter, C.F.: conomics: Essays in Honour of G.L.S. Shackle.   [265]
Halmes, A. : ory: Publication in Honour of Prof.Hans Blomberg.   [717]
Von Luetzow,: sion of Cat Retinal Horizontal Cells.   [1822]
Hoye, R.E. :   Current Status of Hospital Management Information System [793]
Ruotsalainen: ion of the Regional Hospital System with Different Service [1463]
Batchelor, B: for Finding Convex Hulls of Planar Figures.   [121]
Batchelor, B: n Based Upon Convex Hulls of Concavities.   [119]
Boulding, K.:   Human Knowledge as a Special System. [201]
Busch, J.A. : -Type Applicable to Human Groups.   [252]
Camana, P.C.: ion of Feedback for Human Posture Control without Physical [258]
England J.L.: Ashby Hierarchy for Human Action.   [483]
Georgiev, A.: del from Individual Human Growth Curve.   [597]
Johnson, L. : tical Framework for Human Communication.   [849]
Jumarie, G. : c Systems Involving Human Factors.   [862]
Kuhlenbeck, :   The   Human Brain and its Universe. Vol.I: T [993]
Lowen, W.   : Creep-Creep: Out of Human Scale.   [1066]
McPherson, P: lling and Improving Human Systems.   [1142]
Newell, A. :   Human Problem Solving.   [1217]
Nowotny, H. :   Not Quite Human : Science and Utopia.   [1246]
Ruben, B.D. : Systems Theory and Human Communication.   [1455]
Ruffner, M.A: Human Factors in Systems Analysis. [1461]
Schal, R.K. : ning the Proc s of Human Transformation.   [1504]
Waller, R.J.: d the Boundaries of Human Policy Making.   [1835]

Zimmermann, : tent Connectives in Human Decision Making.   [1926]
Oguntade, O.: Pragmatic Theory of Humanistic Systems.   [1258]
Gloria-Botti: Phosphate Locus in Humans .   [625]
Wang, Z.   : Planning of Animal Husbandry System in Farming Region.   [1837]
Troncale, L.:   A Hybrid Systems Method: Tests for Hiera [1747]
Oden, A.S. : of the Behavior of Hydrocarbon Reservoirs.   [1252]
Ciriani, T.A: s for Surface Water Hydrology . (Proc. of the Workshop on M [316]
Overton, W.S: Evolution of a Hydrology Model - an Exercise in Model [1275]
Overton, W.S: ctural Aspects of a Hydrology Model.   [1276]
Jorgenson, S: Hydrophysical and Ecological Modelling [858]
Jorgenson, S: Hydrophysical and Ecological Models of   [859]
Eigen, M.   :   The Hypercycle : a Principle of Natural Sel   [462]
Huxham, C.S.:   Hypergames as a Soft Systems Methodolo   [800]
Broekstra, G: n of Reconstruction Hypotheses by Information Measures.   [226]
Elias, D.   : ting Reconstruction Hypotheses in the Reconstructability A [474]
Hajek, P.   : ration of Inductive Hypotheses .   [702]
Klir, G.J. : ting Reconstruction Hypotheses in the Reconstructability A [947]
Cavallo, R.E: n of Reconstruction Hypothesis .   [283]
Dijkstra, W.:   Testing Auxiliary Hypothesis Behind the Interview.   [421]
Hajek, P.   :   Mechanizing Hypothesis Formationmathematical Found [703]
Hajek, P.   : ethod of Mechanized Hypothesis Formation.   [704]
            : ethod of Mechanized Hypothesis Formation.   [1941]
Batchelor, B: attern Recognition: Ideas in Practice.   [117]
Mulej, M.   : rganize Associates' Ideas about New Products and Programs. [1195]
Rasvan, V. : Some System Theory Ideas Connected with the Stability Pro [1387]
Gasparski, W: Si Duo Dicunt Idem Non Est Idem: Or on Design Method [582]
Gasparski, W: Dicunt Idem Non Est Idem : Or on Design Methodology as Seen [582]
Nugyen, V.V.: ification of Linear Identifiability Concepts.   [1247]
Thowsen, A. :   Identifiability of Dynamic Systems.   [1687]
Aarnes, E. :   On the Problem of Identification in Compartment Analysis   [1]
Broekstra, G: lysis and Structure Identification II.   [223]
Broekstra, G: Representation and Identification of Structure Systems. [224]
Bubnicki, Z.:   Global and Local Identification of Complex Systems with [235]
Czogala, E. :   On Identification of Fuzzy Systems and It [368]
Eisenstein, : xtraction by System Identification .   [466]
El-Sherief, :   Identification and Control of Linear D [470]
El-Sherief, :   Identification and Modeling for Linear [468]
El-Sherief, :   Identification of Multivariable System [469]
El-Sherief, : System Modeling and Identification Algorithms and their Ap [471]
Eyckhoff, P.: Progress in Systems Identification .   [498]
Eyckhoff, P.: Progress in System Identification .   [499]
Fogel, E.   : formation in System Identification : Bounded Noise Case.   [532]
Gaines, B.R.: ential Fuzzy System Identification .   [560]
Gerardy, R. : Finite State System Identification .   [598]
Gerardy, R. : ome Methods for the Identification of Finite-State Systems [599]
Gollmann, D.:   On the Identification of Certain Non-Linear N [633]
Gourlay, A.R: Identification and Analysis of the Sub [650]
Iserman, R. :   Identification and System Parameter Es [815]
Iserman, R. :   System Identification (Papers from an IFAC Sy [816]
Joerres, R. :   Identification of the Temperature Mask [847]
Kalata, P. :   On System Identification with and without Certai [871]
Kellerman, W: Computer-Aided Identification of System Structure: An [899]
Klir, G.J. : ter-Aided Structure Identification : Some Experimental Obse [938]
Sastry, D. : State and Parameter Identification .   [1495]
Sinha, K.K. : ol without Explicit Identification .   [1569]
Tamura, H. :   Identification of a Dynamic Lake Model [1666]
Uyttenhove, : e Behavioral System Identification Problem (Part I).   [1777]
Van Steenkis: Modeling, Identification and Control in Environm [1791]
Haydon, P.G.: rent Computation in Identified Molluscan Neurones.   [742]
Conant, R.C.:   Identifying Relations from Partial Inf [341]
Krippendorff: n the Algorithm for Identifying Structures in Multi-Variat [984]
Conant, R.C.: Efficient Proofs of Identities in N-Dimensional Informatio [340]
Cornelis, A.: ators of Scientific Identity .   [348]
Slawski, C. :   Identity Bargaining: A Policy Systems [1575]

Lilienfeld, :	Systems Theory: an Ideological Analysis.	[1049]
Pla Lopez, R:	mic Transition from Ideological Learning to Scientific Lea	[1345]
Levien, R.E.:	Introduction to IIASA : Applying Systems Analysis in an	[1041]
Casti, J. :	Illusion or Reality?: the Mathematics	[270]
Overton, W.S:	n the Flex Paradigm Illustrated by Structural Aspects of a	[1276]
Tripp, R.S. :	and Development: An Illustration within Air Force Logistic	[1736]
Gonzales, R.:	Digital Image Processing.	[639]
Hall, E.L. :	Computer Image Processing and Recognition.	[714]
Herman, G.T.:	Image Reconstruction From Projections:	[751]
Kovalevsky, :	Image Pattern Recognition.	[975]
Lee, E.T. :	Fuzzy Set Theory to Image Sciences.	[1027]
Melkumian, A:	tortion Measures in Image Data Compression and Transmissio	[1151]
Pichler, F. :	Robotics and Image Processing.	[1338]
Pratt, W.K. :	Digital Image Processing.	[1356]
Simon, J.C. :	Digital Image Processing.	[1565]
Stucki, P. :	Advances in Digital Image Processing.	[1630]
Kaufmann, G.:	Visual Imagery and Its Relation to Problem So	[887]
Collins, D.M:	Electron Density Images from Imperfect Data by Iterativ	[331]
Bergeron, R.:	Imaginary , Fluid, and Dancing Systems.	[161]
Bruni, C. :	Systems Theory in Immunology .	[233]
Desouza, G.R:	c and Environmental Impact Analysis.	[411]
Pipino, L.L.:	Potential Impact of Fuzzy Sets on the Social Sci	[1342]
Steinacker, :	Future, Impacts , and Future Impacts of Artific	[1616]
Steinacker, :	Impacts, and Future Impacts of Artificial Intelligence: A	[1616]
Trappl, R. :	The Impacts of Artificial Intelligence. Sc	[1733]
Trappl, R. :	Impacts of Artificial Intelligence: An	[1730]
Trappl, R. :	Impacts of Artificial Intelligence.	[1727]
Kerner, H. :	Functional versus Imperative Programming - Complement or	[908]
Collins, D.M:	Density Images from Imperfect Data by Iterative Entropy Ma	[331]
Smith, R.B. :	w to Plan, Design & Implement a Bad System.	[1592]
Best, D.P. :	A Case Study in the Implementation of a Cybernetic Organis	[164]
Christensen,:	ns; Vol.3: Computer Implementation ; Vol.4: Applications.	[311]
Ginzberg, M.:	on Model for System Implementation .	[616]
Herman, G.T.:	n From Projections: Implementation and Applications.	[751]
Kellerman, W:	ture: An Attempt at Implementation .	[899]
Meadows, D.L:	Techniques for Implementing Computer-Based Policy Mod	[1143]
Oguntade, O.:	Implementing a Pragmatic Theory of Hum	[1258]
Bohm, D. :	Wholeness and the Implicate Order.	[184]
Bandler, W. :	Semantics of Implication Operators and Fuzzy Relati	[100]
Klir, G.J. :	n Systems Research: Implications for the Social Sciences (	[941]
Koenderink, :	Stretching by Eye: Implications for Robot Vision.	[962]
Scandura, J.:	with Instructional Implications .	[1500]
Windeknecht,:	System Theoretic Implications of Numerical Methods Appl	[1866]
Rushinek, A.:	er Software and Its Importance to the Accounting Informati	[1465]
Yager, R.R. :	uistic Variable for Importance of Fuzzy Sets.	[1892]
Aulin-Ahmava:	The Impossibility of Genuinely Self-Steerin	[74]
Kinston, W. :	Improving Health Care Institutions: An	[927]
McPherson, P:	nding Modelling and Improving Human Systems.	[1142]
Sachs, W. :	Systemic Inactivism : A Modern Laissez-Faire Thi	[1469]
Nicolis, J.S:	Inadequate Communication between Self-	[1225]
Yager, R.R. :	zzy Decision Making Including Unequal Objectives.	[1887]
Haydon, P.G.:	Coherent and Incoherent Computation in Identified M	[742]
De Raadt, J.:	to the Skewness of Income and Output Distributions.	[398]
Chatterjee, :	s with Complete and Incomplete Information .	[295]
Gehrlein, W.:	ethods for Ordering Incomplete Ordinal Data.	[588]
Lipski, W. :	mation Systems with Incomplete Information.	[1057]
Lipski, W. :	On the Logic of Incomplete Information.	[1058]
Zarri, G.P. :	blem of Information Incompleteness .	[1915]
Ben-Eli, M.U:	ulation and Variety Increase in Evolving Systems.	[149]
Wisniewski, :	imate Change due to Increasing Carbon Dioxide.	[1872]
Guida, G. :	Interactivity and Incrementality in Natural Language Und	[680]
Hisdal, E. :	onal Possibilities, Independence and Noninteraction	[761]
Holst, P. :	ation 1951-1976: An Index to the Literature.	[775]
Legget, R.W.:	A Reliability Index for Models.	[1032]

Cornelis, A.:   Epistemological   Indicators   of Scientific Identity.   [348]
Ivan, I.T.   :   System of   Indicators   for Estimating the Software   [817]
Auger, P.   : ween Collective and   Individual   Levels of Organization in a   [68]
Georgiev, A.: ematical Model from   Individual   Human Growth Curve.   [597]
Kindler, E. : iption of Types and   Individual   Typology.   [918]
Scott, W.A. : and Measurement of   Individual   Differences.   [1519]
Bunge, M.   : of Society: Beyond   Individualism and Holism.   [240]
Royce, J.R. : eory in Multifactor   Individuality Theory.   [1453]
Rescher, N. :   Induction .   [1402]
Christensen,:   Foundations of   Inductive   Reasoning.   [310]
Hajek, P.   :   On Generation of   Inductive   Hypotheses.   [702]
Klir, G.J.   : ems Methodology and   Inductive   Reasoning: The Issue of Part   [948]
Michalski, R: tion as Rule-Guided   Inductive   Inference.   [1158]
Stemmer, N. :   The Reliability of   Inductive   Inferences and Our Innate Ca   [1619]
Dimitrova, L: aterial Flow in the   Industrial   Enterprise.   [423]
Knappe, U.   : er-Based System for   Industrial   Energy Control.   [952]
Lauenroth, H: s for Automation of   Industrial   Production and Information   [1017]
Singh, R.   : hes to the Study of   Industrial   Relations.   [1567]
Tong, R.M.   : of Fuzzy Models for   Industrial   Processes - Some Recent Res   [1700]
Drechsler, F: ring (CIM) in Irish   Industry .   [436]
Espejo, R.   : : The Management of   Industry   in Chile 1970-1973.   [490]
Kawamura, K.: he Microelectronics   Industry : A Systems Approach.   [891]
Dial, G.   :   Beta and Shannon's   Inequality .   [420]
Gurdial,   :   A Note on Fano   Inequality   Involving Distortion.   [690]
Pearl, J.   : n the Complexity of   Inexact   Computations.   [1300]
Shapiro, L.G: al Descriptions and   Inexact   Matching.   [1535]
Hakkala, L. :   An   Infeasible Gradient-Type Coordination   [709]
Bandler, W. :   The Four Modes of   Inference   in Fuzzy Expert Systems.   [102]
Feldman, J.A: Complexity and the   Inference   of Bent Programs.   [517]
Levitan, G.L: Use of Deductive   Inference   in Situation Control Systems   [1043]
Michalski, R: le-Guided Inductive   Inference .   [1158]
Mizumoto, M.: r Fuzzy Conditional   Inference .   [1176]
Morris, C.N.: sis and Statistical   Inference .   [1188]
Rosenkrantz,:   Inference , Method and Decision.   [1448]
Turksen, I.B: Bounds for Fuzzy   Inference .   [1761]
Waterman, D.:   Pattern-Directed   Inference   Systems.   [1847]
Yager, R.R. :   An Approach to   Inference   in Approximate Reasoning.   [1893]
Stemmer, N. : bility of Inductive   Inferences   and Our Innate Capacities.   [1619]
Pearl, J.   :   Storage Economy of   Inferential   Question-Answer Systems.   [1301]
Anderson, B.: heory: Finitary and   Infinitary   Conditions.   [29]
Sworder, D.D: Influence   of Sensor Failures on Log Re   [1647]
Adams, W.M. : llabaries: Relative   Information   Content.   [13]
Agrawal, V.D: An   Information   Theoretic Approach to Digi   [18]
Andrew, A.M.:   Information   Transmission on Nerves.   [37]
Apostolico, :   On the Role of   Information   and Hierarchy in the Model   [41]
Arbel, A.   :   Aggregation and   Information   Structuring in Large-Scale   [46]
Arigoni, A.C: c Transformation of   Information .   [51]
Atlan, H.   :   Information   Theory.   [65]
Best, D.P.   : the Application of   Information   Technology.   [165]
Bogart, D.H.: edwithin: Strategic   Information   in Systems.   [181]
Brockett, P.:   Information   Theoretic Analysis of Ques   [222]
Broekstra, G: ction Hypotheses by   Information   Measures.   [226]
Campbell, J.:   Grammatical Man: Information , Entropy, Language, and Li   [259]
Chaitin, G.J: Algorithmic   Information   Theory.   [291]
Chang, Shi-K: Policy Analysis and   Information   Systems, (A Journal Publis   [294]
Chatterjee, : lete and Incomplete   Information .   [295]
Chen, K.   :   Information   Technology in Developing C   [304]
Conant, R.C.: ling Using a Simple   Information   Measure.   [338]
Conant, R.C.: es in N-Dimensional   Information   Theory.   [340]
Conant, R.C.: ations from Partial   Information .   [341]
Cremers, A.B: Orthogonality of   Information   Structures.   [356]
Dedons, A.   :   The   Information   Professional: Survey of an   [401]
Dessauchoy, :   Generalized   Information   Theory and Decomposability   [412]
Dolby, J.L. : ns of Ambiguity and   Information   Loss.   [427]

Donald, A.   :   Management, Information and Systems (Second Editio [434]
Driankow, D.: Making with Ordinal Information .   [439]
Dutton, G.   : apers on Geographic Information Systems.   [450]
Engell, S.   :   Variety, Information and Feedback.   [481]
Espejo, R.   :   Management Information Systems: a System for Desi   [489]
Espejo, R.   :   Management and Information : The Complementarity Contr [492]
Fairhurst, M: Cyclic Activity in Information Processing Systems.   [500]
Fedanzo, A.J: The Origin of Information in Systems.   [514]
Flavin, M.   : amental Concepts of Information Modelling.   [528]
Fogel, E.   :   On the Value of Information in System Identification: [532]
Fowler, T.B. : and the Concept of Information .   [538]
Gage, W.L.   : ditorium Management Information Systems.   [552]
Garfolo, B.T: al Dialysis Medical Information System.   [573]
Gatlin, L.L.:   Information Theory and the Living Syst   [586]
Goffman, W.   :   Scientific Information Systems and the Principle   [627]
Gokhale, D.V: The Information in Contingency Tables.   [629]
Havranek, T.: Approach to Missing Information in the Guha Method.   [738]
Higashi, M.   : of Uncertainty and Information Based on Possibility Distr   [754]
Higashi, M.   : stance Representing Information Closeness: Possibility and   [756]
Hoaglin, D.C: Data for Decisions: Information Strategies for Policymaker   [763]
Hooda, D.S.   : easures of "Useful" Information .   [778]
Hoye, R.E.   : Hospital Management Information Systems.   [793]
Hull, R.   :   Communication and Information : Their Relationship to Com [796]
Hull, R.   : p to Computer-Based Information Systems.   [796]
Hyvarinen, L: Information Theory for Systems Enginee   [806]
Jones, A.H.   : mputerised Business Information Systems.   [853]
Jumarie, G.   : ons of Relativistic Information Fuzzy Sets, Linguistics, R   [861]
Keyes, R.W.   : l Limits in Digital Information Processing.   [909]
Kleyle, R.   : ms in a Generalized Information System.   [936]
Klir, G.J.   :   Possibilistic Information Theory.   [950]
Kobayashi, I: Information and Information Processing   [954]
Kobayashi, I: Information and Information Processing Structure.   [954]
Kornwachs, K: Pragmatic Information and Nonclassical Systems.   [972]
Kornwachs, K: Some Notes on Information and Interaction.   [973]
Kubat, L.   :   Entropy and Information in Science and Philosophy.   [990]
Langefors, L: Information and Data in Systems.   [1009]
Lauenroth, H: rial Production and Information Systems.   [1017]
Lipski, W.   :   Information Systems with Incomplete In   [1057]
Lipski, W.   : ems with Incomplete Information .   [1057]
Lipski, W.   : Logic of Incomplete Information .   [1058]
Lucas, R.G.   : cal Determinants of Information System Configuration.   [1070]
Lundeberg, M: Information Systems Development: A Sys   [1074]
Martin, P.E.: tion and Control of Information Patterns - Further Develop   [1111]
Masuda, Y.   :   The Information Society as Past-Industrial   [1115]
Michie, D.   : Related Topics: An Information Scientist's Weekend Book.   [1161]
Mock, T.J.   : and Organizational Information .   [1177]
Okuda, T.H.   : Problems with Fuzzy Information Using Probability Measures   [1261]
Palm, G.   :   for the Storage of Information in the Brain.   [1278]
Pardo, L.   :   Information Energy of a Fuzzy Event an   [1281]
Pinske, M.S.:   Information and Information Stability   [1341]
Pinske, M.S.:   Information and Information Stability of Random Variab   [1341]
Pippenger, N: Information Theory and the Complexity   [1343]
Potter, J.M.: Partial Prior Information and Decision-Making.   [1355]
Reisig, G.H.: ata Analysis in the Information Domain.   [1399]
Royce, J.R.   : General Systems and Information Theory in Multifactor Indi   [1453]
Rushinek, A.: e to the Accounting Information System.   [1465]
Sage, A.P.   : ns in the Design of Information Systems and Process of Pla   [1475]
Sampson, J.R: Adaptive Information Processing: An Introductor   [1485]
Samuelson, K: Information Systems and Networks.   [1486]
Scott, C.H.   : the Analysis of an Information Theoretic Model of Spacial   [1518]
Silviu, G.   :   Information Theory with Applications.   [1559]
Skvoretz, J.:   Information Diffusion in Formally Stru   [1572]
Skvoretz, J.: red Populations: An Information Processing Approach.   [1572]
Slyadz, N.N.: is of Fuzzy Initial Information in Decision-Making Models.   [1582]

Spoehr, K.T.:   Visual Information   Processing.   [1605]
Sugeno, M. : ning Based on Fuzzy Information .   [1633]
Sultan, L.H.: anization for Fuzzy Information   Processing.   [1636]
Szanto, E. : on of Technological Information .   [1651]
Tou, J.T.   :   Advances in Information   Systems Science, Vol.6. [1705]
Toutenburg, :   Prior Information   in Linear Models.   [1708]
Tsokos, C.P.:   Decision Information .   [1759]
Tussey, C.C.: Kentucky Longevity Information   System.   [1763]
Valdes-Perez: Information   Loss Due to Data Quantizat [1781]
Von Luetzow,: System Analysis of Information   Transmission of Cat Retina [1822]
Wang, P.   : Policy Analysis and Information   Systems.   [1836]
Williams, P.: rinciple of Minimum Information .   [1865]
Wyner, A.D. : Conditional Mutual Information   for Arbitrary Ensembles. [1884]
Wyner, A.D. : ndamental Limits in Information   Theory.   [1885]
Zarri, G.P. : olve the Problem of Information   Incompleteness.   [1915]
            : Policy Analysis and Information   Systems.   [1934]
Szaniawksi, :   Science as an Information-Seeking   Process.   [1650]
Hendel, R.J.: he Grue Paradox: An Information-Theoretic   Solution.   [749]
Kak, S.   : al Homeostasis - an Information-Theoretic   Approach.   [869]
Takatsuji, M: An Information-Theoretical   Approach to a [1662]
Steinberg, J: An Informational   Analysis of the Inter-Ma [1618]
Becker, P.W.: edundancy and Their Inherent   Bounds on Achievable Reliabi [130]
Becker, P.W.: edundancy and their Inherent   Bounds on Achievable Reliabi [129]
De Delara, G: mulation of Lateral Inhibition   in the Retina.   [389]
Slyadz, N.N.:   Analysis of Fuzzy Initial   Information in Decision-Making [1582]
Vallee, R. : System with Random Initial   Conditions.   [1782]
Hellerman, L: Symbolic Initiative   and its Application to Comp [747]
Stemmer, N. : Inferences and Our Innate   Capacities.   [1619]
Lauenroth, H: c Model Systems and Innovation   Strategies for Automation o [1017]
Slawski, C. : Institutionalizing Innovation   in an Organization: A Model [1576]
Trappl, R. :   The Soft Side of Innovation : Medicine - a Hard Case. [1716]
Sahal, D.   : ns of Technological Innovations .   [1481]
De La Sen, M: Avoid Delays in the Input   Generation.   [394]
Walker, C.C.: tion, Priority, and Input   Span in a Class of Fixed-Structu [1830]
Dammasch, I.: Differences between Input-Constant   and Input-Variant Netwo [375]
Bruckmann, G: Input-Output   Approaches in Global Mode [231]
De La Sen, M: rizations in Single Input-Output   Adaptive Control Systems [394]
Maker, P.P :   Input-Output   Relations in Multivariate [1094]
Orava, P.J. : On the Concepts of Input-Output   Model, Consolity, and Sta [1268]
Dammasch, I.: Input-Constant and Input-Variant   Networks.   [375]
Buchberger, :   Input/Output   Codings and Transition Fu [237]
Burns, J.R. :   An Input/Output   Approach to the Structura [250]
Roberts, G.F: An Input/Output   Approach to the Structura [1426]
Ciborra, C. :   A Methodological Inquiry   of Organizational Noise in Soc [315]
Cosier, R.A.: ness of Dialectical Inquiry   Systems.   [349]
Estep, M.L. : d Interdisciplinary Inquiry   in Curriculum Design.   [493]
Rescher, N. :   Empirical Inquiry .   [1403]
Glanville, R: Inside   Every White Box there are Two B [620]
Molloy, K.J.:   Insights   into Complex Organisational P [1182]
Nowakowska, :   Instability , Monopolies, and Alienatio [1242]
Thompson, J.:   The Instability   of Evolving Systems.   [1686]
Slawski, C. :   Institutionalizing   Innovation in an Or [1576]
Kinston, W. : proving Health Care Institutions : An Action Research Appro [927]
Scandura, J.: ical Foundations of Instruction : A Systems Alternative to [1501]
Scandura, J.: ocess Approach with Instructional   Implications.   [1500]
Craine, R. : he Optimal Monetary Instrument : An Empirical Assessment. [354]
Mattessich, :   Instrumental   Reasoning and Systems Met [1126]
Brugger, K. : l Model of Biphasic Insulin   Secretion.   [232]
Binder, Z. : turalization in the Integrated   Automation of Complex Syste [170]
Dayal, R.   :   An Integrated   System of World Models. [385]
Drechsler, F: Computer Integrated   Manufacturing (CIM) in Iris [436]
Gigley, H. : ts Brain Theory. An Integrated   Approach to Simulation Mode [610]
Grubbstrom, :   The Axsater Integrated   Production-Inventory System [676]
Guida, G.   :   An Integrated   Model of Problem Solver. [679]

Mak, K.L.   : mics and Control of   Integrated   Production-Inventory-Shippi [1093]
Sonde, B.S.   : System Design Using   Integrated   Circuits.   [1598]
Brown, J.H.U: Integration and Coordination of Metabo [230]
Estep, M.L.   :   A Systems   Integration of Multidisciplinary and I [493]
Kittler, J.   : on Fast Monte Carlo   Integration of P.D.F. Estimators.   [930]
Ryan, R.M.   : Conceptualizing the   Integration of Social Welfare Delivery [1467]
Fernandez, E: tabase Security and   Integrity .   [518]
Ashby, R.   :   Mechanisms of Intelligence : Ross Ashby's Writings on [55]
Barr, A.   : dbook of Artificial   Intelligence (Vol.1).   [109]
Barr, A.   : dbook of Artificial   Intelligence (Vol.2).   [110]
Boden, M.   :   Artificial   Intelligence and Natural Man.   [180]
Chen, C.N.   : tion and Artificial   Intelligence .   [302]
Cohen, P.R.   : dbook of Artificial   Intelligence (Vol.3).   [327]
Davis, R.   : stems in Artificial   Intelligence .   [383]
Gause, D.C.   : tics and Artificial   Intelligence .   [587]
Gigley, H.   :   Artificial   Intelligence Meets Brain Theory. An In   [610]
Hayes, P.   : rends in Artificial   Intelligence .   [743]
Horn, W.   :   An Artificial   Intelligence System for Medical Decisi   [786]
Hunt, E.B.   :   Artificial   Intelligence .   [797]
Hyde, M.O.   : arch for Artificial   Intelligence .   [805]
Michie, D.   :   Machine   Intelligence and Related Topics: An In   [1161]
Nilsson, N.J: iples of Artificial   Intelligence .   [1230]
Ringle, M.   : tives in Artificial   Intelligence .   [1420]
Steinacker,   : pacts of Artificial   Intelligence : A Bibliography.   [1616]
Trappl, R.   : through Artificial   Intelligence : A Proposal for 3 Project [1732]
Trappl, R.   : pacts of Artificial   Intelligence . Science, Technology, Mil [1733]
Trappl, R.   : pacts of Artificial   Intelligence : An Overview.   [1730]
Trappl, R.   :   Artificial   Intelligence : A One-Hour Course.   [1728]
Trappl, R.   : pacts of Artificial   Intelligence .   [1727]
Webber, B.L.: dings in Artificial   Intelligence .   [1850]
Winston, P.H: Artificial   Intelligence .   [1869]
Winston, P.H: Artificial   Intelligence : An MIT Perspective (2 Vo [1871]
Paritsis, N.: Problems in Natural   Intelligent Systems with Changes in En [1282]
Steinberg, J: nal Analysis of the   Inter-Male   Behavior of the Grasshopper [1618]
Takatsuji, M: oach to a System of   Interacting Elements.   [1662]
Vallee, R.   : " for Observing and   Interacting Subjects.   [1783]
Blunden, M.   : nology and Systemic   Interaction .   [176]
Gelfand, A.E: ensus Level and its   Interaction with Other Management Stra [589]
Horn, W.   : d Real-Time Graphic   Interaction System.   [787]
Kornwachs, K: on Information and   Interaction .   [973]
Scott, C.H.   : ic Model of Spacial   Interaction .   [1518]
Shmulyan, B.:   Interaction of Entropy Systems.   [1548]
Zouwen, J.v.:   Modeling   Interaction Processes in Interviews; a [1930]
Auger, P.   :   Interactions Between Collective and In   [68]
Neck, R.   :   Interactions between the Political and [1206]
Takahara, Y.: Characterization of   Interactions .   [1657]
Baptistella,: y Methodologies for   Interactive Multicriteria Optimization [104]
Darlington,   :   Interactive Validation of Program Spec [376]
Elison, D.I.:   Interactive Simulation on a Computer.   [475]
Fitter, M.   : ards More 'Natural'   Interactive Systems.   [527]
Hazony, Y.   : roblem Solving with   Interactive Graphics.   [745]
Horn, W.   :   Interactive Computer-Aided Planning: E [782]
Koczy, L.T.   :   Interactive Sigma-Algebras and Fuzzy O [959]
Leal, A.   :   An   Interactive Program for Conversational [1025]
Rauch, W.D.   : ic Extracting as an   Interactive Process.   [1391]
Retti, J.   :   SISSY - A Fast Interactive Simulation System.   [1407]
Retti, J.   :   A Framework for Interactive Engineering of Simulation   [1408]
Sakawa, M.   :   An   Interactive Goal Attainment Method for [1483]
Veroy, B.S.   : itecture Synthesis: Interactive Strategy.   [1813]
Wisniewski,   :   An   Interactive Model for Analysing Climat [1872]
Wnuk, A.J.   : The Block-Oriented   Interactive Simulator BORIS.   [1876]
Guida, G.   :   Interactivity and Incrementality in Na   [680]
Nicholson, H: Structure of   Interconnected Systems.   [1221]
Drozen, V.   :   Multilateral Interdependence .   [441]

Altmann, A. :   Interdisciplinary System Analysis.   [27]
Estep, M.L. : ltidisciplinary and   Interdisciplinary Inquiry in Curriculu   [493]
Roth, G.   : anizing Systems: An   Interdisciplinary Approach.   [1451]
Shahane, A.H: Interdisciplinary Models of Water Syst   [1531]
White, A.M. :   Interdisciplinary Teaching.   [1860]
Innocent, P.: oward Self-Adaptive   Interface System.   [813]
Leeming, A.M: he Design of a User   Interface .   [1029]
Bibbero, R.J: roprocessor System,   Interfacing and Applications.   [168]
Viliums, E.R: Methods to Obtain   Interlevel Consistency of Decisions un [1816]
Barenblatt, : Self-Similarity and   Intermediate Asymptotics.   [106]
Bosserman, R: Internal   Security Processes in General   [196]
Bosserman, R: Internal   Security Processes and Subsys   [195]
Favella, L.F: nostic Problems for   Internal   Combustion Engines.   [512]
McIntyre, C.:   Addendum to   Internal   Report 165: Mathematical Docu [1140]
Bensoussan, : n Systems Analysis: International Symposium, Versailles 19   [158]
Bensoussan, :   International Symposium on Systems Opt   [159]
Hajek, P.   : pecial Issue of the   International Journal of Man-Machine S   [704]
Hawkes, N. :   International Seminar on Trends in Mat   [739]
Hawkes, N. :   International Seminar on Trends in Mat   [740]
Klement, E.P: Proceedings of the   International Seminar on Fuzzy Set The   [933]
Ladanyi, O.G: nts Using Published   International Statistics.   [1003]
Levien, R.E.: tems Analysis in an   International Setting.   [1041]
Negoita, C. :   and Systems: a New   International Journal.   [1208]
Trappl, R. :   Reducing   International Tension through Artifici   [1732]
            : edings of the First   International Symposium on Policy Anal   [1934]
            : Metamedicine. A New   International Journal for Philosophy a   [1939]
Slawski, C. :   A Systems Model of   Interparty Negotiation.   [1577]
Slawski, C. :   The   Interpersonal Conflict Episode: A Syst   [1580]
Hanassab, S.:   An   Interpolator Based Controller for Cybe   [719]
Hellwig, Z.H: s Acceptability and   Interpretation .   [748]
Trussel, P. : very, Analysis, and   Interpretation .   [1755]
Varshavskiy,: . I. Definition and   Interpretation .   [1802]
Ragade, R.K.:   Fuzzy   Interpretative Modeling.   [1374]
Auguin, M. : d of Realization of   Interpreted Petri Nets.   [70]
Grubbstrom, : on-Inventory System   Interpreted in Terms of the Theory of   [676]
Toulotte, J.: hod for Decomposing   Interpreted Petri Nets and its Utiliza [1707]
Lesmo, L.   :   An   Interpreter of Fuzzy Production Rules. [1037]
McDowall, D.:   Interrupted Time Series Analysis.   [1136]
Yager, R.R. : bership in the Unit   Interval .   [1890]
Camana, P.C.: ol without Physical   Intervention .   [258]
Dijkstra, W.: pothesis Behind the   Interview .   [421]
Zouwen, J.v.: action Processes in   Interviews ; a Contribution of Systems   [1930]
Garey, M.R. :   Computers and   Intractability : a Guide to NP-Complete   [571]
Hallam, T.G.:   On the   Intrinsic Structure of Differential Eq   [715]
Nobile, A.G.: gulated Growth with   Intrinsic Lower Bounds.   [1234]
Stockmeyer, :   Intrinsically Difficult Problems.   [1621]
Kahne, S.   :   Introducing Systems Concepts to All Un   [865]
Squire, E. :   Introducing Systems Design.   [1607]
Baase, S.   : omputer Algorithms: Introduction to Design and Analysis.   [77]
Beer, S.   :   Introduction : Questions of Quest.   [137]
Bellman, R. :   Introduction to Mathematical Theory of   [143]
Bender, E.A.:   An   Introduction to Mathematical Modelling   [150]
Bennett, S. :   An   Introduction to Multivariate Technique   [153]
Bowen, B.D. :   An   Introduction to Data Analysis.   [202]
Dunn, G.   :   An   Introduction to Mathematical Taxonomy.   [447]
Frank, P.M. :   Introduction to System Sensitivity The   [540]
Golubitsky, :   An   Introduction to Catastrophe Theory and   [635]
Gould, P.   : ge of Structure: An   Introduction for Social Scientists, Ge   [649]
Hall, C.A.S.: ry and Practice: an   Introduction with Case Studies.   [713]
Harrison, M.:   Introduction to Formal Language Theory   [728]
Hopcroft, J.:   Introduction to Automata Theory, Langu   [780]
Jeffers, J.N: An   Introduction to Systems Analysis: with   [842]
Kobayashi, H: ng and Analysis: an   Introduction to System Performance Eva   [953]
Koyayashi, H: ng and Analysis: an   Introduction to System Performance Eva   [977]

Krishana, V.:   An Introduction to Category Theory.   [985]
Levien, R.E.:   Introduction to IIASA: Applying System [1041]
Luenberger, :   Introduction to Dynamic Systems: Theor [1071]
Machtey, M. :   An Introduction to the General Theory of   [1081]
Moraga, C. :   Introduction to Linear p-Adic Invarian [1184]
Morris, C.N.:   Introduction to Data Analysis and Stat [1188]
Pierce, J.R.:   Introduction to Communication Science   [1340]
Rastogi, P.N: An Introduction to Social and Management   [1386]
Ricciardi, L: gy and Medicine: An   Introduction .   [1417]
Sain, M.K. :   Introduction to Algebraic System Theor [1482]
Saunders, P.:   An Introduction to Catastrophe Theory.   [1497]
Sonde, B.S. :   Introduction to System Design Using In [1598]
Srivastava, :   An Introduction to Multivariate Statistic [1609]
Strauss, A. :   An Introduction to Optimal Control Theory [1627]
Taylor, J.R.:   An Introduction to Error Analysis.   [1673]
Techo, R.   : Communications: an   Introduction to Concepts and Design.   [1681]
Tremblay, J.:   An Introduction to Data Structures with A [1735]
Uttenhove, H: Problem Solver): An   Introduction and Guide.   [1775]
Van der Ven,:   Introduction to Scaling.   [1794]
Vemuri, V. : Complex Systems: an   Introduction .   [1808]
Wellstead, P: Introduction to Physical System Modell [1856]
Winkel, D. : Digital Design: An   Introduction to Top-Down Design.   [1867]
Gold, H.J. : logical Systems: an   Introductory Guidebook.   [630]
Sampson, J.R: tion Processing: An   Introductory Survey.   [1485]
Atanassov, K: Intuitionstic L-Fuzzy Sets.   [57]
Sierocki, I.:   Invariance-Relations in General Dynami [1553]
Moraga, C. : on to Linear p-Adic   Invariant Systems.   [1184]
Koenderink, : through Photometric   Invariants .   [961]
Andersson, H: Analysis of an   Inventory Control System Using the The [32]
Van der Zouw: Inventory of Systems Research in the N [1796]
Van der Zouw: Inventory of Systems Research on the N [1797]
Matevosyan, :   The Invers Problem of Karhunen-Loeve.   [1117]
Thissen, W. :   Investigations into World 3 Model: Les [1684]
Van der Zand: gnal Theory: Part I   Investigations on the Foundations.   [1795]
Gurdial,   : on Fano Inequality   Involving Distortion.   [690]
Jumarie, G. : ing Dynamic Systems   Involving Human Factors.   [862]
Jumarie, G. : Random Experiments   Involving Fuzzy Observations.   [863]
Greenspan, D: Modeling of Atoms,   Ions and Molecules. (Part I: Hydrogen; [666]
Drechsler, F: ufacturing (CIM) in   Irish Industry.   [436]
Casti, J.   : ttainable Goals and   Irreducible Uncertainties.   [270]
Venkatesan, : nsitive Coupling in   ISM .   [1809]
Cheng, J.K. :   A Subgraph   Isomorphism Algorithm Using Resolution [306]
Cook, N.D. :   An   Isomorphism of Control in Natural, Soc   [345]
Gati, G.   : Bibliography on the   Isomorphism Disease.   [585]
Miller, G.L.:   Graph Isomorphism , General Remarks.   [1166]
Read, R.C. :   The Graph Isomorphism Disease.   [1393]
Schwartz, D.:   Isomorphisms of Spencer-Brown's Laws o [1515]
Troncale, L.: chy & Links between   Isomorphs .   [1747]
Slawski, C. : ation: Arabs Versus   Israeli .   [1573]
Hajek, P.   :   Special   Issue of the International ournal of   [704]
Klir, G.J. : tive Reasoning: The   Issue of Parts and Wholes.   [948]
Leinhardt, S: Special   Issue of Mathematical Sociology on Soc [1035]
            :   Second Special   Issue on the Guha Method of Mechanized [1941]
            :   Special   Issue on Simulation Modelling and Stat [1943]
            :   Special English   Issue of the Polish Journal "Progress   [1942]
Bailey, N.T.:   Some   Issues in the Cybernetics of Disease C [83]
Farley, A.M.:   Issues in Knowledge-Based Problem Solv [507]
Fick, G.   : on Support Systems: Issues and Challenges.   [521]
Lee, R.M.   : t Systems: Semantic   Issues .   [1028]
Tighe, T.J. : ons and Fundamental   Issues .   [1689]
Collins, D.M: m Imperfect Data by   Iterative Entropy Maximization.   [331]
Mark, J.W. :   Application of   Iterative Algorithms to Adaptive Predi [1105]
            : nd Spring School in   Jadwisin , Poland, April 13-19, 1980   [1942]
Sugiyama, K.: tems Development in   Japan .   [1635]

Scavia, D. : Ecological Model of Lake Ontario.   [1503]
Tamura, H. : cation of a Dynamic Lake Model by the Method of Data Handl [1666]
Tamura, H. : : An Application to Lake Ontario.   [1666]
Van Straaten: sk Force Meeting on Lake Balaton Modelling Veszprem (Hunga [1792]
Jorgenson, S: l Modelling of Deep Lakes and Reservoirs.   [858]
Jorgenson, S: l Models of Shallow Lakes (Summary of an IIASA Workshop).   [859]
Abe, J.   : s-World' in Natural Language Processing.   [2]
Bennett, T.J: Language as a Self-Organizing System.   [154]
Buchberger, : VIE-LANG: A German   Language Understandig System.   [238]
Campbell, J.: formation, Entropy,   Language , and Life.   [259]
Delaney, W. : ry Based Simulation   Language .   [404]
Faribault, M: From Natural Language to a Canonical Representation   [504]
Fum, D.   : Approach to Natural   Language Text Representation and Compr   [547]
Gigley, H. : odelling of Natural   Language Processing.   [610]
Gould, P.   :   Q-Analysis or a Language of Structure: An Introduction   [649]
Guida, G.   : entality in Natural   Language Understanding Systems.   [680]
Guida, G.   :   Natural   Language Access to Online Data Bases.   [681]
Harris, L.R.: Parsing of Natural   Language Data Base Queries.   [727]
Harrison, M.: roduction to Formal   Language Theory.   [728]
Kindler, E. : of Some Simulation   Language Concepts.   [922]
Leinfellner,:   Causality and Language : Phonetics / Phonology, Synta   [1034]
Lozinskii, E: Locality in Natural   Language Processing.   [1069]
Nigro, A.   :   Language Behavioral Feedback.   [1228]
Pedretti, A.: The Cybernetics of   Language .   [1308]
Sembugamoort: LAS, a Paradigmatic   Language Acquisition System.   [1523]
Sernadas, A.: emantics as a Query   Language .   [1528]
Strombach, W: Concepts in German   Language .   [1629]
Sussman, G.J: CONSTRAINTS - A   Language for Expressing Almost Hierarc   [1640]
Trattnig, W.: tware Specification   Language Based on the Data Flow Model.   [1734]
Trost, H.   : he Field of Natural   Language Understanding.   [1754]
Trost, H.   : VIE-LANG: A German   Language Dialogue System.   [1753]
Vaccari, E. : odelling in Natural   Language Processing.   [1779]
Wechler, W. : ess in Automata and   Language Theory.   [1851]
Winograd, T.:   Language as a Cognitive Process (Vol.1   [1868]
Pfeilsticker: en Mathematical and   Language-Oriented Economics.   [1331]
Cole, I.D. : alizing Programming   Languages .   [329]
Denning, P.J: Machines, Languages , and Computation.   [409]
Glushkov, V.:   Algebra,   Languages Programming.   [626]
Hopcroft, J.: to Automata Theory,   Languages , and Computation.   [780]
Kindler, E. : ulation Programming   Languages : I. Declaration of Necessary   [917]
Kindler, E. : ulation Programming   Languages : II. Description of Types an   [918]
Kindler, E. : ulation Programming   Languages .   [921]
Lakshmivarah: r Fuzzifying Formal   Languages and Synthesis of Fuzzy Gramm   [1005]
Metzer, J.R.:   Decision Table Languages and Systems.   [1155]
Ratko, I.   : ions in Programming   Languages .   [1389]
Santos, E.S.: listic Machines and   Languages .   [1493]
Stoy, J.E. : oach to Programming   Languages .   [1625]
Vitanyi, P.M: Systems: Structure,   Languages , and Growth Functions.   [1819]
Fontela, E. : Condensed Forms of   Large Scale Models.   [533]
Franksen, O.: litative Aspects of   Large Scale Systems: Developing Design   [541]
Gaines, B.R.: Methodology in the   Large : Modeling All There Is.   [561]
Golledge, R.: nsional Analysis of   Large Data Sets.   [632]
Gregory, S.A: Large Technological Projects.   [667]
Hafele, W. :   Modelling of Large Scale Energy Systems. (IIASA Pro   [698]
Henize, J. :   Modelling   Large Scale Socio-Economic Systems: Ca   [750]
Klement, E.P: Law of   Large Numbers and Central Limit Theore   [935]
Mahmoud, M.S: Large Scale Systems Modelling.   [1089]
May, R.M.   :   Will a   Large Complex System be Stable?   [1130]
Michel, A.N.: itative Analysis of   Large Scale Systems.   [1159]
Oeren, T.I. : g and Simulation of   Large Scale Systems.   [1254]
Oeren, T.I. :   Rationale for Large Scale System Simulation Software   [1255]
Pospelov, D.:   Large Systems: Situational Control .   [1351]
Roberts, P.C: Modelling   Large Systems.   [1427]
Roberts, P.C: Modelling   Large Systems: Limits to Growth Revisi   [1428]

Grenander, U: Regular Structures: Lectures in Pattern Theory. Vol.3.   [669]
Astola, J. : heme and Bounds for Lee-Codes .   [56]
Astola, J. :   The Lee-Scheme and Bounds for Lee-Codes. [56]
Bailey, K.D.: A Multidisciplinary Legacy .   [82]
Lipset, D. : regory Bateson: The Legacy of a Scientist.   [1056]
Jensen, H.L.: ork for Quantifying Legal Decisions.   [844]
Sadovsky, V.: ystem Principle and Leibnitz 's Theoretical Heritage.   [1471]
Sherlock, R.: tion of Limit Cycle Lengths ; Part 2: the State Cycle Fract [1545]
Clark, W.C. :   Lessons for Ecology Policy Design: a C [318]
Thissen, W. : into World 3 Model: Lessons for Understanding Complicated [1684]
Airenti, G. :   A Two Level Model of Knowledge and Belief.   [23]
Gelfand, A.E: gement by Consensus Level and its Interaction with Other M [589]
Kerefky, P. : anipulation on User Level .   [904]
Newell, A. :   The Knowledge Level .   [1218]
Radecki, T. :   Level Fuzzy Sets.   [1373]
Auger, P.   :   Coupling between N Levels of Observation of a System (Bio [67]
Auger, P.   : tive and Individual Levels of Organization in a Hierarchic [68]
Oldershaw, R: On the Number of Levels in the Self-Similar Hierarchica [1263]
Ruotsalainen: h Different Service Levels .   [1463]
Troncale, L.: ins of Hierarchical Levels : An Emergent Evolutionary Proce [1738]
Troncale, L.:   Are Levels of Complexity in Bio-Systems Re [1739]
Castelfranch: A Lexicon Based Approach to Sentence Und [269]
Gupta, M.M. : ncertain Plants via Liapunov Signal Synthesis Method.   [687]
Glanville, R: inguished and Exact Lies .   [621]
Baltes, P.B.:   Life Span Development and Behavior, Vo [94]
Burghes, D.N: ial, Management and Life Sciences.   [243]
Campbell, J.: ropy, Language, and Life .   [259]
Economos, A.: n a Systems View of Life .   [454]
Kerman, C.E.: eative Tension: the Life and Thought of Kenneth Boulding. [907]
Resconi, G. : ems-Applications to Life Sciences.   [1404]
Caselles, A.: are Theories in the Light of General Systems Theory.   [268]
Klement, E.P: Numbers and Central Limit Theorem for Fuzzy Random Variabl [935]
Sherlock, R.: and Distribution of Limit Cycle Lengths; Part 2: the State [1545]
Keyes, R.W. :   Fundamental Limits in Digital Information Processi [909]
Roberts, P.C: ling Large Systems: Limits to Growth Revisited.   [1428]
Wyner, A.D. :   Fundamental Limits in Information Theory.   [1885]
Vitanyi, P.M: Lindenmayer Systems: Structure, Langua [1819]
Blomberg, H.: y for Multivariable Linear Systems.   [175]
Casti, J.   : Their Applications: Linear Theory.   [271]
Egiazarian, : mard Transforms and Linear Filters.   [459]
El-Sherief, : tion and Control of Linear Discrete Multivariable Systems [470]
El-Sherief, : on and Modeling for Linear Multivariable Discrete-Time Sys [468]
Gupta, M.M. : ic Optimization for Linear Discrete Systems.   [688]
Kucera, V. :   Discrete Linear Control: the Polynomial Equatio [992]
Leigh, J.R. : tional Analysis and Linear Control Theory.   [1033]
Moraga, C. :   Introduction to Linear p-Adic Invariant Systems. [1184]
Nugyen, V.V.: and Unification of Linear Identifiability Concepts.   [1247]
Santis, F.de: alysis and Fisher's Linear Discriminant Analysis.   [1492]
Sengupta, J.: blems in Stochastic Linear Programming.   [1525]
Strejc, V. : Theory of Discrete Linear Control.   [1628]
Takeuchi, A.: of Projection onto Linear Subspeces.   [1664]
Toutenburg, : rior Information in Linear Models.   [1708]
Vallee, R. : lution of Dynamical Linear System with Random Initial Cond [1782]
Wang, Z.   : on Making and Fuzzy Linear Programming in Production and P [1837]
Zeleny, M. : ces and Profits Via Linear Programming.   [1923]
Graham, J.H.:   Linguistic Decision Structures for Hie [653]
Kickert, W.J: An Example of Linguistic Modelling.   [910]
Mamdani, E.H: ate Reasoning Using Linguistic Synthesis.   [1098]
Rao, G.P.   : ping Functions from Linguistic Descriptions in Problems of [1380]
Tong, R.M. :   A Linguistic Approach to Decision Making [1701]
Walichiewicz: Decomposition of Linguistic Rules in the Design of a Mu [1828]
Yager, R.R. :   Linguistic Model and Fuzzy Truths. [1888]
Yager, R.R. :   A Linguistic Variable for Importance of [1892]
Andersson, H: ysis of an Inventory Linguistic ystem Using the Theory of Re [32]

Jumarie, G. : rmation Fuzzy Sets,   Linguistics , Relativistic Sets and Com [861]
Mel'cuk, I.A: Cybernetics and Linguistics : Some Reasons for as Well [1148]
Mel'cuk, I.A: Cybernetics and Linguistics .   [1149]
Yager, R.R. :   Lingustic Hedges: Their Relation to Co [1904]
Rapoport, A.:   Cybernetics as a Link between two Theories of Cognition [1383]
Troncale, L.: n-Made Systems: The   Linkage   Proposition Model.   [1749]
Troncale, L.:   Linkage   Propositions between Fifty Pri [1737]
Troncale, L.:   Linkage   Propositions between Principal [1743]
Troncale, L.: lization of Systems   Linkage   Propositions.   [1746]
Kinston, W. :   Linking   System Research and Delivery S [928]
Troncale, L.: sts for Hierarchy &   Links between Isomorphs.   [1747]
Holst, P.   : 76: An Index to the   Literature .   [775]
Scoltock, J.:   A Survey of the Literature on Cluster Analysis.   [1517]
Verhagen, C.: ther Disciplines; a   Literature Survey.   [1812]
Bosserman, R: rocesses in General   Living Systems.   [196]
Bosserman, R: s and Subsystems in   Living Systems.   [195]
Durkin, J.E.:   Living Groups: Group Psychotherapy and [449]
Ferracin, A.: ganizing Ability in   Living Systems.   [519]
Gatlin, L.L.: tion Theory and the   Living Systems.   [586]
Maturana, H.:   Realization of the   Living .   [1128]
Miller, J.G.:   Living Systems.   [1167]
Slawski, C. : Theory of Community   Living Arrangements and a Practical Ut [1579]
Zeleny, M.   : oiesis: A Theory of   Living Organization.   [1922]
Tracy, L.   :   A Dynamic   Living-Systems Model of Work Motivatio [1709]
Bubnicki, Z.:   Global and   Local Identification of Complex System [235]
Holden, A.V.: roach to Global and   Local Mode of Action of General Aneste [770]
Lozinskii, E: Locality in Natural Language Processin [1069]
Hogeweg, P. :   Locally Synchronised Developmental Sys [769]
Gloria-Botti: cyte Acid Phosphate   Locus in Humans.   [625]
Favella, L.F: Application of   Loeve-Karhunen Expansion to Acoustic D [512]
Sworder, D.D: Sensor Failures on   Log Regulators.   [1647]
Andrew, A.M.:   Logic and Continuity - a Systems Dicho   [36]
Belnap, N.P. :   The   Logic of Questions and Answers.   [144]
Braae, M.   : ameters for a Fuzzy   Logic Controller.   [209]
Farinas-del-: rogramming in Modal   Logic .   [505]
Haack, S.   :   Do We Need 'Fuzzy   Logic '?   [697]
Kowalaski, R: Logic of Problem Solving.   [976]
Kozen, D.   : ositive First-Order   Logic Is NP-Complete.   [978]
Lipski, W.   :   On the   Logic of Incomplete Information.   [1058]
Mamdani, E.H: pplication of Fuzzy   Logic to Approximate Reasoning Using L [1098]
Morein, R.   : alysis of Threshold   Logic Unit Networks.   [1185]
Negoita, C.V: On the   Logic of Discrete Systems Dynamics.   [1211]
Pappis, C.P.:   A Fuzzy   Logic Controller for a Traffic Junctio [1280]
Rao, G.P.   : n Problems of Fuzzy   Logic Applied to Systems Control.   [1380]
Resconi, G. : sis by Mathematical   Logic .   [1405]
Shoesmith, D: Multiple-Conclusion   Logic .   [1549]
Sutherland, : heoretical Decision   Logic .   [1643]
Von Wright, :   The   Logic of Preference.   [1824]
             :   Non-Monotonic Logic (Special Issue).   [1936]
Brightman, H: Problem Solving: A   Logical and Creative Approach.   [219]
Brightman, H: Problem Solving: A   Logical and Creative Approach.   [220]
Dominique, J: Logical Construction of Systems.   [430]
Gaines, B.R.:   Logical Foundations for Database Syste [559]
Gasparski, W: An Outline of the   Logical Theory of Design.   [576]
Kubinski, T.: An Outline of the   Logical Theory of Questions.   [991]
Nowakowska, :   The   Logical Structure of the Development o [1240]
Ratko, I.   : On Evaluating of   Logical Expressions in Programming Lan [1389]
Resconi, G. :   Logical Analysis of Dynamical Systems- [1404]
Resconi, G. : General Systems by   Logical and Algebraic Structures.   [1406]
Svoboda, A. :   Advanced   Logical Circuit Design Techniques.   [1645]
Tjoa, A.M.   : iversal Relation as   Logical View of Databases.   [1692]
Warnier, J.-:   Logical Construction of Systems.   [1842]
Pospelov, D.:   Logical-Linguistic Models in Control S [1352]
Hooker, C.A.: Physical Theory as   Logico-Operational Structure.   [779]

Edmons, E.A.:   Lattice Fuzzy Logics .   [455]
Gaines, B.R.: ability Uncertainty Logics .   [556]
Skala, H.J. :   On Many-Valued Logics , Fuzzy Sets, Fuzzy Logics and T [1570]
Skala, H.J. : , Fuzzy Sets, Fuzzy Logics and Their Applications.   [1570]
            :   Logik , Erkenntnistheorie, Wissenschaft [1935]
Tripp, R.S. : on within Air Force Logistics Command.   [1736]
Millendorfer: tical Approaches to Long Economic Waves.   [1165]
Tussey, C.C.:   The Kentucky Longevity Information System.   [1763]
Erlandson, R: ting Process. A New Look .   [488]
Locker, A. :   A New Look at the Description and Prescripti [1059]
Syski, R.   :   Processes: a First Look .   [1649]
Matsuno, K. : dness, and Feedback Loops in Open Systems.   [1123]
Bandler, W. :   On the Theory of Lorrain-Categories .   [98]
Dolby, J.L. : ity and Information Loss .   [427]
Valdes-Perez: Information Loss Due to Data Quantization in Recon [1781]
Colby, J.A. : Documentation for a Lotic Ecosystem Model.   [328]
McIntyre, C.: erarchical Model of Lotic Ecosystems.   [1139]
McIntyre, C.: Documentation for a Lotic Ecosystem Model.   [1140]
Slawski, C. :   Love , Power and Conflict: A Systems Mo [1577]
Hartl, R.F. : Dynamic Continuous Lovers Under Risk of Being Killed a Ri [729]
Schmutzer, M: Hi & Low - In & Out: Approaches to Social S [1512]
Aracil, J. : ctural Stability of Low-Order System Dynamics Models.   [44]
Nobile, A.G.: owth with Intrinsic Lower Bounds.   [1234]
Rozenberg, G: hematical Theory of LSystems .   [1454]
Beer, S.   : inal: Eighth Annual Ludwig Von Bertalanffy Memorial Lectur [135]
Boulding, K.: stem: Fourth Annual Ludwig von Bertalanffy Memorial Lectur [199]
Wang, Z.   :   M-Fuzzy Numbers and Random Variables. [1838]
Globus, G.G:: here a Ghost in the Machine after all?   [624]
Jakubowski, : Characterization of Machine Parts Shapes.   [830]
Kaiser, J.S.: ithin a Determinate Machine .   [868]
Michie, D. :   Machine Intelligence and Related Topic [1161]
Matsuno, K. : of Self-Reproducing Machineries : Multiplication of Microsy [1120]
Aggarwal, S.: Systems by Ergodic Machines .   [17]
Aulin-Ahmava: uinly Self-Steering Machines : A Fundamental Theorem of Act [74]
Denning, P.J: Machines , Languages, and Computation.   [409]
Laing, R.   :   Machines as Organisms: an Exploration [1004]
Santos, E.S.:   Probabilistic Machines and Languages.   [1493]
Stoklosa, J.: s-Networks of (a,k)- Machines Generating Cyclic Sequences. [1622]
Greenspan, D: icrocosm and in the Macrocosm .   [665]
Neck, R.   : nalysis of a Simple Macroeconomic Model.   [1207]
De Rosnay, J: The Macroscope : a New World Scientific Sys [399]
De Greene, K: a in Sociotechnical Macrosystems : Theories and Models.   [390]
O'Donovan, T: GSPS: Simulation Made Simple.   [1250]
Augustine, N: ugustine's Laws and Major System Development Programs.   [71]
Lasker, G.E.: he Relation between Major World Problems and Systems Learn [1011]
Benedikt, S.: Repetitive Decision Making under Risk.   [151]
Cavallo, R.E: Decision Making in Reconstructability Analysis. [285]
Driankow, D.: objectives Decision Making with Ordinal Information.   [439]
Efstathiou, : iattribute Decision Making Using a Fuzzy Heuristic Approac [457]
Fandel, G. : e Criteria Decision Making Theory and Application. (Proc.   [502]
Hoffer, J.A.: nd Batched Decision Making .   [766]
Hwang, C.L. : Objective Decision Making : Methods and Applications (a St [801]
Hwang, C.L. : Attribute Decision Making : Methods and Applications: A St [803]
Shakun, M.F.: esolution in Policy Making .   [1532]
Shakun, M.F.:   Policy Making and Meaning as Design of Purpos [1533]
Sharif, N. : Models for Decision Making .   [1537]
Starr, M.K. : e Criteria Decision Making .   [1610]
Tong, R.M. : pproach to Decision Making with Fuzzy Sets.   [1701]
Umpleby, S. :   Methods for Making Social Organisation Adaptive.   [1770]
Viliums, E.R: uating and Decision Making Under Uncertainty and Multiple [1817]
Waller, R.J.: ies of Human Policy Making .   [1835]
Wang, Z.   : n of Fuzzy Decision Making and Fuzzy Linear Programming in [1837]
Wildhelm, J.: iobjective Decision Making Under Uncertainty.   [1864]
Yager, R.R. :   Fuzzy Decision Making Including Unequal Objectives.   [1887]

Walker, C.C.: plex Organizations: Management   by Exception, Priority, and [1830]
Probst, G.J.: sitivity Models for Managers , Planners and Politicians in [1364]
Smith, A.W. : Mangagement Systems Analyses and Appli [1587]
Horn, W.   :   The Structure of Manifestations in a Medical Consultati [785]
Kerefky, P. : am Optimization and Manipulation on User Level. [904]
Milgram, M. : ycentric Entropy of Mankov Systems. [1163]
Bryant, D.T.: Manpower Planning and Organization Des [234]
Drechsler, F: Computer Integrated Manufacturing (CIM) in Irish Industry. [436]
Mulej, M.   : istic Creativity of Many (WCM)" Applied to Collect and Org [1195]
Skala, H.J. :   On Many-Valued Logics, Fuzzy Sets, Fuzzy [1570]
Balossino, N: ECG Map Filtering by Means of Spherical Ha [93]
Favella, L.F: Foundations of ECG Map Reconstruction. [510]
Warfield, J.: ntary Relations and Map Reading. [1841]
Martin, J.N.:   On Mapping Real Systems. [1109]
Rao, G.P.   : e Reconstruction of Mapping Functions from Linguistic Desc [1380]
Slawski, C. : Mapping Joint Action. [1578]
Warfield, J.: h₃₀'y and Hierarchy Mapping . [1839]
Batchelor, B: ssification of Pole Maps in the Complex Plane. [118]
Kilmann, R.H: tive Theory and the Maps Technology. [916]
Bolter, M. : uctural Analysis in Marine Ecosystems. [185]
Agassi, J. : Essays in Honour of Mario Bunge. [14]
Mattessich, : Differences Between Mario Bunges's World of the Systems an [1127]
Aulin, A.   : and a Criticism of Marixsm . [73]
Kaynak, E. : olistic Approach in Marketing Planning. [892]
Holler, M.J.: Markets as Self-Policing Quality Enfor [772]
Barfoot, C.B: nditional Absorbing Markov Chains. [107]
Cover, T.M. : ences with Apparent Markov Structure. [353]
Hartley, R. : ent Developments in Markov Decision Processes. [730]
Monahan, G.E: artially Observable Markov Decision Processes Theory, Mode [1183]
Teodorescu, : netical Actions and Markov Processes. [1682]
Barber, M.C.:   A Markovian Model for Ecosystem Flow Ana [105]
Joerres, R. : of the Temperature Masking of the Circadian System of Eug [847]
Hirata, H. :   Utility Efficiency Mass and its Stability. [759]
Hirata, H. :   A Model of Mass and Energy Flow in Ecosystems. [757]
Kryukov, V.I: etastable State. I. Master Equation Approach. [988]
Broekstra, G: MAMA: Management by Matching . A Consistency Model for Orga [227]
Cordier, M.O: on: Another Way for Matching Patterns in KBS. [346]
Shapiro, L.G: iptions and Inexact Matching . [1535]
Varshavskiy,: II. Composition and Matching . [1801]
Cavallo, R.E: Supplementary Material to the SGSR Report on Systems [279]
Dimitrova, L: the Organization of Material Flow in the Industrial Enterp [423]
Kindler, E. :   Systems of Material Flow. [923]
Matsuno, K. : Material Self-Assembly as a Physiochem [1124]
Bavel, Z.   : Math Companion for Computer Science. [126]
Ackerman, E.: Mathematical Models in the Health Scie [7]
Andrews, J.G: Mathematical Modelling. [38]
Aris, R.   : Mathematical Modelling Techniques. [52]
Ballonoff, P: Mathematical Demography of Social Syst [92]
Bellman, R. :   Introduction to Mathematical Theory of Control Process [143]
Bender, E.A.: An Introduction to Mathematical Modelling. [150]
Biswas, A.K.: Mathematical Modelling and Environment [171]
Boyce, W.E. :   Case Studies in Mathematical Modeling. [204]
Bradley, R. :   Case Studies in Mathematical Modelling: A Course Book [210]
Brugger, K. :   A Simple Mathematical Model of Biphasic Insulin [232]
Burghes, D.N: Mathematical Models in the Social, Man [243]
Chytil, M.K.: Mathematical Methods as Cognitive Prob [314]
Ciriani, T.A: Mathematical Models for Surface Water [316]
Coffman, C.V: ctive Approaches to Mathematical Models. [325]
Colby, J.A. : Mathematical Documentation for a Lotic [328]
Collot, F. :   A New Mathematical Concept for General Syste [333]
Dunn, G.   : An Introduction to Mathematical Taxonomy. [447]
Dym, C.L.   : Principles of Mathematical Modeling. [453]
Eisen, M.   : Mathematical Models in Cell Biology an [465]
Favella, L.F: Mathematical Foundations of ECG Map Re [510]

Favella, L.F: Methodological and Mathematical Problems in ECG Analysis.   [513]
Frauenthal, :   Mathematical Modelling in Epidemiology.   [544]
Georgiev, A.:   Nonparametric Mathematical Model from Individual Hum   [597]
Getz, W.M. :   Mathematical Modelling in Biology and   [602]
Gold, H.J. :   Mathematical Modeling of Biological Sy   [630]
Gruska, J. :   Mathematical Foundations of Computer S   [678]
Hawkes, N. : eminar on Trends in   Mathematical Modelling.   [739]
Hawkes, N. : eminar on Trends in   Mathematical Modelling: a UNESCO Repor   [740]
Jacoby, S.L.:   Mathematical Modeling with Computers.   [826]
James, D.J.G: Case Studies in Mathematical Modelling.   [831]
Kaufmann, A.:   Mathematical Models for the Study of t   [886]
Kavakoglu, I: Mathematical Modelling of Energy Syste   [890]
Kuhn, H.W. :   Mathematical Systems Theory and Econom   [994]
Leinhardt, S: Special Issue of Mathematical Sociology on Social Netwo   [1035]
Lester, F.K.:   Mathematical Problem Solving.   [1038]
Lions, J.I. : Some Methods in the   Mathematical Analysis of Systems and T   [1054]
Lions, J.L. : Some Methods in the   Mathematical Analysis of Systems and T   [1055]
Martin, N.F.:   Mathematical Theory of Entropy.   [1110]
McIntosh, J.:   Mathematical Modelling and Computers i   [1138]
McIntyre, C.: nternal Report 165:   Mathematical Documentation for a Lotic   [1140]
Miller, M.C.:   Mathematical Models in Medical Diagnos   [1168]
Molina, F.W.: ve Decomposition in   Mathematical Programming.   [1181]
Morley, D.A.:   Mathematical Modelling in Water and Wa   [1187]
Oden, A.S. :   An Overview of Mathematical Modeling of the Behavior   [1252]
Pfeilsticker: ing the Gap between   Mathematical and Language-Oriented Eco   [1331]
Pichler, F. : tems Algorithms for   Mathematical Systems Theory.   [1339]
Porenta, G. :   A Mathematical Model of the Dopaminergic   [1349]
Resconi, G. : ine Biosynthesis by   Mathematical Logic.   [1405]
Roberts, F.S: Discrete   Mathematical Models.   [1424]
Rozenberg, G: The   Mathematical Theory of LSystems.   [1454]
Schimanovich: c Theory of Winning   Mathematical Games and Economic Action   [1510]
Segel, L.A. :   Mathematical Models in Molecular and C   [1521]
Sellers, P.H: torial Complexes: A   Mathematical Theory of Algorithms.   [1522]
Straszak, A.:   New Trends in Mathematical Modelling.   [1626]
Suzuki, R. :   Mathematical and Phenomenological Stud   [1644]
West, B.J. :   Mathematical Models as a Tool for the   [1859]
Yager, R.R. : nd Specificity in a   Mathematical Theory of Evidence.   [1907]
Zimmermann, : ty and Measuring of   Mathematical Models.   [1927]
Bosserman, R: The Role of Mathematics   in Systems Science.   [194]
Burkhardt, H: The Real World and   Mathematics .   [245]
Casti, J.   : on or Reality?: the   Mathematics   of Attainable Goals and Ir   [270]
Dhrymes, P.J: Mathematics   for Econometrics.   [417]
Dold, A.   :   Lecture Notes in Mathematics , Vol.453   [428]
Greene, D.H.:   Mathematics   for the Analysis of Algori   [662]
Greenspan, D: Arithmetic Applied   Mathematics .   [664]
Kuyk, W.   : Complementarity in   Mathematics .   [995]
Smale, S.   :   The   Mathematics   of Time: Essays on Dynamic   [1583]
Wilder, R.L.:   Mathematics   as a Cultural System.   [1863]
Arbib, M.A. : Theory: The Hankel   Matrix .   [48]
El-Sherief, : e Transfer-Function   Matrix Form.   [469]
Jantzen, J. : ed Fuzzy Relational   Matrix .   [836]
Ryan, R.M. : Delivery Systems: A   Matrix Organization Approach.   [1467]
Iri, M.   :   Use of Matroid Theory in Operations Research,   [814]
Matejko, A. : ia of Social System   Maturity .   [1116]
Collins, D.M: y Iterative Entropy   Maximization .   [331]
Levine, R.D.:   The   Maximum Entropy Formalism.   [1042]
Sperb, R.P. :   Maximum Principles and their Applicati   [1604]
Van Campenho: Maximum Entropy and Conditional Probab   [1786]
Kapur, J.N. :   On Maximum-Entropy Complexity Measures.   [880]
Aplin, J.C. : rnetic Model of the   MBO Process.   [39]
Dammasch, I.: Randomly Connected   McCulloch-Pitts Networks: Differences   [375]
Cerbone, G. :   Mean Variance and Skewness of the Firs   [290]
Gasparski, W: What Does it Mean a Notion of "System Modelling".   [583]
Jahoda, M. :   Wholes and Parts, Meaning and Mechanism.   [827]

Shakun, M.F.:   Policy Making and   Meaning   as Design of Purposeful System [1533]
Strombach, W: alt, System: On the   Meaning   of these Concepts in German La [1629]
Byerly, H. :   eory Mechanisms and   Meanings .   [254]
Balossino, N: CG Map Filtering by   Means   of Spherical Harmonics: A Simple   [93]
Takeuchi, A.: Unified Approach by   Means   of Projection onto Linear Subspe [1664]
Conant, R.C.:   Simple Information   Measure .   [338]
Groen, J.C.F: Constraint Analytic   Measure   for the Distance between Neigh   [671]
Mackenzie, K: A Process Based   Measure   for the Degree of Hierarchy in [1083]
Yager, R.R. :   On the   Measure   of Fuzziness and Negation. Par [1890]
Ghiselli, E.:   Measurement   Theory for the Behavioral   [606]
Mason, R.O. :   Measurement   for Management Decision.   [1114]
Mock, T.J. :   Measurement , Accounting, and Organizat [1177]
Nowakowska, : logical Problems of   Measurement   of Fuzzy Concepts in the S [1239]
Roberts, F.S: Measurement   Theory with Applications t [1425]
Rosen, R.   :   Fundamentals of   Measurement   and Representation of Natu [1441]
Scott, W.A. : ructure: Theory and   Measurement   of Individual Differences. [1519]
Van der Zand: Measurement   Theory and Signal Theory: [1795]
Broekstra, G: eses by Information   Measures .   [226]
De Luca, A. : vergence of Entropy   Measures   of a Fuzzy Set.   [395]
Dubois, D. :   A Class of Fuzzy   Measures   Based on Triangular Norms.   [444]
Goodman, L.A: Measures   of Association for Cross Clas   [642]
Higashi, M. :   On   Measures   of Fuzziness and Fuzzy Comple   [755]
Higashi, M. :   Measures   of Uncertainty and Informatio   [754]
Hoehle, U. : ect to Plausibility   Measures .   [764]
Hooda, D.S. :   Two Generalized   Measures   of "Useful" Information.   [778]
Kapur, J.N. : -Entropy Complexity   Measures .   [880]
Loo, S.G.   :   Measures   of Fuzziness.   [1063]
Melkumian, A: Distortion   Measures   in Image Data Compression and [1151]
Okuda, T.H. : n Using Probability   Measures   of Fuzzy Events.   [1261]
Yager, R.R. :   Measuring   Tranquility and Anxiety in D [1905]
Zimmermann, :   Testability and   Measuring   of Mathematical Models.   [1927]
Labudde, R.A: iscrete Hamiltonian   Mechanics .   [1001]
Jahoda, M. :   Parts, Meaning and   Mechanism .   [827]
Ashby, R.   :   Mechanisms   of Intelligence: Ross Ashby   [55]
Byerly, H. : Evolutionary Theory   Mechanisms   and Meanings.   [254]
Kleyle, R. :   Switching   Mechanisms   in a Generalized Informatio   [936]
Kobsa, A.   : on: A Survey of its   Mechanisms , a Sketch of its Semantics.   [957]
Hajek, P.   :   the Guha Method of   Mechanized   Hypothesis Formation.   [704]
            :   the Guha Method of   Mechanized   Hypothesis Formation.   [1941]
Hajek, P.   :   Mechanizing   Hypothesis Formationmathem   [703]
Garfolo, B.T: The Renal Dialysis   Medical   Information System.   [573]
Horn, W.   : lligence System for   Medical   Decision Support.   [786]
Horn, W.   : Manifestations in a   Medical   Consultation System.   [785]
Kerin, T.   : mance Assessment of   Medical   Radioisotope Equipment.   [906]
Miller, M.C.: hematical Models in   Medical   Diagnosis.   [1168]
Troncale, L.:   the Biological and   Medical   Sciences.   [1750]
Zorbas, Y.G.:   Medical   Monitoring of Man's Reliabilit [1929]
Ricciardi, L: tics in Biology and   Medicine : An Introduction.   [1417]
Trappl, R. :   puter Efficiency in   Medicine .   [1711]
Trappl, R. : Side of Innovation: Medicine - a Hard Case.   [1716]
            : ive Systems Theory.   Medicine .   [1940]
            :   and Methodology of   Medicine .   [1939]
Petrovski, A: Medico-Demographic Models in Health Ca   [1330]
Best, D.P. : tion Structure in a   Medium   Sized UK Company.   [164]
Gheorghe, A.: ransition towards a   Medium   Developed Economy.   [605]
Banathy, B.H: of the 29th Annual   Meeting .   [97]
Van Straaten: TA/IIASA Task Force   Meeting   on Lake Balaton Modelling Vesz [1792]
Gigley, H. : ficial Intelligence   Meets   Brain Theory. An Integrated Appr   [610]
Apple, A.   :   Megasynthesis .   [42]
Yager, R.R. : d Negation. Part 1:   Membership   in the Unit Interval.   [1890]
Yager, R.R. :   Membership   in a Compound Fuzzy Subset. [1908]
Beer, S.   : wig Von Bertalanffy   Memorial   Lecture.   [135]
Boulding, K.: wig von Bertalanffy   Memorial   Lecture.   [199]
Kohonen, T. :   Associative   Memory : a Systems-Theoretical Approach   [964]

Kovalenko, Y: Short-Term Memory   as a Metastable State. II. Simu [974]
Kryukov, V.I: Short-Term Memory   as a Metastable State. I. Maste [988]
Ventriglia, : of Neural Systems: Memory   Effects.   [1810]
Ackoff, R.L.: Art and Science of Mess   Management.   [10]
Brown, J.H.U: and Coordination of Metabolic   Processes: a Systems Approac   [230]
            :   Metamedicine . A New International Jour   [1939]
Chandrasekar: l and Social System Metaphors   for Distributed Problem Solv [292]
            : ssenschaftstheorie,   Metaphysik , Theologie.   [1935]
Barmark, J. :   Perspectives in Metascience .   [108]
Kovalenko, Y: rt-Term Memory as a Metastable   State. II. Simulation Model   [974]
Kryukov, V.I: rt-Term Memory as a Metastable   State. I. Master Equation A   [988]
Kickert, W.J: A Metasystem   Approach to Organizational   [912]
Van Gigch, J: cience, Systems and Metasystem   Paradigms.   [1787]
Auguin, M. :   Systematic Method   of Realization of Interpreted P   [70]
Caselles, A.:   A Method   to Compare Theories in the Ligh   [268]
Cole, I.D. : ized Programming: A Method   of Naturalizing Programming Lan   [329]
Davies, A.J.: The Finite Element Method : A First Approach.   [380]
Gupta, M.M. : ov Signal Synthesis Method .   [687]
Hajek, P.   : Studies on the Guha Method   of Mechanized Hypothesis Format   [704]
Havranek, T.: al Data by the Guha Method : an Example.   [737]
Havranek, T.: rmation in the Guha Method .   [738]
Horvath, T. :   Method   and Systematic Reflections.   [789]
Kerin, T.   :   A Method   for Performance Assessment of M [906]
Mulej, M.   :   Method   "Wholistic Creativity of Many ([1195]
Mumford, E. : sign: Structure and Method .   [1196]
Randers, J. : the System Dynamic Method .   [1379]
Rosenkrantz,:   Inference, Method   and Decision.   [1448]
Rubinstein, : ion and Monte Carlo Method .   [1456]
Sakawa, M. : ive Goal Attainment Method   for Multiobjective Nonconvex Pr   [1483]
Steinacker, :   Focussing as a Method   for Disambiguation and Anaphora [1617]
Tamura, H. : c Lake Model by the Method   of Data Handling: An Applicatio   [1666]
Toulotte, J.:   A Method   for Decomposing Interpreted Pet   [1707]
Troncale, L.:   A Hybrid Systems Method : Tests for Hierarchy & Links be   [1747]
            : l Issue on the Guha Method   of Mechanized Hypothesis Format   [1941]
Blauberg, I.: : Philosophical and Methodological   Problems.   [173]
Ciborra, C. :   A Methodological   Inquiry of Organization [315]
Favella, L.F: Methodological   and Mathematical Proble   [513]
Gasparski, W: Design and System: Methodological   Aspects, Vol.3.   [577]
Gvishiani, J: Systems Research: Methodological   Problems - 1979 Yearboo [694]
Jean, R.V. :   A Methodological   Paradigm in Plant Biolo [841]
Lewicka-Strz: stems Approach as a Methodological   Attitude. An Empirical   [1045]
Mitroff, I.I: Methodological   Approaches to Social Sc   [1172]
Nowakowska, :   Methodological   Problems of Measurement   [1239]
Nowakowska, :   New Theoretical, Methodological , and Empirical Possibil [1244]
Rescher, N. :   Methodological   Pragmatism: a Systems-T   [1400]
Van Gigch, J: A Methodological   Comparison of the Scien [1787]
            :   Systems Research: Methodological   Problems - 1982 Yearboo [1944]
Bahm, A.J. :   Methodologies   of Five Types of Philoso   [80]
Baptistella,:   Fuzzy Methodologies   for Interactive Multicri   [104]
McGowan, C.L: position and Design Methodologies .   [1137]
Oren, T.I. : Advanced Simulation Methodologies .   [1270]
Revell, N. :   Systematic Methodologies   for the Design of Small   [1414]
Adam, A.   :   Discrete Systems Methodology .   [11]
Burton, D.J.:   Methodology   and Epistemology for Secon   [251]
Cavallo, R.E: The Role of Systems Methodology   in Social Science Research   [277]
Cavallo, R.E: lopments in Systems Methodology   for Social Science Researc [278]
Cavallo, R.E: General Systems Methodology   and Large-Scale Systems.   [275]
Cavallo, R.E: Systems Methodology   in Social Science Research   [287]
Furukawa, O.:   A Methodology   for Quality Goal-Seeking a [550]
Gaines, B.R.:   Methodology   in the Large: Modeling All   [561]
Gasparski, W: Problems of Design Methodology .   [574]
Gasparski, W: Studies in Design Methodology   in Poland (Special Issue o   [575]
Gasparski, W: Idem: Or on Design Methodology   as Seen from Neither an En   [582]
Gasparski, W: idem: Or on Design Methodology   as Seen from neither Engli   [584]

Hofferbert, : of General Systems Methodology to the Comparative Study o [767]
Huxham, C.S.: s as a Soft Systems Methodology .   [800]
Jennergren, : a Systems Analysis Methodology .   [843]
Keravnou, E.: Conversation Theory Methodology .   [903]
Klir, G.J. : ems Problem Solving Methodology .   [946]
Klir, G.J. :   On Systems Methodology and Inductive Reasoning: T [948]
Kobayashi, H: formance Evaluation Methodology .   [953]
Koyayashi, H: formance Evaluation Methodology .   [977]
Mattessich, : asoning and Systems Methodology : an Epistemology of the Ap [1126]
Mattessich, : and My Own Systems Methodology .   [1127]
Mihram, G.A.:   Simulation: Methodology for Decision Theorists.   [1162]
Mingers, J.C: ry and Soft Systems Methodology .   [1171]
Nowak, S.   :   Methodology of Sociological Research. [1238]
Prevost, P. :   "Soft" Systems Methodology , Functionalism and the Soc [1359]
Rastogi, P.N: Part I): Theory and Methodology .   [1384]
Sage, A.P. :   Methodology for Large Scale Systems.   [1474]
Shaeffer, D.: A Model Evaluation Methodology Applicable to Environmenta [1530]
Sutherland, :   Societal Systems: Methodology , Modelling and Management. [1642]
Topolski, J.:   Methodology of History.   [1704]
Troncale, L.: l Survey of Systems Methodology . (Proc.of the Twenty-Sixth [1742]
Troncale, L.: l Survey of Systems Methodology .   [1741]
Van den Boga: in Response Surface Methodology .   [1793]
Yeh, R.T.   : ends in Programming Methodology : Vol. IV - Data Structurin [1912]
Zeigler, B.P: Methodology in Systems Modelling and S [1919]
Zouwen, J.v.: ribution of Systems Methodology .   [1930]
            :   for Philosophy and Methodology of Medicine.   [1939]
            :   earch: A Science, a Methodology , a Technology (Proceedings [1933]
Apostolakis,: thesis and Analysis Methods   for Safety and Reliability Stu [40]
Batchelor, B: Two Methods   for Finding Convex Hulls of Pl [121]
Bunn, D.W. :   Formal Methods   in Policy Analysis.   [242]
Casti, J.   :   Topological Methods   for Social and Behavioral Syst [274]
Chow, G.C. : Analysis by Control Methods .   [309]
Chytil, M.K.:   Mathematical Methods   as Cognitive Problem-Solvers. [314]
Desouza, G.R: System Methods   for Socioeconomic and Environm [411]
Dirickx, Y.M: lysis by Multilevel Methods : with Applications for Economi [424]
Dominowski, :   Research Methods .   [431]
Estep, M.L. :   Toward Alternative Methods   of Systems Analysis: The Case [494]
Favella, L.F: lated Computational Methods .   [511]
Fiacco, A.V.:   Extremal Methods   and Systems Analysis.   [520]
Fletcher, R.:   Practical Methods   in Optimization. Vol.1: Uncons [530]
Gasparski, W: onference in Design Methods .   [579]
Gehrlein, W.:   of Simple Counting Methods   for Ordering Incomplete Ordina [588]
Gerardy, R. : periments with Some Methods   for the Identification of Fini [599]
Grant, D,P. :   Design Methods   and Theories: Vol.15, No.2.   [656]
Grant, D,P. :   Design Methods   and Theories: Vol.15, No.3.   [657]
Hanson, O.J.: signing Systems and Methods   of Training Using Systems Tech [722]
Hazlett, B.A: Quantitative Methods   in the Study of Animal Behavio [744]
Hickin, J. : rojective Reduction Methods   to Estimation and Control.   [753]
Hwang, C.L. : ve Decision Making: Methods   and Applications (a State-of-t [801]
Hwang, C.L. : te Decision Making: Methods   and Applications: A State-of-t [803]
Lions, J.I. :   Some Methods   in the Mathematical Analysis o [1054]
Lions, J.L. :   Some Methods   in the Mathematical Analysis o [1055]
Makridakis, :   Forecasting: Methods   and Applications.   [1095]
Marko, H.   :   Methods   of System Theory.   [1106]
McLean, J.M.:   Methods   of System Analysis and Model B [1141]
Nachane, D.M: Optimization Methods   in Multi-Level Systems.   [1199]
Pearl, J.   : c Basis for Certain Methods   of Evaluating Probabilistic Fo [1304]
Protzen, J.-:   Design Methods   and Theories: Vol.10, No.2.   [1366]
Raghaven, V.: heoretic Clustering Methods .   [1377]
Simon, H.A. : Other Topics in the Methods   of Science.   [1562]
Soete, W.   :   Structural Methods   for Decomposition of Large-Sca [1596]
Southwood, T: Ecological Methods   (2nd Edition).   [1601]
Sugiyama, K.:   Methods   for Visual Understanding of Hi [1634]
Swets, J.A. : Diagnostic Systems: Methods   from Signal Detection Theory. [1646]

Trappl, R.   : Computerized Formal Methods   with Computer Efficiency in Me [1711]
Umpleby, S. :   Methods   for Making Social Organisation [1770]
Viliums, E.R: Methods   to Obtain Interlevel Consisten [1816]
Windeknecht,: ations of Numerical   Methods   Applied to the Solution of Ord [1866]
Kaushik, M.L: traints Under a New   Metric .   [888]
Kaushik, M.L: Codes Through a New   Metric .   [889]
Perlis, A.J.:   Software   Metrics : An Analysis and Evaluation.   [1317]
Knappe, U. :   Microcomputer-Based System for Industr   [952]
Kerefky, P. : atient Registers on   Microcomputers .   [905]
Greenspan, D: ete Modeling in the   Microcosm and in the Macrocosm.   [665]
Michie, D. : pert Systems in the   Microelectronic Age.   [1160]
Kawamura, K.:   On Developing the   Microelectronics Industry: A Systems A [891]
Degermendzhy: The Reliability of   Microevolution Process of the Stationa [402]
Blaquire, A.: namical Systems and   Microphysics . (Seminar in Udine/Italy,   [172]
Bibbero, R.J: Microprocessor System, Interfacing and   [168]
Matsuno, K. : ational Description   Microsystems Formation in Probiologica   [1122]
Matsuno, K. : : Multiplication of   Microsystems with Self-Instructing Pol   [1120]
Matsuno, K. : and Reproduction of   Microsystems in Protobiogenesis.   [1125]
Janko, J.   : zech Science in the   Middle of the 19th Century.   [832]
Fairtlough, : ow Systems Thinking   Might Evolve.   [501]
Hirata, H. : cal Ecosystems with   Migration .   [758]
Blunden, M. :   Military Technology and Systemic Inter   [176]
Trappl, R. : cience, Technology,   Military , Economy, Society, Culture, a   [1733]
Elstob, C.M.:   Emergentism and   Mind .   [479]
Kuhlenbeck, : : The Brain and its   Mind .   [993]
Lowen, W.   :   Dichotomies of the   Mind : A Systems Science Model of the M   [1067]
Lowen, W.   : cience Model of the   Mind and Personality.   [1067]
Meldman, M.J: nagement System for   Minicomputers .   [1150]
Christensen,:   Entropy   Minimax Sourcebook. Vol.1: General Des   [311]
Sengupta, J.:   Minimax Solutions in Stochastic Progra   [1526]
Shore, J.E. :   Cross-Entropy   Minimization Given Fully-Decomposable   [1550]
Shore, J.E. : es of Cross-Entropy   Minimization .   [1551]
Van den Boga: Minimizing Wafting Times Using Priorit   [1793]
Watanabe, S.: tion as a Quest for   Minimum Entropy.   [1846]
Williams, P.: nd the Principle of   Minimum Information.   [1865]
Murdick, R.G: MIS : Concepts and Design.   [1197]
Probst, G.J.: Cybernetics against   Mismanagement .   [1363]
Lavorel, P.M: How we Name or   Misname Objects.   [1019]
Havranek, T.: rnative Approach to   Missing Information in the Guha Method   [738]
Winston, P.H: al Intelligence: An   MIT Perspective (2 Vols.).   [1871]
Farinas-del-: for Programming in   Modal Logic.   [505]
Holden, A.V.: to Global and Local   Mode of Action of General Anestetics.   [770]
Achinstein, :   Can There Be a   Model of Explanation?   [6]
Airenti, G. :   A Two Level   Model of Knowledge and Belief.   [23]
Andreae, J.H: A Dual   Model of the Brain.   [34]
Aplin, J.C. :   A Cybernetic   Model of the MBO Process.   [39]
Bandler, W. : uton: An Automation   Model of a Sociological Theory of Valu   [101]
Barber, M.C.:   A Markovian   Model for Ecosystem Flow Analysis.   [105]
Beckett, J.A: of the Disappearing   Model .   [132]
Bogner, S. :   A Cybernetic   Model of Cognitive Processes.   [183]
Bridge, J. :   Beginning   Model Theory.   [217]
Broekstra, G: hing. A Consistency   Model for Organizational Assessment an   [227]
Brown, D.J.H: ee Concept Learning   Model Employing Generalization an Abst   [229]
Brugger, K. : Simple Mathematical   Model of Biphasic Insulin Secretion.   [232]
Chikan, A. :   Analysis of a   Model System.   [307]
Colby, J.A. : r a Lotic Ecosystem   Model .   [328]
Csaki, P.   : General Problems of   Model Description.   [360]
De Malherbe,:   Model Building in Ecology: a Hierarchi   [396]
Genesio, R. :   On the   Model Order for Uncertain System Appro   [590]
Gentile, S. :   A Discrete   Model for the Study of a Lake.   [592]
Georgiev, A.: metric Mathematical   Model from Individual Human Growth Cur   [597]
Ginzberg, M.:   A Prescription   Model for System Implementation.   [616]
Glanville, R: The   Model 's Dimensions: A Form for Argumen   [619]
Greenberg, H: sisted Analysis and   Model Simplification.   [660]

Greenberg, H:	sisted Analysis and	Model	Simplification.	[661]
Guida, G. :	An Integrated	Model	of Problem Solver.	[679]
Guidorzi, R.:	Multistructural	Model	Selection.	[682]
Hirata, H. :	A	Model	of Hierarchical Ecosystems with	[759]
Hirata, H. :	A	Model	of Hierarchical Ecosystems with	[758]
Hirata, H. :	A	Model	of Mass and Energy Flow in Ecosy	[757]
Hossforf, H.:		Model	Analysis of Structures.	[791]
Huber, O. :	tical Analysis of a	Model .		[795]
Ihara, J. :	teria for Selecting	Model	Variables.	[808]
Kontur, I.F.:	with a Random Walk	Model .		[967]
Kovalenko, Y:	ate. II. Simulation	Model .		[974]
Landau, Y.D.:	aptive Control: the	Model	Reference Approach.	[1007]
Lauenroth, H:	Algorithmic	Model	Systems and Innovation Strategie	[1017]
Le Bonniec, :	The Development of	Model	Reasoning.	[1021]
Lowen, W. :	: A Systems Science	Model	of the Mind and Personality.	[1067]
McIntyre, C.:	A Hierarchical	Model	of Lotic Ecosystems.	[1139]
McIntyre, C.:	r a Lotic Ecosystem	Model .		[1140]
McLean, J.M.:	System Analysis and	Model	Building.	[1141]
Mesarovic, M:	Multilevel Computer	Model	of World Development System.	[1153]
Neck, R. :	imple Macroeconomic	Model .		[1207]
O'Muircheart:	Structures. Vol. 2:	Model	Fitting.	[1251]
Orava, P.J. :	pts of Input-Output	Model	, Consolity, and State in the The	[1268]
Overton, W.S:	A Strategy of	Model	Construction.	[1274]
Cverton, W.S:	tion of a Hydrology	Model	- an Exercise in Modelling Strat	[1275]
Overton, W.S:	ting a Hierarchical	Model	in the Flex Paradigm Illustrated	[1276]
Overton, W.S:	ects of a Hydrology	Model .		[1276]
Pedrycz, W. :	A	Model	of Decision-Making in a Fuzzy En	[1312]
Petkoff, B. :	A Cybernetic	Model	of Scientific Research and Cogni	[1328]
Porenta, G. :	A Mathematical	Model	of the Dopaminergic Synapse: Sta	[1349]
Ricci, F.J. :	ion of a Cybernetic	Model	of Man in his Cultural Environme	[1416]
Rivett, P. :		Model	Building for Decision Analysis.	[1422]
Scavia, D. :	An Ecological	Model	of Lake Ontario.	[1503]
Schellhorn, :		Model	Following in Multiple Control: E	[1506]
Scott, C.H. :	formation Theoretic	Model	of Spacial Interaction.	[1518]
Shaeffer, D.:	A	Model	Evaluation Methodology Applicabl	[1530]
Simionescu, :	On a	Model	Neurology.	[1560]
Slawski, C. :	Conflict: A Systems	Model	of Interparty Negotiation.	[1577]
Slawski, C. :	cy Systems Research	Model	of Career Development.	[1575]
Slawski, C. :	Episode: A Systems	Model .		[1580]
Slawski, C. :	an Organization: A	Model	and Case Study.	[1576]
Slawski, C. :	A Systems	Model	of Negotiation: Arabs Versus Isr	[1573]
Sugeno, M. :	A	Model	of Learning Based on Fuzzy Infor	[1633]
Takahara, Y.:	ration of the World	Model .		[1654]
Takeuchi, A.:	d Restle's Learning	Model .		[1663]
Tamura, H. :	n of a Dynamic Lake	Model	by the Method of Data Handling:	[1666]
Thissen, W. :	ations into World 3	Model	: Lessons for Understanding Compl	[1684]
Tracy, L. :	amic Living-Systems	Model	of Work Motivation.	[1709]
Trappl, R. :	A Second Order	Model	of the Optic Generator Potential	[1710]
Trattnig, W.:	ed on the Data Flow	Model .		[1734]
Troncale, L.:	Linkage Proposition	Model .		[1749]
Uccellini, L:	Applying Discrete	Model	Concepts to the Computation of A	[1765]
Wisniewski, :	An Interactive	Model	for Analysing Climate Change due	[1872]
Wolaver, T.G:	haracteristics of a	Model	Ecosystem.	[1877]
Yager, R.R. :	Linguistic	Model	and Fuzzy Truths.	[1888]
Nurmi, H.	egies of Cybernetic	Model-Building .		[1248]
Savage, G.J.:	aph-Theoretic Field	Model-I	- Modelling and Formulations.	[1498]
Apostolico, :	nd Hierarchy in the	Modeling	of Biomolecular Systems.	[41]
Boyce, W.E. :	ies in Mathematical	Modeling .		[204]
Capocelli, R:	Population Dynamics	Modeling .		[260]
De La Sen, M:	Some Aspects of the	Modeling	of Multivariable Aperiodic Sa	[393]
Dym, C.L. :	les of Mathematical	Modeling .		[453]
El-Sherief, :	Identification and	Modeling	for Linear Multivariable Disc	[468]
El-Sherief, :	ultivariable System	Modeling	and Identification Algorithms	[471]
Gaines, B.R.:	ology in the Large:	Modeling	All There Is.	[561]

Galitsky, V.:   On Modeling the Plant Community Dynamics. [562]
Gold, H.J.   :   Mathematical Modeling of Biological Systems: an Int [630]
Greenspan, D: Discrete Modeling in the Microcosm and in the M [665]
Greenspan, D: ter Approach to the   Modeling of Atoms, Ions and Molecules. [666]
Haimes, Y.Y.:   Resources Systems: Modeling and Optimization of Large-Sca [700]
Haimes, Y.Y.: rchical Holographic   Modeling .   [701]
Hall, C.A.S.:   Ecosystem   Modeling in Theory and Practice: an In [713]
Jacoby, S.L.:   Mathematical Modeling with Computers.   [826]
Jarisch, W. :   Stochastic   Modeling of Fetal Heart Rate Variabili [837]
Koyayashi, H: Modeling and Analysis: an Introduction [977]
LeMoigne, J.:   Modeling Complexity: A French History   [1024]
Lendaris, G.:   Structural   Modeling - A Tutorial Guide.   [1036]
Madan, S.   :   Cybernetic   Modeling of Queueing Complex.   [1086]
Oden, A.S. : iew of Mathematical   Modeling of the Behavior of Hydrocarbo [1252]
Peterson, J.:   Net Theory and the   Modeling of Systems.   [1327]
Pritsker, A.:   Modeling and Analysis Using Q-Gert Net [1362]
Ragade, R.K.: uzzy Interpretative   Modeling .   [1374]
Ricciardi, L: ttempts at Neuronal   Modeling .   [1418]
Rissanen, J.:   Universal   Modeling and Coding.   [1421]
Ruotsalainen: Modeling and Simulating the Resource A [1463]
Sauer, C.H. : Systems Performance   Modeling .   [1496]
Sawaragi, Y.:   Modeling , Estimation, and Their Applic [1499]
Scavia, D. : on Lake Eco-System   Modeling .   [1502]
Shigan, E.N.:   Systems   Modeling in Health Care.   [1547]
Soong, T.T. :   Probabilistic Modeling and Analysis in Science and E [1600]
Tazaki, E. :   Structural   Modeling in a Class of Systems Using F [1675]
Tazaki, E. :   Structural   Modeling with Consensus for Societal S [1676]
Toern, A.A. :   A General Tool for   Modeling Simulation Designs.   [1695]
Troncale, L.: lustering Theory to   Modeling Systems Emergence.   [1739]
Umnov, A.   : oach to Distributed   Modeling .   [1769]
Van Steenkis: Modeling , Identification and Control i [1791]
Vemuri, V. :   Modeling of Complex Systems: an Introd [1808]
Wedde, H.F. :   in Flexible Formal   Modeling under Realistic Assumptions.   [1852]
Zouwen, J.v.:   Modeling Interaction Processes in Inte [1930]
Andrews, J.G: Mathematical Modelling .   [38]
Aris, R.   :   Mathematical Modelling Techniques.   [52]
Baldwin, J.F: Modelling Controllers Using Fuzzy Rela   [84]
Bender, E.A.: ion to Mathematical   Modelling .   [150]
Biswas, A.K.:   Mathematical Modelling and Environmental Decision-M [171]
Bradley, R. : ies in Mathematical   Modelling : A Course Book for Engineers [210]
Bruckmann, G: pproaches in Global   Modelling . (Proc.of the Fifth IIASA Sy [231]
Cellier, F.E: Process in   Modelling and Simulation.   [289]
Conant, R.C.: Theoretic Structure   Modelling .   [342]
Conant, R.C.:   Structural   Modelling Using a Simple Information M [338]
Cross, M.   :   Modelling and Simulation in Practice.   [359]
David, F.W. :   Experimental Modelling in Engineering.   [379]
Ditleusen, O: Uncertain   Modelling .   [425]
Duechting, W: Recent Results in   Modelling and Simulation of 3-D Tumor   [446]
Fedra, K.   :   riness in Ecosystem   Modelling : A Lake Modelling Example.   [515]
Fedra, K.   :   m Modelling: A Lake   Modelling Example.   [515]
Flavin, M. :   epts of Information   Modelling .   [528]
Frauenthal, :   Mathematical Modelling in Epidemology.   [544]
Gasparski, W: a Notion of "System   Modelling ".   [583]
Getz, W.M. :   Mathematical Modelling in Biology and Ecology.   [602]
Gigley, H. :   roach to Simulation   Modelling of Natural Language Processi [610]
Hafele, W. :   Modelling of Large Scale Energy System [698]
Hawkes, N. :   nds in Mathematical   Modelling .   [739]
Hawkes, N. :   nds in Mathematical   Modelling : a UNESCO Report.   [740]
Henize, J. :   Modelling Large Scale Socio-Economic S [750]
Jacak, W.   :   Modelling of Evenistic Systems.   [821]
James, D.J.G: ies in Mathematical   Modelling .   [831]
Joergensen, : e Art in Ecological   Modelling .   [846]
Jorgenson, S: ical and Ecological   Modelling of Deep Lakes and Reservoirs [858]
Jorgenson, S: e Art in Ecological   Modelling .   [860]

Jumarie, G. : ivistic Approach to Modelling Dynamic Systems Involving Hu [862]
Kavakoglu, I: Mathematical Modelling of Energy Systems.   [890]
Kickert, W.J: ample of Linguistic Modelling .   [910]
Kindler, J. : gs of a Workshop on Modelling of Water Demands.   [924]
Klir, G.J.   : puter-Aided Systems Modelling .   [945]
Kobayashi, H: Modelling and Analysis: an Introductio [953]
Ladanyi, O.G: Modelling Economic Movements Using Pub [1003]
Linkens, D.A: Biological Systems   Modelling and Control.   [1051]
Maciejowski,:   The Modelling of Systems with Small Observ [1082]
Mahmoud, M.S: Large Scale Systems Modelling .   [1089]
Marchuk, G.I: Modelling and Optimization of Complex [1103]
McIntosh, J.:   Mathematical Modelling and Computers in Endocrinolo [1138]
McPherson, P: On Understanding Modelling and Improving Human Systems. [1142]
Morley, D.A.:   Mathematical Modelling in Water and Wastewater Trea [1187]
Nash, P.   :   Systems Modelling and Optimization.   [1202]
Nicholson, H: Modelling of Dynamical Systems.   [1222]
Nicholson, H: Modelling of Dynamical Systems.   [1223]
Nijssen, G.M: Modelling in Data Base Management Syst [1229]
Norman, M.   : ackage for Economic Modelling .   [1237]
Nowakowska, : eories of Research: Modelling Approaches.   [1241]
Oeren, T.I. :   Cybernetics and Modelling and Simulation of Large Scal [1254]
Overton, W.S: el - an Exercise in Modelling Strategy   [1275]
Retti, J.   :   Representation for Modelling and Simulation.   [1409]
Retti, J.   :   Representation for Modelling and Simulation of Dynamical [1410]
Roberts, P.C: Modelling Large Systems.   [1427]
Roberts, P.C: Modelling Large Systems: Limits to Gro [1428]
Savage, G.J.: tic Field Model-I - Modelling and Formulations.   [1498]
Schwarzenbac: Systems Modelling and Control.   [1516]
Sharif, N.   : nference on Systems Modelling in Developing Countries.   [1536]
Spriet, J.A.:   Computer-Aided Modelling and Simulation.   [1606]
Straszak, A.: nds in Mathematical Modelling .   [1626]
Sutherland, : stems: Methodology, Modelling and Management.   [1642]
Szuecs, E.   :   Similitude and Modelling .   [1652]
Tani, S.T.   :   A Perspective on Modelling in Decision Analysis.   [1668]
Tonella, G. :   Pluralistic Modelling for Development Planning.   [1699]
Tsetlin, M.L: Automata Theory and Modelling of Biological Systems.   [1758]
Ulanowicz, R: The Empirical Modelling of an Ecosystem.   [1766]
Vaccari, E. :   The Role of System Modelling in Natural Language Processi [1779]
Van Straaten,: ing on Lake Balaton Modelling Veszprem (Hungary).   [1792]
Vavrousek, J: An Approach to Modelling in Empirical Sciences.   [1804]
Wellstead, P: to Physical System   Modelling .   [1856]
Wendt, S.   :   Element for System Modelling .   [1857]
Witten, I.H.: on Nondeterministic Modelling of Behaviour Sequences.   [1874]
Zeigler, B.P: hodology in Systems Modelling and Simulation.   [1919]
Zeigler, B.P: r Multifacet System Modelling .   [1920]
             : Issue on Simulation Modelling and Statistical Computing. [1943]
Ackerman, E.:   Mathematical Models in the Health Sciences: a Compu [7]
Andersen, E.: iscrete Statistical Models with Social Science Application [28]
Aracil, J.   : der System Dynamics Models .   [44]
Barto, A.G. : rete and Continuous Models .   [112]
Bekey, G.A. :   Models and Reality: Some Reflections o [140]
Beltrami, E.:   Models for Public Systems Analysis.   [147]
Burghes, D.N: Mathematical Models in the Social, Management and L [243]
Burgin, M.S.: ensional Structured Models of Systems.   [244]
Burks, A.W. :   Models of Deterministic Systems.   [246]
Burns, D.W. : esis of Forecasting Models in Decision Analysis.   [247]
Chatterjee, : tration Procedures: Models with Complete and Incomplete In [295]
Checkland, P: Building Conceptual   Models .   [297]
Ciriani, T.A: Mathematical Models for Surface Water Hydrology. (P [316]
Cliff, A.D. :   Spacial Processes: Models and Applications.   [320]
Coffman, C.V: hes to Mathematical Models .   [325]
Collins, L. :   Use of Models in the Social Sciences.   [332]
Dayal, R.   :   ted System of World Models .   [385]
De Greene, K: stems: Theories and Models .   [390]

Deutsch, K.W: rocessing for World Models .   [414]
Eisen, M.   :   Mathematical Models in Cell Biology and Cancer Chem [465]
Elandt-Johns: Survival Models and Data Analysis.   [473]
Fontela, E. : orms of Large Scale Models .   [533]
Getis, A.   :   Models of Spatial Processes.   [601]
Gottinger, H: eristics Economical Models : a Study of Complexity.   [648]
Gourlay, A.R: ystems Structure of Models .   [650]
Graedel, T.E: Scientific Computer Models .   [652]
Halfon, E. : equacy of Ecosystem Models .   [711]
Hallam, T.G.: fferential Equation Models of Cosystems.   [715]
Halldin, S. : and Energy Exchange Models .(Proc. of an IUFRO Workshop, Up   [716]
Jezek, J.   : gebra and Theory of Models .   [845]
Johnson, P.E: Computer Simulation Models of Expert Reasoning.   [851]
Jorgenson, S: ical and Ecological Models of Shallow Lakes (Summary of an   [859]
Kashyap, R. : Classes of Dynamic Models Using Empirical Data.   [882]
Kaufmann, A.:   Mathematical Models for the Study of the Reliabilit [886]
Kieras, D.E.: te Automata and S-R Models .   [915]
Kobsa, A.   : Representing Belief Models in Semantic Networks.   [958]
Kornai, J. : anatory Theoretical Models of the Socialist Economy: Revie   [971]
Lazarevic, B: Large-Scale System Models . (Proc. of a Seminar in Dubrovn   [1020]
Legget, R.W.: liability Index for Models .   [1032]
Luenberger, : ic Systems: Theory, Models , and Applications.   [1071]
MacDonald, N: Lags in Biological Models .   [1079]
Matis, J.H. : lysis of Ecosystems Models .   [1118]
Maybeck, P.S: Stochastic Models , Estimation, and Control.   [1131]
McClamroch, :   State Models of Dynamic Systems: A Case Stud   [1134]
McClamroch, :   State Models of Dynamic Systems.   [1133]
Meadows, D.L: mputer-Based Policy Models .   [1143]
Metz, H.A.J.:   State Space Models for Animal Behaviour.   [1154]
Miller, M.C.:   Mathematical Models in Medical Diagnosis.   [1168]
Monahan, G.E: n Processes Theory, Models , and Algorithms.   [1183]
Nobile, A.G.: a Class of Discrete Models for Regulated Growth with Intri   [1234]
Oeren, T.I. : ation with Discrete Models .   [1256]
Peay, E.R. :   Structural Models with Qualitative Values.   [1306]
Pedrycz, W. : of Fuzzy Relational Models .   [1313]
Peschel, M. : with Classification Models .   [1321]
Petrovski, A: Medico-Demographic Models in Health Care Systems.   [1330]
Pospelov, D.: Logical-Linguistic Models in Control Systems.   [1352]
Probst, G.J.: sity of Sensitivity Models for Managers, Planners and Poli   [1364]
Pugh, R.E. : f Policy Simulation Models : a Conceptual Approach to Case   [1368]
Rapoport, A.:   Response Models for Detection of Change.   [1382]
Redhead, M. :   Models in Physics.   [1396]
Retti, J.   : ering of Simulation Models of Time-Oriented Systems.   [1408]
Roberts, F.S: screte Mathematical Models .   [1424]
Segel, L.A. :   Mathematical Models in Molecular and Cellular Biolo [1521]
Sengupta, J.:   Bilinear Models in Stochastic Programming.   [1524]
Shaeffer, D.: onmental Assessment Models .   [1530]
Shahane, A.H: Interdisciplinary Models of Water Systems.   [1531]
Sharif, N. :   Systems Models for Decision Making.   [1537]
Simon, H.A. :   Models of Discovery and Other Topics i   [1562]
Simon, H.A. :   Models of Bounded Rationality (2 Volum   [1563]
Slyadz, N.N.: in Decision-Making Models .   [1582]
Smith, M.   :   Models in Ecology.   [1590]
Smithson, M.:   Models for Fuzzy Nominal Data.   [1594]
Subrahmanian: A Generator for Models of Discrete-Event Systems.   [1632]
Thissen, W. : tanding Complicated Models .   [1684]
Thomas, A. : nd Abuse of Systems Models .   [1685]
Toda, M.   : Robot, and Society: Models and Speculations.   [1693]
Tong, R.M. : Synthesis of Fuzzy Models for Industrial Processes - Some   [1700]
Tong, R.M. : Evaluation of Fuzzy Models Derived from Experimental Data.   [1703]
Toutenburg, : formation in Linear Models .   [1708]
Troncale, L.: Testing Hierarchy Models with Data Using Computerized, E   [1745]
Unterberger,: using Articulation Models .   [1772]
Walker, C.C.: of Fixed-Structure Models .   [1830]

West, B.J.   :   Mathematical Models as a Tool for the Social Scienc [1859]
Witten, I.H.: ng Transitive Moore Models . [1873]
Yager, R.R. : of Fuzzy-Linguistic Models . [1889]
Yager, R.R. : g and Possibilistic Models in Classification. [1899]
Yager, R.R. : Based on Regression Models . [1902]
Zeigler, B.P: nization of Partial Models . [1918]
Zimmermann, : ing of Mathematical Models . [1927]
Hellwig, Z.H: Basic Problems of Modern   Econometrics and Some Problems [748]
Launer, R.L.:   Modern   Data Analysis. [1018]
Rose, J.   :   Modern   Trends in Cybernetics and Syste [1436]
Sachs, W.   : temic Inactivism: A Modern   Laissez-Faire Thinking. [1469]
Tighe, T.J. :   Modern   Learning Theory: Foundations an [1689]
Bandler, W. :   The Four Modes of Inference in Fuzzy Expert Sys [102]
Utkin, V.I. : ystems with Sliding Modes . [1774]
Hanson, O.J.:   Assessing and Modifying an Advanced Business Systems [723]
Agusti, J. :   Modularity Grade of the Neural Network [19]
Dalenoort, G: Modularity of Behavior of Networks. [373]
Gruber, H. :   Modularization and Abstracting; a Disc [677]
Wendt, S.   : e Programmed Action Module : an Element for System Modellin [1857]
Matsuno, K. : on in Probiological Molecular Evolution. [1122]
Segel, L.A. : hematical Models in Molecular and Cellular Biology. [1521]
Conrad, M. : of Variability from Molecule to Ecosystem. [343]
Greenspan, D: of Atoms, Ions and Molecules . (Part I: Hydrogen; Part II: [666]
Haydon, P.G.: ation in Identified Molluscan Neurones. [742]
Meyer, A.R. :   Weak Monadic   Second Order Theory of Success [1156]
Caianiello, : A Systemic Study of Monetary   Systems. [255]
Craine, R. :   The Optimal Monetary Instrument: An Empirical Asse [354]
Pau, L.F.   : sis and Performance Monitoring . [1296]
Zorbas, Y.G.:   Medical Monitoring of Man's Reliability during [1929]
Nowakowska, :   Instability, Monopolies , and Alienation in Science. [1242]
Boncina, R. : posed of Non-Active Monotonous Parts in Passive Surroundin [188]
Kittler, J. :   A Comment on Fast Monte   Carlo Integration of P.D.F. Esti [930]
Rubinstein, :   Simulation and Monte   Carlo Method. [1456]
Witten, I.H.: ns Using Transitive Moore   Models. [1873]
Fitter, M. :   Towards More   'Natural' Interactive Systems. [527]
Retti, J.   :   MOSES : A Schema-Based Knowledge Repres [1410]
Ladanyi, O.G: ghties - A World in Motion , A Challenge for Systems Engine [1002]
Labos, E.   : ic and Non-Periodic Motions in Different Classes of Formal [1000]
Tracy, L.   : stems Model of Work Motivation . [1709]
Cavallo, R.E: Sysems Research Movement : Characteristics, Accomplishm [280]
Cavallo, R.E: Systems Research Movement : Characteristics, Accomplishm [281]
Cavallo, R.E: on Systems Research Movement . [279]
Checkland, P: hape of the Systems Movement . [298]
Checkland, P: The Systems Movement and the "Failure" of Manageme [299]
Klir, G.J. : al Systems Research Movement . [942]
Koenderink, : sional Shape: Rigid Movement , Bending and Stretching by Ey [962]
Ladanyi, O.G: Modelling Economic Movements Using Published Internationa [1003]
Jarisch, W. : f Fetal Respiratory Movements-Preliminary Results. [837]
Shepp, L.A. :   On Prediction of Moving-Averaging Processes. [1544]
Van Straaten: of the Second Joint MTA/IIASA Task Force Meeting on Lake B [1792]
Grauer, M. : ly Strategies using Multi-Criteria Optimization. [658]
Atkin, R.H. : e as a Pattern on a Multi-Dimensional Structure. [63]
Hu, N.C.   :   Multi-Dimensional FFT Computation. [794]
Walichiewicz: in the Design of a Multi-Dimensional Fuzzy Control Algori [1828]
Findler, N.V: A Multi-Level Learning Technique Using P [526]
Nachane, D.M: mization Methods in Multi-Level Systems. [1199]
Krippendorff: fying Structures in Multi-Variate Data. [984]
Efstathiou, :   Multiattribute Decision Making Using a [457]
Seo, F.   : Fuzzy Extension of Multiattribute Utility Analysis for Co [1527]
Baptistella,: ies for Interactive Multicriteria Optimization. [104]
Zhukovin, V.:   The Multicriteria Decision Making with Vec [1925]
Atkin, R.   :   Multidimensional Man. [62]
Bose, N.K. :   Multidimensional Systems: Theory and A [192]
Bose, N.K. :   Applied Multidimensional Systems Theory. [193]

Burgin, M.S.: cts of Operators of   Multidimensional   Structured Models of   [244]
Golledge, R.: ce: Problems in the   Multidimensional   Analysis of Large Dat   [632]
Huber, O.   :   Nontransitive   Multidimensional   Preferences: Theoreti   [795]
Nowakowska, : ormal Semiotics and   Multidimensional   Semiotic Systems.   [1243]
Shepard, R.N: Multidimensional   Scaling, Tree-fitting   [1543]
Bailey, K.D.: and Homeostasis: A   Multidisciplinary Legacy.   [82]
De Beaufrand: Process: Towards a   Multidisciplinary Science of Texts.   [386]
Estep, M.L. : tems Integration of   Multidisciplinary and Interdisciplinar   [493]
Bernard, N. :   Multiensembles , a Contribution to the   [162]
Zeigler, B.P: ring Principles for   Multifacet   System Modelling.   [1920]
Royce, J.R. : formation Theory in   Multifactor Individuality Theory.   [1453]
Drozen, V. :   Multilateral   Interdependence.   [441]
Dirickx, Y.M: Systems Analysis by   Multilevel   Methods: with Applications   [424]
Jennergren, :   The   Multilevel   Approach: a Systems Analysi   [843]
Mesarovic, M: Multilevel   Computer Model of World Dev   [1153]
Botez, M.C. : neral Systems: Open   Multimodelling .   [197]
Sakawa, M. : tainment Method for   Multiobjective   Nonconvex Problems.   [1483]
Wildhelm, J.:   Objectives and Multiobjective   Decision Making Under U [1864]
Driankow, D.:   An Approach to   Multiobjectives Decision Making with O [439]
Edwards, J.B: ysis and Control of   Multipass Processes.   [456]
Babloyantz, : ogical Systems with   Multiple   Cellular Contacts.   [78]
Fandel, G. :   Multiple   Criteria Decision Making Theo   [502]
Hwang, C.L. :   Multiple   Objective Decision Making: Me   [801]
Hwang, C.L. :   Multiple   Attribute Decision Making: Me   [803]
Linstone, H.:   The   Multiple   Perspective Concept with Appl   [1052]
MacDonald, B: The Competence of a   Multiple   Context Learning System.   [1078]
Orlovsky, S.: ve Alternatives for   Multiple   Fuzzy Preference Relations.   [1272]
Schellhorn, : Model Following in   Multiple   Control: Exact and Approximat   [1506]
Starr, M.K. :   Multiple   Criteria Decision Making.   [1610]
Vavrousek, J: Multiple   Structure of Systems.   [1803]
Vavrousek, J: Multiple   Structure of Systems.   [1805]
Viliums, E.R: der Uncertainty and   Multiple   Objectives.   [1817]
White, C.   :   A   Multiple   Objective Optimization-Based   [1861]
Zeleny, M. :   Multiple   Criteria Decision Making.   [1924]
Zionts, S. :   Multiple   Criteria Problem Solving.   [1928]
Shoesmith, D: Multiple-Conclusion   Logic.   [1549]
Takatsu, S. :   Multiple-Objective   Satisficing Decisio   [1660]
Matsuno, K. : ducing Machineries: Multiplication   of Microsystems with Se   [1120]
Matsuno, K. :   Self-Sustaining   Multiplication   and Reproduction of Mic [1125]
Guidorzi, R.:   Multistructural   Model Selection.   [682]
Blomberg, H.: lgebraic Theory for   Multivariable   Linear Systems.   [175]
De La Sen, M: of the Modeling of   Multivariable   Aperiodic Sampling Syste   [393]
El-Sherief, : of Linear Discrete   Multivariable   Systems with Process Noi   [470]
El-Sherief, : Modeling for Linear   Multivariable   Discrete-Time Systems: A   [468]
El-Sherief, :   Identification of   Multivariable   Systems in the Transfer- [469]
El-Sherief, :   Recent Advances in   Multivariable   System Modeling and Iden [471]
Anderson, T.:   A Bibliography of   Multivariate   Statistical Analysis.   [31]
Bennett, S. :   An Introduction to   Multivariate   Techniques for Social and [153]
Hawkins, D.M: Topics in Applied   Multivariate   Analysis (2 volumes).   [741]
Korhonen, P.: pwise Procedure for   Multivariate   Clustering.   [970]
Maker, P.P : Output Relations in   Multivariate   Biosystems.   [1094]
Owens, D.H. :   Feedback and   Multivariate   Systems.   [1277]
Patel, R.V. :   Multivariate   System Theory and Design.   [1289]
Sanderson, A: is of Nonstationary   Multivariate   Data.   [1489]
Srivastava, :   An Introduction to   Multivariate   Statistics.   [1609]
Subrahmaniam: Multivariate   Analysis: a Selected and   [1631]
Takeuchi, A.: The Foundations of   Multivariate   Analysis: A Unified Appro   [1664]
Bertoni, A. : s and Compacting of   Musical   Texts.   [163]
Boyd, G.M. : ns in the Design of   Mutual   Control in Education.   [205]
Wyner, A.D. : tion of Conditional   Mutual   Information for Arbitrary Ensem   [1884]
Hatze, H.   :   Myocybernetics - A New Development in   [736]
Conant, R.C.: fs of Identities in   N-Dimensional   Information Theory.   [340]
Lavorel, P.M: How we   Name   or Misname Objects.   [1019]
Athanassov, : Specialists for the   National   Economy.   [58]

```
Balkus, K.    : nal Determinants of   National   Development.                         [87]
Abe, J.       : 'Speakers-World' in   Natural    Language Processing.                   [2]
Basta, D.J. :            Analyzing    Natural    Systems: Analysis for Regional       [114]
Boden, M.     : al Intelligence and   Natural    Man.                                 [180]
Chandrasekar:                         Natural    and Social System Metaphors fo      [292]
Cook, N.D.    : ity: an Analysis of   Natural    Systems.                             [344]
Cook, N.D.    : phism of Control in   Natural    , Social, and Cybernetic System     [345]
Eigen, M.     : cle: a Principle of   Natural    Selforganization.                    [462]
Faribault, M:                 From    Natural    Language to a Canonical Repres       [504]
Fitter, M.    :        Towards More ' Natural    ' Interactive Systems.               [527]
Fum, D.       : e-Based Approach to   Natural    Language Text Representation a       [547]
Gigley, H.    : lation Modelling of   Natural    Language Processing.                 [610]
Guida, G.     : d Incrementality in   Natural    Language Understanding Systems       [680]
Guida, G.     :                       Natural    Language Access to Online Data       [681]
Harris, L.R.: d in the Parsing of     Natural    Language Data Base Queries.          [727]
Janko, J.     : Systems Approach in   Natural    Science Reflecting Orientation       [832]
Kuhlenbeck, : Vol.I: The World of     Natural    Sciences and its Phenomenology       [993]
Lozinskii, E:             Locality in Natural    Language Processing.                [1069]
Paritsis, N.: ptional Problems in     Natural    Intelligent Systems with Chang      [1282]
Rosen, R.     : d Representation of   Natural    Systems.                            [1441]
Scott, W.R.   : izations: Rational,   Natural    and Open Systems.                   [1520]
Troncale, L.:              Knowing    Natural    Systems Enables Better Design       [1749]
Trost, H.     : ion in the Field of   Natural    Language Understanding.             [1754]
Vaccari, E.   : System Modelling in   Natural    Language Processing.                [1779]
Vallet, C.    : for Description of    Natural    Systems.                            [1784]
Trost, H.     :            Outline of a Natural-Language  Understanding System.      [1752]
Cole, I.D.    :                       Naturalized  Programming: A Method of N        [329]
Cole, I.D.    :                       Naturalized  System Development.               [330]
Cole, I.D.    : amming: A Method of   Naturalizing Programming Languages.            [329]
Checkland, P:        The Origins and  Nature   of 'Hard' Systems Thinking.           [296]
Eigen, M.     : : How Principles of   Nature   Govern Chance.                        [463]
Odum, H.T.    : y Basis for Man and   Nature .                                      [1253]
Patten, B.C.:        The Cybernetic   Nature   of Ecosystem.                        [1295]
Mehlhorm, K.:                         Nearly   Optimal Binary Search Trees.         [1147]
Kindler, E.   : : I. Declaration of   Necessary  System Conceptions.                 [917]
Balkus, K.    :                       Necessities  as the Basis of the Social        [88]
Carvallo, M.:           Chance and the Necessity  of the Third Classical Parad       [267]
Giere, R.N. :         Propensity and  Necessity .                                    [609]
Probst, G.J.:                   The   Necessity  of Sensitivity Models for Ma       [1364]
Haack, S.     :               Do We   Need   'Fuzzy Logic'?                          [697]
Yager, R.R. : re of Fuzziness and     Negation . Part 1: Membership in the Un       [1890]
Slawski, C.   : Model of Interp?r'y   Ne?o::?.                                       [····?]
Slawski, C.   : A Systems Model of    Negotiation : Arabs Versus Israeli.           [1573]
Groen, J.C.F: he Distance between     Neighbouring  Structures.                      [671]
Andrew, A.M.: ion Transmission on     Nerves .                                        [37]
Caianiello, : ommunication in the     Nervous  System.                               [256]
Brauer, W.    :                       Net  Theory and Applications. (Proc. of        [213]
Peterson, J.:                  Petri  Net  Theory and the Modeling of Systems       [1327]
Van der Zouw: ems Research in the     Netherlands .                                 [1796]
Van der Zouw: ems Research on the     Netherlands .                                 [1797]
Auguin, M.    : f Interpreted Petri   Nets .                                          [70]
Landweber, L: nd Persistent Petri     Nets .                                        [1008]
Peterson, J.:                  Petri  Nets .                                        [1326]
Toulotte, J.: g Interpreted Petri     Nets  and its Utilization.                    [1707]
Ball, M.O.    :          Complexity of Network  Reliability Computations.             [89]
Gage, W.L.    :          Flexibility in Network  Planning.                           [553]
Hostetter, G:         Fundamentals of Network  Analysis.                             [792]
Kauffman, L.:                         Network  Synthesis and Varela's Calculu        [885]
Kellermayr, : an Optical Bus Area     Network .                                      [900]
Leinhardt, S: Sociology on Social     Network  Research.                            [1035]
Schnakenberg:         Thermodynamic   Network  Analysis of Biological Systems       [1513]
Veroy, B.S. :           Large Scale   Network  Architecture Synthesis: Intera       [1813]
Veroy, B.S. : /Data Communication     Network .                                     [1814]
Palvoelgyi, :                         Network-Modelling  of Learning.               [1279]
```

Agusti, J.   : Grade of the Neural Networks Synthesized with Z-Fivex Form [19]
Atlan, H.   :   Random Boolean Networks .   [64]
Dalenoort, G: rity of Behavior of   Networks .   [373]
Dalenoort, G: eral, and in Neural   Networks in Particular.   [374]
Dammasch, I.: ted McCulloch-Pitts   Networks : Differences between Input-Co   [375]
Dammasch, I.: t and Input-Variant   Networks .   [375]
Findler, N.V: Associative   Networks : Representation and Use of Kn   [524]
Gollmann, D.: Certain Non-Linear   Networks of Automata.   [633]
Hastings, N.:   Decision   Networks .   [734]
Kobsa, A.   : Models in Semantic   Networks .   [958]
Labos, E.   : of Formal Neuronal   Networks and Chaotic Spike Generators. [1000]
Morein, R.   : hreshold Logic Unit Networks .   [1185]
Noguchi, S.   : of Analogue Neural   Networks .   [1235]
Pritsker, A.: alysis Using Q-Gert   Networks .   [1362]
Samuelson, K: rmation Systems and   Networks .   [1486]
Sherlock, R.: of Kauffman Binary   Networks : Part 1: State Space Descript [1545]
Sherlock, R.: Cycle Fraction for   Networks of Different Connectivities. [1545]
Thulasiraman: Graphs,   Networks , and Algorithms.   [1688]
Veelenturf, : ing Learning Neural   Networks .   [1806]
Agusti, J.   : larity Grade of the Neural Networks Synthesized with Z-Fiv [19]
Dalenoort, G: in General, and in   Neural Networks in Particular.   [374]
Erdi, P.   : cal Approach to the Neural Organization: Feed-Forward Cont [484]
Holden, A.V.: Chaotic Activity in   Neural Systems.   [771]
Noguchi, S.   : perties of Analogue Neural Networks.   [1235]
Veelenturf, : Developing Learning   Neural Networks.   [1806]
Ventriglia, :   Kinetic Theory of   Neural Systems: Memory Effects.   [1810]
Ventriglia, :   Kinetic Theory of   Neural Systems: An Attempt Towards a S [1811]
Simionescu, :   On a Model   Neurology .   [1560]
Hatze, H.   :   New Development in Neuromuscular Control.   [736]
Labos, E.   : t Classes of Formal Neuronal Networks and Chaotic Spike Ge [1000]
Ricciardi, L: Classic Attempts at   Neuronal Modeling.   [1418]
Agusti, J.   : with Z-Fivex Formal Neurones .   [19]
Haydon, P.G.: dentified Molluscan   Neurones .   [742]
Trost, H.   : puter Simulation of Neurotic Processes.   [1751]
Bensoussan, :   New Trends in Systems Analysis: Intern   [158]
Boulding, K.:   Ecodynamics: a New Theory of Societal Evolution.   [200]
Collot, F.   :   A New Mathematical Concept for General S   [333]
De Rosnay, J: The Macroscope: a New World Scientific System.   [399]
Dhillon, B.S: eering Reliability: New Techniques and Applications.   [416]
Elton, M.C.J: Evaluating   New Telecommunication Services.   [480]
Erlandson, R: sfacting Process. A   New Look.   [488]
Gengoux, K.G: en Participation: A   New Factor in Social Systems.   [591]
Greenspan, D: A New Computer Approach to the Modeling   [666]
Grindley, K.:   Systematics: a New Approach to Systems Analysis.   [670]
Hatze, H.   : Myocybernetics - A   New Development in Neuromuscular Contr [736]
Innis, G.S.   :   New Directions in the Analysis of Ecol [812]
Jumarie, G.   :   A New Approach to the Transinformation o [863]
Kaushik, M.L: Constraints Under a   New Metric.   [888]
Kaushik, M.L: ing Codes Through a   New Metric.   [889]
Locker, A.   :   A New Look at the Description and Prescr [1059]
Medina, B.F.: System Analysis: A   New Technique.   [1145]
Mulej, M.   : ciates' Ideas about New Products and Programs.   [1195]
Negoita, C.   : Sets and Systems: a New International Journal.   [1208]
Nowakowska, :   New Theoretical, Methodological, and E   [1244]
Oldershaw, R: New Evidence for the Principle of Self   [1265]
Probst, G.J.:   New ('second order') Cybernetics again   [1363]
Resconi, G.   :   A New Formalization of the General Syste [1406]
Rifkin, J.   :   Entropy: a New World View.   [1419]
Straszak, A.:   New Trends in Mathematical Modelling.   [1626]
Tesfatsion, :   A New Approach to Filtering and Adaptive [1683]
Trappl, R.   : , Autonomy, Utopia: New Approaches Towards Complex Systems [1731]
Yashin, A.   :   A New Proff and New Results in Condition [1911]
Yashin, A.   :   A New Proff and New Results in Conditional Gaussian Es [1911]
              : Design Studies. A   New Journal Covering the Study, Resear [1938]

	:	Metamedicine. A	New International Journal for Philosop	[1939]
Ellison, D.	:	Simulation	Newsletter .	[476]
Patten, B.C.:	h to the Concept of	Niche .	[1293]
Patten, B.C.:	and the Ecological	Niche .	[1294]
Tancig, P.	: are Environment for	NLU-Systems .	[1667]
Carvallo, M.:	In Search of the	Noetic Structure of Systems.	[266]
Ciborra, C.	: y of Organizational	Noise in Sociotechnical Systems.	[315]
El-Sherief,	: ystems with Process	Noise .	[470]
Fogel, E.	: tification: Bounded	Noise Case.	[532]
Smithson, M.:	Models for Fuzzy	Nominal Data.	[1594]
Boncina, R.	: System Composed of	Non-Active Monotonous Parts in Passive	[188]
Primas, H.	: heory Reduction and	Non-Boolean Theories.	[1361]
Kodratoff, Y:		Non-Canonical Simplification.	[960]
Trappl, R.	:	Non-Computerized Formal Methods with C	[1711]
Cohen, J.	:	Non-Deterministic Algorithms.	[326]
Nicolis, G.	: elf-Organization in	Non-Equilibrium Systems.	[1224]
Gollmann, D.:	fication of Certain	Non-Linear Networks of Automata.	[633]
	:	Non-Monotonic Logic (Special Issue).	[1936]
Nicolis, J.S:	Bifurcation in	Non-Negotiable Games: A Paradigm for S	[1226]
Banerji, R.B:	tical Approaches to	Non-Numerical Problem Solving.	[103]
Labos, E.	: Periodic and	Non-Periodic Motions in Different Clas	[1000]
Benedikt, S.:		Non-Repetitive Decision Making under R	[151]
Robinson, A.:		Non-Standard Analysis.	[1431]
Kornwachs, K:	tic Information and	Nonclassical Systems.	[972]
Sakawa, M.	: for Multiobjective	Nonconvex Problems.	[1483]
Kintala, C.M:	Amounts of	Nondeterminism in Finite Automata.	[929]
Gini, G.	: Deterministic and	Nondeterministic Programming in Robot	[615]
Witten, I.H.:	e Recent Results on	Nondeterministic Modelling of Behaviou	[1874]
Hisdal, E.	: s, Independence and	Noninteraction	[761]
Aggarwal, J.:		Nonlinear Systems: Stability Analysis.	[16]
Atherton, D.:	Stability of	Nonlinear Systems.	[59]
Atherton, D.:		Nonlinear Control Engineering.	[60]
Bellman, R.	: Control Processes:	Nonlinear Processes.	[143]
Hunt, L.R.	: lability of General	Nonlinear Systems.	[798]
Ibidapo-Obe,:	Error Covariance in	Nonlinear Dynamical Systems.	[807]
Peschel, M.	: lterra Approach for	Nonlinear Systems.	[1325]
Rugh, W.J.	:	Nonlinear Systems Theory: The Volterra	[1462]
De Mottoni,	: urcation of Regular	Nonnegative Stationary Solutions for a	[397]
Georgiev, A.:		Nonparametric Mathematical Model from	[597]
Sanderson, A:	jectory Analysis of	Nonstationary Multivariate Data.	[1489]
Huber, O.	:	Nontransitive Multidimensional Prefere	[795]
Airaksinen, :		Normative Aspects of Some Stabilizing	[22]
Kilmann, R.H:	ial Systems Design:	Normative Theory and the Maps Technolo	[916]
Dubois, D.	: Based on Triangular	Norms .	[444]
Pedrycz, W.	: ons with Triangular	Norms and their Resolutions.	[1310]
Klir, G.J.	: The	North-Holland Series in General System	[943]
Mittermeir, :	CML-Graphs - A	Notation for Systems Development.	[1175]
Barto, A.G.	: A	Note on Pattern Reproduction in Tessel	[111]
Dial, G.	: A	Note on Bayes Risk with Utilities.	[419]
Gurdial,	: A	Note on Fano Inequality Involving Dist	[690]
Mizumoto, M.:		Note on the Arithmetic Rule by Zadeh f	[1176]
Dold, A.	: Lecture	Notes in Mathematics, Vol.453	[428]
Kornwachs, K:	Some	Notes on Information and Interaction.	[973]
Gasparski, W:	What Does it Mean a	Notion of "System Modelling".	[583]
Higashi, M.	: On the	Notion of Distance Representing Inform	[756]
Jack, W.	: tic Approach to the	Notion of System.	[823]
Pedretti, A.:	incides - Coining a	Notion of Cybernetic Description?	[1309]
Dolby, J.L.	: On the	Notions of Ambiguity and Information L	[427]
Kramer, J.T.:	nking: Concepts and	Notions .	[981]
Gardner, M.R:	Predicting	Novel Facts.	[570]
Neal, M.	: Personality in the	Novel .	[1205]
Kozen, D.	: irst-Order Logic Is	NP-Complete .	[978]
Garey, M.R.	: ability: a Guide to	NP-Completeness .	[571]
Janssens, L.:	f Operator Crews of	Nuclear Power Plants.	[833]

Sinha, K.K.   :   Adaptive Nuclear   Reactor Control without Explic [1569]
Oldershaw, R: On the Number   of Levels in the Self-Similar H [1263]
Field, H.H.   :   Science Without Numbers .   [522]
Klement, E.P: Law of Large Numbers   and Central Limit Theorem for   [935]
Wang, Z.   :   M-Fuzzy Numbers   and Random Variables.   [1838]
Borodin, A.   : ty of Algebraic and Numeric   Problems.   [190]
Windeknecht,: tic Implications of Numerical   Methods Applied to the Solut [1866]
Eilbert, R.F: ern Recognition and Numerology .   [464]
Putt, A.M.   : s Theory Applied to Nursing .   [1369]
Trussel, P.   :   Using Nursing   Research: Discovery, Analysis, [1755]
Morris, H.M.: lity in a Conserved Nutrient   Ecosystem.   [1189]
Walker, V.R.:   Criteria for Object   Perception.   [1832]
Bezdek, J.C.: ognition with Fuzzy Objective   Function Algorithms.   [167]
Hwang, C.L.   :   Multiple Objective   Decision Making: Methods and [801]
White, C.   :   A Multiple Objective   Optimization-Based Approach   [1861]
Bell, D.E.   :   Conflicting Objectives   in Decisions.   [142]
Viliums, E.R: tainty and Multiple Objectives .   [1817]
Wildhelm, J.:   Objectives   and Multiobjective Decision [1864]
Yager, R.R.   : g Including Unequal Objectives .   [1887]
Farreny, H.   : Way of Designating Objects   in Sentence Generation.   [508]
Koczy, L.T.   : -Algebras and Fuzzy Objects   of Type N.   [959]
Lavorel, P.M: we Name or Misname Objects .   [1019]
Monahan, G.E: Survey of Partially Observable   Markov Decision Processes T [1183]
Auger, P.   : between N Levels of Observation   of a System (Biological or   [67]
Maciejowski,: Systems with Small Observation   Sets.   [1082]
Conant, R.C.:   Faculty Observations   of Variables and Relation   [337]
Jumarie, G.   : nts Involving Fuzzy Observations .   [863]
Klir, G.J.   : : Some Experimental Observations   and Resulting Guidelines. [938]
Vallee, R.   : Eigen-Elements" for Observing   and Interacting Subjects.   [1783]
Troncale, L.: al Systems Science: Obstaches , Potentials, Case Studies.   [1748]
Viliums, E.R: Methods to Obtain   Interlevel Consistency of Decis [1816]
Franova, M.   : Constructive Proofs Obtained   by Beth's Tableaux.   [542]
Trappl, R.   : ial Intelligence: A One-Hour   Course.   [1728]
Guida, G.   : Language Access to Online   Data Bases.   [681]
Scavia, D.   : gical Model of Lake Ontario .   [1503]
Tamura, H.   : Application to Lake Ontario .   [1666]
Erdi, P.   : ward Control of the Ontogenetic   Development.   [484]
Trappl, R.   :   Ontology   on the Computer.   [1719]
Boncina, R.   :   An Open   System Composed of Non-Active Mon [188]
Botez, M.C.   : of General Systems: Open   Multimodelling.   [197]
Degermendzhy: atory Population in Open   Systems.   [402]
Matsuno, K.   : d Feedback Loops in Open   Systems.   [1123]
Scott, W.R.   : tional, Natural and Open   Systems.   [1520]
Green, M.A.   :   Solar Cells: Operating   Principles, Technology, and [659]
Rus, T.   : Data Structures and Operating   Systems.   [1464]
Haralik, R.M: Reduction Operation   for Constraint Satisfaction. [725]
Ratko, I.   : Waiting for Cardiac Operation .   [1388]
Ackoff, R.L.: cting the Future of Operational   Research.   [8]
Matsuno, K.   :   Operational   Description Microsystems F [1122]
Ackoff, R.L.:   The Future of Operations   Research is Past.   [9]
Ghosal, A.   : s: its Relevance to Operations   Research.   [607]
Iri, M.   : f Matroid Theory in Operations   Research, Circuits and Syst   [814]
Lopes, M.A.   :   Operations   on Problems and Their Solut [1065]
Tchon, K.   :   On Some Operations   Preserving Generic Properti [1679]
Haley, K.B.   :   Applied Operatons   Research in Fishing. (Vol. 1 [710]
Janssens, L.: for the Training of Operator   Crews of Nuclear Power Plants [833]
Rota, G.C.   :   Finite Operator   Calculus.   [1450]
Bandler, W.   : tics of Implication Operators   and Fuzzy Relational Product [100]
Bandler, W.   : uzzy Simplification Operators .   [99]
Burgin, M.S.:   Products of Operators   of Multidimensional Structur [244]
Yager, R.R.   : Fuzzy Set-Theoretic Operators .   [1906]
Trappl, R.   : Order Model of the Optic   Generator Potential and its Rela [1710]
Kellermayr, : ce Simulation of an Optical   Bus Area Network.   [900]
Ben-Dov, Y.   :   Optimal   Testing Procedures for Special [148]

Boltyanskii,:   Optimal   Control of Discrete Systems. ([187]
Boltyanskii,:   Optimal   Control of Discrete Systems.   [186]
Craine, R.   :   The   Optimal   Monetary Instrument: An Empiri   [354]
Gabasov, R. : alitative Theory of   Optimal   Processes. (Translated From Ru [551]
Hartl, R.F. :   Optimal   Seducing Policies for Dynamic   [729]
Mehlhorm, K.:   Nearly   Optimal   Binary Search Trees.   [1147]
Stone, L.D. :   Theory of   Optimal   Search.   [1624]
Strauss, A. : An Introduction to   Optimal   Control Theory.   [1627]
Tesfatsion, : d Adaptive Control: Optimality Results   [1683]
Vincent, T.L: Optimality in Parametric Systems.   [1818]
Baptistella,: ctive Multicriteria   Optimization .   [104]
Bensoussan, : ymposium on Systems   Optimization and Analysis.   [159]
Bensoussan, :   Analysis and Optimization of Systems. (Proc. of a C [160]
Fletcher, R.: ractical Methods in   Optimization . Vol.1: Unconstrained Opt [530]
Fletcher, R.: ol.1: Unconstrained   Optimization ; Vol.2: Constrained Optim [530]
Fletcher, R.:   Vol.2: Constrained   Optimization .   [530]
Gill, P.   :   Practical   Optimization .   [613]
Grauer, M.   : sing Multi-Criteria   Optimization .   [658]
Gupta, M.M. : ierarchical Dynamic   Optimization for Linear Discrete Syste [688]
Haimes, Y.Y.: stems: Modeling and   Optimization of Large-Scale Systems.   [700]
Kerefky, P. :   Program   Optimization and Manipulation on User   [904]
Krzanik, L. : y Projects: Dynamic   Optimization of Delivery Step Structur [989]
Marchuk, G.I: Modelling and   Optimization of Complex Systems. (Proc [1103]
Nachane, D.M: Optimization Methods in Multi-Level Sy [1199]
Nash, P.   : stems Modelling and   Optimization .   [1202]
Sandblom, C.:   Optimization of Economic Policy Using [1487]
Tzafestas, S: Optimization of System Reliability: a   [1764]
White, C.   :   Multiple Objective   Optimization-Based Approach to Choicem [1861]
Fortuny-Amat: tion of Constrained   Optimizing Systems.   [536]
Auger, P.   :   Order , Disorder in Hierarchically Orga   [69]
Bohm, D.   : s and the Implicate   Order .   [184]
Dial, G.   :   Entropy of   Order   Alpha and Type Beta and Shannon' [420]
Genesio, R. :   On the Model Order   for Uncertain System Approximati [590]
Lefebre, V.A: Second   Order   Cybernetics in the Soviet Union [1031]
Meyer, A.R. : Weak Monadic Second   Order   Theory of Successor Is Not Eleme [1156]
Trappl, R.   :   A Second   Order   Model of the Optic Generator Pot [1710]
Zarri, G.P. : ng AI Techniques in   Order   to Solve the Problem of Informat [1915]
Gorbatov, V.: Theory of Partially   Ordered Systems.   [644]
Gehrlein, W.: ounting Methods for   Ordering Incomplete Ordinal Data.   [588]
Lamport, L. : Time, Clock and the   Ordering of Events in Distributed Syst [1006]
Zeleny, M.   :   Spontaneous Social   Orders .   [1921]
Driankow, D.: ecision Making with   Ordinal   Information.   [439]
Gehrlein, W.: Ordering Incomplete   Ordinal   Data.   [588]
Windeknecht,: to the Solution of   Ordinary Differential Equations.   [1866]
Gomez, P.   : hodology in Action: Organic   Problem Solving in a Publishin [638]
Best, D.P.   : ion of a Cybernetic   Organisation Structure in a Medium Siz [164]
Kinston, W. : esearch Approach to   Organisation of Complex Systems.   [927]
Umpleby, S. : s for Making Social   Organisation Adaptive.   [1770]
Molloy, K.J.: sights into Complex   Organisational Problem Solving.   [1182]
Gomez, P.   :   ization in Business   Organisations : Cybernetic Rules for Ef [637]
Butzer, K.W.:   Civilizations: Organisms or Systems?   [253]
Laing, R.   :   Machines as   Organisms : an Exploration of the Relev [1004]
Auger, P.   : ndividual Levels of   Organization in a Hierarchically Organ   [68]
Bryant, D.T.: npower Planning and   Organization Design.   [234]
Davis, C.   :   Organization for Program Management.   [382]
De Greene, K: The Adaptive   Organization - Anticipation and Manage [391]
Dimitrova, L: Cybernetics and the   Organization of Material Flow in the I [423]
Erdi, P.   : roach to the Neural   Organization : Feed-Forward Control of   [484]
Espejo, R.   : s in Management and   Organization .   [491]
Kickert, W.J: Organization of Decision-Making: a Sys [914]
Kickert, W.J: Formal Approach to   Organization Design.   [913]
Mager, P.P. : Biorhythms, System   Organization and Bioactive Compounds. [1088]
Ryan, R.M.   : y Systems: A Matrix   Organization Approach.   [1467]
Slawski, C. : ng Innovation in an   Organization : A Model and Case Study. [1576]

Sultan, L.H.: erations on Systems   Organization   for Fuzzy Information Pro [1636]
Taylor, A.M.: ch to the Political   Organization   of Space.   [1672]
Warfield, J.: ciples of Knowledge   Organization .   [1840]
Wenger, M.S:: etic View of Social   Organization .   [1858]
Zeigler, B.P: Structuring the   Organization   of Partial Models.   [1918]
Zeleny, M. : A Theory of Living   Organization .   [1922]
Acar, W.   :   Bureaucracy as   Organizational   Pathology.   [5]
Balkus, K. :   Organizational   Determinants of Nationa [87]
Broekstra, G: nsistency Model for   Organizational   Assessment and Change. [227]
Ciborra, C. : ological Inquiry of   Organizational   Noise in Sociotechnical [315]
Faucheuz, C.: tion or Autonomy in   Organizational   Design.   [509]
Fox, M.S.   :   An   Organizational   View of Distributed Sys [539]
Gomez, P.   : wn Versus Bottom-Up   Organizational   Design: a Cybernetic Pe [636]
Kaiser, J.S.:   Organizational   Behavior: A Vector of S [868]
Keen, G.P.W.: Support Systems: an   Organizational   Perspective.   [895]
Kickert, W.J: asystem Approach to   Organizational   Decision-Making.   [912]
Mock, T.J. : nt, Accounting, and   Organizational   Information.   [1177]
Sage, A.P. :   Behavioral and   Organizational   Considerations in the D [1475]
Takatsu, S. :   Organizational   Equilibrium under Uncer [1661]
Tripp, R.S. : e of Cybernetics in   Organizational   Design and Development: [1736]
Gelfand, A.E: nagement of Complex   Organizations : Management by Consensus [589]
Peery, N.S. : heory Approaches to   Organizations : Some Problems in Applic [1314]
Scott, W.R. :   Organizations : Rational, Natural and O [1520]
Tomlinson, R: oach to Planning in   Organizations - Developing a Collabora [1697]
Walker, C.C.: nagement of Complex   Organizations : Management by Exception [1830]
Elohim, J.L.: in our Learning to   Organize   Educational Systems.   [477]
Mulej, M.   : lied to Collect and   Organize   Associates' Ideas about New P [1195]
Taylor, W.R.: g Systems Theory to   Organize   Confusion.   [1674]
Auger, P.   : in a Hierarchically   Organized   System.   [68]
Auger, P.   : r in Hierarchically   Organized   Systems.   [69]
La Porte, T.:   Organized   Complexity: Challenge to Pol [998]
Janko, J.   : Science Reflecting   Orientation   Towards Practice: and Epis [832]
Simon, J.C. :   Computer   Oriented   Learning Processes.   [1564]
Fedanzo, A.J: The   Origin of Information in Systems.   [514]
Pask, G.   :   The   Originality of Cybernetics and the Cyb [1286]
Pask, G.   : the Cybernetics of   Originality .   [1286]
Checkland, P: The   Origins   and Nature of 'Hard' Systems T [296]
Christensen,: ol.2: Philosophical   Origins ; Vol.3: Computer Implementatio [311]
Troncale, L.:   Origins   of Hierarchical Levels: An Eme [1738]
Cerbone, G. : assage Time for the   Ornstein-Uhlenbeck   Process.   [290]
Cremers, A.B: Orthogonality of Information Structure   [356]
Degermendzhy: the Stationary and   Oscillatory   Population in Open Systems [402]
Glanville, R: Boxes Trying to Get   Out .   [620]
Lowen, W.   :   Creep-Creep: Out of Human Scale.   [1066]
Schmutzer, M: Hi & Low - In &   Out : Approaches to Social Status.   [1512]
Gasparski, W: An   Outline of the Logical Theory of Desig   [576]
Kubinski, T.:   An   Outline of the Logical Theory of Quest   [991]
Trost, H.   :   Outline of a Natural-Language Understa [1752]
De Raadt, J.: wness of Income and   Output   Distributions.   [398]
Veroy, B.S. :   Overall   Design of Reliable Centralized [1814]
Okeda, M.   :   Overlapping   Decompositions, Expansions [1260]
Cavallo,' R.E: ctability Analysis: Overview   and Bibliography.   [284]
Mohler, R. :   An   Overview of Bilinear System Theory and [1180]
Oden, A.S. :   An   Overview of Mathematical Modeling of t [1252]
Soi, I.M.   :   An   Overview of Computer-Orientied Learnin [1597]
Trappl, R. : al Intelligence: An   Overview .   [1730]
Trost, H.   : Representation: An   Overview   and an Application in the Fie [1754]
Mattessich, : the Systems and My   Own   Systems Methodology.   [1127]
Moraga, C. : roduction to Linear   p-Adic   Invariant Systems.   [1184]
Laszlo, C.A.: nd Computer Program   Package   for Health Hazard Appraisal. [1013]
Norman, R. :   Software   Package   for Economic Modelling.   [1237]
Uyeno, D.H. : te-Event Simulation   Package   for PASCAL.   [1776]
Ciupa, M.   : ernetic Approach to   Packet   Switched Data Communication Sys [317]
Xuemou, W. :   Pansystems   Analysis and Fuzzy Sets.   [1886]

Dutton, G. :   Harvard Papers   on Geographic Information Syste [450]
Gasparski, W: Papers   from the Conference in Design M [579]
Zeeman, E.C.: he Theory: Selected   Papers   1972-1977   [1916]
             : rnetics" Devoted to   Papers   Presented at the Second Spring   [1942]
De Mottoni, : s for a Quasilinear   Parabolic System.   [397]
Battista, J.:   The Holistic   Paradigm and General Systems Theory.   [124]
Carvallo, M.: the Third Classical   Paradigm .   [267]
Jantch, E. : : Toward a Unifying   Paradigm of Physical, Biological and S   [835]
Jean, R.V. :   A Methodological   Paradigm in Plant Biology.   [841]
Keeny, B.P. : ogy: an Alternative   Paradigm for Diagnosis.   [896]
Krallmann, H: larging the Systems   Paradigm : Historical View from 1973 un   [980]
Nicolis, J.S: Negotiable Games: A   Paradigm for Self-Organization in Cogn   [1226]
Overton, W.S: l Model in the Flex   Paradigm Illustrated by Structural Asp   [1276]
Sembugamoort: PLAS, a   Paradigmatic Language Acquisition Syst   [1523]
Floyd, R.W. :   The   Paradigms of Programming.   [531]
Van Gigch, J: tems and Metasystem   Paradigms .   [1787]
Hendel, R.J.:   The Grue   Paradox : An Information-Theoretic Solu   [749]
Hanassab, S.:   Parallel Processors for Cybernetic Sys   [718]
Iserman, R. : fication and System   Parameter Estimation.   [815]
Sastry, D. : ultaneous State and   Parameter Identification.   [1495]
Sawaragi, Y.: ons for Distributed   Parameter Systems.   [1499]
De La Sen, M: Parameterizations in Single Input-Outp   [394]
Braae, M.   :   Selection of   Parameters for a Fuzzy Logic Controlle   [209]
Vincent, T.L: Optimality in   Parametric Systems.   [1818]
Porenta, G. : , and Simulation of   Parkinson 's Disease and Aging Processe   [1349]
Harris, L.R.: onent to Aid in the   Parsing of Natural Language Data Base   [727]
Bluth, B.J. :   Parson 's General Theory of Action. A S   [178]
Conant, R.C.: ying Relations from   Partial Information.   [341]
Kalman, R.E.:   On   Partial Realizations, Transfer Functio   [872]
Lieberherr, :   Complexity of   Partial Satisfaction.   [1048]
Potter, J.M.:   Partial Prior Information and Decision   [1355]
Zeigler, B.P: the Organization of   Partial Models.   [1918]
Gorbatov, V.:   Theory of   Partially Ordered Systems.   [644]
Monahan, G.E: A Survey of   Partially Observable Markov Decision P   [1183]
Vallet, C. :   Use of a   Partially Self-Referent Formalism for   [1784]
Erlandson, R: The   Participant-Observer Role in Systems-M   [487]
Gengoux, K.G: Citizen   Participation : A New Factor in Social   [591]
Mumford, E. :   Participative Systems Design: Structur   [1196]
Dalenoort, G: Neural Networks in   Particular .   [374]
Hisdal, E. : zzy Set Systems and   Particularization .   [762]
Dwyer, R.F. :   Sequential   Partition Detectors.   [451]
Dwyer, R.F. :   Sequential   Partition Detectors with Dependent Sam   [452]
Pardo, L.   : a Fuzzy Event and a   Partition of Fuzzy Events.   [1281]
Bezdek, J.C.:   Fuzzy   Partitions and Relations: an Axiomatic   [166]
Boncina, R. : n-Active Monotonous   Parts   in Passive Surroundings.   [188]
Jahoda, M. :   Wholes and   Parts , Meaning and Mechanism.   [827]
Jakubowski, : rization of Machine   Parts Shapes.   [830]
Klir, G.J. : oning: The Issue of   Parts and Wholes.   [948]
Koffman, E.B: ured Programming in   PASCAL .   [963]
Uyeno, D.H. : ulation Package for   PASCAL .   [1776]
Cerbone, G. : ewness of the First   Passage Time for the Ornstein-Uhlenbec   [290]
Favella, L.F: First   Passage Time Problems and Some Related   [511]
Gericke, M. : Changes of Regional   Passenger-Transport Demands by Society   [600]
Uyeno, D.H. :   PASSIM : A Discrete-Event Simulation Pa   [1776]
Boncina, R. : Monotonous Parts in   Passive Surroundings.   [188]
Ackoff, R.L.: rations Research is   Past .   [9]
Hanson, O.J.: Using Feedback from   Past Students.   [723]
Pszczolowski: y - The Theory with   Past and Future.   [1367]
Masuda, Y. : ormation Society as   Past-Industrial Society.   [1115]
Breitenecker: ion and Analysis of   Pathological Blood Pressure Behaviour   [215]
Acar, W.   : y as Organizational   Pathology .   [5]
Kerefky, P. :   Patient Registers on Microcomputers.   [905]
Breitenecker: r Treadmill Test in   Patients with Coarctation of the Aorta   [215]
Ratko, I.   : ter Registration of   Patients Waiting for Cardiac Operation   [1388]

Roberts, S.L: ms of Critical Care Patients .   [1429]
Atkin, R.H. :   Time as a Pattern   on a Multi-Dimensional Structu   [63]
Barto, A.G. :   A Note on Pattern   Reproduction in Tessellation S   [111]
Batchelor, B: Pattern   Recognition: Ideas in Practice   [117]
Batchelor, B: matic Approaches to   Pattern   Recognition.   [116]
Bozinovski, : A Representation of   Pattern   Classification Teaching.   [208]
Chen, C.N.   :   Pattern   Recognition and Artificial Int   [302]
Drozen, V.   : es - An Approach to Pattern   Analysis.   [442]
Duda, R.O.   :   Pattern   Classification and Scene Analy   [445]
Eilbert, R.F: he Boundary Between   Pattern   Recognition and Numerology.   [464]
Gonzales, R.:   Syntactic Pattern   Recognition.   [640]
Grenander, U: ctures: Lectures in   Pattern   Theory. Vol.3.   [669]
Haken, H.   :   Pattern   Formation by Dynamic Systems a   [707]
Haken, H.   : Dynamic Systems and Pattern   Recognition.   [707]
Kittler, J. :   Pattern   Recognition Theory and Applica   [931]
Klir, G.J.   :   Pattern   Discovery in Activity Arrays.   [939]
Kovalevsky, :   Image Pattern   Recognition.   [975]
Michalski, R: Pattern   Recognition as Rule-Guided Ind   [1158]
Sanderson, A: Pattern   Trajectory Analysis of Nonstat   [1489]
Tsai, W.H.   : tical Approaches to Pattern   Recognition.   [1757]
Verhagen, C.: neral Remarks About   Pattern   Recognition; its Definitions;   [1812]
Watanabe, S.:   Pattern   Recognition as a Quest for Min   [1846]
Young, T.Y. :   Statistical Pattern Classification with Binary Var   [1914]
Waterman, D.:   Pattern-Directed Inference Systems.   [1847]
Bezdek, J.C.:   Patterns   Recognition with Fuzzy Object   [167]
Cordier, M.O: er Way for Matching   Patterns   in KBS.   [346]
Katz, M.J.   : tion of Autogenetic Patterns .   [884]
Martin, P.E.: trol of Information   Patterns   - Further Developments.   [1111]
Sahal, D.   :   Patterns   of Technological Innovations.   [1481]
Koenderink, :   Perception of Solid Shape and Spatial   [961]
Walker, V.R.: Criteria for Object   Perception .   [1832]
Wolkowski, Z: Hierarchical Perception of Time and Periodicity.   [1878]
Sharma, B.D.: r Correcting Binary   Perfect   Codes.   [1538]
Sharma, B.D.: ed Error Correcting   Perfect   Codes.   [1540]
Meyer, J.F. : d-Form Solutions of Performability .   [1157]
Atre, S.   : hniques for Design, Performance and Management.   [66]
Bartolini, G: Development of Performance Adaptive Fuzzy Controllers   [113]
Kellermayr, :   Performance Simulation of an Optical B   [900]
Kerin, T.   :   A Method for Performance Assessment of Medical Radi   [906]
Kobayashi, H: roduction to System   Performance Evaluation Methodology.   [953]
Koyayashi, H: roduction to System   Performance Evaluation Methodology.   [977]
Pau, L.F.   : ilure Diagnosis and Performance Monitoring.   [1296]
Sauer, C.H. :   Computer Systems Performance Modeling.   [1496]
Ivan, I.T.   : mating the Software Performances .   [817]
Beloglavec, : the Properties of a Periodic Discrete String.   [145]
Cushing, J.M: es Competition in a   Periodic Environment.   [367]
Labos, E.   :   Periodic and Non-Periodic Motions in D   [1000]
Wolkowski, Z: ception of Time and   Periodicity .   [1878]
Hwang, C.L. : Sequences of Finite Periods   and a Class of Error Correctin   [802]
Goh, B.S.   : , Vulnerability and Persistence of Complex Ecosystems.   [628]
Landweber, L: f Conflict-Free and   Persistent Petri Nets.   [1008]
Shaw, M.L.G.:   On Becoming a Personal Scientist.   [1541]
Shaw, M.L.G.:   The Personal   Scientist in the Community of   [1542]
Lowen, W.   : del of the Mind and Personality .   [1067]
Neal, M.   : The Construction of Personality in the Novel.   [1205]
Ollero, A.   : ure of Relations in Personnel Management.   [1266]
Gomez, P.   : esign: a Cybernetic Perspective .   [636]
Keen, G.P.W.: : an Organizational   Perspective .   [895]
Keravnou, E.: rt Systems from the   Perspective of Conversation Theory Met   [903]
Linstone, H.:   The Multiple Perspective Concept with Applications   [1052]
Tani, S.T.   :   A Perspective on Modelling in Decision A   [1668]
Urban, H.B. : pment for a Systems Perspective .   [1773]
Winston, P.H: ntelligence: An MIT   Perspective (2 Vols.).   [1871]
Allen, T.F.H: Hierarchy: Perspectives for Ecological Complexity   [24]

Barmark, J. :   Perspectives in Metascience.   [108]
Norman, D.A.:   Perspectives in Cognitive Science. (Pa [1236]
Ringle, M. :   Philosophical Perspectives in Artificial Intelligenc [1420]
Scavia, D. :   Perspectives on Lake Eco-System Modeli [1502]
Auguin, M. : tion of Interpreted Petri Nets.   [70]
Landweber, L: Free and Persistent Petri Nets.   [1008]
Peterson, J.:   Petri Nets.   [1326]
Peterson, J.:   Petri Net Theory and the Modeling of S [1327]
Toulotte, J.: mposing Interpreted Petri Nets and its Utilization.   [1707]
Sharma, B.D.:   Phased Error Correcting Perfect Codes. [1540]
De Greene, K: Fields and Emergent Phenomena in Sociotechnical Macrosyste [390]
Suzuki, R. :   Mathematical and Phenomenological Studies of Biological [1644]
Kuhlenbeck, : al Sciences and its Phenomenology . Vol.II: The Brain and i [993]
Blauberg, I.:   Systems Theory: Philosophical and Methodological Probl [173]
Christensen,: Description; Vol.2: Philosophical Origins; Vol.3: Computer [311]
George, F.H.:   Philosophical Foundations of Cyberneti [595]
Ringle, M. :   Philosophical Perspectives in Artifici [1420]
Rzevski, G. :   Some Philosophical Aspects of System Design [1468]
Agassi, J. :   Scientific Philosophy Today: Essays in Honour of [14]
Aulin, A.   : s a Critical Social Philosophy and a Criticism of Marixsm. [73]
Bahm, A.J. : es of Five Types of Philosophy .   [80]
Bahm, A.J. : ve Types of Systems Philosophy .   [79]
Bennett, R.J: ironmental Systems: Philosophy , Analysis and Control.   [152]
Devries, R.P: tems Theory and the Philosophy of Science.   [415]
George, F.H.:   The Science of Philosophy .   [596]
Kubat, L.   : tion in Science and Philosophy .   [990]
            : ational Journal for Philosophy and Methodology of Medicine [1939]
Leinfellner,: ality and Language: Phonetics / Phonology, Syntax, Semanti [1034]
Leinfellner,: nguage: Phonetics / Phonology , Syntax, Semantics.   [1034]
Gloria-Botti: he Erythrocyte Acid Phosphate Locus in Humans.   [625]
Koenderink, : ial Lay-Out through Photometric Invariants.   [961]
Bell, D.A. :   Physical Record Clustering in Database [141]
Camana, P.C.: ure Control without Physical Intervention.   [258]
Elstob, C.M.:   The Formal and the Physical .   [478]
Hooker, C.A.:   Physical Theory as Logico-Operational [779]
Jantch, E. : nifying Paradigm of Physical , Biological and Sociocultural [835]
Prigogine, I: d Complexity in the Physical Sciences.   [1360]
Schieve, W.C: Applications in the Physical and Social Sciences.   [1507]
Wellstead, P: Introduction to Physical System Modelling.   [1856]
Bogdanski, C: ments of Cybernetic Physics .   [182]
Greenspan, D: pecial Relativistic Physics .   [663]
Redhead, M. :   Models in Physics .   [1396]
Rosen, R.   :   The Physics of Complex Systems.   [1446]
Matsuno, K. : Self-Assembly as a Physiochemical Process.   [1124]
Marmarelis, :   Analysis of Physiological Systems.   [1107]
Nicolis, J.S: esynchronization of Physiological Rhythms.   [1225]
Voyat, G.E. :   Piaget Systematized.   [1825]
Majone, G. :   Pitfalls of Analysis.   [1091]
Presern, S. :   Plan Generating System for a Tactile S [1357]
Smith, R.B. :   How to Plan , Design & Implement a Bad System. [1592]
Batchelor, B: ing Convex Hulls of Planar Figures.   [121]
Batchelor, B: Maps in the Complex Plane .   [118]
Gould, P.   : ts, Geographers and Planners .   [649]
Probst, G.J.: odels for Managers, Planners and Politicians in Health Car [1364]
Bryant, D.T.:   Manpower Planning and Organization Design.   [234]
Gage, W.L. : xibility in Network Planning .   [553]
Gheorghe, A.:   Energy Management Planning Tools for the Transition towa [605]
Horn, W.   : tive Computer-Aided Planning : Estimating Health-Care Resou [782]
Johnson, D.T: A Planning Control Strategy that Allows [848]
Johnson, D.T: ows for the Cost of Planning .   [848]
Kaynak, E. : proach in Marketing Planning .   [892]
Laszlo, E. : ernetics and Global Planning .   [1016]
Rudwick, B.H: Systems Approach to Planning and Control.   [1460]
Sage, A.P. : tems and Process of Planning and Decision Support.   [1475]

Sage, A.P.   : ysis in Systems for   Planning   and Decision Support.   [1476]
Tomlinson, R: Systems Approach to   Planning   in Organizations - Developing [1697]
Tonella, G. : ing for Development   Planning .   [1699]
Wang, Z.   : g in Production and   Planning   of Animal Husbandry System in [1837]
Galitsky, V.:   On Modeling the   Plant   Community Dynamics.   [562]
Jean, R.V. : logical Paradigm in   Plant   Biology.   [841]
Bartolini, G: Continuous Casting   Plants .   [113]
Gupta, M.M. : oller for Uncertain   Plants via Liapunov Signal Synthesis M   [687]
Janssens, L.: ws of Nuclear Power   Plants .   [833]
Sembugamoort: PLAS , a Paradigmatic Language Acquisit   [1523]
Hoehle, U. : ies with Respect to   Plausibility Measures.   [764]
Pernici, B. : lem Solving in Game   Playing : Computer Chess.   [1319]
Tonella, G. :   Pluralistic Modelling for Development   [1699]
Gasparski, W: sign Methodology in   Poland (Special Issue of "Design Metho   [575]
            : School in Jadwisin,   Poland , April 13-19, 1980   [1942]
Batchelor, B: e Classification of   Pole Maps in the Complex Plane.   [118]
Hartl, R.F. :   Optimal Seducing   Policies for Dynamic Continuous Lovers   [729]
Sandblom, C.: luation of Economic   Policies .   [1488]
Bunn, D.W. :   Formal Methods in   Policy Analysis.   [242]
Caputo, R.S.: g a Rational Energy   Policy for Western Europe.   [261]
Chang, Shi-K: Policy Analysis and Information System   [294]
Clark, W.C. : Lessons for Ecology   Policy Design: a Case Study of Ecosyst   [318]
Hofferbert, : ive Study of Public   Policy .   [767]
Krone, R.M. : ystems Analysis and   Policy Sciences: Theory and Practice.   [986]
La Porte, T.: nge to Politics and   Policy .   [998]
Manohar Raho: antitative Economic   Policy .   [1102]
Meadows, D.L: ting Computer-Based   Policy Models.   [1143]
Pugh, R.E. :   Evaluation of   Policy Simulation Models: a Conceptual [1368]
Sandblom, C.: ization of Economic   Policy Using Lagged Controls.   [1487]
Shakun, M.F.: flict Resolution in   Policy Making.   [1532]
Shakun, M.F.:   Policy Making and Meaning as Design of   [1533]
Slawski, C. : ntity Bargaining: A   Policy Systems Research Model of Caree   [1575]
Waller, R.J.: Boundaries of Human   Policy Making.   [1835]
Wang, P.   :   and Application to   Policy Analysis and Information System   [1836]
            : tional Symposium on   Policy Analysis and Information System   [1934]
Hoaglin, D.C: tion Strategies for   Policymakers .   [763]
            : nglish Issue of the   Polish Journal "Progress in Cybernetic [1942]
Blunden, M. : nd Management - The   Political Dimension.   [177]
Lucas, R.G. :   Political Determinants of Information   [1070]
Neck, R.   : actions between the   Political and the Economic System in A [1206]
Taylor, A.M.: ems Approach to the   Political Organization of Space.   [1672]
Probst, G.J.: agers, Planners and   Politicians in Health Care Systems.   [1364]
De Cleris, M: Concepts in Law and   Politics .   [388]
La Porte, T.: exity: Challenge to   Politics and Policy.   [998]
Trappl, R. : ciety, Culture, and   Politics .   [1733]
Atkin, R.   :   Polyhedral Dynamics and the Geometry o   [61]
Casti, J.   : Lake Ecosystems: a   Polyhedral Dynamic Representation.   [273]
Matsuno, K. : th Self-Instructing   Polymerization of Amino Acids.   [1120]
Gloria-Botti: Selective   Polymorphism of the Erythrocyte Acid P   [625]
Kucera, V. : Linear Control: the   Polynomial Equation Approach.   [992]
Peschel, M. :   Application of   Polyoptimization to Processes of Evolu [1320]
Capocelli, R: ernetic Approach to   Population Dynamics Modeling.   [260]
Degermendzhy: ary and Oscillatory   Population in Open Systems.   [402]
Kriegel, U. : nsumption and World   Population .   [982]
Skvoretz, J.: Formally Structured   Populations : An Information Processing   [1572]
Kozen, D.   :   Positive First-Order Logic Is NP-Compl   [978]
Cavallo, R.E: Reconstruction of   Possibilistic Behavior Systems.   [286]
Cumani, A. :   On a   possibilistic Approach to the Analysis   [366]
Klir, G.J. :   Possibilistic Information Theory.   [950]
Yager, R.R. :   Aspects of   Possibilistic Uncertainty.   [1894]
Yager, R.R. : imate Reasoning and   Possibilistic Models in Classification [1899]
Hisdal, E. :   Conditional   Possibilities , Independence and Nonint   [761]
Nowakowska, : ical, and Empirical   Possibilities in Decision Theory.   [1244]
Hagg, C.   :   Possibility and Cost in Decision Analy   [699]

Higashi, M. : nformation Based on Possibility Distributions.   [754]
Higashi, M. : ormation Closeness: Possibility and Probability Distributi [756]
Nguyen, H.T.:   On Conditional Possibility Distributions.   [1220]
Yager, R.R. :   Fuzzy Set and Possibility Theory.   [1903]
Yager, R.R. : ion for a Theory of Possibility .   [1891]
Jacob, F.   :   The   Possible and the Actual.   [824]
Troncale, L.:   On a   Possible Discrimination between Bioevo [1740]
Camana, P.C.: Feedback for Human   Posture Control without Physical Inter [258]
Pipino, L.L.:   Potential Impact of Fuzzy Sets on the [1342]
Trappl, R. : the Optic Generator Potential and its Relation to Stevens' [1710]
Jarisch, W. : is of Visual Evoked Potentials .   [838]
Troncale, L.: Science: Obstaches, Potentials , Case Studies.   [1748]
Bandler, W. :   Fuzzy   Power Sets and Fuzzy Simplification Op [99]
Beer, S.   :   Recursions of   Power .   [138]
Darwish, M. : lity of Large-Scale Power Systems.   [377]
Janssens, L.: or Crews of Nuclear Power Plants.   [833]
Luhman, N. :   Trust and   Power .   [1072]
Slawski, C. :   Love,   Power and Conflict: A Systems Model of [1577]
Trappl, R. :   Power , Autonomy, Utopia: New Approache [1731]
Trappl, R. : elation to Stevens' Power Function.   [1710]
Fletcher, R.:   Practical Methods in Optimization. Vol [530]
Furukawa, O.: ordination, and the Practical Application.   [550]
Gill, P.   :   Practical Optimization.   [613]
Kleijnen, J.: ysis of Simualtion: Practical Statistical Techniques.   [932]
Slawski, C. : Arrangements and a Practical Utopia.   [1579]
Tompa, F.W. :   A   Practical Example of the Specification [1698]
Viliums, E.R: Practical Aspects of Alternatives Eval [1817]
Alter, S.L. : rt Systems: Current Practice and Continuing Challenges.   [26]
Batchelor, B: cognition: Ideas in Practice .   [117]
Checkland, P: s in 'Soft' Systems Practice . Part 1: Systems Diagram - So [297]
Checkland, P: s Thinking, Systems Practice .   [301]
Cross, M.   : g and Simulation in Practice .   [359]
Hall, C.A.S.: eling in Theory and Practice : an Introduction with Case St [713]
Janko, J.   : Orientation Towards Practice : and Episode from the History [832]
Jones, L.M. : stems Boundaries in Practice : Some Proposals and Guideline [857]
Krone, R.M. : ciences: Theory and Practice .   [986]
Linger, R.C.: ramming: Theory and Practice .   [1050]
Siewiorek, D: The Theory and   Practice of Reliable System Design. [1554]
Slawski, C. : ship: In Theory and Practice .   [1574]
White, D.J. : ries of Decision in Practice .   [1862]
Pressman, R.: ware Engineering: A Practitioner 's Approach.   [1358]
Kornwachs, K: Pragmatic Information and Nonclassical [972]
Oguntade, O.:   Implementing a Pragmatic Theory of Humanistic Systems [1258]
Rescher, N. :   Methodological   Pragmatism : a Systems-Theoretical Appr [1400]
Majewska-Tro: Selected   Praxiological Bibliography.   [1090]
Gasparski, W: Praxiology .   [580]
Pszczolowski: Praxiology - The Theory with Past and [1367]
             :   Praxiology (a Yearbook published by th [1937]
Espejo, R. :   Cybernetic   Praxis in Government: The Management o [490]
Furukawa, O.:   Precoordination in Quality Control.   [548]
Reckmeyer, W: esearch and Design: Precursors and Futures.   [1394]
Hassell, M.P: namics of Arthropod Predator-Prey Systems.   [733]
Mark, J.W. : orithms to Adaptive Predicitive Coding.   [1105]
Gardner, M.R: Predicting Novel Facts.   [570]
Shepp, L.A. :   On   Prediction of Moving-Averaging Process [1544]
Yager, R.R. :   Fuzzy   Prediction Based on Regression Models. [1902]
Silvert, W. : n and Evaluation of Predictions .   [1558]
Cover, T.M. :   Compound Bayes   Predictors for Sequences with Apparent [353]
Golledge, R.:   Proximity and Preference : Problems in the Multidimen [632]
Orlovsky, S.: Making with a Fuzzy Preference Relation.   [1271]
Orlovsky, S.: for Multiple Fuzzy   Preference Relations.   [1272]
Von Wright, :   The Logic of   Preference .   [1824]
Zhukovin, V.: g with Vector Fuzzy Preference Realtion.   [1925]
Huber, O.   : ve Multidimensional Preferences : Theoretical Analysis of a [795]

Roedding, W.:   the Aggregation of   Preferences .   [1434]
Katz, M.J.   :   Preformation of Autogenetic Patterns. [884]
Blauberg, I.: Systemic Approach: Prerequisites , Problems and Difficulti [174]
Ginzberg, M.:   A   Prescription Model for System Implemen [616]
Locker, A.   : the Description and   Prescription of Systems.   [1059]
Graedel, T.E: Graphical   Presentation of Results from Scientifi [652]
             : " Devoted to Papers   Presented at the Second Spring School [1942]
Foo, N.Y.   :   Stability   Preservation under Homomorphisis.   [534]
Tchon, K.   : On Some Operations   Preserving Generic Properties of Syste [1679]
Breitenecker: Pathological Blood   Pressure Behaviour after Treadmill Tes [215]
Farkas, J.   : king of Theoretical   Presuppositions of Systems Theory.   [506]
Smetanic, Y.: for Determining the   Primary Structure of Biopolymers.   [1585]
Troncale, L.: tions between Fifty   Principal Systems Concepts.   [1737]
Troncale, L.: ropositions between   Principal Systems Concepts.   [1743]
Eigen, M.   :   The Hypercycle: a   Principle of Natural Selforganization. [462]
Goffman, W. : ion Systems and the   Principle of Selectivity.   [627]
Oldershaw, R: ew Evidence for the   Principle of Self-Similarity.   [1265]
Sadovsky, V.:   The System   Principle and Leibnitz's Theoretical H [1471]
Williams, P.: onalisation and the   Principle of Minimum Information.   [1865]
Dym, C.L.   :   Principles of Mathematical Modeling.   [453]
Eigen, M.   : ws of the Game: How   Principles of Nature Govern Chance.   [463]
Green, M.A. : ar Cells: Operating   Principles , Technology, and System App [659]
Janssens, L.: ation of Cybernetic   Principles for the Training of Operato [833]
Laszlo, E.   : GST: Prospects and   Principles .   [1014]
Nilsson, N.J: Principles of Artificial Intelligence. [1230]
Ryabinin, I.: ngineering Systems: Principles and Analysis.   [1466]
Sahal, D.   :   Principles of Regulation and Control. [1477]
Sperb, R.P. :   Maximum   Principles and their Applications.   [1604]
Takahara, Y.: Theory of Decision   Principles .   [1656]
Takahara, Y.: f Rational Decision   Principles .   [1658]
Varela, F.J.:   Principles of Biological Autonomy. (Se [1799]
Warfield, J.:   Some   Principles of Knowledge Organization. [1840]
Witten, I.H.:   Principles of Computer Speech.   [1875]
Zeigler, B.P: Structuring   Principles for Multifacet System Model [1920]
Potter, J.M.:   Partial   Prior Information and Decision-Making. [1355]
Toutenburg, :   Prior Information in Linear Models.   [1708]
Van den Boga: Wafting Times Using   Priority Classes: A Case Study in Resp [1793]
Walker, C.C.: ement by Exception,   Priority , and Input Span in a Class of [1830]
Aggarwal, S.: c Representation of   Probabilistic Systems by Ergodic Machi   [17]
Domotor, Z. :   Probabilistic and Causal Dependence St [432]
Gerardy, R. :   Probabilistic Finite State System Iden [598]
Pearl, J.   : marizing Data Using   Probabilistic Assertions.   [1303]
Pearl, J.   : thods of Evaluating   Probabilistic Forecasts.   [1304]
Santos, E.S.:   Probabilistic Machines and Languages. [1493]
Soong, T.T. :   Probabilistic Modeling and Analysis in [1600]
Srinivasan, :   Probabilistic Analysis of Redundant Sy [1608]
Witten, I.H.:   Probabilistic Behaviour-Structure Tran [1873]
Yager, R.R. :   Fuzzy Sets,   Probabilities , and Decision.   [1895]
Gaines, B.R.:   Fuzzy and   Probability Uncertainty Logics.   [556]
Higashi, M. : ss: Possibility and   Probability Distributions.   [756]
Okuda, T.H. : y Information Using   Probability Measures of Fuzzy Events. [1261]
Van Campenho: opy and Conditional   Probability .   [1786]
Von Plato, J: Probability and Determinism.   [1823]
Yakowitz, S.:   Computational   Probability and Simulation.   [1910]
Matsuno, K. : ystems Formation in   Probiological Molecular Evolution.   [1122]
Aarnes, E.   :   On the   Problem of Identification in Compartme   [1]
Agazzi, E.   : stem Theory and the   Problem of Reductionism.   [15]
Averkin, A.N: lized Strategies in   Problem Solving.   [75]
Banerji, R.B: es to Non-Numerical   Problem Solving.   [103]
Branin, F.H.:   Problem Analysis in Science and Engine [212]
Brightman, H: Problem Solving: A Logical and Creativ [219]
Brightman, H: Problem Solving: A Logical and Creativ [220]
Carlsson, C.: h to Handling Fuzzy   Problem Structures.   [264]
Cavallo, R.E: ndation for Systems   Problem Solving.   [276]

Chandrasekar: ors for Distributed Problem Solving. [292]
Chen, D.T.-W: ogical Reasoning in Problem Solving by Computers. [303]
Dreben, B. : The Decision Problem . [435]
Farley, A.M.: in Knowledge-Based Problem Solving. [507]
Gomez, P. : in Action: Organic Problem Solving in a Publishing Compan [638]
Grabowski, J: Sequencing Problem with Resource Constraints. [651]
Grogono, P. : Problem Solving and Computer Programmi [672]
Guida, G. : Integrated Model of Problem Solver. [679]
Hazony, Y. : Problem Solving with Interactive Graph [745]
Kaufmann, G.: and Its Relation to Problem Solving. [887]
Kekes, J. : The Centrality of Problem Solving. [898]
King, L.T. : Problem Solving in a Project Environme [926]
King, L.T. : Problem Solving in a Project Environme [925]
Klir, G.J. : On the Problem of Computer-Aided Structure Id [938]
Klir, G.J. : General Systems Problem Solving Methodology. [946]
Koffman, E.B: Problem Solving and Structured Program [963]
Kowalaski, R: Logic of Problem Solving. [976]
Lester, F.K.: Mathematical Problem Solving. [1038]
Matevosyan, : The Invers Problem of Karhunen-Loeve. [1117]
Miller, R.B.: pes of Man-Computer Problem Solving. [1169]
Molloy, K.J.: plex Organisational Problem Solving. [1182]
Newell, A. : Human Problem Solving. [1217]
Ohlsson, S. : tomated Learning of Problem Solving Rules. [1259]
Pernici, B. : Problem Solving in Game Playing: Compu [1319]
Rasvan, V. : with the Stability Problem . [1387]
Robertshaw, : Problem Solving: a Systems Approach. [1430]
Sanderson, M: Successful Problem Management. [1491]
Scandura, J.: Problem Solving: a Structural Process [1500]
Smith, R.G. : tion in Distributed Problem Solving. [1593]
Stoklosa, J.: On the Reachability Problem for Some Dynamical Systems. [1623]
Ulrich, W. : The Design of Problem Solving Systems. [1767]
Uyttenhove, : stem Identification Problem (Part I). [1777]
Uyttenhove, : he Systems Approach Problem Solver and a Stockmarket Examp [1778]
Zarri, G.P. : Order to Solve the Problem of Information Incompleteness. [1915]
Zionts, S. : Multiple Criteria Problem Solving. [1928]
Chytil, M.K.: ethods as Cognitive Problem-Solvers . [314]
Callebaut, W: ibution Approach to Problem-Solving in Science: Prospects [257]
Gladun, V.P.: Adaptive Problem-Solving Systems. [617]
Wade, J.W. : l Design as a Basic Problem-Solving Process. [1826]
Blauberg, I.: and Methodological Problems . [173]
Blauberg, I.: ach: Prerequisites, Problems and Difficulties. [174]
Borodin, A. : gebraic and Numeric Problems . [190]
Carlsson, C.: A System of Problems and How to Deal with It. [262]
Carlsson, C.: A System of Problems and a Way Treating it. [263]
Csaki, P. : ach to Some General Problems of Model Description. [360]
Czogala, E. : lication in Control Problems . [368]
Czogala, E. : Control Problems in Fuzzy Systems. [369]
El-Shirbeeny: ined Control Design Problems . [472]
Everitt, B.S: Unresolved Problems in Cluster Analysis. [497]
Favella, L.F: First Passage Time Problems and Some Related Computationa [511]
Favella, L.F: Acoustic Diagnostic Problems for Internal Combustion Engin [512]
Favella, L.F: al and Mathematical Problems in ECG Analysis. [513]
Gasparski, W: Problems of Design Methodology. [574]
Golledge, R.: ity and Preference: Problems in the Multidimensional Analy [632]
Gottinger, H: s and Environmental Problems . [645]
Gvishiani, J: rch: Methodological Problems - 1979 Yearbook. [694]
Hattiangadi,: The Structure of Problems . [735]
Hellwig, Z.H: Basic Problems of Modern Econometrics and So [748]
Hellwig, Z.H: onometrics and Some Problems of its Acceptability and Inte [748]
Lasker, G.E.: between Major World Problems and Systems Learning. [1011]
Lopes, M.A. : Operations on Problems and Their Solution Spaces. [1065]
Nowakowska, : Methodological Problems of Measurement of Fuzzy Conce [1239]
Okuda, T.H. : n of Fuzzy Decision Problems with Fuzzy Information Using [1261]
Paritsis, N.: Adaptional Problems in Natural Intelligent System [1282]

Peery, N.S. : Organizations: Some Problems in Application.   [1314]
Petrov, B.N.:   Some   Problems in the Theory of Searchless S [1329]
Rao, G.P.   : tic Descriptions in Problems of Fuzzy Logic Applied to Sys [1380]
Roberts, S.L: ssessing Behavioral   Problems of Critical Care Patients.   [1429]
Rudwick, B.H: Solving Management   Problems : a Systems Approach to Planni [1460]
Sakawa, M.   : objective Nonconvex Problems .   [1483]
Sengupta, J.: ting and Validation   Problems in Stochastic Linear Programm [1525]
Stockmeyer, : insically Difficult   Problems .   [1621]
Takatsu, S. : atisficing Decision Problems .   [1660]
Tamburrini, : ome Epistemological   Problems ?   [1665]
Tzafestas, S: bility: a Survey of   Problems and Techniques.   [1764]
Veloso, P.A.: ities Among General   Problems .   [1807]
Wade, J.W.   :   Architecture, Problems , and Purposes: Architectural [1826]
Watt, K.E.F.: ate - Environmental   Problems , Growth, and Culture.   [1849]
             : rch: Methodological Problems - 1982 Yearbook.   [1944]
Fortuny-Amat: A Solution Procedure for the Hierarchical Coordin   [536]
Korhonen, P.:   A Stepwise Procedure for Multivariate Clustering.   [970]
Ben-Dov, Y. :   Optimal Testing Procedures for Special Structures of C   [148]
Chatterjee, : ison of Arbitration   Procedures : Models with Complete and I   [295]
Elias, D.   :   Procedures for Generating Reconstructi   [474]
Klir, G.J.   :   Procedures for Generating Reconstructi   [947]
Smith, P.M. : red Knowledge about Procedures Used in Fault Diagnosis Tas   [1591]
Trappl, R.   : Simple Forecasting Procedures in Health Care.   [1714]
Yager, R.R. :   Some   Procedures for Selecting Fuzzy Set-The [1906]
Yashin, A.   : Gaussian Estimation Procedures .   [1911]
Banathy, B.H: Proceedings of the 29th Annual Meeting   [97]
Kindler, J. :   Proceedings of a Workshop on Modelling   [924]
Klement, E.P: Proceedings of the International Semin   [933]
Sharif, N.   :   Proceedings of the Int. Conference on   [1536]
Van Straaten: Proceedings of the Second Joint MTA/II [1792]
             :   Proceedings of the First International [1934]
             :   Proceedings of the Workshop on Applica [1940]
Aplin, J.C. : ic Model of the MBO Process .   [39]
Cellier, F.E: Process in Modelling and Simulation.   [289]
Cerbone, G. : Ornstein-Uhlenbeck Process .   [290]
Day, R.H.   : ment as an Adaptive Process .   [384]
De Beaufrand: ext, Discourse, and   Process : Towards a Multidisciplinary S [386]
De Beaugrand: ext, Discourse, and   Process .   [387]
Degermendzhy: y of Microevolution   Process of the Stationary and Oscillat [402]
El-Sherief, : riable Systems with   Process Noise.   [470]
Erlandson, R: The Satisficating Process . A New Look.   [488]
Franta, W.R.:   Process View of Simulation.   [543]
King, L.T.   : nment: A Consulting Process .   [926]
Mackenzie, K: A Process Based Measure for the Degree o [1083]
Mackenzie, K: eadership as a Task   Process Uncertainty Control Process. [1084]
Mackenzie, K: Uncertainty Control   Process .   [1084]
Matsuno, K. : as a Physiochemical Process .   [1124]
Rauch, W.D. : g as an Interactive Process .   [1391]
Ray, W.H.   :   Advanced Process Control.   [1392]
Roffel, B.   :   Process Dynamics, Control and Protecti [1435]
Sage, A.P.   : rmation Systems and Process of Planning and Decision Suppo [1475]
Scandura, J.: lving: a Structural   Process Approach with Instructional Im [1500]
Schal, R.K. : ists Concerning the Process of Human Transformation.   [1504]
Szaniawksi, : Information-Seeking   Process .   [1650]
Troncale, L.: ergent Evolutionary   Process Based on Systems Concepts.   [1738]
Van Navta, L: ter Applications to   Process Control.   [1789]
Wade, J.W.   : sic Problem-Solving Process .   [1826]
Winograd, T.: uage as a Cognitive Process (Vol.1: Syntax).   [1868]
Bellman, R. : l Theory of Control Processes : Nonlinear Processes.   [143]
Bellman, R. : rocesses: Nonlinear Processes .   [143]
Bogner, S.   : Model of Cognitive Processes .   [183]
Bosserman, R: Internal Security Processes in General Living Systems.   [196]
Bosserman, R: Internal Security Processes and Subsystems in Living Sys [195]
Brown, J.H.U: nation of Metabolic   Processes : a Systems Approach to Endoc [230]

Cliff, A.D. :   Spacial Processes : Models and Applications.   [320]
Das, J.P.   : uccessive Cognitive Processes .   [378]
De Witt, C.T: ation of Ecological Processes (2nd Edition, Revised and Ex   [400]
Demillo, R.A: Social Processes and Proofs of Theorems and P   [407]
Edwards, J.B: ontrol of Multipass Processes .   [456]
Gabasov, R. : e Theory of Optimal Processes . (Translated From Russian by   [551]
Getis, A.   :   Models of Spatial Processes .   [601]
Gihman, I.I.: ntrolled Stochastic Processes .   [611]
Gupta, M.M. : tomata and Decision Processes .   [685]
Hartley, R. : in Markov Decision Processes .   [730]
Honkasalo, A: Entropic Processes and Economic Systems.   [777]
Jachymczyk, : ain Class of Random Processes .   [822]
Kandel, A. :   Fuzzy Processes .   [875]
Karlin, S. : ourse in Stochastic Processes .   [881]
Monahan, G.E: ble Markov Decision Processes Theory, Models, and Algorith   [1183]
Murphy, R.E.:   Adaptive Processes in Economic Systems.   [1198]
Peschel, M. : Polyoptimization to Processes of Evolution.   [1320]
Pinske, M.S.: andom Variables and Processes .   [1341]
Porenta, G. : s Disease and Aging Processes .   [1349]
Shepp, L.A. : of Moving-Averaging Processes .   [1544]
Simon, J.C. : r Oriented Learning Processes .   [1564]
Smale, S.   : l Systems, Economic Processes and Related Topics.   [1583]
Soi, I.M.   : -Orientied Learning Processes .   [1597]
Syski, R.   :   Random Processes : a First Look.   [1649]
Teodorescu, : Actions and Markov Processes .   [1682]
Tong, R.M. : dels for Industrial Processes - Some Recent Results.   [1700]
Trost, H.   : ulation of Neurotic Processes .   [1751]
Varshavskiy,:   Asynchronous Processes . I. Definition and Interpret [1802]
Varshavskiy,:   Asynchronous Processes II. Composition and Matching [1801]
Zouwen, J.v.: odeling Interaction Processes in Interviews; a Contributio   [1930]
Abe, J.   : in Natural Language Processing .   [2]
Deutsch, K.W: Complexity in Data Processing for World Models.   [414]
Fairhurst, M: vity in Information Processing Systems.   [500]
Gigley, H. : of Natural Language Processing .   [610]
Gonzales, R.:   Digital Image Processing .   [639]
Hall, E.L. :   Computer Image Processing and Recognition.   [714]
Havranek, T.:   Processing Sociological Data by the Gu   [737]
Keyes, R.W. : Digital Information Processing .   [909]
Kobayashi, I: ion and Information Processing Structure.   [954]
Lozinskii, E: in Natural Language Processing .   [1069]
Pearl, J.   :   A Framework for Processing Value Judgements.   [1302]
Pichler, F. : Robotics and Image Processing .   [1338]
Pratt, W.K. :   Digital Image Processing .   [1356]
Sampson, J.R: daptive Information Processing : An Introductory Survey.   [1485]
Simon, J.C. :   Digital Image Processing .   [1565]
Skvoretz, J.: ons: An Information Processing Approach.   [1572]
Spoehr, K.T.: Visual Information Processing .   [1605]
Stucki, P. : es in Digital Image Processing .   [1630]
Sultan, L.H.: r Fuzzy Information Processing .   [1636]
Vaccari, E. : in Natural Language Processing .   [1779]
Hanassab, S.:   Parallel Processors for Cybernetic Systems.   [718]
Bottaci, L. : ct Specification of Production System Programs Used in Exp   [198]
Findler, N.V: ing Technique Using Production Systems.   [526]
Lauenroth, H: ation of Industrial Production and Information Systems.   [1017]
Lesmo, L.   : nterpreter of Fuzzy Production Rules.   [1037]
Wang, Z.   : near Programming in Production and Planning of Animal Husb [1837]
Grubbstrom, : Axsater Integrated Production-Inventory System Interprete   [676]
Mak, K.L.   : ntrol of Integrated Production-Inventory-Shipping Systems. [1093]
Shakun, M.F.:   Effectiveness, Productivity and Design of Purposeful [1534]
Bandler, W. : nd Fuzzy Relational Products .   [100]
Burgin, M.S.:   Products of Operators of Multidimensio   [244]
Mulej, M.   : es' Ideas about New Products and Programs.   [1195]
Dedons, A. :   The Information Professional : Survey of an Emerging Fi   [401]
Yashin, A. :   A New Proff and New Results in Conditional G [1911]

Shakun, M.F.: oseful Systems: The   Profit-Making   Case.   [1534]
Zeleny, M.   : ng of Resources and   Profits   Via Linear Programming.   [1923]
Darlington, : ctive Validation of   Program   Specifications.   [376]
Davis, C.   :   Organization for Program   Management.   [382]
Franova, M. :   Program   Synthesis and Constructive Pro   [542]
Gams, M.   : A Circuit Analysis   Program   that Explains its Reasoning.   [565]
Greibach, S.:   Theory of   Program   Structures: Schemes, Semantics   [668]
Hoernig, K.M: spects of Automatic   Program   Construction.   [765]
Kerefky, P. :   Program   Optimization and Manipulation   [904]
Laszlo, C.A.: System and Computer   Program   Package for Health Hazard Appr   [1013]
Leal, A.   :   An Interactive   Program   for Conversational Elicitation   [1025]
Trappl, R.   : th Hazard Appraisal   Program   for Austria.   [1717]
Batchelor, B: Experimental and   Programmatic   Approaches to Pattern Rec   [116]
Wendt, S.   :   The   Programmed   Action Module: an Element f   [1857]
Baldwin, J.F: Dynamic   Programming   for Fuzzy Systems with Fuz   [85]
Cole, I.D.   :   Naturalized   Programming : A Method of Naturalizing   [329]
Cole, I.D.   : hod of Naturalizing   Programming   Languages.   [329]
Dreyfus, S.E: Dynamic   Programming   and the Calculus of Variat   [437]
Dreyfus, S.E: d Theory of Dynamic   Programming .   [438]
Farinas-del-:   Prolegomena for   Programming   in Modal Logic.   [505]
Floyd, R.W. :   The Paradigms of   Programming .   [531]
Gini, G.   : nd Nondeterministic   Programming   in Robot Systems.   [615]
Glushkov, V.:   Algebra, Languages   Programming .   [626]
Grogono, P. : olving and Computer   Programming .   [672]
Kerner, H.   : l versus Imperative   Programming   - Complement or Contradict   [908]
Kindler, E. : ation of Simulation   Programming   Languages: I. Declaration   [917]
Kindler, E. : ation of Simulation   Programming   Languages: II. Description   [918]
Kindler, E. :   Simulation   Programming   Languages.   [921]
Koffman, E.B: ving and Structured   Programming   in PASCAL.   [963]
Linger, R.C.:   Structured   Programming : Theory and Practice.   [1050]
Molina, F.W.: ion in Mathematical   Programming .   [1181]
Ratko, I.   : ical Expressions in   Programming   Languages.   [1389]
Sengupta, J.: tions in Stochastic   Programming .   [1526]
Sengupta, J.: n Stochastic Linear   Programming .   [1525]
Sengupta, J.: odels in Stochastic   Programming .   [1524]
Stoy, J.E.   : trachey Approach to   Programming   Languages.   [1625]
Wang, Z.   : ng and Fuzzy Linear   Programming   in Production and Planning   [1837]
Yeh, R.T.   :   Current Trends in Programming   Methodology: Vol. IV - Dat   [1912]
Zeleny, M.   :   Profits Via Linear Programming .   [1923]
Anderson, R.:   Proving   Programs   Correct.   [30]
Augustine, N: System Development   Programs .   [71]
Bottaci, L. : f Production System   Programs   Used in Expert Systems.   [198]
Demillo, R.A: ofs of Theorems and   Programs .   [407]
Feldman, J.A: e Inference of Bent   Programs .   [517]
Kobsa, A.   :   On Regarding AI Programs   as Theories.   [956]
Mulej, M.   : ut New Products and   Programs .   [1195]
Aulin, A.   : etic Laws of Social   Progress : Towards a Critical Social Ph   [73]
Eyckhoff, P.:   Trends and Progress   in Systems Identification.   [498]
Eyckhoff, P.:   Trends and Progress   in Systems Identification.   [499]
Pichler, F. :   Progress   in Cybernetics and Systems Re   [1336]
Pichler, F. :   Progress   in Cybernetics and Systems Re   [1335]
Rosen, R.   :   Progress   in Theoretical Biology, Vol.6   [1445]
Trappl, R.   :   Progress   in Cybernetics and Systems Re   [1712]
Trappl, R.   :   Progress   in Cybernetics and Systems Re   [1713]
Trappl, R.   :   Progress   in Cybernetics and Systems Re   [1721]
Trappl, R.   :   Progress   in Cybernetics and Systems Re   [1720]
Trappl, R.   :   Progress   in Cybernetics and Systems Re   [1723]
Trappl, R.   :   Progress   in Cybernetics and Systems Re   [1724]
Trappl, R.   :   Progress   in Cybernetics and Systems Re   [1722]
             : the Polish Journal "   Progress   in Cybernetics" Devoted to Pa   [1942]
King, L.T.   : roblem Solving in a   Project   Environment: A Consulting Proc   [926]
King, L.T.   : roblem Solving in a   Project   Environment.   [925]
Schellenberg: Project   Management, Key to Successful   [1505]
Zarri, G.P. :   The RESEDA   Project : Using AI Techniques in Order   [1915]

Takeuchi, A.: pproach by Means of Projection onto Linear Subspeces.   [1664]
Herman, G.T.: Reconstruction From Projections : Implementation and Applic   [751]
Katz, M.J.   : Reconstruction from Projections .   [883]
Hickin, J.   :   Applications of Projective Reduction Methods to Estima [753]
Gregory, S.A: Large Technological Projects .   [667]
Krzanik, L. :   Evolutionary Projects : Dynamic Optimization of Deli   [989]
Schellenberg: Successful Computer Projects .   [1505]
Trappl, R. : e: A Proposal for 3 Projects .   [1732]
Farinas-del-:   Prolegomena for Programming in Modal L   [505]
Conant, R.C.:   Efficient Proofs of Identities in N-Dimensional   [340]
Demillo, R.A: ocial Processes and Proofs of Theorems and Programs.   [407]
Franova, M. : is and Constructive Proofs Obtained by Beth's Tableaux.   [542]
Cordier, M.O: Propagation : Another Way for Matching   [346]
Patten, B.C.:   Propagation of Cause in Ecosystems.   [1290]
Giere, R.N. :   Propensity and Necessity.   [609]
Arigoni, A.C: Synthesis of Formal Properties .   [50]
Beloglavec, : ory Approach to the Properties of a Periodic Discrete Stri   [145]
Dammasch, I.:   On the Properties of Randomly Connected McCul   [375]
Foo, N.Y.   :   Closure Properties and Homomorphisms of Timeva   [535]
Landweber, L: Properties of Conflict-Free and Persis   [1008]
Matsuda, T. :   Algebraic Properties of Satisficing Decision.   [1119]
Noguchi, S. :   Properties of Analogue Neural Networks   [1235]
Reichle, D.E: Dynamic Properties of Forest Ecosystems.   [1398]
Shore, J.E. :   Properties of Cross-Entropy Minimizati   [1551]
Tchon, K.   : Preserving Generic Properties of Systems.   [1679]
Tong, R.M. :   Some Properties of Fuzzy Feedback Systems.   [1702]
Yager, R.R. :   Some Properties of Fuzzy Relationships.   [1901]
Arigoni, A.C: ebraic Structure of Property Spaces - On the Synthesis of   [50]
Parks, M.S. :   Dichotic Property and Teleogenesis.   [1283]
Nevitt, B. :   ABC of Prophecy : Understanding the Environmen [1216]
Trappl, R. : ial Intelligence: A Proposal for 3 Projects.   [1732]
Jones, L.M. : s in Practice: Some Proposals and Guidelines.   [857]
Troncale, L.: ystems: The Linkage Proposition Model.   [1749]
Troncale, L.:   Linkage Propositions between Fifty Principal S   [1737]
Troncale, L.:   Linkage Propositions between Principal Systems   [1743]
Troncale, L.: of Systems Linkage Propositions .   [1746]
Rosen, R.   : s in Retrospect and Prospect .   [1443]
Callebaut, W: Solving in Science: Prospects for GSM.   [257]
Laszlo, E. :   GST: Prospects and Principles.   [1014]
LeMoigne, J.: nch History and its Prospects .   [1024]
Kohout, L.J.: of a General System Protection Design.   [966]
Roffel, B. : namics, Control and Protection .   [1435]
Walker, B.J.: mputer Security and Protection Structures.   [1829]
Rosen, R.   :   Protein Folding: a Prototype for Contr   [1444]
Matsuno, K. : of Microsystems in Protobiogenesis .   [1125]
Rosen, R.   : Protein Folding: a Prototype for Control of Complex Syste   [1444]
Gyarfas, F. :   CODEX: Prototypes Driven Backward and Forward   [695]
Yager, R.R. :   Prototypical Values for Fuzzy Subsets.   [1900]
Anderson, R.:   Proving Programs Correct.   [30]
Golledge, R.:   Proximity and Preference: Problems in   [632]
Suppes, P. :   PSA 1976.   [1639]
Slawski, C. : Synthesis of Social Psychological "Theories".   [1581]
Kantor, J.R.: ture and Scientific Psychology .   [878]
Pask, G.   :   Cybernetics in Psychology and Education.   [1287]
Scandura, J.: native to Cognitive Psychology .   [1501]
Cliffe, M.J.: s to the Science of Psychopathology .   [322]
Durkin, J.E.: iving Groups: Group Psychotherapy and General System Theor   [449]
Trappl, R. :   Computer Psychotherapy : Is it Acceptable, Feasi   [1715]
Beltrami, E.:   Models for Public Systems Analysis.   [147]
Hofferbert, : omparative Study of Public Policy.   [767]
Halmes, A. : in Systems Theory: Publication in Honour of Prof.Hans Blo   [717]
Ladanyi, O.G: mic Movements Using Published International Statistics.   [1003]
Gomez, P.   : roblem Solving in a Publishing Company.   [638]
Ulrich, W. :   A Critique of Pure Cybernetic Reason: The Chilean Ex [1768]

Resconi, G.   :   Analysis of the   Purine   Biosynthesis by Mathematical Lo [1405]
Tancig, P.   :   SOVA - General   Purpose   Software Environment for NLU-S [1667]
Shakun, M.F.: eaning as Design of   Purposeful   Systems.   [1533]
Shakun, M.F.: ivity and Design of   Purposeful   Systems: The Profit-Making   [1534]
Wade, J.W.   : ture, Problems, and   Purposes : Architectural Design as a Ba   [1826]
Gould, P.   :   Q-Analysis   or a Language of Structure: [649]
Sugiyama, K.:   Visual   Q-Analysis   (I): A Case of Future Compu   [1635]
Toda, M.   :   Visual   Q-Analysis   (II): Strategy Analysis for   [1694]
Pritsker, A.: and Analysis Using   Q-Gert   Networks.   [1362]
Eveland, S.   :   The Use of the   QBE   Relational Database System for a G [496]
Estep, M.L. : alysis: The Case of   Qualitative   Knowing.   [494]
Franksen, O.:   Qualitative   Aspects of Large Scale Sys   [541]
Gabasov, R.   :   The   Qualitative   Theory of Optimal Processe   [551]
Goodman, L.A: Analyzing   Qualitative   / Categorical Data.   [641]
Hofmann, Th.:   Qualitative   Terms for Quantity.   [768]
Michel, A.N.:   Qualitative   Analysis of Large Scale Sy   [1159]
Peay, E.R.   : uctural Models with   Qualitative   Values.   [1306]
Basta, D.J. : duals-Environmental   Quality   Management.   [114]
Furukawa, O.:   Refinement of   Quality   Control Systems.   [549]
Furukawa, O.: Precoordination in   Quality   Control.   [548]
Furukawa, O.:   A Methodology for   Quality   Goal-Seeking and Coordination,   [550]
Holler, M.J.: ts as Self-Policing   Quality   Enforcement Systems.   [772]
Jensen, H.L.: tical Framework for   Quantifying   Legal Decisions.   [844]
Cliff, A.D.   : patial Structure: a   Quantitative   Approach.   [319]
Hazlett, B.A: Quantitative   Methods in the Study of A   [744]
Manohar Raho: Control Systems and   Quantitative   Economic Policy.   [1102]
Wallenmaier,:   Towards a   Quantitative   Concept of Teleological S   [1834]
Hofmann, Th.: alitative Terms for   Quantity .   [768]
Valdes-Perez: on Loss Due to Data   Quantization   in Reconstructability Ana   [1781]
De Mottoni, : ary Solutions for a   Quasilinear   Parabolic System.   [397]
Harris, L.R.: Language Data Base   Queries .   [727]
Sernadas, A.: and Semantics as a   Query   Language.   [1528]
Beer, S.   : ction: Questions of   Quest .   [137]
Watanabe, S.: rn Recognition as a   Quest   for Minimum Entropy.   [1846]
Pearl, J.   : nomy of Inferential   Question-Answer   Systems.   [1301]
Brockett, P.: eoretic Analysis of   Questionnaire   Data.   [222]
Quade, E.   : he State-of-the-Art   Questionnaire   on Applied Systems Analy   [1370]
Beer, S.   :   Introduction: Questions   of Quest.   [137]
Belnap, N.P.:   The Logic of   Questions   and Answers.   [144]
Boyd, G.M.   : tic Aesthetics: Key   Questions   in the Design of Mutual Cont   [205]
Katz, M.J.   :   Questions   of Uniqueness and Resolution   [883]
Kubinski, T.: e Logical Theory of   Questions .   [991]
Sadovskiy, L: Questions   Regarding Simulation of Hier   [1470]
Troncale, L.: ome Key, Unanswered   Questions   about Hierarchies.   [1744]
Courtois, P.:   Decomposability: Queueing   and Computer System Applicati [352]
Gross, D.   :   Fundamentals of   Queueing   Theory.   [674]
Madan, S.   : ernetic Modeling of   Queueing   Complex.   [1086]
Madan, S.   : tment of Cybernetic   Queueing   Systems.   [1087]
Yager, R.R.   :   Fuzzy Thinking as   Quick   and Efficient.   [1897]
Kaindl, H.   : amic Control of the   Quiescence   Search in Computer Chess.   [867]
Nowotny, H.   :   Not   Quite   Human: Science and Utopia.   [1246]
Gaines, B.R.: l Systems Research: Quo   Vadis?   [557]
Kerin, T.   : sessment of Medical   Radioisotope   Equipment.   [906]
Atlan, H.   :   Random   Boolean Networks.   [64]
Granero-Pora: Random   Ecological Systems with Structu   [655]
Jachymczyk, : or Certain Class of   Random   Processes.   [822]
Jumarie, G. : Transinformation of   Random   Experiments Involving Fuzzy Obs   [863]
Klement, E.P: t Theorem for Fuzzy   Random   Variables.   [935]
Kontur, I.F.: f Simulation with a   Random   Walk Model.   [967]
Kwakernaak, :   Fuzzy   Random   Variables. Part 1: Definitions   [996]
Pinske, M.S.: mation Stability of   Random   Variables and Processes.   [1341]
Syski, R.   :   Random   Processes: a First Look.   [1649]
Vallee, R.   :   Linear System with   Random   Initial Conditions.   [1782]
Wang, Z.   : M-Fuzzy Numbers and   Random   Variables.   [1838]

Dammasch, I.: n the Properties of Randomly Connected McCulloch-Pitts Net [375]
Tao, K.M.   : ith Applications: A Rapprochement with Systems Theory (Par [1669]
Tao, K.M.   : ith Applications: A Rapprochement with Systems Theory (Par [1670]
Jarisch, W. : ling of Fetal Heart Rate Variability with Applications to [837]
Caputo, R.S.:   Establishing a Rational Energy Policy for Western Eur [261]
Scott, W.R. :   Organizations: Rational , Natural and Open Systems.   [1520]
Takahara, Y.:   A Structure of Rational Decision Principles.   [1658]
Oeren, T.I. :   Rationale for Large Scale System Simul [1255]
Dompere, K.K: and Decision-Choice Rationality .   [433]
Simon, H.A. :   Models of Bounded Rationality (2 Volumes).   [1563]
Hanson, O.J.:   Review and Re-Design of an Advanced Systems Analy [724]
Stoklosa, J.:   On the Reachability Problem for Some Dynamica [1623]
Sinha, K.K. :   Adaptive Nuclear Reactor Control without Explicit Ident [1569]
Warfield, J.: y Relations and Map Reading .   [1841]
Webber, B.L.:   Readings in Artificial Intelligence.   [1850]
Burkhardt, H: The Real World and Mathematics.   [245]
Martin, J.N.:   On Mapping Real Systems.   [1109]
Troncale, L.: xity in Bio-Systems Real ? Applications of Clustering Theor [1739]
Horn, W.   :   A Frame-Based Real-Time Graphic Interaction System.   [787]
Wedde, H.F. : rmal Modeling under Realistic Assumptions.   [1852]
Bekey, G.A. :   Models and Reality : Some Reflections on the Art a [140]
Casti, J.   :   Illusion or Reality ?: the Mathematics of Attainabl [270]
Kent, W.   :   Data and Reality .   [902]
Von Glasersf: on of "Others" and " Reality ": A Study of Self-Regulation. [1821]
Auguin, M. : ystematic Method of Realization of Interpreted Petri Nets. [70]
Maturana, H.:   and Cognition: The Realization of the Living.   [1128]
Nishimura, H: Realization of Dependency Structures. [1232]
Yager, R.R. :   Their Experimental Realization .   [1904]
Zeigler, B.P: Canonical Realization of General Time Systems. [1917]
Kalman, R.E.:   On Partial Realizations , Transfer Functions, and [872]
Takahara, Y.: or and its Concrete Realizations .   [1659]
Zhukovin, V.: or Fuzzy Preference Realtion .   [1925]
Ulrich, W. :   of Pure Cybernetic Reason : The Chilean Experience with Cy [1768]
Chen, D.T.-W: Toward Analogical Reasoning in Problem Solving by Comput [303]
Christensen,: ations of Inductive Reasoning .   [310]
Gams, M.   : m that Explains its Reasoning .   [565]
Johnson, P.E: on Models of Expert Reasoning .   [851]
Klir, G.J. : ology and Inductive Reasoning : The Issue of Parts and Whol [948]
Le Bonniec,: evelopment of Model Reasoning .   [1021]
Mamdani, E.H: ogic to Approximate Reasoning Using Linguistic Synthesis. [1098]
Mamdani, E.M: Fuzzy Reasoning and Its Applications.   [1099]
Mattessich, :   Instrumental Reasoning and Systems Methodology: an [1126]
Reynolds, J.:   Reasoning About Arrays.   [1415]
Winston, P.H: Learning and Reasoning by Analogy.   [1870]
Yager, R.R. :   Approximate Reasoning and Possibilistic Models in [1899]
Yager, R.R. : ence in Approximate Reasoning .   [1893]
Mel'cuk, I.A: d Linguistics: Some Reasons for as Well as Some Consequenc [1148]
Cavallo, R.E: Recent Developments in Systems Methodo [278]
Cavallo, R.E: l Science Research: Recent Developments.   [287]
Duechting, W: Recent Results in Modelling and Simula [446]
El-Sherief, :   Recent Advances in Multivariable Syste [471]
Hartley, R. :   Recent Developments in Markov Decision [730]
Laing, R.   : of the Relevance of Recent Results.   [1004]
Mohler, R. :   Recent Developments in Variable Struct [1179]
Tong, R.M. : al Processes - Some Recent Results.   [1700]
Witten, I.H.:   Some Recent Results on Nondeterministic Mod [1874]
Wolaver, T.G: , Connectivity, and Recipient Control on the Sensitivity C [1877]
Batchelor, B: Pattern Recognition : Ideas in Practice.   [117]
Batchelor, B: proaches to Pattern Recognition .   [116]
Bezdek, J.C.:   Patterns Recognition with Fuzzy Objective Funct [167]
Chen, C.N. :   Pattern Recognition and Artificial Intelligenc [302]
Dixon, N.R. : Speech and Speaker Recognition .   [426]
Eilbert, R.F: ary Between Pattern Recognition and Numerology.   [464]
Gonzales, R.:   Syntactic Pattern Recognition .   [640]

Haken, H.   : Systems and Pattern Recognition .   [707]
Hall, E.L.   : mage Processing and   Recognition .   [714]
Kittler, J. :   Pattern   Recognition   Theory and Applications. [931]
Kovalevsky, :   Image Pattern   Recognition .   [975]
Michalski, R: Pattern   Recognition   as Rule-Guided Inductive I [1158]
Tsai, W.H.   : proaches to Pattern   Recognition .   [1757]
Unterberger,:   Automatic   Recognition   of Voiced Signals using Ar [1772]
Verhagen, C.: marks About Pattern   Recognition ; its Definitions; its Rela [1812]
Watanabe, S.:   Pattern   Recognition   as a Quest for Minimum Ent [1846]
Cavallo, R.E: Reconstructability   Analysis: Evaluatio [283]
Cavallo, R.E: Reconstructability   Analysis: Overview   [284]
Cavallo, R.E: Decision Making in   Reconstructability   Analysis.   [285]
Elias, D.   : n Hypotheses in the   Reconstructability   Analysis: APL Versi [474]
Klir, G.J.   : ms: a Basis for the   Reconstructability   Analysis.   [944]
Klir, G.J.   : n Hypotheses in the   Reconstructability   Analysis.   [947]
Valdes-Perez: ata Quantization in   Reconstructability   Analysis.   [1781]
Cavallo, R.E: The Structure of Reconstructable Relations: a Comprehen   [282]
Broekstra, G: n and Evaluation of   Reconstruction   Hypotheses by Informati [226]
Cavallo, R.E: ysis: Evaluation of   Reconstruction   Hypothesis.   [283]
Cavallo, R.E: Reconstruction   of Possibilistic Behavi [286]
Elias, D.   : ures for Generating   Reconstruction   Hypotheses in the Recon [474]
Favella, L.F: ndations of ECG Map   Reconstruction .   [510]
Herman, G.T.:   Image   Reconstruction   From Projections: Imple [751]
Jones, B.   :   Determination of   Reconstruction   Families.   [854]
Katz, M.J.   : s and Resolution in   Reconstruction   from Projections.   [883]
Klir, G.J.   : ures for Generating   Reconstruction   Hypotheses in the Recon [947]
Rao, G.P.   :   Approximate   Reconstruction   of Mapping Functions fr [1380]
Bell, D.A.   :   Physical   Record Clustering in Databases.   [141]
Gumovski, I.:   Recurrences and Discrete Dynamic Syste   [683]
Beer, S.   :   Recursions of Power.   [138]
Meyer, A.R. : r Is Not Elementary   Recursive .   [1156]
Veloso, P.A.:   On   Reducibilities Among General Problems.   [1807]
Millendorfer: Growth   Reducing Factors in Complex Systems.   [1164]
Trappl, R.   :   Reducing International Tension through   [1732]
Deutsch, K.W: of Simplification: Reduction of Complexity in Data Proces   [414]
Haralik, R.M: Reduction Operation for Constraint Sat   [725]
Hickin, J.   : tions of Projective   Reduction Methods to Estimation and Co   [753]
Primas, H.   :   Theory   Reduction and Non-Boolean Theories.   [1361]
Agazzi, E.   : and the Problem of   Reductionism .   [15]
Becker, P.W.:   On Systems with   Redundancy and Their Inherent Bounds o   [130]
Becker, P.W.:   On Systems with   Redundancy and their Inherent Bounds o   [129]
Becker, P.W.: ms' Reliability and   Redundancy .   [131]
Srinivasan, : ilistic Analysis of   Redundant Systems.   [1608]
Rauch, W.D. :   Automatic   Refereeing and Excerpting.   [1390]
Landau, Y.D.: Control: the Model   Reference Approach.   [1007]
Furukawa, O.:   Refinement of Quality Control Systems.   [549]
Janko, J.   : in Natural Science   Reflecting Orientation Towards Practic   [832]
Bekey, G.A. : s and Reality: Some   Reflections on the Art and Science of   [140]
Horvath, T. : thod and Systematic   Reflections .   [789]
Laszlo, E.   :   Some   Reflections on Systems Theory's Critic [1015]
Kobsa, A.   :   On   Regarding AI Programs as Theories.   [956]
Sadovskiy, L: Questions   Regarding Simulation of Hierarchical S [1470]
Wang, Z.   : y System in Farming   Region .   [1837]
Basta, D.J. : stems: Analysis for   Regional Residuals-Environmental Quali   [114]
Gericke, M. :   Changes of   Regional Passenger-Transport Demands b   [600]
Ruotsalainen: e Allocation of the   Regional Hospital System with Differen [1463]
Gollmann, D.: y of Bilinear Shift   Registers .   [634]
Kerefky, P. :   Patient   Registers on Microcomputers.   [905]
Ratko, I.   :   Computer   Registration of Patients Waiting for C [1388]
Schilderinck: Regression and Factor Analysis in Econ [1509]
Yager, R.R. : Prediction Based on   Regression Models.   [1902]
De Mottoni, :   Bifurcation of Regular Nonnegative Stationary Solutio   [397]
Grenander, U: Regular Structures: Lectures in Patter   [669]
Koppelaar, H: Regular Career Systems II.   [968]

Nobile, A.G.: Discrete Models for   Regulated   Growth with Intrinsic Lower   [1234]
Ben-Eli, M.U: Amplifying   Regulation and Variety Increase in Evo   [149]
Sahal, D.   :   Principles of   Regulation and Control.   [1477]
Sworder, D.D: sor Failures on Log   Regulators .   [1647]
Pot, John S.: c Relevance and the   Rehabilitation of the Goal Concept.   [1354]
Bozinovski, : tem Using Secondary   Reinforcement .   [207]
Kalaba, R.E.: niques for Adaptive   Reinvestment : A Small Sample Compariso   [870]
Favella, L.F: e Problems and Some   Related   Computational Methods.   [511]
Gaines, B.R.: Systems and Closely   Related   Topics.   [555]
Michie, D. : ne Intelligence and   Related   Topics: An Information Scienti   [1161]
Smale, S.   : nomic Processes and   Related   Topics.   [1583]
DiNola, A. : ions of Alpha-Fuzzy   Relation   Equations on Finite Spaces.   [418]
Haralik, R.M: erization of Binary   Relation   Homomorphisms.   [726]
Kaufmann, G.: ual Imagery and Its   Relation   to Problem Solving.   [887]
Lasker, G.E.:   The   Relation   between Major World Problems   [1011]
Orlovsky, S.: a Fuzzy Preference   Relation .   [1271]
Tjoa, A.M. :   The Universal   Relation   as Logical View of Databases.   [1692]
Trappl, R. : r Potential and its   Relation   to Stevens' Power Function.   [1710]
Verhagen, C.: ts Definitions; its   Relation   with Other Disciplines; a Lit   [1812]
Yager, R.R. : ustic Hedges: Their   Relation   to Context and Their Experime   [1904]
Aho, A.V.   :   Theory of Joins in   Relational Data Bases.   [20]
Bandler, W. : Operators and Fuzzy   Relational Products.   [100]
Bottaci, L. :   A   Relational Scheme for the Abstract Spe   [198]
Eveland, S. :   The Use of the QBE   Relational Database System for a Gothi   [496]
Hutt, A.T.F.:   A   Relational Data Base Management System   [799]
Jantzen, J. : A Generalized Fuzzy   Relational Matrix.   [836]
Kannellakis,: lity Constraints in   Relational Databases.   [877]
Lozinskii, E: ion of Relations in   Relational Databases.   [1068]
Meldman, M.J: Riss: a   Relational Data Base Management System   [1150]
Pedrycz, W. :   Fuzzy   Relational Equations with Triangular N   [1310]
Pedrycz, W. : al Aspects of Fuzzy   Relational Equations in Systems Analys   [1311]
Pedrycz, W. : nstruction of Fuzzy   Relational Models.   [1313]
Phillips, R.: P: an Architectural   Relational Data Base.   [1333]
Sanderson, J: A   Relational Theory of Computing.   [1490]
Baldwin, J.F: rollers Using Fuzzy   Relations .   [84]
Bezdek, J.C.: uzzy Partitions and   Relations : an Axiomatic Basis for Clus   [166]
Brewer, J.N.: nd Cause and Effect   Relations   in Social System Simulations   [216]
Cavallo, R.E: of Reconstructable   Relations : a Comprehensive Study.   [282]
Conant, R.C.: ns of Variables and   Relations .   [337]
Conant, R.C.:   Identifying   Relations   from Partial Information.   [341]
Jaron, J.   :   Relations   and Topologies in the Goal-S   [840]
Loo, S.G.   :   Fuzzy   Relations   in the Social and Behavioral   [1064]
Lozinskii, E: Construction of   Relations   in Relational Databases.   [1068]
Maker, P.P :   Input-Output   Relations   in Multivariate Biosystems.   [1094]
Matsuno, K. :   Dynamics of   Relations , Sequential Connectedness, a   [1123]
Neuwirth, E.: Fuzzy Similiarity   Relations   and Distance Coefficients in   [1215]
Ollero, A. :   The Structure of   Relations   in Personnel Management.   [1266]
Orlovsky, S.: le Fuzzy Preference   Relations .   [1272]
Singh, R.   : Study of Industrial   Relations .   [1567]
Warfield, J.:   Complementary   Relations   and Map Reading.   [1841]
Granero-Pora: tability-Complexity   Relationship .   [655]
Hull, R.   : Information: Their   Relationship to Computer-Based Informa   [796]
Slawski, C. : loping the Learning   Relationship : In Theory and Practice.   [1574]
Faribault, M: responding Semantic   Relationships .   [504]
Yager, R.R. : Properties of Fuzzy   Relationships .   [1901]
Adams, W.M. : takana Syllabaries: Relative Information Content.   [13]
Andersson, H: Using the Theory of   Relatively Closed Systems.   [32]
Grubbstrom, :   Theory of   Relatively Closed Systems and Applicat   [675]
Grubbstrom, : ms of the Theory of   Relatively Closed Systems.   [676]
Greenspan, D: assical and Special   Relativistic Physics.   [663]
Jumarie, G. : cal Applications of   Relativistic Information Fuzzy Sets, L   [861]
Jumarie, G. : Sets, Linguistics,   Relativistic Sets and Communication.   [861]
Jumarie, G. :   A   Relativistic Approach to Modelling Dyn   [862]
Ghosal, A. : ed Cybernetics: its   Relevance to Operations Research.   [607]

Ghosal, A.   :   Relevance   of Cybernetics in the Study   [608]
Laing, R.   : Exploration of the Relevance   of Recent Results.   [1004]
Palm, G.   : c Changes and their Relevance   for the Storage of Informati [1278]
Pot, John S.:   Scientific   Relevance   and the Rehabilitation of th [1354]
Colombano, S: e Unstable. Is this   Relevant   to Ecology?   [334]
Apostolakis,: hods for Safety and   Reliability   Studies (NATO Advanced Stu   [40]
Ball, M.O.   : mplexity of Network Reliability   Computations.   [89]
Becker, P.W.: ounds on Achievable   Reliability .   [130]
Becker, P.W.: ounds on Achievable   Reliability .   [129]
Becker, P.W.:   On Systems'   Reliability   and Redundancy.   [131]
Conant, R.C.: iable: Satisfying a   Reliability   Criterion.   [336]
Degermendzhy: The   Reliability   of Microevolution Process   [402]
Dhillon, B.S: Engineering   Reliability : New Techniques and Applic   [416]
Gopal, K.   : nsion Algorithm for Reliability   Evaluation in Complex Syst [643]
Kaufmann, A.: or the Study of the   Reliability   of Systems.   [886]
Legget, R.W.:   A   Reliability   Index for Models.   [1032]
Rees, R.K.D.:   Software   Reliability   (2 Vols).   [1397]
Ryabinin, I.:   Reliability   of Engineering Systems: Pr [1466]
Stemmer, N. :   The   Reliability   of Inductive Inferences an [1619]
Tzafestas, S: imization of System   Reliability : a Survey of Problems and   [1764]
Zorbas, Y.G.: Monitoring of Man's   Reliability   during Cosmic Flights.   [1929]
Siewiorek, D: ory and Practice of   Reliable   System Design.   [1554]
Veroy, B.S. :   Overall Design of Reliable   Centralized Voice/Data Commun [1814]
Hoehle, U.   :   A Remark   on Entropies with Respect to Pl [764]
Miller, G.L.: somorphism, General   Remarks .   [1166]
Suppes, P.   :   Some   Remarks   About Complexity.   [1638]
Verhagen, C.:   Some General Remarks   About Pattern Recognition; its [1812]
Garfolo, B.T: The   Renal   Dialysis Medical Information Sys   [573]
Mergenthaler: The Total   Repair   Cost in a Defective, Coherent B [1152]
Cavallo, R.E: aterial to the SGSR   Report   on Systems Research Movement.   [279]
Hawkes, N. : Modelling: a UNESCO   Report .   [740]
McIntyre, C.: ddendum to Internal   Report   165: Mathematical Documentation [1140]
Quade, E.   : Systems Analysis: a   Report   on the Response.   [1370]
Abe, J.   :   Schema   Representation   of 'Speakers-World' in   [2]
Aggarwal, S.:   Deterministic Representation   of Probabilistic System [17]
Bobrow, D.G.:   Representation   and Understanding: Stud   [179]
Botez, M.C. :   Toward a Systemic Representation   of General Systems: Ope   [197]
Bozinovski, :   A   Representation   of Pattern Classificati   [208]
Broekstra, G: On the   Representation   and Identification of S   [224]
Broekstra, G: is of C-Structures: Representation   and Evaluation of Recon   [226]
Casti, J.   :   Polyhedral Dynamic   Representation .   [273]
Dalenoort, G: On the Representation   of Control in Systems i   [374]
Faribault, M: uage to a Canonical   Representation   of the Corresponding Se   [504]
Findler, N.V: sociative Networks: Representation   and Use of Knowledge by   [524]
Fum, D.   : tural Language Text   Representation   and Comprehension.   [547]
Kobsa, A.   :   Knowledge   Representation : The Representation of   [955]
Kobsa, A.   : Representation: The   Representation   of Knowledge in the Com [955]
Kobsa, A.   :   Knowledge   Representation : A Survey of its Mechan   [957]
Malec, J.   : l for Dynamic Scene   Representation .   [1096]
Mattessich, :   Axiomatic   Representation   of the Systems Framewor [1127]
Retti, J.   : -Oriented Knowledge   Representation   for Modelling and Simul [1409]
Retti, J.   : ema-Based Knowledge   Representation   for Modelling and Simul [1410]
Rosen, R.   : of Measurement and   Representation   of Natural Systems.   [1441]
Shieh, L.   :   Representation   of Continuous Time Stat [1546]
Trost, H.   : clarative Knowledge   Representation : An Overview and an App   [1754]
Chomsky, N. :   Rules and   Representations .   [308]
Higashi, M. : Notion of Distance   Representing   Information Closeness: Po   [756]
Kobsa, A.   :   Representing   Belief Models in Semantic [958]
Barto, A.G. :   A Note on Pattern   Reproduction   in Tessellation Structure [111]
Matsuno, K. :   Multiplication and   Reproduction   of Microsystems in Protob [1125]
Horn, W.   : ealth-Care Resource   Requirements .   [782]
Ackoff, R.L.: uture of Operations   Research   is Past.   [9]
Ackoff, R.L.: ture of Operational   Research .   [8]
Cavallo, R.E: y in Social Science   Research .   [277]

Cavallo, R.E: for Social Science   Research .   [278]
Cavallo, R.E: Sysems   Research Movement: Characteristics, Ac [280]
Cavallo, R.E: Systems   Research Movement: Characteristics, Ac [281]
Cavallo, R.E: R Report on Systems   Research Movement.   [279]
Cavallo, R.E: y in Social Science   Research : Recent Developments.   [287]
Dominowski, :   Research Methods.   [431]
Fosko, P.D. : the Management of/   Research and Development.   [537]
Gaines, B.R.: for General Systems   Research .   [558]
Gaines, B.R.:   General Systems   Research : Quo Vadis?   [557]
Gasparski, W: Theory and Systems   Research and Science and Technology.   [578]
Ghosal, A.   : vance to Operations   Research .   [607]
Gvishiani, J: Systems   Research : Methodological Problems - 19 [694]
Haley, K.B. :   Applied Operatons   Research in Fishing. (Vol. 10 in NATO [710]
Iri, M.   : heory in Operations   Research , Circuits and Systems Theory. [814]
Kaposi, A.   :   Research into the Assessment of Comple [879]
Kinston, W. : itutions: An Action   Research Approach to Organisation of C [927]
Kinston, W. :   Linking System   Research and Delivery Systems. The Exa [928]
Klir, G.J.   : rontiers in Systems   Research : Implications for the Social [941]
Klir, G.J.   : The General Systems   Research Movement.   [942]
Klir, G.J.   : in General Systems   Research .   [943]
Klir, G.J.   : ied General Systems   Research .   [940]
Klir, G.J.   : ied General Systems   Research : A Bibliography.   [937]
Kornai, J.   : conomy: Review of a   Research Direction.   [971]
Leinhardt, S: y on Social Network   Research .   [1035]
Lusk, E.J.   : alidity in Systemic   Research .   [1075]
Naughton, J.: nowledge in Systems   Research : The Rules of the Game.   [1203]
Nowak, S.   : ogy of Sociological   Research .   [1238]
Nowakowska, :   Theories of   Research : Modelling Approaches.   [1241]
Petkoff, B. : Model of Scientific   Research and Cognition.   [1328]
Pichler, F. : rnetics and Systems   Research , Vol.VI.   [1336]
Pichler, F. : rnetics and Systems   Research , Vol.VII.   [1335]
Ragade, R.R.: for General Systems   Research .   [1376]
Reckmeyer, W: General Systems   Research and Design: Precursors and Fu [1394]
Slawski, C. : g: A Policy Systems   Research Model of Career Development. [1575]
Steinacker, :   Computer-Aided   Research in Learning.   [1614]
Tomlinson, R: ing a Collaborative   Research Study.   [1697]
Trappl, R.   : rnetics and Systems   Research , Vol.III.   [1712]
Trappl, R.   : rnetics and Systems   Research , Vol.IV.   [1713]
Trappl, R.   : rnetics and Systems   Research , Vol.VII.   [1721]
Trappl, R.   : rnetics and Systems   Research , Vol.IX.   [1720]
Trappl, R.   : rnetics and Systems   Research , Vol.X.   [1723]
Trappl, R.   : rnetics and Systems   Research , Vol.XI.   [1724]
Trappl, R.   : rnetics and Systems   Research .   [1718]
Trappl, R.   : rnetics and Systems   Research , Vol.II.   [1726]
Trappl, R.   : ied General Systems   Research : A Bibliography, 1977-1984.   [1729]
Trappl, R.   : rnetics and Systems   Research , Vol.VIII.   [1722]
Trussel, P. :   Using Nursing   Research : Discovery, Analysis, and Int [1755]
Unger, C.K. : ology and Ecosystem   Research .   [1771]
Van der Zouw: nventory of Systems   Research in the Netherlands.   [1796]
Van der Zouw: nventory of Systems   Research on the Netherlands.   [1797]
Van Rootsela: Annals of Systems   Research , Vol. 7   [1790]
Wasniowski, :   Systems in Futures   Research .   [1844]
             : Covering the Study,   Research and Techniques of All Design [1938]
             :   Systems   Research : Methodological Problems - 19 [1944]
             :   General Systems   Research : A Science, a Methodology, a [1933]
Zarri, G.P. :   The   RESEDA Project: Using AI Techniques in [1915]
Jorgenson, S: g of Deep Lakes and   Reservoirs .   [858]
Oden, A.S.   : vior of Hydrocarbon   Reservoirs .   [1252]
Basta, D.J. : alysis for Regional   Residuals-Environmental Quality Manage [114]
Holling, C.S: Resilience and Stability of Ecological [773]
Cheng, J.K. : ism Algorithm Using   Resolution .   [306]
Katz, M.J.   : s of Uniqueness and   Resolution in Reconstruction from Proj [883]
Shakun, M.F.: ormalizing Conflict   Resolution in Policy Making.   [1532]
Steinacker, : uation and Anaphora   Resolution .   [1617]

Pedrycz, W. : lar Norms and their   Resolutions .   [1310]
Grabowski, J: encing Problem with   Resource Constraints.   [651]
Horn, W.   : imating Health-Care   Resource Requirements.   [782]
Molina, F.W.:   A Survey of   Resource Directive Decomposition in Ma [1181]
Ruotsalainen: and Simulating the   Resource Allocation of the Regional Ho [1463]
Haimes, Y.Y.: l Analysis of Water   Resources Systems: Modeling and Optimi [700]
Zeleny, M.   : the Squandering of   Resources and Profits Via Linear Progr [1923]
Hoehle, U.   : k on Entropies with   Respect to Plausibility Measures.   [764]
Jarisch, W. : Detection of Fetal   Respiratory Movements-Preliminary Resu [837]
Quade, E.   : is: a Report on the   Response .   [1370]
Rapoport, A.:   Response Models for Detection of Chang [1382]
Van den Boga : es: A Case Study in   Response Surface Methodology.   [1793]
Takeuchi, A.: mation Behavior and   Restle 's Learning Model.   [1663]
Maruyama, M.: an Epistemological   Restructuring of Biological and Social [1112]
Auger, P.   : ogical or Physical)   Resulting in Creation of Structures. [67]
Klir, G.J.   : al Observations and   Resulting Guidelines.   [938]
Balossino, N: e Approximation and   Results .   [93]
Duechting, W: Recent   Results in Modelling and Simulation of [446]
Graedel, T.E: cal Presentation of   Results from Scientific Computer Model [652]
Jarisch, W. : vements-Preliminary   Results .   [837]
Laing, R.   : Relevance of Recent   Results .   [1004]
Sandblom, C.:   Simulation   Results on the Evaluation of Economic [1488]
Tesfatsion, : Control: Optimality   Results   [1683]
Tong, R.M.   : esses - Some Recent   Results .   [1700]
Witten, I.H.:   Some Recent   Results on Nondeterministic Modelling [1874]
Yashin, A.   : A New Proff and New   Results in Conditional Gaussian Estima [1911]
Ackoff, R.L.:   Resurrecting the Future of Operational [8]
Checkland, P: Rethinking a Systems Approach.   [300]
Farkas, J.   :   Rethinking of Theoretical Presuppositi [506]
De Delara, G: l Inhibition in the   Retina .   [389]
Von Luetzow,: Transmission of Cat   Retinal Horizontal Cells.   [1822]
Rosen, R.   : cipatory Systems in   Retrospect and Prospect.   [1443]
Toffoli, T. : ion Universality of   Reversible Cellular Automata.   [1696]
Hanson, O.J.:   Review and Re-Design of an Advanced Sy [724]
Hwang, C.L. : ex/Large Systems: A   Review .   [804]
Kornai, J.   : Socialist Economy: Review of a Research Direction.   [971]
McGowan, C.L: A   Review of Decomposition and Design Met [1137]
Nugyen, V.V.:   Review and Unification of Linear Ident [1247]
Nagel, E.   :   Teleology   Revisited .   [1200]
Roberts, P.C: s: Limits to Growth   Revisited .   [1428]
Nicolis, J.S: on of Physiological   Rhythms .   [1225]
Suzuki, R.   : udies of Biological   Rhythms .   [1644]
Koenderink, : -Dimensional Shape: Rigid Movement, Bending and Stretching [962]
Lilienfeld, :   The   Rise of Systems Theory: an Ideological [1049]
Benedikt, S.: cision Making under   Risk .   [151]
Dial, G.   :   A Note on Bayes Risk with Utilities.   [419]
Hartl, R.F. : inuous Lovers Under   Risk of Being Killed a Rival.   [729]
Rowe, W.D.   :   An Anatomy of Risk .   [1452]
Jarisch, W. : cations to Clinical   Risk-Assessment and Detection of Fetal [837]
Meldman, M.J: Riss : a Relational Data Base Managemen [1150]
Hartl, R.F. : k of Being Killed a   Rival .   [729]
Gini, G.   : stic Programming in   Robot Systems.   [615]
Koenderink, : e: Implications for   Robot Vision.   [962]
Toda, M.   :   Man,   Robot , and Society: Models and Specula [1693]
Hirzinger, G: Robot-Teaching via Force-Torque-Sensor [760]
Rose, J.   :   Robotica (Int.Journal of Information, [1438]
Pichler, F. :   Robotics and Image Processing.   [1338]
Apostolico, :   On the   Role of Information and Hierarchy in t [41]
Bosserman, R: The   Role of Mathematics in Systems Science [194]
Cavallo, R.E: The   Role of Systems Methodology in Social [277]
Erlandson, R: articipant-Observer   Role in Systems-Methodologies.   [487]
Hanson, O.J.:   The   Role of Systems Analysis in Designing [722]
Royce, J.R. :   The   Role of General Systems and Informatio [1453]
Vaccari, E. :   The   Role of System Modelling in Natural La [1779]

White, D.J.   :   Role   and Effectiveness of Theories of   [1862]
Smyth, D.S.   : h: the Structure of   Root   Definitions.   [1595]
Ashby, R.   : ms of Intelligence: Ross   Ashby's Writings on Cybernetics.   [55]
Mizumoto, M.: e on the Arithmetic   Rule   by Zadeh for Fuzzy Conditional In   [1176]
Fum, D.   :   A Rule-Based Approach to Natural Languag   [547]
Michalski, R: tern Recognition as   Rule-Guided Inductive Inference.   [1158]
Chomsky, N.   :   Rules   and Representations.   [308]
Czogala, E.   :   Fuzzy Rules   Generation for Fuzzy Control.   [370]
Franksen, O.: : Developing Design   Rules   Using APL.   [541]
Gomez, P.   : sations: Cybernetic   Rules   for Effective Management.   [637]
Lesmo, L.   : of Fuzzy Production   Rules .   [1037]
Naughton, J.: stems Research: The   Rules   of the Game.   [1203]
Ohlsson, S.   : of Problem Solving   Rules .   [1259]
Palm, G.   :   Rules   for Synaptic Changes and their R   [1278]
Walichiewicz: ition of Linguistic   Rules   in the Design of a Multi-Dimensi   [1828]
Batchelor, B: Converting Run   Code to Chain Code.   [122]
Kontur, I.F.:   Runoff Simulation with a Random Walk M   [967]
Kieras, D.E.: Finite Automata and   S-R Models.   [915]
Apostolakis,: nalysis Methods for   Safety   and Reliability Studies (NATO A   [40]
Brown, D.B.   : ysis and Design for   Safety .   [228]
Kalaba, R.E.: investment: A Small   Sample   Comparison.   [870]
De La Sen, M: ivariable Aperiodic   Sampling Systems.   [393]
Dwyer, R.F.   : tors with Dependent   Sampling .   [452]
Uttenhove, H: SAPS   (Systems Approach Problem Solver)   [1775]
Erlandson, R: The Satisfacting Process. A New Look.   [488]
Haralik, R.M: tion for Constraint   Satisfaction .   [725]
Lieberherr, : mplexity of Partial   Satisfaction .   [1048]
Matsuda, T. : braic Properties of   Satisficing Decision.   [1119]
Takatsu, S.   : Multiple-Objective   Satisficing Decision Problems.   [1660]
Conant, R.C.: a Binary Variable: Satisfying a Reliability Criterion.   [336]
Schiffman, Y: r Estimating Energy   Savings   in Communities.   [1508]
Fontela, E.   : nsed Forms of Large   Scale   Models.   [533]
Franksen, O.: ve Aspects of Large   Scale   Systems: Developing Design Rules   [541]
Hafele, W.   : Modelling of Large   Scale   Energy Systems. (IIASA Proceedin   [698]
Henize, J.   :   Modelling Large   Scale   Socio-Economic Systems: Can We B   [750]
Lowen, W.   : Creep: Out of Human   Scale .   [1066]
Mahmoud, M.S: Large   Scale   Systems Modelling.   [1089]
Michel, A.N.: e Analysis of Large   Scale   Systems.   [1159]
Oeren, T.I.   : Simulation of Large   Scale   Systems.   [1254]
Oeren, T.I.   : Rationale for Large   Scale   System Simulation Software Based   [1255]
Sage, A.P.   : thodology for Large   Scale   Systems.   [1474]
Singh, M.G.   : Handbook of Large   Scale   Systems Engineering Applications   [1566]
Tacker, E.C.: Estimation on Large   Scale   Systems.   [1653]
Tilti, A.   :   Large   Scale   System Theory and Applications ([1690]
Veroy, B.S.   :   Large   Scale   Network Architecture Synthesis: [1813]
Shepard, R.N: Multidimensional   Scaling , Tree-fitting, and Clustering.   [1543]
Van der Ven,:   Introduction to   Scaling .   [1794]
Malec, J.   :   Scenarios as a Tool for Dynamic Scene   [1096]
Duda, R.O.   : Classification and   Scene Analysis.   [445]
Malec, J.   : a Tool for Dynamic   Scene Representation.   [1096]
Dempster, A.: stic and Stochastic   Scheduling (NATO Advanced Study Instit   [408]
Abe, J.   :   Schema   Representation of 'Speakers-Wor   [2]
Retti, J.   :   MOSES: A Schema-Based Knowledge Representation   [1410]
Burns, J.R.   : graphs to Forrester   Schematics .   [249]
Bolter, M.   :   A Statistical   Scheme   for Structural Analysis in Mari   [185]
Bottaci, L.   :   A Relational   Scheme   for the Abstract Specification   [198]
Greibach, S.: Program Structures: Schemes , Semantics, Verification.   [668]
             : t the Second Spring   School   in Jadwisin, Poland, April 13-1   [1942]
Ackoff, R.L.:   The Art and   Science of Mess Management.   [10]
Andersen, E.: Models with Social   Science Applications.   [28]
Banathy, B.H: Systems   Science and Science. (Proc. 24th Annua   [96]
Banathy, B.H: Systems Science and   Science . (Proc. 24th Annual Meeting of   [96]
Bavel, Z.   : panion for Computer   Science .   [126]
Bayraktar, B: ducation in Systems   Science .   [127]

Bekey, G.A. : ions on the Art and   Science of Simulation.   [140]
Bobrow, D.G.: tudies in Cognitive   Science .   [179]
Bosserman, R: hematics in Systems   Science .   [194]
Branin, F.H.: Problem Analysis in   Science and Engineering.   [212]
Bubnicki, Z.:   Systems Science . (A Journal Published Quarterl   [236]
Callebaut, W: Problem-Solving in   Science : Prospects for GSM.   [257]
Cavallo, R.E: thodology in Social   Science Research.   [277]
Cavallo, R.E: hodology for Social   Science Research.   [278]
Cavallo, R.E: thodology in Social   Science Research: Recent Developments.   [287]
Checkland, P: lure" of Management   Science .   [299]
Cliffe, M.J.: Cybernetics to the   Science of Psychopathology.   [322]
Coblentz, A.:   Systems Science in Health Care.   [323]
Cramer, F. : ncept in Biological   Science and Beyond.   [355]
De Beaufrand: a Multidisciplinary   Science of Texts.   [386]
Devries, R.P: d the Philosophy of   Science .   [415]
Field, H.H. :   Science Without Numbers.   [522]
Gasparski, W: Science , Technology, and Systems.   [581]
Gasparski, W: ystems Research and   Science and Technology.   [578]
George, F.H.:   The Science of Philosophy.   [596]
Gruska, J. : dations of Computer   Science .   [678]
Janko, J.   : Approach in Natural   Science Reflecting Orientation Towards   [832]
Janko, J.   : he History of Czech   Science in the Middle of the 19th Cent   [832]
Kefalas, A.G: dom: On the Art and   Science of Anticipatory Management.   [897]
Kraft, R.W. :   Symbols, Systems, Science & Survival.   [979]
Kubat, L.   :   and Information in Science and Philosophy.   [990]
Levy, L.S. : uctures of Computer   Science .   [1044]
Loo, S.G.   :   cial and Behavioral Science .   [1064]
Lowen, W.   : the Mind: A Systems Science Model of the Mind and Personal   [1067]
Mitroff, I.I: pproaches to Social   Science .   [1172]
Norman, D.A.: ctives in Cognitive   Science . (Papers from a Meeting in La   [1236]
Nowakowska, : , and Alienation in   Science .   [1242]
Nowotny, H. :   Not Quite Human: Science and Utopia.   [1246]
Phillips, D.: c Thought in Social   Science .   [1332]
Pierce, J.R.: on to Communication   Science and Systems.   [1340]
Shaw, M.L.G.: in the Community of   Science .   [1542]
Simon, H.A. : s in the Methods of   Science .   [1562]
Smith, C.S. :   Selected Essays on Science , Art and History.   [1588]
Sonnier, I.L: cation: Teaching of   Science in the Affective Domain.   [1599]
Soong, T.T. : ing and Analysis in   Science and Engineering.   [1600]
Szaniawksi, :   Science as an Information-Seeking Proc   [1650]
Tamburrini, : ybernetics, Systems   Science and Fuzzy Sets Share Some Epis   [1665]
Tou, J.T.   : Information Systems   Science , Vol.6.   [1705]
Trappl, R. : icial Intelligence.   Science , Technology, Military, Economy   [1733]
Troncale, L.: of General Systems   Science : Obstaches, Potentials, Case S   [1748]
Van Gigch, J: l Comparison of the   Science , Systems and Metasystem Paradi   [1787]
Van Gigch, J: Taxonomy of Systems   Science .   [1788]
Wulf, W.A. : uctures of Computer   Science .   [1883]
            : Systems Research: A   Science , a Methodology, a Technology ([1933]
Ackerman, E.: odels in the Health   Sciences : a Computer Approach.   [7]
Airaksinen, : ms Theory in Social   Sciences .   [21]
Bennett, S. : cial and Behavioral   Sciences .   [153]
Benseler, F.: ystem in the Social   Sciences .   [157]
Burghes, D.N: Management and Life   Sciences .   [243]
Collins, L. : odels in the Social   Sciences .   [332]
Ghiselli, E.: for the Behavioral   Sciences .   [606]
Gill, A.   : ra for the Computer   Sciences .   [612]
Gyula, P.   :   Common Laws of   Sciences and Systems.   [696]
Klir, G.J. : ions for the Social   Sciences (a New Book Series) .   [941]
Krone, R.M. : Analysis and Policy   Sciences : Theory and Practice.   [986]
Kuhlenbeck, : he World of Natural   Sciences and its Phenomenology. Vol.II   [993]
Lee, E.T.   : Set Theory to Image   Sciences .   [1027]
Maruyama, M.: ological and Social   Sciences .   [1112]
Mattessich, : Applied and Social   Sciences .   [1126]
Nowakowska, : cepts in the Social   Sciences .   [1239]

Nowakowska, : etics in the Social   Sciences .   [1245]
Pipino, L.L.: Sets on the Social   Sciences .   [1342]
Prevost, P. : lism and the Social   Sciences .   [1359]
Prigogine, I: ity in the Physical   Sciences .   [1360]
Resconi, G. : pplications to Life   Sciences .   [1404]
Roberts, F.S: ity, and the Social   Sciences .   [1425]
Schieve, W.C: Physical and Social   Sciences .   [1507]
Sussmann, H.: cial and Biological   Sciences : a Critique.   [1641]
Troncale, L.: logical and Medical   Sciences .   [1750]
Vavrousek, J: elling in Empirical   Sciences .   [1804]
West, B.J. : Tool for the Social   Sciences .   [1859]
Agassi, J. :   Scientific   Philosophy Today: Essays in   [14]
Cornelis, A.: in the Dynamics of   Scientific   Learning.   [347]
Cornelis, A.: gical Indicators of   Scientific   Identity.   [348]
Csanyi, V. : siderations for the   Scientific   Formulation of Epistemology   [361]
De Rosnay, J: oscope: a New World   Scientific   System.   [399]
Goffman, W. :   Scientific   Information Systems and the   [627]
Graedel, T.E: ion of Results from   Scientific   Computer Models.   [652]
Gurel, O.   : and Applications in   Scientific   Disciplines.   [691]
Kantor, J.R.: ystem Structure and   Scientific   Psychology.   [878]
Nowakowska, : he Development of a   Scientific   Discipline.   [1240]
Petkoff, B. : Cybernetic Model of   Scientific   Research and Cognition.   [1328]
Pla Lopez, R: logical Learning to   Scientific   Learning.   [1345]
Pot, John S.:   Scientific   Relevance and the Rehabilit   [1354]
Suppe, F.   :   The Structure of   Scientific   Theories (Second Ed.)   [1637]
Lipset, D. : on: The Legacy of a   Scientist .   [1056]
Michie, D. : ics: An Information   Scientist 's Weekend Book.   [1161]
Shaw, M.L.G.: Becoming a Personal   Scientist .   [1541]
Shaw, M.L.G.:   The Personal   Scientist in the Community of Science.   [1542]
Bradley, R. : k for Engineers and   Scientists .   [210]
Gilmore, R. : astrophe Theory for   Scientists and Engineers.   [614]
Gould, P.   : oduction for Social   Scientists , Geographers and Planners.   [649]
Bowler, T.D.: stems Thinking: Its   Scope   and Applicability.   [203]
Della Riccia: oximation of Factor   Scores   in a Factor Analysis Algorithm.   [405]
Stoy, J.E. : onal Semantics: the   Scott-Strachey Approach to Programming   [1625]
Bernard, N. : Contribution to the   Search   of Fuzzy Sets Theory.   [162]
Carvallo, M.:   In   Search   of the Noetic Structure of Syst   [266]
Hyde, M.O. : rs that Think?: The   Search   for Artificial Intelligence.   [805]
Kaindl, H. : l of the Quiescence   Search   in Computer Chess.   [867]
Mehlhorm, K.: arly Optimal Binary   Search   Trees.   [1147]
Smith, C.S. :   A   Search   for Structure: Selected Essays   [1588]
Stefanski, J: A Heuristic   Search   for the Best Hierarchical Struc   [1611]
Stone, L.D. :   Theory of Optimal Search .   [1624]
Wu, W.T.   : o Chinese-Character   Search .   [1881]
Petrov, B.N.: ms in the Theory of   Searchless Self-Adaptive Systems, I.   [1329]
Karlin, S. :   A   Second   Course in Stochastic Processes.   [881]
Lefebre, V.A: Second   Order Cybernetics in the Soviet   [1031]
Meyer, A.R. :   Weak Monadic   Second   Order Theory of Successor Is No   [1156]
Trappl, R. :   A   Second   Order Model of the Optic Genera   [1710]
Van Straaten: Proceedings of the   Second   Joint MTA/IIASA Task Force Meet   [1792]
            :   Second   Special Issue on the Guha Metho   [1941]
            : rs Presented at the   Second   Spring School in Jadwisin, Pola   [1942]
Burton, D.J.: nd Epistemology for   Second-Order Cybernetics.   [251]
Bozinovski, : arning System Using   Secondary Reinforcement.   [207]
Brugger, K. : of Biphasic Insulin   Secretion .   [232]
Burns, J.R. :   A   Sector Approach to the Formulation of   [248]
Bosserman, R: Internal   Security Processes in General Living S   [196]
Bosserman, R: Internal   Security Processes and Subsystems in L   [195]
Fernandez, E: Database   Security and Integrity.   [518]
Walker, B.J.:   Computer   Security and Protection Structures.   [1829]
Hartl, R.F. :   Optimal   Seducing Policies for Dynamic Continuo   [729]
Gasparski, W: sign Methodology as   Seen   from Neither an English nor a Ger   [582]
Gasparski, W: sign Methodology as   Seen   from neither English nor German S   [584]
Majewska-Tro: Selected Praxiological Bibliography.   [1090]

Smith, C.S. : arch for Structure: Selected   Essays on Science, Art and Hi [1588]
Subrahmaniam: variate Analysis: a   Selected   and Abstracted Bibliography,   [1631]
Zeeman, E.C.: Catastrophe Theory: Selected   Papers 1972-1977   [1916]
Ihara, J.   : sis of Criteria for   Selecting Model Variables.   [808]
Yager, R.R. : Some Procedures for   Selecting Fuzzy Set-Theoretic Operator   [1906]
Braae, M.   :   Selection of Parameters for a Fuzzy Lo   [209]
Guidorzi, R.: ltistructural Model   Selection .   [682]
Gloria-Botti: Selective Polymorphism of the Erythroc   [625]
Goffman, W. : nd the Principle of   Selectivity .   [627]
Innocent, P.:   Toward   Self-Adaptive Interface System.   [813]
Petrov, B.N.: heory of Searchless   Self-Adaptive Systems, I.   [1329]
Matsuno, K. :   Material   Self-Assembly as a Physiochemical Proc   [1124]
Matsuno, K. : f Microsystems with   Self-Instructing Polymerization of Ami   [1120]
Bozinovski, :   A   Self-Learning System Using Secondary R   [207]
Andrew, A.M.:   Autopoiesis and   Self-Organization .   [35]
Babloyantz, :   Self-Organization in Biological System   [78]
Balkus, K. : Basis of the Social   Self-Organization .   [88]
Nicolis, G. :   Self-Organization in Non-Equilibrium S   [1224]
Nicolis, J.S: mes: A Paradigm for   Self-Organization in Cognitive Systems [1226]
Sahal, D.   : A Unified Theory of   Self-Organization .   [1480]
Schieve, W.C: Self-Organization and Dissipative Stru   [1507]
Bennett, T.J: Language as a Self-Organizing System.   [154]
Ferracin, A.:   Self-Organizing Ability in Living Syst   [519]
Jantch, E. :   The   Self-Organizing Universe.   [834]
Nicolis, J.S: mmunication between   Self-Organizing Systems and Desynchron   [1225]
Roth, G.   :   Self-Organizing Systems: An Interdisci   [1451]
Saridis, G.N: Self-Organizing Control of Stochastic   [1494]
Holler, M.J.:   Markets as   Self-Policing Quality Enforcement Syst   [772]
Pedretti, A.: ogy, Semantics, and   Self-Reference .   [1307]
Schwartz, D.: rela's Calculus for   Self-Reference .   [1515]
Vallet, C. : Use of a Partially   Self-Referent Formalism for Descriptio [1784]
Von Glasersf: eality": A Study of   Self-Regulation .   [1821]
Matsuno, K. : artmentalization of   Self-Reproducing Machineries: Multipli [1120]
Oldershaw, R: Foundations of the   Self-Similar Hierarchical Cosmology.   [1262]
Oldershaw, R: er of Levels in the   Self-Similar Hierarchical Cosmology.   [1263]
Barenblatt, :   Similarity, Self-Similarity and Intermediate Asymp   [106]
Oldershaw, R: retical Support for   Self-Similarity between Atomic and Ste   [1264]
Oldershaw, R: or the Principle of   Self-Similarity .   [1265]
Aulin-Ahmava: ibility of Genuinly   Self-Steering Machines: A Fundamental   [74]
Matsuno, K. :   Self-Sustaining Multiplication and Rep   [1125]
Eigen, M.   : rinciple of Natural   Selforganization .   [462]
Faribault, M: f the Corresponding   Semantic   Relationships.   [504]
Harris, L.R.: the Data Base as a   Semantic   Component to Aid in the Parsi   [727]
Kobsa, A.   : ng Belief Models in   Semantic   Networks.   [958]
Lee, R.M.   : nt Support Systems: Semantic   Issues.   [1028]
Bandler, W. :   Semantics of Implication Operators and   [100]
Greibach, S.: tructures: Schemes,   Semantics , Verification.   [668]
Kobsa, A.   : ms, a Sketch of its   Semantics .   [957]
Leinfellner,: Phonology, Syntax,   Semantics .   [1034]
Pedretti, A.:   Epistemology, Semantics , and Self-Reference.   [1307]
Sernadas, A.: ICS: Its Syntax and   Semantics as a Query Language.   [1528]
Stoy, J.E. :   Denotational   Semantics : the Scott-Strachey Approach   [1625]
Trattnig, W.:   Denotational   Semantics of a Software Specification   [1734]
Hawkes, N. :   International   Seminar on Trends in Mathematical Mode   [739]
Hawkes, N. :   International   Seminar on Trends in Mathematical Mode   [740]
Klement, E.P: f the International   Seminar on Fuzzy Set Theory.   [933]
Arigoni, A.C: Correctness of the   Semiotic Transformation of Information   [51]
Nowakowska, : nd Multidimensional   Semiotic Systems.   [1243]
Nowakowska, :   Formal   Semiotics and Multidimensional Semioti [1243]
El-Shirbeeny: A   Sensitivity Augmented Conjugate Gradie   [472]
Frank, P.M. : roduction to System   Sensitivity Theory.   [540]
Haruna, K. :   Structural   Sensitivity Analysis and its Applicati   [732]
Porenta, G. : apse: Stability and   Sensitivity Analyses, and Simulation o [1349]
Probst, G.J.:   The Necessity of Sensitivity Models for Managers, Plann   [1364]

Sage, A.P.   :   Sensitivity   Analysis in Systems for Pl [1476]
Wolaver, T.G: ient Control on the   Sensitivity   Characteristics of a Model [1877]
Presern, S. : ystem for a Tactile   Sensor .   [1357]
Sworder, D.D: Influence of Sensor Failures on Log Regulators.   [1647]
Castelfranch: n Based Approach to   Sentence Understanding.   [269]
Farreny, H. : ignating Objects in   Sentence Generation.   [508]
Cover, T.M. : ayes Predictors for   Sequences with Apparent Markov Structu [353]
Hwang, C.L. : mposition of Binary   Sequences of Finite Periods and a Clas [802]
Stoklosa, J.: s Generating Cyclic   Sequences .   [1622]
Witten, I.H.: elling of Behaviour   Sequences .   [1874]
Grabowski, J: Sequencing   Problem with Resource Const [651]
Dwyer, R.F. :   Sequential   Partition Detectors.   [451]
Dwyer, R.F. :   Sequential   Partition Detectors with De [452]
Gaines, B.R.:   Sequential   Fuzzy System Identification [560]
Gruber, H.   : ion on the Basis of   Sequential   Systems.   [677]
Hoffer, J.A.: ative Advantages of   Sequential   and Batched Decision Making [766]
Matsuno, K. : amics of Relations,   Sequential   Connectedness, and Feedback [1123]
Iijima, J.   :   Serial   Decomposition of General System [809]
Brillinger, :   Time   Series : Data Analysis and Theory.   [221]
Kedem, B.   :   Binary Time Series .   [893]
Klir, G.J.   :   The North-Holland   Series in General Systems Research.   [943]
McDowall, D.:   Interrupted Time   Series Analysis.   [1136]
Otnes, R.K. :   Applied Time Series Analysis. Vol. 1: Basic Techniq [1273]
Robinson, E.: Foundations of Time   Series Analysis. Vol. 1: The Box-Jenki [1433]
Vagnucci, A.: of Biological Time   Series .   [1780]
Ruotsalainen: stem with Different   Service Levels.   [1463]
Elton, M.C.J: w Telecommunication   Services .   [480]
De Luca, A. : Measures of a Fuzzy   Set .   [395]
Gupta, M.M. :   Advances in Fuzzy   Set Theory and Applications.   [684]
Gupta, M.M. :   Advances in Fuzzy   Set Theory and Applications.   [686]
Gupta, M.M. : plications of Fuzzy   Set Theory: A Survey.   [689]
Hisdal, E.   :   Generalized Fuzzy   Set Systems and Particularization.   [762]
Klement, E.P: al Seminar on Fuzzy   Set Theory.   [933]
Lee, E.T.   : plications of Fuzzy   Set Theory to Image Sciences.   [1027]
Leung, Y.   :   A Fuzzy   Set Analysis of Sociometric Structure. [1039]
Pfeilsticker: Approach and Fuzzy   Set Theory Bridging the Gap between Ma [1331]
Tazaki, E.   : Systems Using Fuzzy   Set Theory.   [1676]
Yager, R.R. :   Fuzzy   Set and Possibility Theory.   [1903]
Conant, R.C.:   Set-Theoretic Structure Modelling.   [342]
Yager, R.R. : for Selecting Fuzzy   Set-Theoretic Operators.   [1906]
Pogonowski, :   Set-Theoretical Approach to General Sy [1348]
Atanassov, K: tuitionstic L-Fuzzy   Sets .   [57]
Bandler, W. :   Fuzzy Power   Sets and Fuzzy Simplification Operator [99]
Bernard, N. : the Search of Fuzzy   Sets Theory.   [162]
Dubois, D.   :   Fuzzy   Sets and Systems: Theory and Applicati [443]
Freeling, A.:   Fuzzy   Sets and Decision Analysis.   [545]
Golledge, R.: lysis of Large Data   Sets .   [632]
Jumarie, G. : c Information Fuzzy   Sets , Linguistics, Relativistic Sets a [861]
Jumarie, G. : stics, Relativistic   Sets and Communication.   [861]
Kandel, A.   :   Fuzzy   Sets , Fuzzy Algebra, and Fuzzy Statist [873]
Klement, E.P: ardinality of Fuzzy   Sets .   [934]
Maciejowski,: h Small Observation   Sets .   [1082]
Negoita, C. :   Fuzzy   Sets and Systems: a New International [1208]
Pipino, L.L.: ial Impact of Fuzzy   Sets on the Social Sciences.   [1342]
Poch, F.A.   : plications of Fuzzy   Sets to Statistics .   [1347]
Radecki, T. :   Level Fuzzy   Sets .   [1373]
Skala, H.J. : alued Logics, Fuzzy   Sets , Fuzzy Logics and Their Applicati [1570]
Tamburrini, : s Science and Fuzzy   Sets Share Some Epistemological Proble [1665]
Tazaki, E.   : Systems Using Fuzzy   Sets Theory.   [1675]
Tong, R.M.   : n Making with Fuzzy   Sets .   [1701]
Van Ameronge: Designed with Fuzzy   Sets .   [1785]
Xuemou, W.   : Analysis and Fuzzy   Sets .   [1886]
Yager, R.R. :   Fuzzy   Sets , Probabilities, and Decision.   [1895]
Yager, R.R. : Importance of Fuzzy   Sets .   [1892]

De Delara, G: Computer Simulation of Lateral Inhibition in th [389]
De Witt, C.T: Simulation of Ecological Processes (2n [400]
Dekker, L. :   Simulation of Systems.   [403]
Delaney, W. : System Theory Based Simulation Language.   [404]
Dent, J.B. :   Systems Simulation in Agriculture.   [410]
Duechting, W: ts in Modelling and Simulation of 3-D Tumor Growth.   [446]
El-Fattah, Y: Systems: Decision, Simulation , and Control.   [467]
Elison, D.I.:   Interactive Simulation on a Computer.   [475]
Ellison, D. :   Simulation Newsletter.   [476]
Franta, W.R.:   Process View of Simulation .   [543]
Gericke, M. : ctivities - A Fuzzy Simulation Approach.   [600]
Gigley, H. : egrated Approach to Simulation Modelling of Natural Langua [610]
Holst, P.   :   Computer Simulation 1951-1976: An Index to the [775]
Jakubowski, : lex Dynamic Systems Simulation .   [829]
Johnson, P.E: Validating Computer Simulation Models of Expert Reasoning. [851]
Kellermayr, :   Performance Simulation of an Optical Bus Area Netw [900]
Kindler, E. :   Classification of Simulation Programming Languages: I. D [917]
Kindler, E. : stems and Theory of Simulation .   [919]
Kindler, E. :   Classification of Simulation Programming Languages: II. [918]
Kindler, E. :   Simulation Programming Languages.   [921]
Kindler, E. : rmalization of Some Simulation Language Concepts.   [922]
Kontur, I.F.:   Runoff Simulation with a Random Walk Model. [967]
Kovalenko, Y: tastable State. II. Simulation Model.   [974]
Maryanski, F: Digital Computer Simulation .   [1113]
Mihram, G.A.:   Simulation : Methodology for Decision T [1162]
O'Donovan, T: GSPS: Simulation Made Simple.   [1250]
Oeren, T.I. : s and Modelling and Simulation of Large Scale Systems. [1254]
Oeren, T.I. : Large Scale System Simulation Software Based on Cyberneti [1255]
Oeren, T.I. :   Simulation with Discrete Models.   [1256]
Oren, T.I. :   Simulation - As it Has Been, Is, and S [1269]
Oren, T.I. : ncepts for Advanced Simulation Methodologies.   [1270]
Porenta, G. : ivity Analyses, and Simulation of Parkinson's Disease and [1349]
Pugh, R.E. : valuation of Policy Simulation Models: a Conceptual Approa [1368]
Retti, J.   : A Fast Interactive Simulation System.   [1407]
Retti, J.   : n for Modelling and Simulation .   [1409]
Retti, J.   : tive Engineering of Simulation Models of Time-Oriented Sys [1408]
Retti, J.   : n for Modelling and Simulation of Dynamical Systems.   [1410]
Ricci, F.J. :   Computer Simulation of a Cybernetic Model of Ma [1416]
Rubinstein, :   Simulation and Monte Carlo Method.   [1456]
Sadovskiy, L: Questions Regarding Simulation of Hierarchical Systems. [1470]
Sandblom, C.:   Simulation Results on the Evaluation o [1488]
Spriet, J.A.: Aided Modelling and Simulation .   [1606]
Sydow, A.   :   Systems and Simulation 1980   [1648]
Toern, A.A. :   Simulation Graphs: A General Tool for [1695]
Toern, A.A. : l Tool for Modeling Simulation Designs.   [1695]
Trost, H.   :   Computer Simulation of Neurotic Processes.   [1751]
Uyeno, D.H. : M: A Discrete-Event Simulation Package for PASCAL.   [1776]
Ventriglia, : n Attempt Towards a Simulation of Cerebellar Cortex Activi [1811]
Yakowitz, S.: nal Probability and Simulation .   [1910]
Zeigler, B.P: stems Modelling and Simulation .   [1919]
            :   Special Issue on Simulation Modelling and Statistical C [1943]
Brewer, J.N.: ns in Social System Simulations .   [216]
Comstock, F.: o Grading of Flight Simulator Students.   [335]
Wnuk, A.J. : riented Interactive Simulator BORIS.   [1876]
Das, J.P.   :   Simultaneous and Successive Cognitive [378]
Sastry, D. : l Algorithm for the Simultaneous State and Parameter Ident [1495]
De La Sen, M: arameterizations in Single Input-Output Adaptive Control S [394]
Kaushik, M.L: Single Error and Burst Error Correctin [889]
Kaiser, J.S.: havior: A Vector of Single-Valued Transformations within a [868]
Retti, J.   :   SISSY - A Fast Interactive Simulation [1407]
Levitan, G.L: uctive Inference in Situation Control Systems.   [1043]
Pospelov, D.:   Large Systems: Situational Control .   [1351]
Best, D.P. : ructure in a Medium Sized UK Company.   [164]
Kobsa, A.   : f its Mechanisms, a Sketch of its Semantics.   [957]

Cerbone, G. :   Mean Variance and Skewness of the First Passage Time for [290]
De Raadt, J.:   Ashby's Law to the   Skewness of Income and Output Distribu [398]
Utkin, V.I. : ucture Systems with   Sliding   Modes.   [1774]
Lasota, A.   :   The Extinction of   Slowly Evolving Dynamical Systems.   [1012]
Kalaba, R.E.:   ive Reinvestment: A   Small   Sample Comparison.   [870]
Maciejowski,:   ing of Systems with   Small   Observation Sets.   [1082]
Revell, N.   : s for the Design of   Small   Database Systems.   [1414]
Airaksinen, :   Systems Theory in   Social Sciences.   [21]
Airaksinen, :   of Some Stabilizing   Social Systems.   [22]
Andersen, E.:   istical Models with   Social Science Applications.   [28]
Aulin, A.   :   Cybernetic Laws of   Social Progress: Towards a Critical So [73]
Aulin, A.   :   Towards a Critical   Social Philosophy and a Criticism of M [73]
Balkus, K.   : as the Basis of the   Social Self-Organization.   [88]
Ballonoff, P: tical Demography of   Social Systems, II.   [92]
Bennett, S. : iate Techniques for   Social and Behavioral Sciences.   [153]
Benseler, F.:   ietic System in the   Social Sciences.   [157]
Brewer, J.N.:   Effect Relations in   Social System Simulations.   [216]
Burghes, D.N: tical Models in the   Social , Management and Life Sciences. [243]
Casti, J.   :   logical Methods for   Social and Behavioral Systems.   [274]
Cavallo, R.E: tems Methodology in   Social Science Research.   [277]
Cavallo, R.E: ems Methodology for   Social Science Research.   [278]
Cavallo, R.E: tems Methodology in   Social Science Research: Recent Develo [287]
Chandrasekar: Natural and Social System Metaphors for Distribute [292]
Chen, K.   :   alysis in Different   Social Settings.   [305]
Collins, L. : se of Models in the   Social Sciences.   [332]
Cook, N.D.   :   Control in Natural,   Social , and Cybernetic Systems.   [345]
Demillo, R.A: Social Processes and Proofs of Theorem [407]
Fararo, T.J.:   Social Activity and Social Structure: [503]
Fararo, T.J.:   Social Activity and   Social Structure: A Contribution to th [503]
Fararo, T.J.:   on to the Theory of   Social Systems.   [503]
Gengoux, K.G: on: A New Factor in   Social Systems.   [591]
Gould, P.   :   An Introduction for   Social Scientists, Geographers and Pla [649]
Hanken, A.F.:   Social Systems and Learning Systems.   [721]
Hanken, A.F.:   ety: An Analysis of   Social Systems.   [720]
Jutila, S.T.:   Social Gaming and Control by Dislocati [864]
Kilmann, R.H: Social Systems Design: Normative Theor [916]
Klir, G.J.   : mplications for the   Social Sciences (a New Book Series) . [941]
Leinhardt, S: atical Sociology on   Social Network Research.   [1035]
Loo, S.G.   :   zy Relations in the   Social and Behavioral Science.   [1064]
Luhman, N.   :   World Society as a   Social System.   [1073]
Maruyama, M.:   g of Biological and   Social Sciences.   [1112]
Matejko, A. : uctural Criteria of   Social System Maturity.   [1116]
Mattessich, :   of the Applied and   Social Sciences.   [1126]
Mingers, J.C: ards an Appropriate   Social Theory for Applied Systems Thin [1171]
Mitroff, I.I: gical Approaches to   Social Science.   [1172]
Nowakowska, :   zzy Concepts in the   Social Sciences.   [1239]
Nowakowska, :   Cybernetics in the   Social Sciences.   [1245]
Phillips, D.:   Holistic Thought in   Social Science.   [1332]
Pipino, L.L.:   f Fuzzy Sets on the   Social Sciences.   [1342]
Prevost, P. : nctionalism and the   Social Sciences.   [1359]
Rastogi, P.N: An Introduction to   Social and Management Cybernetics.   [1386]
Roberts, F.S: g, Utility, and the   Social Sciences.   [1425]
Ryan, R.M.   :   the Integration of   Social Welfare Delivery Systems: A Mat [1467]
Schieve, W.C: in the Physical and   Social Sciences.   [1507]
Schmutzer, M: Out: Approaches to   Social Status.   [1512]
Slawski, C. :   Social Actualization: A Theory of Comm [1579]
Slawski, C. : ards a Synthesis of   Social Psychological "Theories".   [1581]
Sussmann, H.:   y as Applied to the   Social and Biological Sciences: a Crit [1641]
Umpleby, S. :   Methods for Making   Social Organisation Adaptive.   [1770]
Wenger, M.S.:   Cybernetic View of   Social Organization.   [1858]
West, B.J.   :   s as a Tool for the   Social Sciences.   [1859]
Zeleny, M.   :   es, and Spontaneous   Social Orders.   [1921]
Kornai, J.   :   tical Models of the   Socialist Economy: Review of a Researc [971]
Boulding, K.:   cs: a New Theory of   Societal Evolution.   [200]

Rastogi, P.N: Cybernetic Study of Societal Systems (Part I): Theory and [1384]
Rastogi, P.N: The Behaviour of Societal Systems. [1385]
Sutherland, : Societal Systems: Methodology, Modelli [1642]
Tazaki, E. : with Consensus for Societal Systems Using Fuzzy Set Theor [1676]
Arbib, M.A. : and the Cybernetic Society . [47]
Arbib, M.A. : and the Cybernetic Society . [49]
Benseler, F.: Communications, and Society : The Theory of Autopoietic Sys [157]
Bunge, M. : Systems Concept of Society : Beyond Individualism and Holi [240]
Gaines, B.R.: ms. Yearbook of the Society for General Systems Research. [558]
Hanken, A.F.: Cybernetics and Society : An Analysis of Social Systems [720]
Krippendorff: tion and Control in Society . [983]
Luhman, N. : The World Society as a Social System. [1073]
Masuda, Y. : The Information Society as Past-Industrial Society. [1115]
Masuda, Y. : as Past-Industrial Society . [1115]
Ragade, R.R.: ms: Yearbook of the Society for General Systems Research. [1376]
Toda, M. : Man, Robot, and Society : Models and Speculations. [1693]
Trappl, R. : Military, Economy, Society , Culture, and Politics. [1733]
Gericke, M. : ransport Demands by Society-Controlled Activities - A Fuzz [600]
Henize, J. : delling Large Scale Socio-Economic Systems: Can We Begin t [750]
Peschel, M. : Socio-Economic Consequences of the Vol [1325]
Balkus, K. : Systems in Sociocultural Development. [86]
Jantch, E. : cal, Biological and Sociocultural Evolution (AAAS Selected [835]
Geyer, R.F. : Sociocybernetics , Vol.II. [603]
Desouza, G.R: System Methods for Socioeconomic and Environmental Impact [411]
Bandler, W. : tomation Model of a Sociological Theory of Value Genesis. [101]
Havranek, T.: Processing Sociological Data by the Guha Method: [737]
Nowak, S. : Methodology of Sociological Research. [1238]
England J.L.: ierarchy Theory and Sociology . [482]
Leinhardt, S: sue of Mathematical Sociology on Social Network Research. [1035]
Leung, Y. : zzy Set Analysis of Sociometric Structure. [1039]
Ciborra, C. : nizational Noise in Sociotechnical Systems. [315]
De Greene, K: ergent Phenomena in Sociotechnical Macrosystems: Theories [390]
Beck, M.B. : Hard or Soft Environmental Systems? [128]
Checkland, P: Techniques in ' Soft ' Systems Practice. Part 1: System [297]
Huxham, C.S.: Hypergames as a Soft Systems Methodology. [800]
Mingers, J.C: Critical Theory and Soft Systems Methodology. [1171]
Prevost, P. : " Soft " Systems Methodology, Functionali [1359]
Trappl, R. : The Soft Side of Innovation: Medicine - a [1716]
Ivan, I.T. : for Estimating the Software Performances. [817]
Norman, M. : Software Package for Economic Modellin [1237]
Oeren, T.I. : e System Simulation Software Based on Cybernetic and Gener [1255]
Perlis, A.J.: Software Metrics: An Analysis and Eval [1317]
Pressman, R.: Software Engineering: A Practitioner's [1358]
Rees, R.K.D.: Software Reliability (2 Vols). [1397]
Rushinek, A.: nalysis of Computer Software and Its Importance to the Acc [1465]
Tancig, P. : A - General Purpose Software Environment for NLU-Systems. [1667]
Trattnig, W.: onal Semantics of a Software Specification Language Based [1734]
Green, M.A. : Solar Cells: Operating Principles, Tec [659]
Koenderink, : Perception of Solid Shape and Spatial Lay-Out throug [961]
Fortuny-Amat: A Solution Procedure for the Hierarchica [536]
Hendel, R.J.: formation-Theoretic Solution . [749]
Kalaba, R.E.: Two Solution Techniques for Adaptive Reinv [870]
Lopes, M.A. : Problems and Their Solution Spaces. [1065]
Windeknecht,: hods Applied to the Solution of Ordinary Differential Equa [1866]
De Mottoni, : negative Stationary Solutions for a Quasilinear Parabolic [397]
DiNola, A. : On the Fuzziness of Solutions of Alpha-Fuzzy Relation Equa [418]
Meyer, J.F. : Closed-Form Solutions of Performability. [1157]
Sengupta, J.: Minimax Solutions in Stochastic Programming. [1526]
Zarri, G.P. : hniques in Order to Solve the Problem of Information Incom [1915]
Guida, G. : ed Model of Problem Solver . [679]
Uyttenhove, : ms Approach Problem Solver and a Stockmarket Example. [1778]
Averkin, A.N: rategies in Problem Solving . [75]
Banerji, R.B: n-Numerical Problem Solving . [103]
Brightman, H: Problem Solving : A Logical and Creative Approa [219]

Brightman, H: Problem Solving : A Logical and Creative Approa [220]
Cavallo, R.E: for Systems Problem Solving .   [276]
Chandrasekar: Distributed Problem Solving .   [292]
Chen, D.T.-W: easoning in Problem Solving by Computers.   [303]
El-Shirbeeny: dient Algorithm for Solving Constrained Control Design Pro   [472]
Farley, A.M.: ledge-Based Problem Solving .   [507]
Gomez, P.   : on: Organic Problem Solving in a Publishing Company.   [638]
Grogono, P. :   Problem Solving and Computer Programming.   [672]
Hazony, Y. :   Problem Solving with Interactive Graphics.   [745]
Kaufmann, G.: Relation to Problem   Solving .   [887]
Kekes, J.   : ntrality of Problem   Solving .   [898]
King, L.T. :   Problem Solving in a Project Environment: A Co   [926]
King, L.T. :   Problem Solving in a Project Environment.   [925]
Klir, G.J. : ral Systems Problem   Solving Methodology.   [946]
Koffman, E.B: Problem Solving and Structured Programming in   [963]
Kowalaski, R: Logic of Problem   Solving .   [976]
Lester, F.K.: athematical Problem   Solving .   [1038]
Miller, R.B.: an-Computer Problem   Solving .   [1169]
Molloy, K.J.: anisational Problem   Solving .   [1182]
Newell, A. :   Human Problem   Solving .   [1217]
Ohlsson, S. : Learning of Problem   Solving Rules.   [1259]
Pernici, B. :   Problem Solving in Game Playing: Computer Ches   [1319]
Robertshaw, :   Problem Solving : a Systems Approach.   [1430]
Rudwick, B.H: Solving Management Problems: a Systems   [1460]
Scandura, J.:   Problem Solving : a Structural Process Approach   [1500]
Smith, R.G. : Distributed Problem   Solving .   [1593]
Ulrich, W. : e Design of Problem   Solving Systems.   [1767]
Zionts, S. : le Criteria Problem   Solving .   [1928]
Christensen,:   Entropy Minimax Sourcebook . Vol.1: General Description   [311]
Tancig, P. :   SOVA - General Purpose Software Enviro   [1667]
Lefebre, V.A: lysis of Wester and   Soviet Ethical Systems.   [1030]
Lefebre, V.A: Cybernetics in the   Soviet Union and the West.   [1031]
Niemeyer, H.:   Soziale   Beziehungsgeflechte.   [1227]
Feintuch, A.: m Theory: A Hilbert   Space Approach.   [516]
Metz, H.A.J.:   State Space Models for Animal Behaviour.   [1154]
Sherlock, R.: orks: Part 1: State   Space Description and Distribution of   [1545]
Strejc, V. :   State Space Theory of Discrete Linear Contro   [1628]
Taylor, A.M.: cal Organization of   Space .   [1672]
Arigoni, A.C: ructure of Property   Spaces - On the Synthesis of Formal Pr   [50]
DiNola, A. : Equations on Finite   Spaces .   [418]
Drozen, V. :   Combinatory   Spaces - An Approach to Pattern Analys   [442]
Gardiner, P.:   Decision   Spaces .   [569]
Lopes, M.A. : and Their Solution   Spaces .   [1065]
Cliff, A.D. :   Spacial   Processes: Models and Applicat   [320]
Scott, C.H. : Theoretic Model of   Spacial   Interaction.   [1518]
Baltes, P.B.:   Life   Span Development and Behavior, Vol.I.   [94]
Walker, C.C.: Priority, and Input   Span in a Class of Fixed-Structure Mod   [1830]
Cliff, A.D. :   Elements of   Spatial   Structure: a Quantitative Appr   [319]
Getis, A.   :   Models of   Spatial   Processes.   [601]
Koenderink, : of Solid Shape and   Spatial   Lay-Out through Photometric In   [961]
Dixon, N.R. : utomatic Speech and   Speaker   Recognition.   [426]
Abe, J.   : Representation of '   Speakers-World ' in Natural Language Pr   [2]
Gasparski, W: nglish nor a German   Speaking   Country.   [582]
Gasparski, W: English nor German   Speaking   Country.   [584]
Ben-Dov, Y. : ting Procedures for   Special   Structures of Coherent Systems   [148]
Boulding, K.: uman Knowledge as a   Special   System.   [201]
Greenspan, D: on of Classical and   Special   Relativistic Physics.   [663]
Hajek, P.   :   Special   Issue of the International Jou   [704]
Kohout, L.J.: herapy Adviser as a   Special   Case of a General System Prote   [966]
Leinhardt, S: Special   Issue of Mathematical Sociolog   [1035]
          :   Second   Special   Issue on the Guha Method of Me   [1941]
          :   Special   Issue on Simulation Modelling   [1943]
          :   Special   English Issue of the Polish Jo   [1942]
Cushing, J.M: Two Species Competition in a Periodic Envi   [367]

Bottaci, L. : me for the Abstract Specification of Production System Pro [198]
Demarco, T. : Analysis and System Specification . [406]
Tompa, F.W. : ical Example of the Specification of Abstract Data Types. [1698]
Trattnig, W.: ntics of a Software Specification Language Based on the Da [1734]
Darlington, : lidation of Program Specifications . [376]
Yager, R.R. :   Entropy and Specificity in a Mathematical Theory o [1907]
Arakelian, A: On an Algorithm of Spectral Analysis. [45]
Toda, M.   : Society: Models and Speculations . [1693]
Dixon, N.R. :   Automatic Speech and Speaker Recognition. [426]
Witten, I.H.: nciples of Computer Speech . [1875]
Kohout, L.J.: ic Structure of the Spencer-Brown and Varela Calculi. [965]
Schwartz, D.:   Isomorphisms of Spencer-Brown 's Laws of Form and Varel [1515]
Balossino, N: ltering by Means of Spherical Harmonics: A Simple Approxim [93]
Labos, E.   : etworks and Chaotic Spike Generators. [1000]
Zeleny, M. : ive Structures, and Spontaneous Social Orders. [1921]
Koenderink, :   On Spotting Three-Dimensional Shape: Rigi [962]
            : ented at the Second Spring School in Jadwisin, Poland, Apr [1942]
Zeleny, M. :   On the Squandering of Resources and Profits V [1923]
Acalugaritei: s of Evolutions and Stabilities . [4]
Aggarwal, J.: Nonlinear Systems: Stability Analysis. [16]
Aracil, J. :   Structural Stability of Low-Order System Dynamics [44]
Atherton, D.:   Stability of Nonlinear Systems. [59]
Brayton, R.K: Stability of Dynamical Systems: a Cons [214]
Cook, N.D. :   Stability and Flexibility: an Analysis [344]
Darwish, M. :   On the Stability of Large-Scale Power Systems [377]
Egardt, B. :   Stability of Adaptive Controllers. [458]
Foo, N.Y.   :   Stability Preservation under Homomorph [534]
Goh, B.S.   :   Stability , Vulnerability and Persisten [628]
Hirata, H. : ciency Mass and its Stability . [759]
Holling, C.S: Resilience and Stability of Ecological Systems. [773]
Kania, A.   :   On Stability of Formal Fuzziness Systems. [876]
LaSalle, J.P: The Stability of Dynamical Systems. [999]
Morris, H.M.:   Critical Stability in a Conserved Nutrient Ecos [1189]
Moylan, P.J.:   Stability Criteria for Large-Scale Sys [1192]
Perlis, J.H.:   Stability Analysis: An Approach to the [1318]
Pinske, M.S.: ion and Information Stability of Random Variables and Proc [1341]
Porenta, G. : paminergic Synapse: Stability and Sensitivity Analyses, an [1349]
Raghaven, V.: A Comparison of the Stability Characteristics of Some Grap [1377]
Rasvan, V. : Connected with the Stability Problem. [1387]
Granero-Pora: ems with Structure: Stability-Complexity Relationship. [655]
Binder, Z. :   Stabilization , Co-ordination and Struc [170]
Airaksinen, : ive Aspects of Some Stabilizing Social Systems. [22]
May, R.M.   : e Complex System be Stable ? [1130]
Weinberg, G.:   On the Design of Stable Systems. [1853]
Gerardy, R. : robabilistic Finite State System Identification. [598]
Joergensen, :   State of the Art in Ecological Modelli [846]
Jorgenson, S: State of the Art in Ecological Modelli [860]
Kovalenko, Y: ory as a Metastable State . II. Simulation Model. [974]
Kryukov, V.I: ory as a Metastable State . I. Master Equation Approach. [988]
McClamroch, :   State Models of Dynamic Systems: A Cas [1134]
McClamroch, :   State Models of Dynamic Systems. [1133]
Metz, H.A.J.:   State Space Models for Animal Behaviou [1154]
Orava, P.J. : del, Consolity, and State in the Theory of Dynamical Syste [1268]
Sastry, D. : or the Simultaneous State and Parameter Identification. [1495]
Sherlock, R.: y Networks: Part 1: State Space Description and Distributi [1545]
Sherlock, R.: engths; Part 2: the State Cycle Fraction for Networks of D [1545]
Shieh, L.   : of Continuous Time State Equations by Discrete-Time State [1546]
Shieh, L.   : ns by Discrete-Time State Equations. [1546]
Strejc, V. :   State Space Theory of Discrete Linear [1628]
Tacker, E.C.: ized Structures for State Estimation on Large Scale System [1653]
Watt, K.E.F.:   The Unsteady State - Environmental Problems, Growth [1849]
Hwang, C.L. : and Applications: A State-of-the-Art Survey. [803]
Quade, E.   :   The State-of-the-Art Questionnaire on Appl [1370]
Takahara, Y.:   A Stationarization Functor and its Concr [1659]

Ihara, J.   :   A Structural Analysis of Criteria for Se [808]
Lendaris, G.:   Structural Modeling - A Tutorial Guide [1036]
Matejko, A. :   The Structural Criteria of Social System M [1116]
Matsuno, K. : s Systems and Their Structural Transformation.   [1121]
Overton, W.S: digm Illustrated by Structural Aspects of a Hydrology Mode [1276]
Peay, E.R. :   Structural Models with Qualitative Val [1306]
Roberts, G.F: put Approach to the Structural Analysis of Digraphs.   [1426]
Scandura, J.:   Problem Solving: a Structural Process Approach with Instr [1500]
Shapiro, L.G: Structural Descriptions and Inexact Ma [1535]
Soete, W.   :   Structural Methods for Decomposition o [1596]
Tazaki, E. :   Structural Modeling in a Class of Syst [1675]
Tazaki, E. :   Structural Modeling with Consensus for [1676]
Stegmuller, :   The Structuralist View of Theories.   [1613]
Binder, Z. : , Co-ordination and Structuralization in the Integrated Au [170]
Arigoni, A.C: Algebraic Structure of Property Spaces - On the [50]
Atkin, R.H. : a Multi-Dimensional Structure .   [63]
Best, D.P. : rnetic Organisation Structure in a Medium Sized UK Company [164]
Brewer, J.N.:   Structure and Cause and Effect Relatio [216]
Broekstra, G: traint Analysis and Structure Identification II.   [223]
Broekstra, G: d Identification of Structure Systems.   [224]
Bubnicki, Z.: ystems with Cascade Structure .   [235]
Carvallo, M.: earch of the Noetic Structure of Systems.   [266]
Cavallo, R.E: The Structure of Reconstructable Relations [282]
Caws, P.   : erence, System, and Structure .   [288]
Cliff, A.D. : Elements of Spatial Structure : a Quantitative Approach. [319]
Conant, R.C.:   Set-Theoretic Structure Modelling.   [342]
Cover, T.M. : ith Apparent Markov Structure .   [353]
Fararo, T.J.: Activity and Social Structure : A Contribution to the Theor [503]
Garbade, K. : riation in the Term Structure of Treasury Bills.   [568]
Gould, P.   : is or a Language of Structure : An Introduction for Social [649]
Gourlay, A.R: s of the Subsystems Structure of Models.   [650]
Granero-Pora: ogical Systems with Structure : Stability-Complexity Relati [655]
Hallam, T.G.:   On the Intrinsic Structure of Differential Equation Mod [715]
Hattiangadi,:   The Structure of Problems.   [735]
Hooker, C.A.: Logico-Operational Structure .   [779]
Horn, W.   :   The Structure of Manifestations in a Medic [785]
Ibidapo-Obe,: ximate Differential Structure for Error Covariance in Nonl [807]
Kantor, J.R.:   System Structure and Scientific Psychology.   [878]
Kellerman, W: ification of System Structure : An Attempt at Implementatio [899]
Klir, G.J. : m of Computer-Aided Structure Identification: Some Experim [938]
Klir, G.J. :   Architecture of Structure Systems: a Basis for the Rec [944]
Kobayashi, I: ormation Processing Structure .   [954]
Kohout, L.J.:   The Algebraic Structure of the Spencer-Brown and Var [965]
Krzanik, L. : on of Delivery Step Structure .   [989]
Leung, Y.   : ysis of Sociometric Structure .   [1039]
Mohler, R. : cations of Variable Structure Systems.   [1178]
Mohler, R. : opments in Variable Structure Systems, Economics and Biolo [1179]
Mumford, E. : ive Systems Design: Structure and Method.   [1196]
Nelson, R.J.:   Structure of Complex Systems.   [1213]
Netchiporenk: Structure Analysis of Systems.   [1214]
Newton-Smith: The Structure of Time.   [1219]
Nicholson, H: Structure of Interconnected Systems. [1221]
Nowakowska, :   The Logical Structure of the Development of a Scie [1240]
Ollero, A. :   The Structure of Relations in Personnel Ma [1266]
Scott, W.A. :   Cognitive Structure : Theory and Measurement of I [1519]
Smetanic, Y.: rmining the Primary Structure of Biopolymers.   [1585]
Smith, C.S. :   A Search for Structure : Selected Essays on Science, [1588]
Smyth, D.S. : stems Approach: the Structure of Root Definitions.   [1595]
Stefanski, J: e Best Hierarchical Structure .   [1611]
Stegmuller, :   The Structure and Dynamics of Theories.   [1612]
Steward, D.V: sis and Management: Structure , Strategy and Design.   [1620]
Suppe, F.   :   The Structure of Scientific Theories (Seco [1637]
Takahara, Y.:   A Structure of Rational Decision Princip [1658]
Tao, K.M.   :   On the Structure of Directed Graphs with Appl [1669]

```
Tao, K.M.    :              On the  Structure  of Directed Graphs with Appl [1670]
Utkin, V.I. :            Variable  Structure  Systems with Sliding Modes.    [1774]
Vavrousek, J:            Multiple  Structure  of Systems.                    [1803]
Vavrousek, J:            Multiple  Structure  of Systems.                    [1805]
Vitanyi, P.M: indenmayer Systems:  Structure , Languages, and Growth Funct  [1819]
Wolaver, T.G:   Effect of Systems  Structure , Connectivity, and Recipient  [1877]
Atre, S.     :          Data Base:  Structured  Techniques for Design, Perf    [66]
Burgin, M.S.: of Multidimensional  Structured  Models of Systems.            [244]
Koffman, E.B: Problem Solving and  Structured  Programming in PASCAL.        [963]
Linger, R.C.:                       Structured  Programming: Theory and Pra  [1050]
Majster, M.E: hs, a Formalism for  Structured  Data and Data Structures.     [1092]
Medina, B.F.:                       Structured  System Analysis: A New Tech  [1145]
Skvoretz, J.: ffusion in Formally  Structured  Populations: An Information  [1572]
Auger, P.    : ting in Creation of  Structures .                              [67]
Barto, A.G.  : ion in Tessellation  Structures .                             [111]
Ben-Dov, Y.  : cedures for Special  Structures  of Coherent Systems.         [148]
Carlsson, C.: dling Fuzzy Problem  Structures .                             [264]
Conant, R.C.: lysis of Dependency  Structures .                             [339]
Cremers, A.B: lity of Information  Structures .                             [356]
Domotor, Z.  : d Causal Dependence  Structures .                             [432]
Drommenhoek,:           Biological  Structures  (translated from Dutch).     [440]
Graham, J.H.: Linguistic Decision  Structures  for Hierarchical Systems.     [653]
Greibach, S.:   Theory of Program  Structures : Schemes, Semantics, Verifi  [668]
Grenander, U:             Regular  Structures : Lectures in Pattern Theory  [669]
Groen, J.C.F: etween Neighbouring  Structures .                             [671]
Horowitz, E.: undamentals of Data  Structures .                             [788]
Hossforf, H.:    Model Analysis of  Structures .                             [791]
Krippendorff: thm for Identifying  Structures  in Multi-Variate Data.        [984]
Leal, A.     : itation of Decision  Structures .                            [1025]
Levy, L.S.   :           Discrete  Structures  of Computer Science.         [1044]
Majster, M.E: tured Data and Data  Structures .                            [1092]
Nishimura, H: ation of Dependency  Structures .                            [1232]
O'Muircheart: . 1: Exploring Data  Structures . Vol. 2: Model Fitting.      [1251]
Resconi, G.  : gical and Algebraic  Structures .                            [1406]
Rus, T.      :               Data  Structures  and Operating Systems.       [1464]
Schieve, W.C: ion and Dissipative  Structures : Applications in the Physic  [1507]
Schkolnick,  : hm for Hierarchical  Structures .                            [1511]
Sugiyama, K.: Hierarchical System  Structures .                            [1634]
Tacker, E.C.:       Decentralized  Structures  for State Estimation on Lar  [1653]
Tremblay, J.: ntroduction to Data  Structures  with Applications.           [1735]
Walker, B.J.: rity and Protection  Structures .                            [1829]
Wulf, W.A.   :          Fundamental  Structures  of Computer Science.        [1883]
Zeleny, M.   : oiesis, Dissipative  Structures , and Spontaneous Social Ord  [1921]
Arbel, A.    : ion and Information  Structuring  in Large-Scale Dynamic Sys   [46]
Yeh, R.T.    : ogy: Vol. IV - Data  Structuring .                           [1912]
Zeigler, B.P:                       Structuring  the Organization of Partia  [1918]
Zeigler, B.P:                       Structuring  Principles for Multifacet  [1920]
Comstock, F.: of Flight Simulator  Students .                              [335]
Hanson, O.J.:  Feedback from Past  Students .                              [723]
Kahne, S.'   : s to All University  Students .                              [865]
Apostolakis,: ety and Reliability  Studies  (NATO Advanced Study Institute   [40]
Bobrow, D.G.: and Understanding:  Studies  in Cognitive Science.           [179]
Boyce, W.E.  :               Case  Studies  in Mathematical Modeling.       [204]
Bradley, R.  :               Case  Studies  in Mathematical Modelling: A C  [210]
Gasparski, W:                       Studies  in Design Methodology in Polan  [575]
Hajek, P.    : rnal of Man-Machine  Studies  on the Guha Method of Mechaniz  [704]
Halfon, E.   : : Advances and Case  Studies .                              [712]
Hall, C.A.S.: roduction with Case  Studies .                              [713]
James, D.J.G:               Case  Studies  in Mathematical Modelling.       [831]
Suzuki, R.   : nd Phenomenological  Studies  of Biological Rhythms.         [1644]
Troncale, L.: s, Potentials, Case  Studies .                             [1748]
             :              Design  Studies . A New Journal Covering the St  [1938]
Best, D.P.   :             A Case  Study  in the Implementation of a Cyber   [164]
Caianiello,  :          A Systemic  Study  of Monetary Systems.             [255]
```

Cavallo, R.E: ns: a Comprehensive   Study .   [282]
Clark, W.C. : licy Design: a Case   Study   of Ecosystem Management.   [318]
Csendes, T. :   A Simulation   Study   on the Chemoton.   [363]
Gentile, S. : crete Model for the   Study   of a Lake.   [592]
Ghosal, A.   :   Cybernetics in the   Study   of Developing Countries.   [608]
Gottinger, H: conomical Models: a   Study   of Complexity.   [648]
Hazlett, B.A: tive Methods in the   Study   of Animal Behavior.   [744]
Hofferbert, : to the Comparative   Study   of Public Policy.   [767]
Kaufmann, A.: ical Models for the   Study   of the Reliability of Systems.   [886]
Leeming, A.M: A Systems   Study   of the Design of a User Interfac   [1029]
McClamroch, : mic Systems: A Case   Study   Approach.   [1134]
Pugh, R.E.   : al Approach to Case   Study .   [1368]
Rastogi, P.N: Cybernetic   Study   of Societal Systems (Part I): Th   [1384]
Singh, R.   : s Approaches to the   Study   of Industrial Relations.   [1567]
Slawski, C. : n: A Model and Case   Study .   [1576]
Tomlinson, R: laborative Research   Study .   [1697]
Van den Boga: ity Classes: A Case   Study   in Response Surface Methodology.   [1793]
Von Glasersf: s" and "Reality": A   Study   of Self-Regulation.   [1821]
            : ournal Covering the   Study , Research and Techniques of All   [1938]
Cheng, J.K. :   A   Subgraph   Isomorphism Algorithm Using R   [306]
Vallee, R. : ing and Interacting   Subjects .   [1783]
Shore, J.E. :   Fully-Decomposable   Subset and Aggregate Constraints.   [1550]
Yager, R.R. : in a Compound Fuzzy   Subset .   [1908]
Plotkin, A.A: archical Systems of   Subsets .   [1346]
Yager, R.R. : al Values for Fuzzy   Subsets .   [1900]
Yager, R.R. :   Fuzzy   Subsets   of Type II in Decisions.   [1896]
Andersson, H: ng the Theory of Rel   Subsets   Closed Systems.   [32]
Takeuchi, A.: jection onto Linear   Subspeces .   [1664]
Bosserman, R: urity Processes and   Subsystems   in Living Systems.   [195]
Gourlay, A.R: and Analysis of the   Subsystems   Structure of Models.   [650]
Sanderson, M: Successful   Problem Management.   [1491]
Schellenberg: Management, Key to   Successful   Computer Projects.   [1505]
Gutierrez, L: Ecosystem   Succession .   [693]
Das, J.P.   :   Simultaneous and   Successive   Cognitive Processes.   [378]
Meyer, A.R. : ond Order Theory of   Successor   Is Not Elementary Recursive.   [1156]
Pearl, J.   :   On   Summarizing   Data Using Probabilistic A   [1303]
Bluth, B.J. : Theory of Action. A   Summary   of the Basic Theory.   [178]
Medgyessy, P: Decomposition of   Superpositions   of Density Functions an   [1144]
Cavallo, R.E: Supplementary   Material to the SGSR Rep   [279]
Grauer, M. : of Efficient Energy   Supply   Strategies using Multi-Criteria   [658]
Alter, S.L. :   Decision   Support   Systems: Current Practice and   [26]
Bonczek, R.H: dations of Decision   Support   Systems.   [189]
Fick, G.   :   Decision   Support   Systems: Issues and Challenges   [521]
Horn, W.   : or Medical Decision   Support .   [786]
Keen, G.P.W.:   Decision   Support   Systems: an Organizational Per   [895]
Keen, G.P.W.:   Decision   Support   Systems.   [894]
Lee, R.M.   : pert vs. Management   Support   Systems: Semantic Issues.   [1028]
Oldershaw, R: cal and Theoretical   Support   for Self-Similarity between At   [1264]
Sage, A.P. : anning and Decision   Support .   [1475]
Sage, A.P. : anning and Decision   Support .   [1476]
Ciriani, T.A: ematical Models for   Surface   Water Hydrology. (Proc. of the   [316]
Van den Boga: e Study in Response   Surface   Methodology.   [1793]
Boncina, R. : us Parts in Passive   Surroundings .   [188]
Dedons, A. : ation Professional: Survey   of an Emerging Field.   [401]
Duran, B.S. : Cluster Analysis: a   Survey .   [448]
El-Sherief, : ete-Time Systems: A   Survey .   [468]
Gupta, M.M. : Fuzzy Set Theory: A   Survey .   [689]
Hwang, C.L. :   A State-of-the-Art   Survey .   [803]
Kobsa, A.   : e Representation: A   Survey   of its Mechanisms, a Sketch of   [957]
Molina, F.W.:   A   Survey   of Resource Directive Decomposi   [1181]
Monahan, G.E: A   Survey   of Partially Observable Markov   [1183]
O'Muircheart: The Analysis of   Survey   Data, Vol. 1: Exploring Data St   [1251]
Sampson, J.R: ng: An Introductory   Survey .   [1485]
Scoltock, J.:   A   Survey   of the Literature on Cluster An   [1517]

Troncale, L.:   A General Survey   of Systems Methodology. (Proc.o [1742]
Troncale, L.:   A General Survey   of Systems Methodology.   [1741]
Tzafestas, S: stem Reliability: a Survey   of Problems and Techniques.   [1764]
Verhagen, C.: lines; a Literature Survey .   [1812]
Elandt-Johns: Survival Models and Data Analysis.   [473]
Kraft, R.W. : Systems, Science & Survival .   [979]
Ninneman, J.:   The Survival of U.S.Business: A General Sy   [1231]
Ciupa, M.   : Approach to Packet Switched Data Communication Systems.   [317]
Davio, M.   :   Discrete and Switching Functions.   [381]
Kandel, A. :   Fuzzy Switching and Automata: Theory and App   [874]
Kleyle, R. :   Switching Mechanisms in a Generalized   [936]
Pearce, J.G.: Telecommunications Switching .   [1299]
Adams, W.M. : ragana and Katakana Syllabaries : Relative Information Cont   [13]
Hellerman, L: Symbolic Initiative and its Applicatio   [747]
Kraft, R.W. :   Symbols , Systems, Science & Survival.   [979]
Moser, J.   :   Hidden Symmetrics in Dynamical Systems.   [1190]
Arora, P.N. : vergence and Cyclic Symmetry .   [54]
Bathe, K.J. : alysis: U.S.-German Symposium .   [123]
Bensoussan, : ysis: International Symposium , Versailles 1976.   [158]
Bensoussan, :   International Symposium on Systems Optimization and   [159]
            : First International Symposium on Policy Analysis and Infor   [1934]
Porenta, G. : of the Dopaminergic Synapse : Stability and Sensitivity Ana   [1349]
Palm, G.   :   Rules for Synaptic Changes and their Relevance f   [1278]
Hogeweg, P. :   Locally Synchronised Developmental Systems: Co   [769]
Haken, H.   :   Dynamics of Synergetic Systems.   [708]
Haken, H.   :   Synergetics .   [706]
Haken, H.   :   Synergetics - A Workshop.   [705]
Gonzales, R.:   Syntactic Pattern Recognition.   [640]
Jakubowski, :   Syntactic Characterization of Machine   [830]
Tsai, W.H. : Tool for Combining Syntactic and Statistical Approaches t   [1757]
Leinfellner,: netics / Phonology, Syntax , Semantics.   [1034]
Sernadas, A.:   SYSTEMATICS: Its Syntax and Semantics as a Query Langua   [1528]
Apostolakis,:   Synthesis and Analysis Methods for Saf   [40]
Arigoni, A.C: rty Spaces - On the Synthesis of Formal Properties.   [50]
Burns, D.W. :   The Synthesis of Forecasting Models in Dec   [247]
Franova, M. :   Program Synthesis and Constructive Proofs Obta   [542]
Gupta, M.M. : via Liapunov Signal Synthesis Method.   [687]
Kauffman, L.:   Network Synthesis and Varela's Calculus.   [885]
Lakshmivarah: ormal Languages and Synthesis of Fuzzy Grammars.   [1005]
Mamdani, E.H: ng Using Linguistic Synthesis .   [1098]
Porter, B. :   Synthesis of Dynamical Systems.   [1350]
Prokop, A. : ystems Analysis and Synthesis in Biology and Biotechnology   [1365]
Slawski, C. :   Towards a Synthesis of Social Psychological "The   [1581]
Tong, R.M. :   Synthesis of Fuzzy Models for Industri   [1700]
Veroy, B.S. : etwork Architecture Synthesis : Interactive Strategy.   [1813]
Agusti, J. : the Neural Networks Synthesized with Z-Fivex Formal Neuron   [19]
Freuder, E.C: Synthesizing Constraint Expressions.   [546]
Cavallo, R.E: Sysems Research Movement: Characterist   [280]
Agazzi, E. :   System Theory and the Problem of Reduc   [15]
Altmann, A. :   Interdisciplinary System Analysis.   [27]
Andersson, H: n Inventory Control System Using the Theory of Relatively   [32]
Aracil, J. : bility of Low-Order System Dynamics Models.   [44]
Arbib, M.A. :   Foundations of System Theory: The Hankel Matrix.   [48]
Athanassov, :   A System for Training of EDP-Specialists   [58]
Auger, P.   : of Observation of a System (Biological or Physical) Result   [67]
Auger, P.   : rchically Organized System .   [68]
Augustine, N: ne's Laws and Major System Development Programs.   [71]
Ballance, R.: its Application to System Evaluation.   [90]
Bennett, T.J: s a Self-Organizing System .   [154]
Benseler, F.: eory of Autopoietic System in the Social Sciences.   [157]
Bibbero, R.J: Microprocessor System , Interfacing and Applications.   [168]
Boncina, R. :   An Open System Composed of Non-Active Monotono   [188]
Bottaci, L. : ation of Production System Programs Used in Expert Systems   [198]
Boulding, K.: iverse as a General System : Fourth Annual Ludwig von Berta   [199]

Boulding, K.: wledge as a Special   System .   [201]
Bozinovski, :   A Self-Learning   System Using Secondary Reinforcement.   [207]
Brewer, J.N.: Relations in Social   System Simulations.   [216]
Buchberger, : nguage Understandig   System .   [238]
Caianiello, : tion in the Nervous   System .   [256]
Carlsson, C.:   A   System of Problems and How to Deal wit [262]
Carlsson, C.:   A   System of Problems and a Way Treating   [263]
Caws, P.   :   Coherence,   System , and Structure.   [288]
Chandrasekar: Natural and Social   System Metaphors for Distributed Probl [292]
Chen, K.   :   System Analysis in Different Social Se [305]
Chikan, A. : Analysis of a Model   System .   [307]
Cole, I.D. :   Naturalized   System Development.   [330]
Couger, J.D.:   Advanced   System Development/Feasibility Techniq [351]
Courtois, P.: eueing and Computer   System Applications.   [352]
Dayal, R.   :   An Integrated   System of World Models.   [385]
De Cleris, M: Certain   System Concepts in Law and Politics.   [388]
De Jong, K. :   Adaptive   System Design: A Genetic Approach.   [392]
De Mottoni, : asilinear Parabolic   System .   [397]
De Rosnay, J: ew World Scientific   System .   [399]
Delaney, W. :   A   System Theory Based Simulation Languag [404]
Demarco, T. : ctural Analysis and   System Specification.   [406]
Desouza, G.R: System Methods for Socioeconomic and E [411]
Destouches, :   Basic Concepts of   System Theory.   [413]
Durkin, J.E.: therapy and General   System Theory.   [449]
Eisenstein, : ature Extraction by   System Identification.   [466]
El-Sherief, : es in Multivariable   System Modeling and Identification Alg [471]
Espejo, R. : ormation Systems: a   System for Design.   [489]
Eveland, S. : Relational Database   System for a Gothic Ethymological Dict [496]
Eyckhoff, P.: nds and Progress in   System Identification.   [499]
Feintuch, A.:   System Theory: A Hilbert Space Approac [516]
Fogel, E.   : e of Information in   System Identification: Bounded Noise C [532]
Frank, P.M. :   Introduction to   System Sensitivity Theory.   [540]
Gaines, B.R.:   Sequential Fuzzy   System Identification.   [560]
Garfolo, B.T: Medical Information   System .   [573]
Gasparski, W: Design and   System : Methodological Aspects, Vol.3. [577]
Gasparski, W: t Mean a Notion of "   System Modelling".   [583]
Genesio, R. : Order for Uncertain   System Approximation.   [590]
Gerardy, R. : listic Finite State   System Identification.   [598]
Ginzberg, M.: scription Model for   System Implementation.   [616]
Glaser, F.B.:   A   System Approach to Alcohol Treatment.   [622]
Gottinger, H: ications of Dynamic   System Theory.   [646]
Green, M.A. : es, Technology, and   System Application.   [659]
Grubbstrom, : roduction-Inventory   System Interpreted in Terms of the The [676]
Gyarfas, F. : d Diagnostic Expert   System .   [695]
Horn, W.   : ficial Intelligence   System for Medical Decision Support.   [786]
Horn, W.   : edical Consultation   System .   [785]
Horn, W.   : Graphic Interaction   System .   [787]
Hutt, A.T.F.: ata Base Management   System .   [799]
Innocent, P.: -Adaptive Interface   System .   [813]
Iserman, R. :   Identification and   System Parameter Estimation.   [815]
Iserman, R. :   System Identification (Papers from an   [816]
Ivan, I.T. :   System of Indicators for Estimating th [817]
Jack, W.   : ch to the Notion of   System .   [823]
Jaron, J.   : the Goal-Space of a   System .   [840]
Joerres, R. : ng of the Circadian   System of Euglena Gracilis.   [847]
Kalata, P. :   On   System Identification with and without [871]
Kantor, J.R.:   System Structure and Scientific Psycho [878]
Kellerman, W: d Identification of   System Structure: An Attempt at Implem [899]
Kindler, E. : ration of Necessary   System Conceptions.   [917]
Kinston, W. :   Linking   System Research and Delivery Systems.   [928]
Kleyle, R. : ralized Information   System .   [936]
Knappe, U. : Microcomputer-Based   System for Industrial Energy Control.   [952]
Kobayashi, H: an Introduction to   System Performance Evaluation Methodol [953]
Kohout, L.J.: l Case of a General   System Protection Design.   [966]

Koyayashi, H: an Introduction to System Performance Evaluation Methodol [977]
Laszlo, C.A.:   Champ - A System and Computer Program Package fo [1013]
Lazarevic, B: bal and Large-Scale System Models. (Proc. of a Seminar in [1020]
Lucas, R.G. : ants of Information System Configuration.   [1070]
Luhman, N.   : Society as a Social System .   [1073]
MacDonald, B: le Context Learning System .   [1078]
Mager, P.P. :   Biorhythms, System Organization and Bioactive Comp [1088]
Marko, H.   :   Methods of System Theory.   [1106]
Matejko, A. : Criteria of Social System Maturity.   [1116]
May, R.M.   : ill a Large Complex System be Stable?   [1130]
McCorkell, C: al Adaptive Control System .   [1135]
McLean, J.M.:   Methods of System Analysis and Model Building.   [1141]
Medina, B.F.:   Structured System Analysis: A New Technique.   [1145]
Meldman, M.J: ata Base Management System for Minicomputers.   [1150]
Mergenthaler: ve, Coherent Binary System .   [1152]
Mesarovic, M: f World Development System .   [1153]
Mohler, R.   : verview of Bilinear System Theory and Applications.   [1180]
Neck, R.   : al and the Economic System in Austria.   [1206]
Oeren, T.I. : ale for Large Scale System Simulation Software Based on Cy [1255]
Patel, R.V. :   Multivariate System Theory and Design.   [1289]
Patten, B.C.:   System Theory and the Ecological Niche [1294]
Perlis, J.H.: o the Evaluation of System Design.   [1318]
Presern, S. :   Plan Generating System for a Tactile Sensor.   [1357]
Randers, J. :   Elements of the System Dynamic Method.   [1379]
Rasvan, V.   :   Some System Theory Ideas Connected with the [1387]
Retti, J.   : eractive Simulation System .   [1407]
Ruotsalainen: e Regional Hospital System with Different Service Levels. [1463]
Rushinek, A.: ounting Information System .   [1465]
Rzevski, G. : sophical Aspects of System Design.   [1468]
Sadovsky, V.:   The System Principle and Leibnitz's Theore [1471]
Sain, M.K.   : uction to Algebraic System Theory.   [1482]
Sembugamoort: anguage Acquisition System .   [1523]
Siewiorek, D: ractice of Reliable System Design.   [1554]
Smith, R.B. : n & Implement a Bad System .   [1592]
Sonde, B.S. :   Introduction to System Design Using Integrated Circuit [1598]
Strombach, W: Wholeness, Gestalt, System : On the Meaning of these Concep [1629]
Sugiyama, K.: ing of Hierarchical System Structures.   [1634]
Takahara, Y.:   A System Theoretic Consideration of the [1654]
Takatsuji, M: tical Approach to a System of Interacting Elements.   [1662]
Tilti, A.   :   Large Scale System Theory and Applications (Proc. [1690]
Trost, H.   : guage Understanding System .   [1752]
Trost, H.   : n Language Dialogue System .   [1753]
Tussey, C.C.: ngevity Information System .   [1763]
Tzafestas, S: Optimization of System Reliability: a Survey of Proble [1764]
Uyttenhove, : y in the Behavioral System Identification Problem (Part I) [1777]
Vaccari, E. :   The Role of System Modelling in Natural Language P [1779]
Vallee, R.   : of Dynamical Linear System with Random Initial Conditions. [1782]
Von Luetzow,:   A System Analysis of Information Transmi [1822]
Walker, C.C.:   A System Theoretic Approach to the Manag [1830]
Wang, Z.   : of Animal Husbandry System in Farming Region.   [1837]
Weir, M.   : for a Goal-Directed System .   [1855]
Wellstead, P: duction to Physical System Modelling.   [1856]
Wendt, S.   : ule: an Element for System Modelling.   [1857]
Wilder, R.L.: atics as a Cultural System .   [1863]
Windeknecht,:   System Theoretic Implications of Numer [1866]
Wuensch, G. :   System Theorie.   [1882]
Zeigler, B.P: ples for Multifacet System Modelling.   [1920]
Hwang, C.L. :   System-Reliability Evaluation Techniqu [804]
Neck, R.   :   A System-Theoretic Analysis of a Simple [1207]
Erdi, P.   :   System-Theoretical Approach to the Neu [484]
Sutherland, :   Towards a System-Theoretical Decision Logic.   [1643]
Tjoa, A.M.   :   A System-Theoretical Approach to Databas [1691]
Cugno, F.   : alitarian Utopia: A System-Theory Approach.   [364]
Busch, J.A. :   Cybernetics IV: A System-Type Applicable to Human Groups [252]

Millendorfer: Systemanalytical Approaches to Long Ec [1165]
Gall, J.   :   Systemantics : How Systems Work and Esp [563]
Auguin, M.   :   Systematic Method of Realization of In [70]
Horvath, T. :   Method and   Systematic Reflections.   [789]
Lundeberg, M: tems Development: A   Systematic Approach.   [1074]
Revell, N.   :   Systematic Methodologies for the Desig [1414]
Koppelaar, H: Systematica .   [969]
Grindley, K.:   Systematics : a New Approach to Systems [670]
Sernadas, A.:   SYSTEMATICS : Its Syntax and Semantics [1528]
Rescher, N. :   Cognitive   Systematization : a Systems-Theoretic A [1401]
Voyat, G.E. :   Piaget   Systematized .   [1825]
Le Moigne, J: Analyse de   Systeme : Bibliographie et Commentaires [1022]
Le Moigne, J: Analyse de   Systeme .   [1023]
Blauberg, I.:   The   Systemic Approach: Prerequisites, Prob [174]
Blunden, M. : tary Technology and   Systemic Interaction.   [176]
Botez, M.C. :   Toward a   Systemic Representation of General Sys [197]
Caianiello, :   A   Systemic Study of Monetary Systems.   [255]
Jain, V.   :   A   Systemic Assessment and Application of [828]
Lusk, E.J. : xternal Validity in   Systemic Research.   [1075]
Pla Lopez, R: Systemic Transition from Ideological L [1345]
Sachs, W.   :   Systemic Inactivism: A Modern Laissez- [1469]
Sahal, D.   :   Systemic Similitude.   [1478]
Seton, F.   :   Economic Flows and   Systemic Efficiency.   [1529]
Small, M.G. :   Systemic and Global Learning.   [1584]
Adam, A.   :   Discrete   Systems   Methodology.   [11]
Aggarwal, J.:   Nonlinear   Systems : Stability Analysis.   [16]
Aggarwal, S.: on of Probabilistic   Systems   by Ergodic Machines.   [17]
Airaksinen, :   Systems   Theory in Social Sciences.   [21]
Airaksinen, :   Stabilizing Social   Systems .   [22]
Alter, S.L. :   Decision Support   Systems : Current Practice and Continui [26]
Anderson, B.:   Foundations of   Systems   Theory: Finitary and Infinitar [29]
Andersson, H: f Relatively Closed   Systems .   [32]
Anderton, R.:   Systems   in the Seventies: an Emerging [33]
Andrew, A.M.:   and Continuity - a   Systems   Dichotomy.   [36]
Apostolico, : ing of Biomolecular   Systems .   [41]
Arbel, A.   : Large-Scale Dynamic   Systems .   [46]
Atherton, D.: bility of Nonlinear   Systems .   [59]
Atkin, R.   : and the Geometry of   Systems .   [61]
Auger, P.   : rchically Organized   Systems .   [69]
Aulin, A.   :   the Foundations of   Systems   Theory.   [72]
Babloyantz, : ation in Biological   Systems   with Multiple Cellular Contact [78]
Bahm, A.J. :   Five Types of   Systems   Philosophy.   [79]
Baldwin, J.F: ogramming for Fuzzy   Systems   with Fuzzy Environment.   [85]
Balkus, K.   :   Systems   in Sociocultural Development. [86]
Ballonoff, P: emography of Social   Systems , II.   [92]
Banathy, B.H: Developing a   Systems   View of Education.   [95]
Banathy, B.H: Systems   Science and Science. (Proc. 24 [96]
Bandler, W. : nce in Fuzzy Expert   Systems .   [102]
Basta, D.J. :   Analyzing Natural   Systems : Analysis for Regional Residua [114]
Battista, J.: aradigm and General   Systems   Theory.   [124]
Bayraktar, B: Education in   Systems   Science.   [127]
Beck, M.B. :   Soft Environmental   Systems ?   [128]
Becker, P.W.:   On   Systems   with Redundancy and Their Inhe [130]
Becker, P.W.:   On   Systems   with Redundancy and their Inhe [129]
Becker, P.W.:   On   Systems ' Reliability and Redundancy. [131]
Bednarek, A.:   Dynamical   Systems   II.   [133]
Beishon, J. :   Learning about   Systems .   [139]
Beloglavec, :   A General   Systems   Theory Approach to the Propert [145]
Beltrami, E.:   Models for Public   Systems   Analysis.   [147]
Ben-Dov, Y. : uctures of Coherent   Systems .   [148]
Ben-Eli, M.U: ncrease in Evolving   Systems .   [149]
Bennett, R.J: Environmental   Systems : Philosophy, Analysis and Cont [152]
Benningfield: Circuit and   Systems   Theory.   [155]
Bensoussan, :   New Trends in   Systems   Analysis: International Sympos [158]

Bensoussan, : tional Symposium on	Systems	Optimization and Analysis.	[159]
Bensoussan, : and Optimization of	Systems	. (Proc. of a Conference in Ver	[160]
Bergeron, R.: Fluid, and Dancing	Systems	.	[161]
Binder, Z. : tomation of Complex	Systems	.	[170]
Blaquire, A.: Dynamical	Systems	and Microphysics. (Seminar in	[172]
Blauberg, I.:	Systems	Theory: Philosophical and Meth	[173]
Blomberg, H.: ultivariable Linear	Systems	.	[175]
Blunden, M. :	Systems	and Management - The Political	[177]
Bogart, D.H.: egic Information in	Systems	.	[181]
Boltyanskii,: Control of Discrete	Systems	. (Translated From Russian by R	[187]
Boltyanskii,: Control of Discrete	Systems	.	[186]
Bonczek, R.H: of Decision Support	Systems	.	[189]
Bose, N.K. : Multidimensional	Systems	: Theory and Applications.	[192]
Bose, N.K. : ed Multidimensional	Systems	Theory.	[193]
Bosserman, R: s in General Living	Systems	.	[196]
Bosserman, R: e of Mathematics in	Systems	Science.	[194]
Bosserman, R: ubsystems in Living	Systems	.	[195]
Botez, M.C. : entation of General	Systems	: Open Multimodelling.	[197]
Bottaci, L. : rams Used in Expert	Systems	.	[198]
Bowler, T.D.: General	Systems	Thinking: Its Scope and Applic	[203]
Brayton, R.K: bility of Dynamical	Systems	: a Constructive Approach.	[214]
Broekstra, G: cation of Structure	Systems	.	[224]
Brown, D.B. :	Systems	Analysis and Design for Safety	[228]
Brown, J.H.U: abolic Processes: a	Systems	Approach to Endocrinology.	[230]
Bruni, C. :	Systems	Theory in Immunology.	[233]
Bubnicki, Z.: fication of Complex	Systems	with Cascade Structure.	[235]
Bubnicki, Z.:	Systems	Science. (A Journal Published	[236]
Buchberger, : ctions in Effective	Systems	.	[237]
Bunge, M. : General	Systems	and Holism.	[239]
Bunge, M. : A	Systems	Concept of Society: Beyond Ind	[240]
Bunge, M. : Analogy between	Systems	.	[241]
Burgin, M.S.: tructured Models of	Systems	.	[244]
Burks, A.W. : ls of Deterministic	Systems	.	[246]
Burns, J.R. : the Formulation of	Systems	Dynamics.	[248]
Butzer, K.W.: tions: Organisms or	Systems	?	[253]
Caianiello, : c Study of Monetary	Systems	.	[255]
Carvallo, M.: Noetic Structure of	Systems	.	[266]
Caselles, A.: he Light of General	Systems	Theory.	[268]
Casti, J. : Dynamical	Systems	and Their Applications: Linear	[271]
Casti, J. : ophe in Large-Scale	Systems	.	[272]
Casti, J. : cial and Behavioral	Systems	.	[274]
Cavallo, R.E: tual Foundation for	Systems	Problem Solving.	[276]
Cavallo, R.E: The Role of	Systems	Methodology in Social Science	[277]
Cavallo, R.E: ent Developments in	Systems	Methodology for Social Science	[278]
Cavallo, R.E: General	Systems	Methodology and Large-Scale Sy	[275]
Cavallo, R.E: ogy and Large-Scale	Systems	.	[275]
Cavallo, R.E: sibilistic Behavior	Systems	.	[286]
Cavallo, R.E:	Systems	Research Movement: Characteris	[281]
Cavallo, R.E: the SGSR Report on	Systems	Research Movement.	[279]
Cavallo, R.E:	Systems	Methodology in Social Science	[287]
Chang, L.C. : of Data Management	Systems	.	[293]
Chang, Shi-K: sis and Information	Systems	, (A Journal Published Semi-Anu	[294]
Checkland, P: nd Nature of 'Hard'	Systems	Thinking.	[296]
Checkland, P: echniques in 'Soft'	Systems	Practice. Part 1: Systems Diag	[297]
Checkland, P: s Practice. Part 1:	Systems	Diagram - Some Tentative Guide	[297]
Checkland, P: The Shape of the	Systems	Movement.	[298]
Checkland, P: Rethinking a	Systems	Approach.	[300]
Checkland, P:	Systems	Thinking, Systems Practice.	[301]
Checkland, P: Systems Thinking,	Systems	Practice.	[301]
Checkland, P: The	Systems	Movement and the "Failure" of	[299]
Churchman, C: The	Systems	Approach and its Enemies.	[312]
Ciborra, C. : e in Sociotechnical	Systems	.	[315]
Ciupa, M. : Data Communication	Systems	.	[317]
Coblentz, A.:	Systems	Science in Health Care.	[323]

Cochin, I.   : d Design of Dynamic   Systems .   [324]
Collot, F.   : Concept for General   Systems   Theory.   [333]
Colombano, S: Complex   Systems   Tend to be Unstable. Is this R   [334]
Comstock, F.:   A   Systems   Approach to Grading of Flight   [335]
Cook, N.D.   : Analysis of Natural   Systems .   [344]
Cook, N.D.   : ial, and Cybernetic   Systems .   [345]
Cosier, R.A.: Dialectical Inquiry   Systems .   [349]
Cotterman, W: Systems   Analysis and Design: A Foundat   [350]
Cumani, A.   : s of Fuzzy Feedback   Systems .   [366]
Czogala, E.   : tification of Fuzzy   Systems   and Its Application in Control   [368]
Czogala, E.   : l Problems in Fuzzy   Systems .   [369]
Dale, M.B.   :   Systems   Analysis and Ecology.   [372]
Dalenoort, G: ation of Control in   Systems   in General, and in Neural Netw   [374]
Darwish, M.   : f Large-Scale Power   Systems .   [377]
Davis, R.   :   Knowledge-Based   Systems   in Artificial Intelligence.   [383]
De La Sen, M: Aperiodic Sampling   Systems .   [393]
De La Sen, M: ut Adaptive Control   Systems   to Avoid Delays in the Input G   [394]
Degermendzhy: Population in Open   Systems .   [402]
Dekker, L.   :   Simulation of   Systems .   [403]
Dent, J.B.   :   Systems   Simulation in Agriculture.   [410]
Dessauchoy, : Decomposability of   Systems .   [412]
Devries, R.P: Systems   Theory and the Philosophy of S   [415]
Dimirivski, :   Categories in   Systems   and Control Theory.   [422]
Dirickx, Y.M: Systems   Analysis by Multilevel Methods   [424]
Dominique, J: cal Construction of   Systems .   [430]
Donald, A.   : nt, Information and   Systems   (Second Edition).   [434]
Dubois, D.   :   Fuzzy Sets and   Systems : Theory and Applications.   [443]
Dutton, G.   : graphic Information   Systems .   [450]
Economos, A.:   Thoughts on a   Systems   View of Life.   [454]
Ehrenkrantz,:   The   Systems   Approach to Design.   [460]
El-Fattah, Y: Learning   Systems : Decision, Simulation, and Con   [467]
El-Sherief, : crete Multivariable   Systems   with Process Noise.   [470]
El-Sherief, : iable Discrete-Time   Systems : A Survey.   [468]
El-Sherief, : on of Multivariable   Systems   in the Transfer-Function Matri   [469]
Elohim, J.L.: rganize Educational   Systems .   [477]
Erickson, S.:   A   Systems   View of the Foundations of Edu   [485]
Ericson, R.F: The General   Systems   Challenge.   [486]
Espejo, R.   : agement Information   Systems : a System for Design.   [489]
Estep, M.L.   :   A   Systems   Integration of Multidisciplina   [493]
Estep, M.L.   : ernative Methods of   Systems   Analysis: The Case of Qualitat   [494]
Eugene, J.   : e General Theory of   Systems .   [495]
Eyckhoff, P.: nds and Progress in   Systems   Identification.   [498]
Fairhurst, M: ormation Processing   Systems .   [500]
Fairtlough, :   How   Systems   Thinking Might Evolve.   [501]
Fararo, T.J.: he Theory of Social   Systems .   [503]
Farkas, J.   :   Presuppositions of   Systems   Theory.   [506]
Fedanzo, A.J: n of Information in   Systems .   [514]
Ferracin, A.: g Ability in Living   Systems .   [519]
Fiacco, A.V.: xtremal Methods and   Systems   Analysis.   [520]
Fick, G.   :   Decision Support   Systems : Issues and Challenges.   [521]
Findeisen, W: ion in Hierarchical   Systems   (Vol.9 in the Wiley IIASA/ Int   [523]
Findler, N.V: ue Using Production   Systems .   [526]
Fitter, M.   : atural' Interactive   Systems .   [527]
Fleddermann,:   Teaching the   Systems   Approach with a Group Thesis.   [529]
Foo, N.Y.   : isms of Timevarying   Systems .   [535]
Fortuny-Amat: strained Optimizing   Systems .   [536]
Fosko, P.D.   : ications of General   Systems   Theory and Cybernetics to the   [537]
Fox, M.S.   : View of Distributed   Systems .   [539]
Franksen, O.: ects of Large Scale   Systems : Developing Design Rules Using   [541]
Furukawa, O.: of Quality Control   Systems .   [549]
Gage, W.L.   : agement Information   Systems .   [552]
Gaines, B.R.: bliography of Fuzzy   Systems   and Closely Related Topics.   [555]
Gaines, B.R.: ations for Database   Systems .   [559]
Gaines, B.R.:   General   Systems . Yearbook of the Society for G   [558]

Gaines, B.R.: Society for General   Systems   Research.   [558]
Gaines, B.R.:   General   Systems   Research: Quo Vadis?   [557]
Gall, J.   : Systemantics: How   Systems   Work and Especially How They F   [563]
Ganascia, J.: ities for Diagnosis   Systems .   [566]
Gane, C.   :   Structural   Systems   Analysis: Tools and Techniques   [567]
Gasparski, W: ce, Technology, and   Systems .   [581]
Gasparski, W: On   Systems   Theory and Systems Research an   [578]
Gasparski, W: Systems Theory and   Systems   Research and Science and Techn   [578]
Gatlin, L.L.: eory and the Living   Systems .   [586]
Gengoux, K.G: ew Factor in Social   Systems .   [591]
Gerardy, R. : ion of Finite-State   Systems .   [599]
Geyer, R.F. : Theories: a General   Systems   Approach   [604]
Gini, G.   : rogramming in Robot   Systems .   [615]
Gladun, V.P.: ive Problem-Solving   Systems .   [617]
Goffman, W. : entific Information   Systems   and the Principle of Selectivi   [627]
Gold, H.J. : eling of Biological   Systems : an Introductory Guidebook.   [630]
Golemanov, L: ity of Hierarchical   Systems .   [631]
Gopal, K.   : aluation in Complex   Systems .   [643]
Gorbatov, V.: f Partially Ordered   Systems .   [644]
Gottinger, H: Systems   Approaches and Environmental P   [645]
Gottinger, H: mplexity in Dynamic   Systems .   [647]
Graham, J.H.: es for Hierarchical   Systems .   [653]
Granero-Pora: Random Ecological   Systems   with Structure: Stability-Comp   [655]
Grindley, K.: : a New Approach to   Systems   Analysis.   [670]
Grollmann, J: ription of Feedback   Systems   via Approximation by Delayed S   [673]
Grollmann, J: ximation by Delayed   Systems .   [673]
Grubbstrom, : f Relatively Closed   Systems   and Applications.   [675]
Grubbstrom, : f Relatively Closed   Systems .   [676]
Gruber, H. : Basis of Sequential   Systems .   [677]
Guida, G.   : guage Understanding   Systems .   [680]
Gumovski, I.: nd Discrete Dynamic   Systems .   [683]
Gupta, M.M. : for Linear Discrete   Systems .   [688]
Gutenbaum, J: Applications of Systems   Theory to Economics, Managemen   [692]
Gvishiani, J: Systems   Research: Methodological Probl   [694]
Gyula, P.   : aws of Sciences and   Systems .   [696]
Hafele, W. : Large Scale Energy   Systems . (IIASA Proceedings Series, Vo   [698]
Haimes, Y.Y.: of Water Resources   Systems : Modeling and Optimization of   [700]
Haimes, Y.Y.: tion of Large-Scale   Systems .   [700]
Haken, H.   : ormation by Dynamic   Systems   and Pattern Recognition.   [707]
Haken, H.   : amics of Synergetic   Systems .   [708]
Hakkala, L. : rithm for Dynamical   Systems .   [709]
Halfon, E. :   Theoretical   Systems   Ecology: Advances and Case Stu   [712]
Halmes, A. :   Topics in   Systems   Theory: Publication in Honour   [717]
Hanassab, S.: sors for Cybernetic   Systems .   [718]
Hanassab, S.: ller for Cybernetic   Systems .   [719]
Hanken, A.F.:   Social   Systems   and Learning Systems.   [721]
Hanken, A.F.: ystems and Learning   Systems .   [721]
Hanken, A.F.: Analysis of Social   Systems .   [720]
Hanson, O.J.:   The Role of   Systems   Analysis in Designing Systems   [722]
Hanson, O.J.: alysis in Designing   Systems   and Methods of Training Using   [722]
Hanson, O.J.: s of Training Using   Systems   Techniques.   [722]
Hanson, O.J.: n Advanced Business   Systems   Analysis Course Using Feedback   [723]
Hanson, O.J.: sign of an Advanced   Systems   Analysis Course.   [724]
Hartnett, W.:   Systems : Approaches, Theories, Applica   [731]
Hassell, M.P: ropod Predator-Prey   Systems .   [733]
Heim, R.   : proaches to Complex   Systems .   [746]
Henize, J. : cale Socio-Economic   Systems : Can We Begin to Do Better.   [750]
Hermann, R. : gebraic Geometry to   Systems   Theory, Part 1   [752]
Hisdal, E. : neralized Fuzzy Set   Systems   and Particularization.   [762]
Hofferbert, : lication of General   Systems   Methodology to the Comparative   [767]
Hogeweg, P. : nised Developmental   Systems : Concrete Advantages of Discre   [769]
Holden, A.V.: Activity in Neural   Systems .   [771]
Holler, M.J.: Quality Enforcement   Systems .   [772]
Holling, C.S: ility of Ecological   Systems .   [773]

Honkasalo, A: cesses and Economic   Systems .   [777]
Hoye, R.E.   : agement Information   Systems .   [793]
Hull, R.   : r-Based Information   Systems .   [796]
Hunt, L.R.   : f General Nonlinear   Systems .   [798]
Huxham, C.S.: ypergames as a Soft   Systems Methodology.   [800]
Hwang, C.L. : s for Complex/Large   Systems : A Review.   [804]
Hyvarinen, L: ormation Theory for   Systems Engineers.   [806]
Ibidapo-Obe,: Nonlinear Dynamical   Systems .   [807]
Iijima, J.   : position of General   Systems .   [809]
Innis, G.S. : lysis of Ecological   Systems .   [812]
Iri, M.   : earch, Circuits and   Systems Theory.   [814]
Jacak, W.   :   Events and   Systems .   [819]
Jacak, W.   :   On Goal-Oriented   Systems .   [820]
Jacak, W.   : elling of Evenistic   Systems .   [821]
Jakubowski, : of Complex Dynamic   Systems Simulation.   [829]
Janko, J.   :   Systems Approach in Natural Science Re   [832]
Jaron, J.   : iented Cybernetical   Systems .   [839]
Jeffers, J.N: An Introduction to   Systems Analysis: with Ecological Appl   [842]
Jennergren, : tilevel Approach: a   Systems Analysis Methodology.   [843]
Johnson, W.A: Applications of   Systems Theory to Ecology.   [852]
Jones, A.H. : usiness Information   Systems .   [853]
Jones, B.   :   General   Systems Theory and Algorithm Theory.   [855]
Jones, D.D. : plied to Ecological   Systems .   [856]
Jones, L.M. :   Defining   Systems Boundaries in Practice: Some P   [857]
Jumarie, G. : o Modelling Dynamic   Systems Involving Human Factors.   [862]
Kahne, S.   :   Introducing   Systems Concepts to All University Stu   [865]
Kania, A.   : of Formal Fuzziness   Systems .   [876]
Kaufmann, A.:   the Reliability of   Systems .   [886]
Kavakoglu, I: Modelling of Energy   Systems .   [890]
Kawamura, K.: tronics Industry: A   Systems Approach.   [891]
Keen, G.P.W.:   Decision Support   Systems : an Organizational Perspective   [895]
Keen, G.P.W.:   Decision Support   Systems .   [894]
Keravnou, E.:   Design of Expert   Systems from the Perspective of Conver   [903]
Kilmann, R.H: Social   Systems Design: Normative Theory and t   [916]
Kindler, E. :   Simula 67 and   Systems Analysis.   [920]
Kindler, E. :   Dynamic   Systems and Theory of Simulation.   [919]
Kindler, E. :   Systems of Material Flow.   [923]
Kinston, W. : nisation of Complex   Systems .   [927]
Kinston, W. : search and Delivery   Systems . The Example of Health Care.   [928]
Klir, G.J.   :   Frontiers in   Systems Research: Implications for the   [941]
Klir, G.J.   :   The General   Systems Research Movement.   [942]
Klir, G.J.   : d Series in General   Systems Research.   [943]
Klir, G.J.   : ecture of Structure   Systems : a Basis for the Reconstructab   [944]
Klir, G.J.   :   Computer-Aided   Systems Modelling.   [945]
Klir, G.J.   :   General   Systems Problem Solving Methodology.   [946]
Klir, G.J.   :   Applied General   Systems Research.   [940]
Klir, G.J.   : and Applied General   Systems Research: A Bibliography.   [937]
Klir, G.J.   :   On   Systems Methodology and Inductive Reas   [948]
Klir, G.J.   :   General   Systems Concepts.   [949]
Kloeden, P.E: Fuzzy Dynamical   Systems .   [951]
Koppelaar, H: Regular Career   Systems II.   [968]
Kornwachs, K: on and Nonclassical   Systems .   [972]
Kraft, R.W. :   Symbols,   Systems , Science & Survival.   [979]
Krallmann, H: Enlarging the   Systems Paradigm: Historical View from   [980]
Kramer, J.T.:   Systems Thinking: Concepts and Notions   [981]
Krone, R.M. :   Systems Analysis and Policy Sciences: [986]
Kuhn, H.W.   :   Mathematical   Systems Theory and Economics.   [994]
La Patra, J.:   Applying the   Systems Approach to Urban Development.   [997]
Ladanyi, O.G: on, A Challenge for   Systems Engineering.   [1002]
Lamport, L. : ents in Distributed   Systems .   [1006]
Langefors, L: rmation and Data in   Systems .   [1009]
LaSalle, J.P: bility of Dynamical   Systems .   [999]
Lasker, G.E.:   Applied   Systems and Cybernetics (6 Vols).   [1010]
Lasker, G.E.: World Problems and   Systems Learning.   [1011]

Lasota, A.   : Evolving Dynamical   Systems .   [1012]
Laszlo, E.   : Some Reflections on   Systems Theory's Critics.   [1015]
Lauenroth, H: Algorithmic Model   Systems and Innovation Strategies for   [1017]
Lauenroth, H: ion and Information   Systems .   [1017]
Lee, E.B.   : orms for Time Delay   Systems .   [1026]
Lee, R.M.   : Management Support   Systems : Semantic Issues.   [1028]
Leeming, A.M: A   Systems Study of the Design of a User   [1029]
Lefebre, V.A: and Soviet Ethical   Systems .   [1030]
Levien, R.E.: to IIASA: Applying   Systems Analysis in an International S   [1041]
Levitan, G.L: n Situation Control   Systems .   [1043]
Lewicka-Strz: The   Systems Approach as a Methodological A   [1045]
Lewin, D.   : of Digital Computer   Systems .   [1046]
Lilienfeld, :   The Rise of   Systems Theory: an Ideological Analysi   [1049]
Linkens, D.A: Biological   Systems Modelling and Control.   [1051]
Lions, J.I.   : matical Analysis of   Systems and Their Control.   [1054]
Lions, J.L.   : matical Analysis of   Systems and Their Control.   [1055]
Lipski, W.   :   Information   Systems with Incomplete Information.   [1057]
Locker, A.   : and Prescription of   Systems .   [1059]
Loefgren, L.: nability of Complex   Systems .   [1060]
Lowen, W.   : mies of the Mind: A   Systems Science Model of the Mind and   [1067]
Luenberger, : oduction to Dynamic   Systems : Theory, Models, and Applicati   [1071]
Lundeberg, M: Information   Systems Development: A Systematic Appr   [1074]
M'Pherson, P: Systems Engineering: An Approach to Wh   [1076]
M'Pherson, P: A Framework for   Systems Engineering Design.   [1077]
Maciejowski,:   The Modelling of   Systems with Small Observation Sets.   [1082]
Madan, S.   : Cybernetic Queueing   Systems .   [1087]
Mahmoud, M.S: Large Scale   Systems Modelling.   [1089]
Mak, K.L.   : -Inventory-Shipping   Systems .   [1093]
Mann, K.H.   : f Coastal Waters: A   Systems Approach (Studies in Ecology,   [1101]
Manohar Raho: Control   Systems and Quantitative Economic Poli   [1102]
Marchuk, G.I: mization of Complex   Systems . (Proc. of a Conference in Nov   [1103]
Margolis, D.: ome Classic Dynamic   Systems .   [1104]
Marmarelis, : is of Physiological   Systems .   [1107]
Marshall, J.: ntrol of Time-Delay   Systems .   [1108]
Martin, J.N.:   On Mapping Real   Systems .   [1109]
Matsuno, K.   : amics of Autonomous   Systems and Their Structural Transform   [1121]
Matsuno, K.   : dback Loops in Open   Systems .   [1123]
Mattessich, : ental Reasoning and   Systems Methodology: an Epistemology o   [1126]
Mattessich, : presentation of the   Systems Framework: Similarities and Di   [1127]
Mattessich, : nges's World of the   Systems and My Own Systems Methodology   [1127]
Mattessich, : Systems and My Own   Systems Methodology.   [1127]
McAleer, W.E: Systems : A Concept for Business and Ma   [1132]
McClamroch, : e Models of Dynamic   Systems : A Case Study Approach.   [1134]
McClamroch, : e Models of Dynamic   Systems .   [1133]
McPherson, P: and Improving Human   Systems .   [1142]
Mees, A.L.   : ynamics of Feedback   Systems .   [1146]
Melkumian, A: on and Transmission   Systems .   [1151]
Metzer, J.R.: Table Languages and   Systems .   [1155]
Michel, A.N.: ysis of Large Scale   Systems .   [1159]
Michie, D.   :   Expert   Systems in the Microelectronic Age.   [1160]
Milgram, M.   : c Entropy of Mankov   Systems .   [1163]
Millendorfer: Factors in Complex   Systems .   [1164]
Miller, J.G.:   Living   Systems .   [1167]
Mingers, J.C: Theory for Applied   Systems Thinking: Critical Theory and   [1171]
Mingers, J.C: cal Theory and Soft   Systems Methodology.   [1171]
Mitroff, I.I: Dialectical Belief   Systems .   [1173]
Mitsch, W.J.:   Energetics and Systems .   [1174]
Mittermeir, : hs - A Notation for   Systems Development.   [1175]
Mohler, R.   : Variable Structure   Systems .   [1178]
Mohler, R.   : Variable Structure   Systems , Economics and Biology.   [1179]
Moraga, C.   : ar p-Adic Invariant   Systems .   [1184]
Moser, J.   : etrics in Dynamical   Systems .   [1190]
Moylan, P.J.: ria for Large-Scale   Systems .   [1192]
Mulej, M.   :   Dialectical   Systems Theory.   [1193]

Mulej, M.   :   Dialectical Systems Theory and Work Simplification [1194]
Mumford, E. :   Participative Systems Design: Structure and Method.   [1196]
Murphy, R.E.: ocesses in Economic   Systems .   [1198]
Nachane, D.M: hods in Multi-Level   Systems .   [1199]
Nash, P.   :   Systems Modelling and Optimization.   [1202]
Naughton, J.: Craft Knowledge in   Systems Research: The Rules of the Gam [1203]
Negoita, C. :   Fuzzy Sets and   Systems : a New International Journal.   [1208]
Negoita, C. : ent Applications of   Systems Theory.   [1209]
Negoita, C. :   Fuzzy   Systems .   [1210]
Negoita, C.V: Fuzzy   Systems .   [1212]
Negoita, C.V: e Logic of Discrete   Systems Dynamics.   [1211]
Nelson, R.J.: tructure of Complex   Systems .   [1213]
Netchiporenk: ructure Analysis of   Systems .   [1214]
Nicholson, H: elling of Dynamical   Systems .   [1222]
Nicholson, H: e of Interconnected   Systems .   [1221]
Nicholson, H: elling of Dynamical   Systems .   [1223]
Nicolis, G. : in Non-Equilibrium   Systems .   [1224]
Nicolis, J.S: een Self-Organizing   Systems and Desynchronization of Physi [1225]
Nicolis, J.S: zation in Cognitive   Systems .   [1226]
Nijssen, G.M: ata Base Management   Systems .   [1229]
Ninneman, J.: Business: A General   Systems Approach.   [1231]
Nitecki, Z. : Theory of Dynamical   Systems .   [1233]
Nowakowska, : imensional Semiotic   Systems .   [1243]
Oeren, T.I. : tion of Large Scale   Systems .   [1254]
Oeren, T.I. : ernetic and General   Systems Theories.   [1255]
Oguntade, O.: heory of Humanistic   Systems .   [1258]
Okeda, M.   : ractions of Dynamic   Systems .   [1260]
Oldershaw, R: Atomic and Stellar   Systems .   [1264]
Optner, S.L.:   Systems Analysis.   [1267]
Orava, P.J. : Theory of Dynamical   Systems and Control.   [1268]
Owens, D.H. : ck and Multivariate   Systems .   [1277]
Paritsis, N.: Natural Intelligent   Systems with Changes in Environmental   [1282]
Patten, B.C.:   Systems Approach to the Concept of Env   [1292]
Patten, B.C.:   Systems Approach to the Concept of Nic   [1293]
Patten, B.C.:   Systems Analysis in Ecology.   [1291]
Paulre, B.E.:   Systems Dynamics and the Analysis of C   [1297]
Pearl, J.   : ial Question-Answer   Systems .   [1301]
Pedrycz, W. : tional Equations in   Systems Analysis.   [1311]
Peery, N.S. :   General   Systems Theory Approaches to Organizat   [1314]
Peixoto, M.M: Dynamical   Systems .   [1315]
Peschel, M. : logical Approach to   Systems Analysis Based on Volterra Equ   [1322]
Peschel, M. : logical Approach to   Systems Analysis Based on Volterra Equ   [1323]
Peschel, M. : roach for Nonlinear   Systems .   [1325]
Peterson, J.: and the Modeling of   Systems .   [1327]
Petrov, B.N.: hless Self-Adaptive   Systems , I.   [1329]
Petrovski, A: dels in Health Care   Systems .   [1330]
Pfeilsticker: The   Systems Approach and Fuzzy Set Theory   [1331]
Pichler, F. : of General Dynamic   Systems .   [1334]
Pichler, F. : in Cybernetics and   Systems Research, Vol.VI.   [1336]
Pichler, F. : in Cybernetics and   Systems Research, Vol.VII.   [1335]
Pichler, F. :   Dynamical   Systems Theory.   [1337]
Pichler, F. :   General   Systems Algorithms for Mathematical Sy   [1339]
Pichler, F. : ms for Mathematical   Systems Theory.   [1339]
Pierce, J.R.: ication Science and   Systems .   [1340]
Plotkin, A.A: Hierarchical Systems of Subsets.   [1346]
Pogonowski, : Approach to General   Systems Theory.   [1348]
Porter, B.   : thesis of Dynamical   Systems .   [1350]
Pospelov, D.:   Large   Systems : Situational Control .   [1351]
Pospelov, D.: c Models in Control   Systems .   [1352]
Prevost, P. :   "Soft"   Systems Methodology, Functionalism and   [1359]
Probst, G.J.: ians in Health Care   Systems .   [1364]
Prokop, A.   :   Systems Analysis and Synthesis in Biol   [1365]
Putt, A.M.   :   General   Systems Theory Applied to Nursing.   [1369]
Quade, E.   : ionnaire on Applied   Systems Analysis: a Report on the Resp   [1370]

Ragade, R.K.:   General Systems Yearbook.   [1375]
Ragade, R.R.:   General Systems : Yearbook of the Society for G [1376]
Ragade, R.R.: Society for General Systems Research.   [1376]
Rao, G.P.   : zy Logic Applied to Systems Control.   [1380]
Rao, G.V.   : lex Digital Control Systems .   [1381]
Rastogi, P.N: c Study of Societal Systems (Part I): Theory and Methodolo [1384]
Rastogi, P.N: haviour of Societal Systems .   [1385]
Reckmeyer, W: General Systems Research and Design: Precursor [1394]
Resconi, G. : tion of the General Systems by Logical and Algebraic Struc [1406]
Retti, J.   : ls of Time-Oriented Systems .   [1408]
Retti, J.   : lation of Dynamical Systems .   [1410]
Reusch, B. :   Continuous Systems with Digital Behavior.   [1411]
Reusch, B. : ation of Continuous Systems that are Digital.   [1412]
Reuver, H.A.: Context of General   Systems Theory.   [1413]
Revell, N. : n of Small Database Systems .   [1414]
Roberts, P.C: Modelling Large Systems .   [1427]
Roberts, P.C: Modelling Large Systems : Limits to Growth Revisited.   [1428]
Roberts, S.L: Systems Approach to Assessing Behavior [1429]
Robertshaw, : Problem Solving: a   Systems Approach.   [1430]
Rose, J.   : in Cybernetics and Systems (3 Vols.).   [1436]
Rose, J.   : in Cybernetics and Systems (Proc.Fourth Mt.Congress on Cy [1437]
Rosen, R.   : entation of Natural Systems .   [1441]
Rosen, R.   : Control of Complex Systems .   [1444]
Rosen, R.   :   Anticipatory Systems in Retrospect and Prospect.   [1443]
Rosen, R.   :   On Anticipatory Systems .   [1442]
Rosen, R.   : forwards and Global Systems Failure.   [1440]
Rosen, R.   : Physics of Complex Systems .   [1446]
Rosenberg, R: ynamics of Discrete Systems .   [1447]
Roth, G.   :   Self-Organizing Systems : An Interdisciplinary Approach [1451]
Royce, J.R. : The Role of General Systems and Information Theory in Mult [1453]
Ruben, B.D. :   General Systems Theory and Human Communication [1455]
Rudall, B.H.: Automated Teaching   Systems .   [1457]
Rudwick, B.H: agement Problems: a Systems Approach to Planning and Contr [1460]
Ruffner, M.A: Human Factors in Systems Analysis.   [1461]
Rugh, W.J. :   Nonlinear   Systems Theory: The Volterra/Wiener Ap [1462]
Rus, T.   : tures and Operating Systems .   [1464]
Ryabinin, I.: lity of Engineering Systems : Principles and Analysis.   [1466]
Ryan, R.M. : al Welfare Delivery Systems : A Matrix Organization Approac [1467]
Sadovskiy, L: ion of Hierarchical Systems .   [1470]
Saeks, R.   : rge-Scale Dynamical Systems .   [1472]
Sage, A.P. : ogy for Large Scale Systems .   [1474]
Sage, A.P. : sign of Information Systems and Process of Planning and De [1475]
Sage, A.P. : itivity Analysis in Systems for Planning and Decision Supp [1476]
Salzberg, P.: traction in General Systems .   [1484]
Samuelson, K: Information   Systems and Networks.   [1486]
Saridis, G.N: ntrol of Stochastic Systems .   [1494]
Sauer, C.H. :   Computer   Systems Performance Modeling.   [1496]
Sawaragi, Y.: stributed Parameter Systems .   [1499]
Scandura, J.: s of Instruction: A Systems Alternative to Cognitive Psych [1501]
Schal, R.K. : Thoughts of General Systems Theorists Concerning the Proce [1504]
Schiffman, Y: A   Systems Approach for Estimating Energy [1508]
Schnakenberg: lysis of Biological Systems (Second Edition).   [1513]
Schwarzenbac: Systems Modelling and Control.   [1516]
Scott, W.R. : l, Natural and Open Systems .   [1520]
Shahane, A.H: ary Models of Water Systems .   [1531]
Shakun, M.F.: esign of Purposeful Systems .   [1533]
Shakun, M.F.: esign of Purposeful Systems : The Profit-Making Case.   [1534]
Sharif, N. : Int. Conference on Systems Modelling in Developing Countr [1536]
Sharif, N. :   Systems Models for Decision Making.   [1537]
Shigan, E.N.:   Systems Modeling in Health Care.   [1547]
Shmulyan, B.: eraction of Entropy Systems .   [1548]
Shugart, H.H: Systems Ecology.   [1552]
Sierocki, I.: in General Dynamic Systems .   [1553]
Siljak, D.D.: Large-Scale Dynamic Systems .   [1555]

Silvern, L.C: Systems   Engineering Applied to Trainin [1556]
Silvern, L.C: Systems   Engineering of Education I: th [1557]
Silvern, L.C: I: the Evolution of   Systems   Thinking in Education.   [1557]
Simon, H.A. : Complex Are Complex   Systems   ?   [1561]
Singh, M.G. : book of Large Scale   Systems   Engineering Applications.   [1566]
Singh, R.   :   Systems   Approaches to the Study of Ind [1567]
Sklenar, J. : ription of Discrete   Systems   .   [1571]
Slawski, C. : wer and Conflict: A   Systems   Model of Interparty Negotiatio [1577]
Slawski, C. : argaining: A Policy   Systems   Research Model of Career Devel [1575]
Slawski, C. : Conflict Episode: A   Systems   Model.   [1580]
Slawski, C. :   A   Systems   Model of Negotiation: Arabs Ve [1573]
Smale, S.   : Essays on Dynamical   Systems   , Economic Processes and Relate [1583]
Smith, A.W. :   Mangagement   Systems   Analyses and Applications.   [1587]
Smyth, D.S. :   Using a   Systems   Approach: the Structure of Roo [1595]
Soete, W.   : tion of Large-Scale   Systems   .   [1596]
Spagovich, V: Theory of Adaptive   Systems   .   [1602]
Spedding, C.:   Agricultural   Systems   .   [1603]
Squire, E. :   Introducing   Systems   Design.   [1607]
Srinivasan, : alysis of Redundant   Systems   .   [1608]
Steward, D.V: Systems   Analysis and Management: Struc [1620]
Stoklosa, J.: for Some Dynamical   Systems   .   [1623]
Subrahmanian: s of Discrete-Event   Systems   .   [1632]
Sugiyama, K.: of Future Computer   Systems   Development in Japan.   [1635]
Sultan, L.H.: e Considerations on   Systems   Organization for Fuzzy Informa [1636]
Sutherland, :   Societal   Systems   : Methodology, Modelling and Ma [1642]
Swets, J.A. : ation of Diagnostic   Systems   : Methods from Signal Detection [1646]
Sydow, A.   :   Systems   and Simulation 1980   [1648]
Tacker, E.C.: tion on Large Scale   Systems   .   [1653]
Takahara, Y.: of General Dynamic   Systems   .   [1655]
Tamburrini, :   Do Cybernetics,   Systems   Science and Fuzzy Sets Share S [1665]
Tao, K.M.   : Rapprochement with   Systems   Theory (Part I).   [1669]
Tao, K.M.   : Rapprochement with   Systems   Theory (Part II).   [1670]
Taschdjian, : rnetics of Stressed   Systems   .   [1671]
Taylor, A.M.:   A   Systems   Approach to the Political Orga [1672]
Taylor, W.R.:   Using   Systems   Theory to Organize Confusion. [1674]
Tazaki, E. : eling in a Class of   Systems   Using Fuzzy Sets Theory.   [1675]
Tazaki, E. : sensus for Societal   Systems   Using Fuzzy Set Theory.   [1676]
Tchon, K.   : proach to Evenistic   Systems   .   [1677]
Tchon, K.   : neric Properties of   Systems   .   [1679]
Tchon, K.   : Global Analysis of   Systems   .   [1680]
Thomas, A. : he Use and Abuse of   Systems   Models.   [1685]
Thompson, J.: ability of Evolving   Systems   .   [1686]
Thowsen, A. : iability of Dynamic   Systems   .   [1687]
Tjoa, A.M. : pproach to Database   Systems   .   [1691]
Toda, M.   : for Future Computer   Systems   Development in Japan.   [1694]
Tomlinson, R: A   Systems   Approach to Planning in Organi [1697]
Tong, R.M. : s of Fuzzy Feedback   Systems   .   [1702]
Tou, J.T.   : nces in Information   Systems   Science, Vol.6.   [1705]
Trappl, R. : in Cybernetics and   Systems   Research, Vol.III.   [1712]
Trappl, R. : in Cybernetics and   Systems   Research, Vol.IV.   [1713]
Trappl, R. : in Cybernetics and   Systems   Research, Vol.VII.   [1721]
Trappl, R. : in Cybernetics and   Systems   Research, Vol.IX.   [1720]
Trappl, R. : in Cybernetics and   Systems   Research, Vol.X.   [1723]
Trappl, R. : in Cybernetics and   Systems   Research, Vol.XI.   [1724]
Trappl, R. :   Cybernetics and   Systems   Research.   [1718]
Trappl, R. :   Cybernetics and   Systems   Research, Vol.II.   [1726]
Trappl, R. : and Applied General   Systems   Research: A Bibliography, 1977 [1731]
Trappl, R. : hes Towards Complex   Systems   .   [1731]
Trappl, R. : in Cybernetics and   Systems   Research, Vol.VIII.   [1722]
Troncale, L.: A General Survey of   Systems   Methodology. (Proc.of the Twen [1742]
Troncale, L.:   A Hybrid   Systems   Method: Tests for Hierarchy & [1747]
Troncale, L.:   Knowing Natural   Systems   Enables Better Design of Man-M [1749]
Troncale, L.:   Design of Man-Made   Systems   : The Linkage Proposition Model [1749]
Troncale, L.:   Future of General   Systems   Science: Obstaches, Potentials [1748]

Troncale, L.:	een Fifty Principal	Systems	Concepts.	[1737]
Troncale, L.:	ry Process Based on	Systems	Concepts.	[1738]
Troncale, L.:	Theory to Modeling	Systems	Emergence.	[1739]
Troncale, L.:	ion and a Theory of	Systems	Emergence.	[1740]
Troncale, L.:	A General Survey of	Systems	Methodology.	[1741]
Troncale, L.:	s between Principal	Systems	Concepts.	[1743]
Troncale, L.:	a Formalization of	Systems	Linkage Propositions.	[1746]
Troncale, L.:		Systems	Theory Applied to the Biologic	[1750]
Trussel, P. :	tructs from General	Systems	Theory to a Health History.	[1756]
Tsetlin, M.L:	lling of Biological	Systems .		[1758]
Tsypkin, Y.Z:	aptive and Learning	Systems .		[1760]
Ulrich, W. :	of Problem Solving	Systems .		[1767]
Urban, H.B. :	f Development for a	Systems	Perspective.	[1773]
Utkin, V.I. :	Variable Structure	Systems	with Sliding Modes.	[1774]
Uyttenhove, :	The	Systems	Approach Problem Solver and a	[1778]
Vallet, C. :	cription of Natural	Systems .		[1784]
Van der Zouw:	Inventory of	Systems	Research in the Netherlands.	[1796]
Van der Zouw:	Inventory of	Systems	Research on the Netherlands.	[1797]
Van Gigch, J:	son of the Science,	Systems	and Metasystem Paradigms.	[1787]
Van Gigch, J:	A Taxonomy of	Systems	Science.	[1788]
Van Rootsela:	Annals of	Systems	Research, Vol. 7	[1790]
Van Steenkis:	ol in Environmental	Systems .		[1791]
Vavrousek, J:	ltiple Structure of	Systems .		[1803]
Vavrousek, J:	ltiple Structure of	Systems .		[1805]
Vemuri, V. :	Modeling of Complex	Systems	: an Introduction.	[1808]
Ventriglia, :	ic Theory of Neural	Systems	: Memory Effects.	[1810]
Ventriglia, :	ic Theory of Neural	Systems	: An Attempt Towards a Simulati	[1811]
Vickers, G. :	Education in	Systems	Thinking.	[1815]
Vincent, T.L:	ality in Parametric	Systems .		[1818]
Vitanyi, P.M:	Lindenmayer	Systems	: Structure, Languages, and Gro	[1819]
Von Bertalan:		Systems	View of Man (edited by Paul La	[1820]
Walker, J.A.:	Dynamical	Systems	and Evolution Equations: Theor	[1831]
Wallenmaier,:	ept of Teleological	Systems .		[1834]
Wang, P. :	Fuzzy	Systems	: Theory and Application to Pol	[1836]
Wang, P. :	sis and Information	Systems .		[1836]
Warnier, J.-:	cal Construction of	Systems .		[1842]
Wasniowski, :	puter-Communication	Systems	in Futures Research.	[1844]
Waterman, D.:	-Directed Inference	Systems .		[1847]
Weinberg, G.:	he Design of Stable	Systems .		[1853]
Weingartner,:	Analogy Among	Systems .		[1854]
Wolaver, T.G:	Effect of	Systems	Structure, Connectivity, and R	[1877]
Wood-Harper,:	rrent Approaches to	Systems	Analysis.	[1879]
Zeigler, B.P:	Methodology in	Systems	Modelling and Simulation.	[1919]
Zeigler, B.P:	ion of General Time	Systems .		[1917]
Zouwen, J.v.:	; a Contribution of	Systems	Methodology.	[1930]
	: sis and Information	Systems .		[1934]
	: cations of Adaptive	Systems	Theory. Medicine.	[1940]
	:	Systems	Research: Methodological Probl	[1944]
	: General	Systems	Research: A Science, a Methodo	[1933]
Grams, R.R. :		Systems-Analysis	Workbook.	[654]
Resconi, G. :	alysis of Dynamical	Systems-Applications	to Life Sciences.	[1404]
Erlandson, R:	nt-Observer Role in	Systems-Methodologies		[487]
Gomez, P. :		Systems-Methodology	in Action: Organic	[638]
Stoklosa, J.:	s of Some Dynamical	Systems-Networks	of (a,k)-Machines Gen	[1622]
Gelfand, A.E:	A	Systems-Theoretic	Approach to the Mana	[589]
Rescher, N. :	Systematization: a	Systems-Theoretic	Approach to a Cohere	[1401]
Kickert, W.J:	Decision-Making: a	Systems-Theoretical	Approach.	[914]
Kohonen, T. :	sociative Memory: a	Systems-Theoretical	Approach.	[964]
Rescher, N. :	gical Pragmatism: a	Systems-Theoretical	Approach to the Th	[1400]
Metzer, J.R.:	Decision	Table	Languages and Systems.	[1155]
Franova, M. :	Obtained by Beth's	Tableaux .		[542]
Gokhale, D.V:	tion in Contingency	Tables .		[629]
Presern, S. :	rating System for a	Tactile	Sensor.	[1357]
Mackenzie, K:	Leadership as a	Task	Process Uncertainty Control Proce	[1084]

Van Straaten: ond Joint MTA/IIASA	Task Force Meeting on Lake Balaton Mod	[1792]
Brown, D.J.H: A	Task-Free Concept Learning Model Emplo	[229]
Honkasalo, A: Design of Work	Tasks and Education of Labor Force.	[776]
Smith, P.M. : in Fault Diagnosis	Tasks .	[1591]
Dunn, G. : ion to Mathematical	Taxonomy .	[447]
Van Gigch, J: A	Taxonomy of Systems Science.	[1788]
Wood-Harper,: A	Taxonomy of Current Approaches to Syst	[1879]
Bozinovski, : tern Classification	Teaching .	[208]
Fleddermann,:	Teaching the Systems Approach with a G	[529]
Rudall, B.H.: eveloping Automated	Teaching Systems.	[1457]
Sonnier, I.L: Holistic Education:	Teaching of Science in the Affective D	[1599]
White, A.M. : Interdisciplinary	Teaching .	[1860]
Jumarie, G. : Some	Technical Applications of Relativistic	[861]
Findler, N.V: ulti-Level Learning	Technique Using Production Systems.	[526]
Medina, B.F.: tem Analysis: A New	Technique .	[1145]
Aris, R. : hematical Modelling	Techniques .	[52]
Atre, S. : ta Base: Structured	Techniques for Design, Performance and	[66]
Bennett, S. : ion to Multivariate	Techniques for Social and Behavioral S	[153]
Brown, D.J.H: tion an Abstraction	Techniques .	[229]
Checkland, P:	Techniques in 'Soft' Systems Practice.	[297]
Couger, J.D.: lopment/Feasibility	Techniques .	[351]
Dhillon, B.S: ng Reliability: New	Techniques and Applications.	[416]
Gane, C. : Analysis: Tools and	Techniques .	[567]
Hanson, O.J.: ining Using Systems	Techniques .	[722]
Hwang, C.L. : iability Evaluation	Techniques for Complex/Large Systems:	[804]
Kalaba, R.E.: Two Solution	Techniques for Adaptive Reinvestment:	[870]
Kleijnen, J.: actical Statistical	Techniques .	[932]
Meadows, D.L:	Techniques for Implementing Computer-B	[1143]
Otnes, R.K. : ysis. Vol. 1: Basic	Techniques .	[1273]
Robinson, D.: igraphs: Theory and	Techniques .	[1432]
Svoboda, A. : ical Circuit Design	Techniques .	[1645]
Tzafestas, S: vey of Problems and	Techniques .	[1764]
Zarri, G.P. : A Project: Using AI	Techniques in Order to Solve the Probl	[1915]
: Study, Research and	Techniques of All Design (Published Qu	[1938]
Gregory, S.A: Large	Technological Projects.	[667]
Sahal, D. : Patterns of	Technological Innovations.	[1481]
Szanto, E. : Generation of	Technological Information.	[1651]
Best, D.P. : tion of Information	Technology .	[165]
Blunden, M. : Military	Technology and Systemic Interaction.	[176]
Boroush, M.A:	Technology Assessment: Creative Future	[191]
Chen, K. : Information	Technology in Developing Countries.	[304]
Gasparski, W: Science,	Technology , and Systems.	[581]
Gasparski, W: rch and Science and	Technology .	[578]
Green, M.A. : erating Principles,	Technology , and System Application.	[659]
Gutenbaum, J: cs, Management, and	Technology .	[692]
Kilmann, R.H: Theory and the Maps	Technology .	[916]
Linstone, H.: ith Applications to	Technology and Other Decision Areas.	[1052]
Sahal, D. : ory of Evolution of	Technology .	[1479]
Trappl, R. : elligence. Science,	Technology , Military, Economy, Society	[1733]
: e, a Methodology, a	Technology (Proceedings of the 1979 No	[1933]
Elton, M.C.J: Evaluating New	Telecommunication Services.	[480]
Pearce, J.G.:	Telecommunications Switching.	[1299]
Jaburek, W.J:	Teleconferencing via Videotex.	[818]
Parks, M.S. : chotic Property and	Teleogenesis .	[1283]
Wallenmaier,: titative Concept of	Teleological Systems.	[1834]
Byerly, H. :	Teleology and Evolutionary Theory Mech	[254]
Nagel, E. :	Teleology Revisited.	[1200]
Joerres, R. : entification of the	Temperature Masking of the Circadian S	[847]
Colombano, S: Complex Systems	Tend to be Unstable. Is this Relevant	[334]
Kerman, C.E.: Creative	Tension : the Life and Thought of Kenne	[907]
Thomas, A. : Generating	Tension for Constructive Change: The U	[1685]
Trappl, R. : ucing International	Tension through Artificial Intelligenc	[1732]
Checkland, P: tems Diagram - Some	Tentative Guidelines. Part 2: Building	[297]
Garbade, K. : me Variation in the	Term Structure of Treasury Bills.	[568]

Grubbstrom, : stem Interpreted in   Terms   of the Theory of Relatively Clos [676]
Hofmann, Th.:   Qualitative   Terms   for Quantity.   [768]
Barto, A.G. : ern Reproduction in   Tessellation Structures.   [111]
Breitenecker: our after Treadmill   Test in Patients with Coarctation of t   [215]
Zimmermann, :   Testability and Measuring of Mathemati   [1927]
Agrawal, V.D: ch to Digital Fault   Testing .   [18]
Ben-Dov, Y. :   Optimal   Testing Procedures for Special Structu   [148]
Dijkstra, W.:   Testing Auxiliary Hypothesis Behind th   [421]
Sengupta, J.:   Testing and Validation Problems in Sto   [1525]
Troncale, L.:   Testing Hierarchy Models with Data Usi   [1745]
Troncale, L.: rid Systems Method: Tests for Hierarchy & Links between Is   [1747]
De Beaufrand: Text , Discourse, and Process: Towards   [386]
De Beaugrand: Text , Discourse, and Process.   [387]
Fum, D.   : to Natural Language   Text Representation and Comprehension.   [547]
Bertoni, A. : mpacting of Musical   Texts .   [163]
De Beaufrand: iplinary Science of   Texts .   [386]
            : heorie, Metaphysik,   Theologie .   [1935]
Aulin-Ahmava: ines: A Fundamental   Theorem of Actor-Systems.   [74]
Klement, E.P: s and Central Limit   Theorem for Fuzzy Random Variables.   [935]
Demillo, R.A: esses and Proofs of   Theorems and Programs.   [407]
Kwakernaak, : 1: Definitions and   Theorems . Part 2: Algorithms and Examp   [996]
Agrawal, V.D: An Information Theoretic Approach to Digital Fault Te   [18]
Brockett, P.:   Information   Theoretic Analysis of Questionnaire Da   [222]
Raghaven, V.: stics of Some Graph   Theoretic Clustering Methods.   [1377]
Scott, C.H. : s of an Information   Theoretic Model of Spacial Interaction   [1518]
Takahara, Y.:   A System   Theoretic Consideration of the World M   [1654]
Walker, C.C.:   A System   Theoretic Approach to the Management o   [1830]
Windeknecht,:   System   Theoretic Implications of Numerical Me   [1866]
Banerji, R.B: Theoretical Approaches to Non-Numerica   [103]
Farkas, J. :   Rethinking of Theoretical Presuppositions of Systems   [506]
Halfon, E. :   Theoretical Systems Ecology: Advances   [712]
Heim, R.   :   Theoretical Approaches to Complex Syst   [746]
Huber, O.   : sional Preferences: Theoretical Analysis of a Model.   [795]
Jensen, H.L.:   A   Theoretical Framework for Quantifying   [844]
Johnson, L. :   A   Theoretical Framework for Human Commun   [849]
Kornai, J. : riptive-Explanatory   Theoretical Models of the Socialist Ec   [971]
Madan, S.   : sic Concepts on the   Theoretical Treatment of Cybernetic Qu   [1087]
Nowakowska, :   New Theoretical , Methodological, and Empir   [1244]
Oldershaw, R: Empirical and Theoretical Support for Self-Similarit   [1264]
Rosen, R.   :   Progress in   Theoretical Biology, Vol.6.   [1445]
Sadovsky, V.: iple and Leibnitz's   Theoretical Heritage.   [1471]
Scandura, J.:   Theoretical Foundations of Instruction   [1501]
Wuensch, G. :   System   Theorie .   [1882]
Caselles, A.: A Method to Compare   Theories in the Light of General Syste   [268]
De Greene, K: nical Macrosystems: Theories and Models.   [390]
Geyer, R.F. :   Alienation   Theories : a General Systems Approach   [604]
Grant, D,P. : Design Methods and   Theories : Vol.15, No.2.   [656]
Grant, D,P. : Design Methods and   Theories ; Vol.15, No.3.   [657]
Hartnett, W.: ystems: Approaches,   Theories , Applications.   [731]
Johnson, L. : Cognitive Coherence   Theories .   [850]
Kickert, W.J: Fuzzy   Theories on Decision-Making. (Vol. 3 i   [911]
Kobsa, A.   : ding AI Programs as   Theories .   [956]
Nowakowska, :   Theories of Research: Modelling Approa   [1241]
Oeren, T.I. : and General Systems   Theories .   [1255]
Primas, H. : ion and Non-Boolean   Theories .   [1361]
Protzen, J.-: Design Methods and   Theories : Vol.10, No.2.   [1366]
Quine, W.V. :   Theories and Things.   [1371]
Rapoport, A.: a Link between two   Theories of Cognition.   [1383]
Slawski, C. : cial Psychological "   Theories ".   [1581]
Stegmuller, : ure and Dynamics of   Theories .   [1612]
Stegmuller, : ructuralist View of   Theories .   [1613]
Suppe, F.   : cture of Scientific   Theories (Second Ed.)   [1637]
White, D.J. : nd Effectiveness of   Theories of Decision in Practice.   [1862]
Mihram, G.A.: dology for Decision   Theorists .   [1162]

Schal, R.K.   :   of General Systems   Theorists Concerning the Process of Hu [1504]
Agazzi, E.   :   System   Theory and the Problem of Reductionism   [15]
Aho, A.V.   :   The   Theory of Joins in Relational Data Bas   [20]
Airaksinen,   :   Systems   Theory in Social Sciences.   [21]
Anderson, B.: ndations of Systems   Theory : Finitary and Infinitary Condit   [29]
Andersson, H: ol System Using the   Theory of Relatively Closed Systems.   [32]
Arbib, M.A.   : undations of System   Theory : The Hankel Matrix.   [48]
Atlan, H.   :   Information   Theory .   [65]
Aulin, A.   : ndations of Systems   Theory .   [72]
Bandler, W.   : l of a Sociological   Theory of Value Genesis.   [101]
Bandler, W.   :   On the   Theory of Lorrain-Categories.   [98]
Battista, J.: and General Systems   Theory .   [124]
Bellman, R.   : ion to Mathematical   Theory of Control Processes: Nonlinear   [143]
Beloglavec,   :   A General Systems   Theory Approach to the Properties of a   [145]
Benningfield: Circuit and Systems   Theory .   [155]
Benseler, F.: s, and Society: The   Theory of Autopoietic System in the So   [157]
Bernard, N.   : earch of Fuzzy Sets   Theory .   [162]
Blauberg, I.:   Systems   Theory : Philosophical and Methodologic [173]
Blomberg, H.:   Algebraic   Theory for Multivariable Linear System   [175]
Bluth, B.J.   :   Parson's General   Theory of Action. A Summary of the Bas   [178]
Bluth, B.J.   : ummary of the Basic   Theory .   [178]
Bose, N.K.   : imensional Systems: Theory and Applications.   [192]
Bose, N.K.   : dimensional Systems   Theory .   [193]
Boulding, K.: Ecodynamics: a New   Theory of Societal Evolution.   [200]
Brams, S.J.   :   Applied Game   Theory .   [211]
Brauer, W.   :   Net   Theory and Applications. (Proc. of the   [213]
Bridge, J.   :   Beginning Model   Theory .   [217]
Brillinger,   : : Data Analysis and   Theory .   [221]
Bruni, C.   :   Systems   Theory in Immunology.   [233]
Byerly, H.   : gy and Evolutionary   Theory Mechanisms and Meanings.   [254]
Caselles, A.: of General Systems   Theory .   [268]
Casti, J.   : pplications: Linear   Theory .   [271]
Chaitin, G.J: rithmic Information   Theory .   [291]
Collot, F.   : for General Systems   Theory .   [333]
Conant, R.C.: nsional Information   Theory .   [340]
Csanyi, V.   :   General   Theory of Evolution.   [362]
Delaney, W.   :   A System   Theory Based Simulation Language.   [404]
Dessauchoy,   : ralized Information   Theory and Decomposability of Systems.   [412]
Destouches,   : Concepts of System   Theory .   [413]
Deutsch, K.W: Toward a   Theory of Simplification: Reduction of   [414]
Devries, R.P: Systems   Theory and the Philosophy of Science.   [415]
Dimirivski,   : Systems and Control   Theory .   [422]
Dreyfus, S.E: The Art and   Theory of Dynamic Programming.   [438]
Dubois, D.   : y Sets and Systems: Theory and Applications.   [443]
Durkin, J.E.: and General System   Theory .   [449]
Ehrig, H.   :   Universal   Theory of Automata.   [461]
England J.L.:   Hierarchy   Theory and Sociology.   [482]
Eugene, J.   : ects of the General   Theory of Systems.   [495]
Fandel, G.   : ria Decision Making   Theory and Application. (Proc. of a co   [502]
Fararo, T.J.: Contribution to the   Theory of Social Systems.   [503]
Farkas, J.   : ositions of Systems   Theory .   [506]
Feintuch, A.:   System   Theory : A Hilbert Space Approach.   [516]
Fosko, P.D.   : of General Systems   Theory and Cybernetics to the Manageme   [537]
Frank, P.M.   : System Sensitivity   Theory .   [540]
Gabasov, R.   :   The Qualitative   Theory of Optimal Processes. (Translat   [551]
Gasparski, W: line of the Logical   Theory of Design.   [576]
Gasparski, W: On Systems   Theory and Systems Research and Scienc   [578]
Gatlin, L.L.:   Information   Theory and the Living Systems.   [586]
Ghiselli, E.:   Measurement   Theory for the Behavioral Sciences.   [606]
Gigley, H.   : ligence Meets Brain   Theory . An Integrated Approach to Simu [610]
Gilmore, R.   :   Catastrophe   Theory for Scientists and Engineers.   [614]
Gollmann, D.:   On the   Theory of Bilinear Shift Registers.   [634]
Golubitsky,   : tion to Catastrophe   Theory and its Applications.   [635]
Gorbatov, V.:   Theory of Partially Ordered Systems.   [644]

Gottinger, H: s of Dynamic System	Theory	.	[646]
Gottinger, H: Toward an Algebraic	Theory	of Complexity in Dynamic System	[647]
Greibach, S.:	Theory	of Program Structures: Schemes,	[668]
Grenander, U: Lectures in Pattern	Theory	. Vol.3.	[669]
Gross, D. : mentals of Queueing	Theory	.	[674]
Grubbstrom, :	Theory	of Relatively Closed Systems an	[675]
Grubbstrom, : ted in Terms of the	Theory	of Relatively Closed Systems.	[676]
Gupta, M.M. : vances in Fuzzy Set	Theory	and Applications.	[684]
Gupta, M.M. : vances in Fuzzy Set	Theory	and Applications.	[686]
Gupta, M.M. : ations of Fuzzy Set	Theory	: A Survey.	[689]
Gurel, O. : Bifurcation	Theory	and Applications in Scientific	[691]
Gutenbaum, J: ications of Systems	Theory	to Economics, Management, and T	[692]
Hajek, P. : ations of a General	Theory	.	[703]
Hall, C.A.S.: osystem Modeling in	Theory	and Practice: an Introduction w	[713]
Halmes, A. : Topics in Systems	Theory	: Publication in Honour of Prof.	[717]
Harrison, M.: to Formal Language	Theory	.	[728]
Hermann, R. : Geometry to Systems	Theory	, Part 1	[752]
Hooker, C.A.: Physical	Theory	as Logico-Operational Structure	[779]
Hopcroft, J.: duction to Automata	Theory	, Languages, and Computation.	[780]
Hyvarinen, L: Information	Theory	for Systems Engineers.	[806]
Ingemarsson,: Toward a	Theory	of Unknown Functions.	[810]
Iri, M. : Use of Matroid	Theory	in Operations Research, Circuit	[814]
Iri, M. : ircuits and Systems	Theory	.	[814]
Jezek, J. : iversal Algebra and	Theory	of Models.	[845]
Johnson, L. : Conversation	Theory	and Cognitive Coherence Theorie	[850]
Johnson, W.A: ications of Systems	Theory	to Ecology.	[852]
Jones, B. : General Systems	Theory	and Algorithm Theory.	[855]
Jones, B. : heory and Algorithm	Theory	.	[855]
Jones, D.D. : Catastrophe	Theory	Applied to Ecological Systems.	[856]
Kandel, A. : ching and Automata:	Theory	and Applications.	[874]
Keravnou, E.: ive of Conversation	Theory	Methodology.	[903]
Kilmann, R.H: s Design: Normative	Theory	and the Maps Technology.	[916]
Kindler, E. : Dynamic Systems and	Theory	of Simulation.	[919]
Kittler, J. : Pattern Recognition	Theory	and Applications.	[931]
Klement, E.P: eminar on Fuzzy Set	Theory	.	[933]
Klir, G.J. : ilistic Information	Theory	.	[950]
Krishana, V.: duction to Category	Theory	.	[985]
Krone, R.M. : nd Policy Sciences:	Theory	and Practice.	[986]
Kubinski, T.: line of the Logical	Theory	of Questions.	[991]
Kuhn, H.W. : athematical Systems	Theory	and Economics.	[994]
Laszlo, E. : lections on Systems	Theory	's Critics.	[1015]
Lee, E.T. : ations of Fuzzy Set	Theory	to Image Sciences.	[1027]
Leigh, J.R. : and Linear Control	Theory	.	[1033]
Lewin, D. :	Theory	and Design of Digital Computer	[1046]
Lilienfeld, : The Rise of Systems	Theory	: an Ideological Analysis.	[1049]
Linger, R.C.: ctured Programming:	Theory	and Practice.	[1050]
Luenberger, : to Dynamic Systems:	Theory	, Models, and Applications.	[1071]
Machtey, M. : tion to the General	Theory	of Algorithms.	[1081]
Marko, H. : Methods of System	Theory	.	[1106]
Martin, N.F.: Mathematical	Theory	of Entropy.	[1110]
Meyer, A.R. : onadic Second Order	Theory	of Successor Is Not Elementary	[1156]
Mingers, J.C: Appropriate Social	Theory	for Applied Systems Thinking: C	[1171]
Mingers, J.C: Thinking: Critical	Theory	and Soft Systems Methodology.	[1171]
Mohler, R. :	Theory	and Applications of Variable St	[1178]
Mohler, R. : of Bilinear System	Theory	and Applications.	[1180]
Monahan, G.E: Decision Processes	Theory	, Models, and Algorithms.	[1183]
Mulej, M. : Dialectical Systems	Theory	.	[1193]
Mulej, M. : Dialectical Systems	Theory	and Work Simplification.	[1194]
Naylor, A.W.: On Decomposition	Theory	: Generalized Dependence.	[1204]
Negoita, C. : ications of Systems	Theory	.	[1209]
Nitecki, Z. : Global	Theory	of Dynamical Systems.	[1233]
Nowakowska, : ilities in Decision	Theory	.	[1244]
Ogborn, J.M.: Conversation	Theory	.	[1257]
Oguntade, O.: menting a Pragmatic	Theory	of Humanistic Systems.	[1258]

Orava, P.J. : y, and State in the   Theory of Dynamical Systems and Contro [1268]
Pask, G.   : t in Conversational   Theory - Part 1   [1285]
Patel, R.V. : Multivariate System   Theory and Design.   [1289]
Patten, B.C.:   System   Theory and the Ecological Niche.   [1294]
Peery, N.S. :   General Systems   Theory Approaches to Organizations: So [1314]
Peterson, J.:   Petri Net   Theory and the Modeling of Systems.   [1327]
Petrov, B.N.: ome Problems in the   Theory of Searchless Self-Adaptive Sys [1329]
Pfeilsticker: roach and Fuzzy Set   Theory Bridging the Gap between Mathem   [1331]
Pichler, F. :   Dynamical Systems   Theory .   [1337]
Pichler, F. : athematical Systems   Theory .   [1339]
Pippenger, N: Information   Theory and the Complexity of Boolean F [1343]
Pippenger, N: Complexity   Theory .   [1344]
Pogonowski, :   to General Systems   Theory .   [1348]
Poston, T. :   Catastrophe   Theory and its Applications.   [1353]
Primas, H. :   Theory Reduction and Non-Boolean Theor [1361]
Pszczolowski: Praxiology - The   Theory with Past and Future.   [1367]
Putt, A.M. :   General Systems   Theory Applied to Nursing.   [1369]
Ralescu, D. :   Toward a General   Theory of Fuzzy Variables.   [1378]
Rastogi, P.N: l Systems (Part I):   Theory and Methodology.   [1384]
Rasvan, V. :   Some System   Theory Ideas Connected with the Stabil [1387]
Reddy, N.P. :   A Thermodynamic   Theory of Knowledge.   [1395]
Rescher, N. : cal Approach to the   Theory of Knowledge.   [1400]
Rescher, N. : roach to a Coherent   Theory of Knowledge.   [1401]
Reuver, H.A.: of General Systems   Theory .   [1413]
Roberts, F.S: Measurement   Theory with Applications to Decision-M [1425]
Robinson, D.:   Digraphs: Theory and Techniques.   [1432]
Rosen, R.   :   Similarity and the   Theory of Biological Transformations.   [1439]
Royce, J.R. : ems and Information   Theory in Multifactor Individuality Th [1453]
Royce, J.R. : actor Individuality   Theory .   [1453]
Rozenberg, G: The Mathematical   Theory of LSystems.   [1454]
Ruben, B.D. :   General Systems   Theory and Human Communication.   [1455]
Rugh, W.J. :   Nonlinear Systems   Theory : The Volterra/Wiener Approach.   [1462]
Sahal, D.   :   A   Theory of Evolution of Technology.   [1479]
Sahal, D.   :   A Unified   Theory of Self-Organization.   [1480]
Sain, M.K. : to Algebraic System   Theory .   [1482]
Salzberg, P.:   A Unified   Theory of Attraction in General System [1484]
Sanderson, J: A Relational   Theory of Computing.   [1490]
Saunders, P.: tion to Catastrophe   Theory .   [1497]
Schimanovich: An Axiomatic   Theory of Winning Mathematical Games a   [1510]
Scott, W.A. : ognitive Structure: Theory and Measurement of Individual D [1519]
Sellers, P.H: xes: A Mathematical   Theory of Algorithms.   [1522]
Siewiorek, D: The   Theory and Practice of Reliable System [1554]
Silviu, G. :   Information   Theory with Applications.   [1559]
Sinha, D.K. :   Catastrophe   Theory and Applications.   [1568]
Slawski, C. : ng Relationship: In   Theory and Practice.   [1574]
Slawski, C. : al Actualization: A   Theory of Community Living Arrangement [1579]
Smith, J.M. :   Evolution and the   Theory of Games.   [1589]
Spagovich, V: Theory of Adaptive Systems.   [1602]
Stone, L.D. :   Theory of Optimal Search.   [1624]
Strauss, A. : to Optimal Control   Theory .   [1627]
Strejc, V. :   State Space   Theory of Discrete Linear Control.   [1628]
Sussmann, H.:   Catastrophe   Theory as Applied to the Social and Bi [1641]
Swets, J.A. : om Signal Detection   Theory .   [1646]
Takahara, Y.:   A Unified   Theory of Decision Principles.   [1656]
Tao, K.M.   : hement with Systems   Theory (Part I).   [1669]
Tao, K.M.   : hement with Systems   Theory (Part II).   [1670]
Taylor, W.R.:   Using Systems   Theory to Organize Confusion.   [1674]
Tazaki, E. : ms Using Fuzzy Sets   Theory .   [1675]
Tazaki, E. : ems Using Fuzzy Set   Theory .   [1676]
Tchon, K.   :   Towards a Formal   Theory of Description.   [1678]
Tighe, T.J. :   Modern Learning   Theory : Foundations and Fundamental Is [1689]
Tilti, A.   :   Large Scale System   Theory and Applications (Proc. of a Sy [1690]
Trappl, R. :   Cybernetics -   Theory and Applications.   [1725]
Troncale, L.: tions of Clustering   Theory to Modeling Systems Emergence.   [1739]

Vagnucci, A.: neity of Biological   Time   Series.   [1780]
Wolkowski, Z: hical Perception of   Time   and Periodicity.   [1878]
Zeigler, B.P: lization of General   Time   Systems.   [1917]
Zwart, P.J. :   About   Time .   [1931]
Marshall, J.:   Control of   Time-Delay Systems.   [1108]
Horn, W.   : rm: A Comparison of   Time-Efficiency .   [784]
Retti, J.   : imulation Models of   Time-Oriented Systems.   [1408]
Van den Boga: Minimizing Wafting   Times Using Priority Classes: A Case S [1793]
Foo, N.Y.   : nd Homomorphisms of   Timevarying Systems.   [535]
Agassi, J.   : ientific Philosophy   Today : Essays in Honour of Mario Bunge [14]
Malec, J.   :   Scenarios as a   Tool   for Dynamic Scene Representation. [1096]
Toern, A.A. : n Graphs: A General   Tool   for Modeling Simulation Designs. [1695]
Tsai, W.H.   : ributed Grammar - A   Tool   for Combining Syntactic and Stati [1757]
West, B.J.   : matical Models as a   Tool   for the Social Sciences.   [1859]
Gane, C.   : l Systems Analysis: Tools and Techniques.   [567]
Gheorghe, A.: Management Planning   Tools for the Transition towards a Med [605]
Gomez, P.   :   Top-Down Versus Bottom-Up Organization [636]
Winkel, D.   :   An Introduction to Top-Down Design.   [1867]
Adam, N.R.   :   Current   Topics in Computer Simulation.   [12]
Gaines, B.R.: and Closely Related   Topics .   [555]
Halmes, A.   :   Topics in Systems Theory: Publication [717]
Hawkins, D.M: Topics in Applied Multivariate Analysi [741]
Michie, D.   : ligence and Related   Topics : An Information Scientist's Wee [1161]
Rose, J.   :   Current   Topics in Cybernetics and Systems (Pro [1437]
Simon, H.A. : Discovery and Other   Topics in the Methods of Science.   [1562]
Smale, S.   : ocesses and Related   Topics .   [1583]
Casti, J.   :   Topological Methods for Social and Beh [274]
Jaron, J.   :   Relations and Topologies in the Goal-Space of a Syst [840]
Feldman, J.A: Total Complexity and the Inference of [517]
Mergenthaler: The   Total Repair Cost in a Defective, Cohe [1152]
Pappis, C.P.: ic Controller for a   Traffic Junction.   [1280]
Athanassov, :   A System for Training of EDP-Specialists for the Na [58]
Hanson, O.J.: tems and Methods of   Training Using Systems Techniques.   [722]
Janssens, L.: Principles for the   Training of Operator Crews of Nuclear [833]
Silvern, L.C: ineering Applied to   Training .   [1556]
Uccellini, L: tion of Atmospheric   Trajectories .   [1765]
Sanderson, A: Pattern   Trajectory Analysis of Nonstationary M [1489]
Yager, R.R. :   Measuring   Tranquility and Anxiety in Decision Ma [1905]
Kalman, R.E.: rtial Realizations,   Transfer Functions, and Canonical Form [872]
El-Sherief, : able Systems in the   Transfer-Function Matrix Form.   [469]
Brigham, E.O: The Fast Fourier   Transform .   [218]
Horn, W.   : sh vs. Fast Fourier   Transform : A Comparison of Time-Effici [784]
Jachymczyk, :   Karhunen-Loewe   Transform Approximation for Certain Cl [822]
Nussbaumer, :   Fast Fourier   Transform and Convolution Algorithms. [1249]
Arigoni, A.C: ess of the Semiotic   Transformation of Information.   [51]
Matsuno, K. : nd Their Structural   Transformation .   [1121]
Schal, R.K. : he Process of Human   Transformation .   [1504]
Cornelis, A.:   Anthropological Transformations in the Dynamics of Sci [347]
Kaiser, J.S.: or of Single-Valued   Transformations within a Determinate M [868]
Rosen, R.   : heory of Biological   Transformations .   [1439]
Stoklosa, J.:   Transformations of Some Dynamical Syst [1622]
Witten, I.H.: Behaviour-Structure   Transformations Using Transitive Moore [1873]
Egiazarian, : eneralized Hadamard   Transforms and Linear Filters.   [459]
Jumarie, G. : New Approach to the   Transinformation of Random Experiments [863]
Buchberger, : /Output Codings and   Transition Functions in Effective Syst [237]
Gheorghe, A.: nning Tools for the   Transition towards a Medium Developed [605]
Pla Lopez, R: Systemic   Transition from Ideological Learning t [1345]
Venkatesan, : ternate Approach to   Transitive Coupling in ISM.   [1809]
Witten, I.H.: ansformations Using   Transitive Moore Models.   [1873]
Andrew, A.M. :   Information Transmission on Nerves.   [37]
Melkumian, A: ata Compression and   Transmission Systems.   [1151]
Von Luetzow,: ysis of Information   Transmission of Cat Retinal Horizontal [1822]
Tchon, K.   :   A '   Transversal ' Approach to Evenistic Sys [1677]
Breitenecker: ure Behaviour after   Treadmill Test in Patients with Coarct [215]

Garbade, K. : e Term Structure of Treasury Bills.   [568]
Carlsson, C.: Problems and a Way Treating it.   [263]
Glaser, F.B.: Approach to Alcohol Treatment .   [622]
Madan, S.   : on the Theoretical Treatment of Cybernetic Queueing Syste   [1087]
Morley, D.A.: ater and Wastewater Treatment .   [1187]
Shepard, R.N: imensional Scaling, Tree-fitting , and Clustering.   [1543]
Batchelor, B: Labeling Concavity Trees .   [120]
Mehlhorm, K.: timal Binary Search Trees .   [1147]
Bensoussan, :   New Trends in Systems Analysis: Internatio   [158]
Eyckhoff, P.:   Trends and Progress in Systems Identif   [498]
Eyckhoff, P.:   Trends and Progress in System Identifi   [499]
Hawkes, N. : national Seminar on Trends in Mathematical Modelling.   [739]
Hawkes, N. : national Seminar on Trends in Mathematical Modelling: a UN   [740]
Hayes, P.   :   Trends in Artificial Intelligence.   [743]
Rose, J.   : :   Modern Trends in Cybernetics and Systems (3 V [1436]
Straszak, A.:   New Trends in Mathematical Modelling.   [1626]
Yeh, R.T.   :   Current Trends in Programming Methodology: Vol   [1912]
Dubois, D. : y Measures Based on Triangular Norms.   [444]
Pedrycz, W. : onal Equations with Triangular Norms and their Resolutions [1310]
Luhman, N. :   Trust and Power.   [1072]
Yager, R.R. : tic Model and Fuzzy Truths .   [1888]
Glanville, R: are Two Black Boxes Trying to Get Out.   [620]
Duechting, W: d Simulation of 3-D Tumor Growth.   [446]
Lendaris, G.: ctural Modeling - A Tutorial Guide.   [1036]
Airenti, G. :   A Two   Level Model of Knowledge and Belie   [23]
Batchelor, B: Two   Methods for Finding Convex Hulls o   [121]
Cushing, J.M: Two   Species Competition in a Periodic   [367]
Glanville, R: White Box there are Two   Black Boxes Trying to Get Out.   [620]
Hooda, D.S. :   Two   Generalized Measures of "Useful" I   [778]
Kalaba, R.E.:   Two   Solution Techniques for Adaptive R   [870]
Uyttenhove, :   On the Two-dimensionality in the Behavioral S   [1777]
Sastry, D. :   A Two-Level Algorithm for the Simultaneo   [1495]
Dial, G.   : of Order Alpha and Type Beta and Shannon's Inequality.   [420]
Koczy, L.T. : nd Fuzzy Objects of Type N.   [959]
Yager, R.R. :   Fuzzy Subsets of Type II in Decisions.   [1896]
Bahm, A.J. : thodologies of Five Types of Philosophy.   [80]
Bahm, A.J. :   Five Types of Systems Philosophy.   [79]
Kindler, E. : II. Description of Types and Individual Typology.   [918]
Tompa, F.W. : on of Abstract Data Types .   [1698]
Jain, V.   : tion of the Ekistic Typology .   [828]
Kindler, E. : ypes and Individual Typology .   [918]
Best, D.P. : e in a Medium Sized UK Company.   [164]
Wallace, M.R: Applications of Ultrastability .   [1833]
Troncale, L.:   Some Key, Unanswered Questions about Hierarchies   [1744]
Ditleusen, O: Uncertain Modelling.   [425]
Genesio, R. : the Model Order for Uncertain System Approximation.   [590]
Gupta, M.M. : tive Controller for Uncertain Plants via Liapunov Signal S [687]
Casti, J.   : als and Irreducible Uncertainties .   [270]
Carter, C.F.:   Uncertainty and Expectations in Econom   [265]
Fedra, K.   :   Uncertainty and Arbitrariness in Ecosy   [515]
Gaines, B.R.: zzy and Probability Uncertainty Logics.   [556]
Higashi, M. :   Measures of Uncertainty and Information Based on P   [754]
Kahneman, D.:   Judgement Under Uncertainty : Heuristics and Biases.   [866]
Mackenzie, K: p as a Task Process Uncertainty Control Process.   [1084]
Takatsu, S. : l Equilibrium under Uncertainty .   [1661]
Viliums, E.R: of Decisions under Uncertainty .   [1816]
Viliums, E.R: cision Making Under Uncertainty and Multiple Objectives.   [1817]
Wildhelm, J.: cision Making Under Uncertainty .   [1864]
Yager, R.R. : ts of Possibilistic Uncertainty .   [1894]
Fletcher, R.: ptimization. Vol.1: Unconstrained Optimization; Vol.2: Con [530]
Buchberger, : : A German Language Understandig System.   [238]
Bobrow, D.G.:   Representation and Understanding : Studies in Cognitive Sc [179]
Castelfranch: pproach to Sentence Understanding .   [269]
Guida, G.   : in Natural Language Understanding Systems.   [680]

McPherson, P: On Understanding Modelling and Improving [1142]
Nevitt, B.   :   ABC of Prophecy: Understanding the Environment.   [1216]
Sugiyama, K.: Methods for Visual Understanding of Hierarchical System S [1634]
Thissen, W. : Model: Lessons for Understanding Complicated Models.   [1684]
Trost, H.   : a Natural-Language Understanding System.   [1752]
Trost, H.   : of Natural Language Understanding .   [1754]
Pelzmann, L.: tribution-Model for Unemployed Workers?   [1316]
Yager, R.R. : on Making Including Unequal Objectives.   [1887]
Hawkes, N.   : atical Modelling: a UNESCO Report.   [740]
Nugyen, V.V.:   Review and Unification of Linear Identifiability [1247]
Sahal, D.   :   A Unified Theory of Self-Organization.   [1480]
Salzberg, P.:   A Unified Theory of Attraction in Genera   [1484]
Takahara, Y.:   A Unified Theory of Decision Principles.   [1656]
Takeuchi, A.: variate Analysis: A Unified Approach by Means of Projectio [1664]
Jantch, E.   : ry Vision: Toward a Unifying Paradigm of Physical, Biologi [835]
Lefebre, V.A: etics in the Soviet Union and the West.   [1031]
Katz, M.J.   :   Questions of Uniqueness and Resolution in Reconstru [883]
Morein, R.   : of Threshold Logic Unit Networks.   [1185]
Yager, R.R. : : Membership in the Unit Interval.   [1890]
Ehrig, H.   :   Universal Theory of Automata.   [461]
Jezek, J.   :   Universal Algebra and Theory of Models [845]
Rissanen, J.:   Universal Modeling and Coding.   [1421]
Tjoa, A.M.   :   The Universal Relation as Logical View of [1692]
Toffoli, T. : on and Construction Universality of Reversible Cellular Au [1696]
Boulding, K.:   The Universe as a General System: Fourth A [199]
Jantch, E.   : The Self-Organizing Universe .   [834]
Kuhlenbeck, : Human Brain and its Universe . Vol.I: The World of Natural [993]
Kahne, S.   : ems Concepts to All University Students.   [865]
Ingemarsson,: Toward a Theory of Unknown Functions.   [810]
Everitt, B.S: Unresolved Problems in Cluster Analysi [497]
Colombano, S: Systems Tend to be Unstable . Is this Relevant to Ecology? [334]
Watt, K.E.F.:   The Unsteady State - Environmental Problem [1849]
La Patra, J.: Systems Approach to Urban Development.   [997]
Collins, L. :   Use of Models in the Social Sciences.   [332]
Eveland, S. :   The Use of the QBE Relational Database Sys [496]
Findler, N.V: Representation and Use of Knowledge by Computers.   [524]
Iri, M.   :   Use of Matroid Theory in Operations Re [814]
Levitan, G.L: Use of Deductive Inference in Situatio [1043]
Morein, R.   :   The Use of the Calculus of Variations and [1185]
Thomas, A.   : ructive Change: The Use and Abuse of Systems Models.   [1685]
Tripp, R.S. :   The Use of Cybernetics in Organizational D [1736]
Vallet, C.   :   Use of a Partially Self-Referent Forma [1784]
Wasniowski, :   The Use of Computer-Communication Systems [1844]
Bottaci, L. : ion System Programs Used in Expert Systems.   [198]
Smith, P.M. : ge about Procedures Used in Fault Diagnosis Tasks.   [1591]
Hooda, D.S. : alized Measures of " Useful " Information.   [778]
Kerefky, P. : and Manipulation on User Level.   [904]
Leeming, A.M: of the Design of a User Interface.   [1029]
Dial, G.   : on Bayes Risk with Utilities .   [419]
Hirata, H.   : cal Ecosystems with Utility Efficiency Mass and its Stabil [759]
Roberts, F.S: to Decision-Making, Utility , and the Social Sciences.   [1425]
Seo, F.   : n of Multiattribute Utility Analysis for Collective Choice [1527]
Toulotte, J.: Petri Nets and its Utilization .   [1707]
Cugno, F.   : lamy's Equalitarian Utopia : A System-Theory Approach.   [364]
Nowotny, H. : Human: Science and Utopia .   [1246]
Slawski, C. : nts and a Practical Utopia .   [1579]
Trappl, R.   : Power, Autonomy, Utopia : New Approaches Towards Complex [1731]
Gaines, B.R.: stems Research: Quo Vadis ?   [557]
Johnson, P.E: Validating Computer Simulation Models [851]
Darlington, :   Interactive Validation of Program Specifications.   [376]
Sengupta, J.:   Testing and Validation Problems in Stochastic Line [1525]
Yager, R.R. :   Validation of Fuzzy-Linguistic Models.   [1889]
Lusk, E.J.   : cussion of External Validity in Systemic Research.   [1075]
Belova, N.   :   Valuation and Control.   [146]

Bandler, W. : iological Theory of Value   Genesis.   [101]
Fogel, E.   :   On the Value   of Information in System Identif [532]
Pearl, J.   : work for Processing Value   Judgements.   [1302]
Peay, E.R.   : ls with Qualitative Values .   [1306]
Yager, R.R. :   Prototypical Values for Fuzzy Subsets.   [1900]
Bandler, W. :   The Valuton : An Automation Model of a Soci [101]
Kauffman, L.: twork Synthesis and Varela 's Calculus.   [885]
Kohout, L.J.: e Spencer-Brown and Varela Calculi.   [965]
Schwartz, D.: 's Laws of Form and Varela 's Calculus for Self-Reference.   [1515]
Conrad, M.   : The Significance of Variability from Molecule to Ecosystem [343]
Jarisch, W. : of Fetal Heart Rate Variability with Applications to Clini [837]
Conant, R.C.: Entropy of a Binary Variable : Satisfying a Reliability Cri [336]
Mohler, R.   : and Applications of Variable   Structure Systems.   [1178]
Mohler, R.   : ent Developments in Variable   Structure Systems, Economics [1179]
Utkin, V.I. :   Variable   Structure Systems with Slidin [1774]
Yager, R.R. :   A Linguistic Variable for Importance of Fuzzy Sets. [1892]
Conant, R.C.: lty Observations of Variables and Relations.   [337]
Ihara, J.   : for Selecting Model Variables .   [808]
Klement, E.P: em for Fuzzy Random Variables .   [935]
Kwakernaak, :   Fuzzy Random Variables . Part 1: Definitions and The [996]
Nahmias, S. :   Fuzzy Variables .   [1201]
Pinske, M.S.: Stability of Random Variables and Processes.   [1341]
Ralescu, D. : ral Theory of Fuzzy Variables .   [1378]
Wang, Z.   : Numbers and Random Variables .   [1838]
Andersson, H: Inventory Control Sy Variables g the Theory of Relatively Cl [32]
Young, T.Y. : ication with Binary Variables .   [1914]
Cerbone, G. :   Mean Variance   and Skewness of the First Pas [290]
Garbade, K. :   Time Variation in the Term Structure of Tre [568]
Dreyfus, S.E: and the Calculus of Variations .   [437]
Morein, R.   : of the Calculus of Variations and Functionals in the Anal [1185]
Ben-Eli, M.U: ying Regulation and Variety   Increase in Evolving Systems.   [149]
Engell, S.   :   Variety , Information and Feedback.   [481]
Paritsis, N.: es in Environmental Variety .   [1282]
Kaiser, J.S.: ational Behavior: A Vector   of Single-Valued Transformation [868]
Zhukovin, V.: ecision Making with Vector   Fuzzy Preference Realtion.   [1925]
Greibach, S.: Schemes, Semantics, Verification .   [668]
Bensoussan, : national Symposium, Versailles 1976.   [158]
Elias, D.   : ility Analysis: APL Version .   [474]
Van Straaten: e Balaton Modelling Veszprem (Hungary).   [1792]
Jaburek, W.J: eleconferencing via Videotex .   [818]
Maurer, H.   : ustrian Approach to Videotex .   [1129]
Buchberger, :   VIE-LANG : A German Language Understand [238]
Trost, H.   :   VIE-LANG : A German Language Dialogue S [1753]
Banathy, B.H: eveloping a Systems View   of Education.   [95]
Economos, A.: oughts on a Systems View   of Life.   [454]
Erickson, S.:   A Systems View   of the Foundations of Education.   [485]
Fox, M.S.   :   An Organizational View   of Distributed Systems.   [539]
Franta, W.R.:   Process View   of Simulation.   [543]
Krallmann, H: aradigm: Historical View   from 1973 until 1983.   [980]
Rifkin, J.   : ntropy: a New World View .   [1419]
Stegmuller, :   The Structuralist View   of Theories.   [1613]
Tjoa, A.M.   : Relation as Logical View   of Databases.   [1692]
Von Bertalan: Systems View   of Man (edited by Paul La Violett [1820]
Wenger, M.S:: hts on a Cybernetic View   of Social Organization.   [1858]
Steinberg, J: shopper Chortophaga Viridifaciata .   [1618]
Ballard, D.H: Computer Vision .   [91]
Jantch, E.   :   The Evolutionary Vision : Toward a Unifying Paradigm of   [835]
Koenderink, : lications for Robot Vision .   [962]
Jarisch, W. : Signal Analysis of Visual   Evoked Potentials.   [838]
Kaufmann, G.:   Visual   Imagery and Its Relation to Pro [887]
Spoehr, K.T.:   Visual   Information Processing.   [1605]
Sugiyama, K.:   Methods for Visual   Understanding of Hierarchical S [1634]
Sugiyama, K.:   Visual   Q-Analysis (I): A Case of Futur [1635]
Toda, M.   :   Visual   Q-Analysis (II): Strategy Analy [1694]

BIBLIOGRAPHICAL LIST

[1] Aarnes, E.: On the Problem of Identification in Compartment Analysis. Modeling, Identification and Control, 1, No.2, 1980, pp.93-103.

[2] Abe, J.: Schema Representation of 'Speakers-World' in Natural Language Processing. In: Trappl, R.(ed.): Cybernetics and Systems Research. North-Holland, Amsterdam, and New York, 1982, pp.885-890.

[3] Abraham, R., Shaw, C.: Dynamics: The Geometry of Behavior. Aerial Press, Santa Cruz, Calif., 1982.

[4] Acalugaritei, G.: Hierarchies of Evolutions and Stabilities. Cybernetica, 20, No.2, 1977, pp.147-166.

[5] Acar, W., Aupperle, K.E.: Bureaucracy as Organizational Pathology. Systems Research, 1, No.3, 1984, pp.157-166.

[6] Achinstein, P.: Can There Be a Model of Explanation? Theory and Decision, 13, No.3, Sept.1981, pp.275-292.

[7] Ackerman, E., Gatewood, L.C.: Mathematical Models in the Health Sciences: a Computer Approach. Univ. of Minnesota Press, Minneapolis, 1979, XIV+358 pp.

[8] Ackoff, R.L.: Resurrecting the Future of Operational Research. J. of the Operations Research Society, 30, No.3, 1979, pp.189-199.

[9] Ackoff, R.L.: The Future of Operations Research is Past. J. of the Operations Research Society, 30, No.2, 1979, pp.93-104 (also General Systems Yearbook, 24, 1979, pp.241-252).

[10] Ackoff, R.L.: The Art and Science of Mess Management. Interfaces, 11, No.1, 1981, pp.20-26.

[11] Adam, A., Beran, H.: Discrete Systems Methodology. In: Trappl, R.(ed.): Cybernetics and Systems Research. North-Holland, Amsterdam, and New York, 1982, pp.141-146.

[12] Adam, N.R., Dogramaci, A.(eds.): Current Topics in Computer Simulation. Academic Press, London, and New York, 1979, XX+292 pp.

[13] Adams, W.M.: Hiragana and Katakana Syllabaries: Relative Information Content. Cybernetics and Systems, 11, No.1-2, 1980, pp.131-141.

[14] Agassi, J., Cohen, R.S.(eds.): Scientific Philosophy Today: Essays in Honour of Mario Bunge. D.Reidel, Dordrecht, Holland, and Boston, 1981, 475pp.

[15] Agazzi, E.: System Theory and the Problem of Reductionism. Erkenntnis, 12, No.3, 1978, pp.339-358.

[16] Aggarwal, J.K., Vidyasagar, M.(eds.): Nonlinear Systems: Stability Analysis. Dowden, Hutchinson and Ross, Stroudsburg, Penn., 1977.

[17] Aggarwal, S.: Deterministic Representation of Probabilistic Systems by Ergodic Machines. Mathematical Systems Theory, 10, No.4, 1977, pp.345-361.

[18] Agrawal, V.D.: An Information Theoretic Approach to Digital Fault Testing. IEEE Trans. on Computers, C-30, No.8, August 1981, pp.582-587.

[19] Agusti, J., Villanueva, J.J., Aguilo, J.: Modularity Grade of the Neural Networks Synthesized with Z-Fivex Formal Neurones. In: Trappl, R.(ed.): Cybernetics and Systems Research. North-Holland, Amsterdam, and New York, 1982, pp.333-338.

[20] Aho, A.V., Beeri, C., Ullman, J.D.: The Theory of Joins in Relational Data Bases. ACM Trans. on Database Systems, 4, No.3, 1979, pp.297-314.

[21] Airaksinen, T.: Systems Theory in Social Sciences. In: Trappl, R.(ed.): Cybernetics and Systems Research. North-Holland, Amsterdam, and New York, 1982, pp.489-494.

[22] Airaksinen, T.: Normative Aspects of Some Stabilizing Social Systems. In: Trappl, R.(ed.): Cybernetics and Systems Research, Vol.II. North-Holland, Amsterdam, and

New York, 1984, pp.465-470.

[23] Airenti, G., Bara, B.G., Colombetti, M.: A Two Level Model of Knowledge and Belief. In: Trappl, R.(ed.): Cybernetics and Systems Research. North-Holland, Amsterdam, and New York, 1982, pp.881-884.

[24] Allen, T.F.H., Starr, T.S.: Hierarchy: Perspectives for Ecological Complexity. Univ. of Chicago Press, Chicago, 1982, XVI+310pp.

[25] Alspach, B., Hell, P., Muller, D.J.(eds.): Algorithmic Aspects of Combinatorics. North-Holland, Amsterdam, and New York, 1978, VI+245 pp.

[26] Alter, S.L.: Decision Support Systems: Current Practice and Continuing Challenges. Addison-Wesley, Reading, Mass., 1980, XVI+316 pp.

[27] Altmann, A.: Interdisciplinary System Analysis. Springer-Verlag, Berlin, FRG, and New York, 1982, XV+294pp. [German]

[28] Andersen, E.B.: Discrete Statistical Models with Social Science Applications. North-Holland, Amsterdam, and New York, 1980, XIV+384 pp.

[29] Anderson, B.D.O., Arbib, M.A., Manes, E.G.: Foundations of Systems Theory: Finitary and Infinitary Conditions. Springer-Verlag, Berlin, FRG, and New York, 1976, VII+93 pp.

[30] Anderson, R.B.: Proving Programs Correct. John Wiley, Chichester and New York, 1979, VIII+184 pp.

[31] Anderson, T.W., (et al.): A Bibliography of Multivariate Statistical Analysis. John Wiley, Chichester and New York, 1972.

[32] Andersson, H.: Analysis of an Inventory Control System Using the Theory of Relatively Closed Systems. Kybernetes, 7, No.4, 1978, pp.291-296.

[33] Anderton, R.: Systems in the Seventies: an Emerging Discipline? J. of Applied Systems Analysis, 5, No.2, May 1978, pp.149-156.

[34] Andreae, J.H., Dowd, R.B., Webb, O.J.: A Dual Model of the Brain. Dept. El.Eng., Univ. of Canterbury, Christchurch, New Zealand, 1978, 36pp. [IFSR-Depository]

[35] Andrew, A.M.: Autopoiesis and Self-Organization. J. of Cybernetics, 9, No.4, 1979, pp.359-367.

[36] Andrew, A.M.: Logic and Continuity - a Systems Dichotomy. In: Trappl, R.(ed.): Cybernetics and Systems Research. North-Holland, Amsterdam, and New York, 1982, pp.19-22.

[37] Andrew, A.M.: Information Transmission on Nerves. In: Trappl, R.(ed.): Cybernetics and Systems Research, Vol.II. North-Holland, Amsterdam, and New York, 1984, pp.257-260.

[38] Andrews, J.G., McLone, R.R.(eds.): Mathematical Modelling. Butterworths, London, 1976, 260 pp.

[39] Aplin, J.C., Schoderbek, P.P.: A Cybernetic Model of the MBO Process. J. of Cybernetics, 10, No.1-3, 1980, pp.19-28.

[40] Apostolakis, G., Garribba, S., Volta, G.(eds.): Synthesis and Analysis Methods for Safety and Reliability Studies (NATO Advanced Study Institute). Plenum Press, New York and London, 1980, IX+463 pp.

[41] Apostolico, A.: On the Role of Information and Hierarchy in the Modeling of Biomolecular Systems. J. of Cybernetics, 8, No.3-4, 1978, pp.223-236.

[42] Apple, A.: Megasynthesis. Univ. of Alberta, Edmonton, Canada, 1980, 511pp.

[43] Apter, M.J.: On the Concept of Bistability. Int. J. of General Systems, 6, No.4, 1981, pp.225-232.

[44] Aracil, J.: Structural Stability of Low-Order System

Dynamics Models. Int. J. of Systems Science, 12, No.4,
April 1981, pp.423-441.

[45] Arakelian, A., Agaian, S.: On an Algorithm of Spectral
Analysis. In: Trappl, R.(ed.): Cybernetics and Systems
Research, Vol.II. North-Holland, Amsterdam, and New York,
1984, pp.273-276.

[46] Arbel, A., Tse, E.: Aggregation and Information
Structuring in Large-Scale Dynamic Systems. IEEE Trans.
on Systems, Man, and Cybernetics, SMC-10, No.11, 1980,
pp.723-729.

[47] Arbib, M.A.: Computers and the Cybernetic Society.
Academic Press, London, and New York, 1977.

[48] Arbib, M.A., Manes, E.G.: Foundations of System Theory:
The Hankel Matrix. J. of Computer and Systems Sciences,
20, No.3, June 1980, pp.330-378.

[49] Arbib, M.A.: Cooperative Computation and the Cybernetic
Society. In: Trappl, R.(ed.): Cybernetics - Theory and
Applications. Hemisphere, Washington, D.C., 1983,
pp.361-372.

[50] Arigoni, A.C.: Algebraic Structure of Property Spaces -
On the Synthesis of Formal Properties. In: Trappl,
R.(ed.): Cybernetics and Systems Research.
North-Holland, Amsterdam, and New York, 1982, pp.705-710.

[51] Arigoni, A.C.: Correctness of the Semiotic Transformation
of Information. In: Trappl, R.(ed.): Cybernetics and
Systems Research, Vol.II. North-Holland, Amsterdam, and
New York, 1984, pp.15-20.

[52] Aris, R.: Mathematical Modelling Techniques. Pitman,
London and Boston, 1978, 191 pp.

[53] Arora, P.N., Chowdhary, S.: Shannon's Entropy and
Cyclicity. Cybernetics and Systems, 13, No.4, 1982,
pp.345-356.

[54] Arora, P.N., Chowdhary, S.: Directed Divergence and
Cyclic Symmetry. Cybernetics and Systems, 15, No.1-2,
1984, pp.127-144.

[55] Ashby, R., Conant, R.C.(eds.): Mechanisms of
Intelligence: Ross Ashby's Writings on Cybernetics.
Intersystems, Seaside, Ca., 1981.

[56] Astola, J.: The Lee-Scheme and Bounds for Lee-Codes.
Cybernetics and Systems, 13, No.4, 1982, pp.331-343.

[57] Atanassov, K., Stoeva, S.: Intuitionstic L-Fuzzy Sets.
In: Trappl, R.(ed.): Cybernetics and Systems Research,
Vol.II. North-Holland, Amsterdam, and New York, 1984,
pp.539-540.

[58] Athanassov, A., Bekjarov, E.: A System for Training of
EDP-Specialists for the National Economy. In: Trappl,
R.(ed.): Cybernetics and Systems Research, Vol.II.
North-Holland, Amsterdam, and New York, 1984, pp.381-386.

[59] Atherton, D.P.: Stability of Nonlinear Systems. Research
Studies Press (A division of John Wiley), New York, 1981,
XII+231 pp.

[60] Atherton, D.R.: Nonlinear Control Engineering. Van
Nostrand Reinhold, New York, 1982, 470pp.

[61] Atkin, R., Casti, J.: Polyhedral Dynamics and the
Geometry of Systems. IIASA, Laxenburg, Austria, 1977,
IV+36 pp.

[62] Atkin, R.: Multidimensional Man. Penguin Books, London,
1981, 199 pp.

[63] Atkin, R.H.: Time as a Pattern on a Multi-Dimensional
Structure. J. of Social and Biological Structures, 1,
No.3, 1978.

[64] Atlan, H., Fogelman-Soulie, F., Salomon, J., Weisbuch, G.:
Random Boolean Networks. Cybernetics and Systems, 12,
No.1-2, 1981, pp.103-121.

[65] Atlan, H.: Information Theory. In: Trappl, R.(ed.): Cybernetics - Theory and Applications. Hemisphere, Washington, D.C., 1983, pp.9-42.

[66] Atre, S.: Data Base: Structured Techniques for Design, Performance and Management. John Wiley, Chichester and New York, 1980, XVI+442 pp.

[67] Auger, P.: Coupling between N Levels of Observation of a System (Biological or Physical) Resulting in Creation of Structures. Int. J. of General Systems, 6, No.2, 1980, pp.83-100.

[68] Auger, P.: Interactions Between Collective and Individual Levels of Organization in a Hierarchically Organized System. Laboratoire de Biomathematiques, Faculte de Med. d'Angers, Angers, France, 1982, 18pp. [IFSR-Depository]

[69] Auger, P.: Order, Disorder in Hierarchically Organized Systems. Int. J. of General Systems, 8, No.2, 1982, pp.109-114.

[70] Auguin, M., Boeri, F., Andre, C.: Systematic Method of Realization of Interpreted Petri Nets. Digital Processes, 8, No.1, 1980, pp.55-68.

[71] Augustine, N.R.: Augustine's Laws and Major System Development Programs. Concepts, 5, No.1, 1982.

[72] Aulin, A.: On the Foundations of Systems Theory. Univ. of Tampere, Finland, 1982, 37pp. [IFSR-Depository]

[73] Aulin, A.: The Cybernetic Laws of Social Progress: Towards a Critical Social Philosophy and a Criticism of Marixsm. Pergamon Press, Oxford and New York, 1982, VII+218pp.

[74] Aulin-Ahmavaara, A.: The Impossibility of Genuinly Self-Steering Machines: A Fundamental Theorem of Actor-Systems. Kybernetes, 10, No.2, 1981, pp.113-121.

[75] Averkin, A.N., (et al.): Generalized Strategies in Problem Solving. Engineering Cybernetics, 16, No.5, 1978, pp.55-63.

[76] Baase, S.: Computer Algorithms. Addison-Wesley, Reading, Mass., 1978, 286 pp.

[77] Baase, S.: Computer Algorithms: Introduction to Design and Analysis. Addison-Wesley, Reading, Mass., 1978, 384 pp.

[78] Babloyantz, A., Kaczmarek, L.K.: Self-Organization in Biological Systems with Multiple Cellular Contacts. Bulletin of Mathematical Biology, 41, No.2, 1977, pp.193-201.

[79] Bahm, A.J.: Five Types of Systems Philosophy. Int. J. of General Systems, 6, No.4, 1981, pp.233-238.

[80] Bahm, A.J.: Methodologies of Five Types of Philosophy. Systems Trends, 4, No.3, March 1982, pp.8-12.

[81] Bahm, A.J.: Holons: Three Conceptions. Systems Research, 1, No.2, 1984, pp.145-150.

[82] Bailey, K.D.: Equilibrium, Entropy and Homeostasis: A Multidisciplinary Legacy. Systems Research, 1, No.1, 1984, pp.25-44.

[83] Bailey, N.T.J.: Some Issues in the Cybernetics of Disease Control. In: Trappl, R.(ed.): Cybernetics and Systems Research, Vol.II. North-Holland, Amsterdam, and New York, 1984, pp.499-504.

[84] Baldwin, J.F., Guild, N.C.F.: Modelling Controllers Using Fuzzy Relations. Kybernetes, 9, No.3, 1980, pp.223-229.

[85] Baldwin, J.F., Pilsworth, B.W.: Dynamic Programming for Fuzzy Systems with Fuzzy Environment. J. of Mathematical Analysis and Applications, 85, No.1, 1982, pp.1-23.

[86] Balkus, K.(ed.): Systems in Sociocultural Development. Proc. of SGSR 1978 Southeastern Meeting, Tallahassee, Florida, 1978.

[87] Balkus, K.:   Organizational   Determinants   of   National
     Development.   In: Trappl,   R.(ed.):   Cybernetics   and
     Systems Research.   North-Holland, Amsterdam, and New York,
     1982, pp.477-484.
[88] Balkus, K.:   Necessities   as   the   Basis   of   the   Social
     Self-Organization.   In: Trappl, R.(ed.):   Cybernetics and
     Systems Research, Vol.II.   North-Holland, Amsterdam, and
     New York, 1984, pp.359-366.
[89] Ball, M.O.:   Complexity   of   Network   Reliability
     Computations.   Networks, 10, 1980, pp.153-165.
[90] Ballance, R.A., Meyer, J.F.:   Functional   Dependence   and
     its   Application   to   System   Evaluation.   Johns   Hopkins
     Univ., Baltimore, Proc. 1978 Conf. on Information Sciences
     and Systems, March 29-31, 1978.
[91] Ballard, D.H.,   Brown,   C.M.:   Computer   Vision.
     Prentice-Hall, Englewood Cliffs, New Jersey, 1982, 523 pp.
[92] Ballonoff, P.A.:   Mathematical   Demography   of   Social
     Systems, II.   In: Trappl, R.(ed.):   Cybernetics   and
     Systems Research.   North-Holland, Amsterdam, and New York,
     1982, pp.555-560.
[93] Balossino, N., Favella, L.F., Reineri, M.T.:   ECG   Map
     Filtering by Means   of   Spherical   Harmonics: A   Simple
     Approximation and Results. Cybernetics and Systems, 15,
     No.1-2, 1984, pp.1-40.
[94] Baltes, P.B.(ed.):   Life Span   Development   and   Behavior,
     Vol.I.   Academic Press, London, and New York, 1978.
[95] Banathy, B.H.:   Developing a Systems   View   of   Education.
     Intersystems, Seaside, Ca., 1973, VI+90 pp.
[96] Banathy, B.H.(ed.):   Systems Science and Science.   (Proc.
     24th Annual   Meeting   of   the   Society for General Systems
     Research).   SGSR, Louisville, Kentucky, 1980, 660 pp.
[97] Banathy, B.H.(ed.):   Proceedings   of   the   29th   Annual
     Meeting.   SGSR, Louisville, Kentucky, 1985.
[98] Bandler,   W.,   Johnson,   J.H.:   On   the   Theory   of
     Lorrain-Categories.   Univ.   of Essex, Colchester, England,
     1977, 119pp.   [IFSR-Depository]
[99] Bandler, W., Kohout, L.J.:   Fuzzy   Power   Sets   and   Fuzzy
     Simplification Operators.   Fuzzy   Sets   and   Systems,   4,
     No.1, July 1980, pp.13-30.
[100] Bandler, W., Kohout,   L.J.:   Semantics   of   Implication
     Operators   and   Fuzzy   Relational   Products.   Int.   J.   of
     Man-Machine Studies, 12, No.1, Jan.1980, pp.89-116.
[101] Bandler, W.,   Schmutzer,   M.E.A.:   The   Valuton: An
     Automation   Model   of   a   Sociological   Theory   of   Value
     Genesis.   Cybernetics   and   Systems,   11,   No.3,   1980,
     pp.267-292.
[102] Bandler, W., Kohout, L.J.:   The Four Modes of Inference in
     Fuzzy Expert Systems.   In: Trappl, R.(ed.):   Cybernetics
     and Systems   Research,   Vol.II.   North-Holland, Amsterdam,
     and New York, 1984, pp.581-588.
[103] Banerji,   R.B.,   Mesarovic,   M.D.(eds.):   Theoretical
     Approaches   to   Non-Numerical   Problem   Solving.
     Springer-Verlag, Berlin, FRG, and New York, 1970.
[104] Baptistella, L.F.B., Ollero, A.:   Fuzzy   Methodologies   for
     Interactive Multicriteria   Optimization.   IEEE   Trans. on
     Systems, Man, and Cybernetics, SMC-10,   No.7,   July   1980,
     pp.355-365.
[105] Barber, M.C.:   A   Markovian   Model   for   Ecosystem   Flow
     Analysis.   Ecological   Modelling,   5,   No.3,   Sept.1978,
     pp.193-206.
[106] Barenblatt,   G.I.:   Similarity,   Self-Similarity   and
     Intermediate Asymptotics.   Plenum   Press,   New   York   and
     London, 1979, XVIII+218 pp.
[107] Barfoot, C.B.:   Aggregation   of   Conditional   Absorbing

Markov Chains. In: Trappl, R.(ed.): Cybernetics and Systems Research. North-Holland, Amsterdam, and New York, 1982, pp.215-218.

[108] Barmark, J.(ed.): Perspectives in Metascience. Kugl. Vetenskapsoch Vitterhets-Samhallet, Goteborg, Sweden, 1979, 199 pp.

[109] Barr, A., Feigenbaum, E.A.(eds.): The Handbook of Artificial Intelligence (Vol.1). William Kaufmann, Los Altos, Calif., 1981, XIV+409 pp.

[110] Barr, A., Feigenbaum, E.A.(eds.): The Handbook of Artificial Intelligence (Vol.2). William Kaufmann, Los Altos, Calif., 1982, XIV + 428pp.

[111] Barto, A.G.: A Note on Pattern Reproduction in Tessellation Structures. J. of Computer and Systems Sciences, 16, No.3, June 1978, pp.445-454.

[112] Barto, A.G.: Discrete and Continuous Models. Int. J. of General Systems, 4, No.3, 1978, pp.163-178.

[113] Bartolini, G., Casalino, G., Davoli, F., Mastretta, M., Minciardi, R., Morten, M.: Development of Performance Adaptive Fuzzy Controllers with Application to Continuous Casting Plants. In: Trappl, R.(ed.): Cybernetics and Systems Research. North-Holland, Amsterdam, and New York, 1982, pp.721-728.

[114] Basta, D.J., Bower, B.T.(eds.): Analyzing Natural Systems: Analysis for Regional Residuals-Environmental Quality Management. John Hopkins University Press, Baltimore, 1982, XVIII+546 pp.

[115] Batchelder, W.H., Narens, L.: A Critical Examination of the Analysis of Dichotomous Data. Philosophy of Science, 44, No.1, March 1977, pp.113-135.

[116] Batchelor, B.G.: Experimental and Programmatic Approaches to Pattern Recognition. Kybernetes, 7, No.4, 1978, pp.269-278.

[117] Batchelor, B.G.(ed.): Pattern Recognition: Ideas in Practice. Plenum Press, New York and London, 1978.

[118] Batchelor, B.G.: The Classification of Pole Maps in the Complex Plane. J. of Cybernetics, 8, No.3-4, 1978, pp.237-252.

[119] Batchelor, B.G.: Hierarchical Shape Description Based Upon Convex Hulls of Concavities. J. of Cybernetics, 10, No.1-3, 1980, pp.205-210.

[120] Batchelor, B.G.: Shape Description for Labeling Concavity Trees. J. of Cybernetics, 110, No.1-3, 1980, pp.233-238.

[121] Batchelor, B.G.: Two Methods for Finding Convex Hulls of Planar Figures. Cybernetics and Systems, 11, No.1-2, 1980, pp.105-113.

[122] Batchelor, B.G., Marlow, B.K.: Converting Run Code to Chain Code. Cybernetics and Systems, 12, No.3, 1981, pp.237-246.

[123] Bathe, K.J., (et al.)(eds.): Formulations and Computational Algorithms in Finite Element Analysis: U.S.-German Symposium. M.I.T.Press, Cambridge, Mass., 1977, 1091 pp.

[124] Battista, J.R.: The Holistic Paradigm and General Systems Theory. General Systems Yearbook, 22, 1977, pp.73-83.

[125] Bauer, M.A.: Defining Structural Descriptions. Kybernetes, 9, No.3, 1980, pp.207-216.

[126] Bavel, Z.: Math Companion for Computer Science. Reston/Prentice-Hall, Reston, Virginia, 1982, XXI+362pp.

[127] Bayraktar, B.A., (et al.)(eds.): Education in Systems Science. Taylor & Francis, London, 1979, 369 pp.

[128] Beck, M.B.: Hard or Soft Environmental Systems? Ecological Modelling, 1!, No.4, Feb.1981, pp.233-251.

[129] Becker, P.W.: On Systems with Redundancy and their

Inherent Bounds on Achievable Reliability. Electr. Inst.,
Techn.Univ. of Denmark, Lyngby, 1978, 8pp.
[IFSR-Depository]

[130] Becker, P.W.: On Systems with Redundancy and Their
Inherent Bounds on Achievable Reliability. IEEE Trans. on
Systems, Man, and Cybernetics, SMC-11, No.5, May 1981,
pp.387-390.

[131] Becker, P.W., Hansen, E.B.: On Systems' Reliability and
Redundancy. Int. J. of General Systems, 9, No.4, 1983,
pp.261-266.

[132] Beckett, J.A.: Management: The Case of the Disappearing
Model. In: Trappl, R.(ed.): Cybernetics and Systems
Research, Vol.II. North-Holland, Amsterdam, and New York,
1984, pp.403-412.

[133] Bednarek, A.R., Cesari, L.(eds.): Dynamical Systems II.
Academic Press, London, and New York, 1982, 664 pp.

[134] Beer, S.: Brain of the Firm (second edition). John
Wiley, Chichester and New York, 1981, XIII+417 pp.

[135] Beer, S.: Death is Equifinal: Eighth Annual Ludwig Von
Bertalanffy Memorial Lecture. Behavioral Science, 26,
No.3, July 1981, pp.185-196.

[136] Beer, S.: The Heart of Enterprise. John Wiley,
Chichester and New York, 1981.

[137] Beer, S.: Introduction: Questions of Quest. In:
Trappl, R.(ed.): Cybernetics - Theory and Applications.
Hemisphere, Washington, D.C., 1983, pp.1-8.

[138] Beer, S.: Recursions of Power. In: Trappl, R.(ed.):
Power, Autonomy, Utopia: New Approaches Towards Complex
Systems. Plenum Press, New York and London, 1985.

[139] Beishon, J.: Learning about Systems. Cybernetics and
Systems, 11, No.4, 1980, pp.297-316.

[140] Bekey, G.A.: Models and Reality: Some Reflections on the
Art and Science of Simulation. Simulation, 29, No.5,
November 1977, pp.161-164.

[141] Bell, D.A.: Physical Record Clustering in Databases.
Kybernetes, 13, No.1, 1984, pp.31-38.

[142] Bell, D.E., Keeney, R.L., Raiffa, H.(eds.): Conflicting
Objectives in Decisions. John Wiley, Chichester and New
York, 1978, X+442 pp.

[143] Bellman, R.: Introduction to Mathematical Theory of
Control Processes: Nonlinear Processes. Academic Press,
London, and New York, 1971.

[144] Belnap, N.P., Steel, T.B.: The Logic of Questions and
Answers. Yale Univ. Press, New Haven, Conn., 1976.

[145] Beloglavec, E., Ribaric, M.: A General Systems Theory
Approach to the Properties of a Periodic Discrete String.
In: Trappl, R.(ed.): Cybernetics and Systems Research.
North-Holland, Amsterdam, and New York, 1982, pp.69-72.

[146] Belova, N., Belov, K.: Valuation and Control. In:
Trappl, R.(ed.): Cybernetics and Systems Research,
Vol.II. North-Holland, Amsterdam, and New York, 1984,
pp.93-96.

[147] Beltrami, E.J.: Models for Public Systems Analysis.
Academic Press, London, and New York, 1977, XV+218 pp.

[148] Ben-Dov, Y.: Optimal Testing Procedures for Special
Structures of Coherent Systems. Management Science, 27,
No.12, Dec.1981, pp.1410 - 1420.

[149] Ben-Eli, M.U.: Amplifying Regulation and Variety Increase
in Evolving Systems. J. of Cybernetics, 9, No.3, 1979,
pp.285-296.

[150] Bender, E.A.: An Introduction to Mathematical Modelling.
John Wiley, Chichester and New York, 1978, 256 pp.

[151] Benedikt, S.: Non-Repetitive Decision Making under Risk.
In: Trappl, R.(ed.): Cybernetics and Systems Research,

Vol.II. North-Holland, Amsterdam, and New York, 1984, pp.183-188.

[152] Bennett, R.J., Chorley, R.J.: Environmental Systems: Philosophy, Analysis and Control. Princeton University Press, Princeton, N.J., 1978, XII+624 pp.

[153] Bennett, S., Bowers, D.: An Introduction to Multivariate Techniques for Social and Behavioral Sciences. John Wiley, Chichester and New York, 1976.

[154] Bennett, T.J.A.: Language as a Self-Organizing System. Cybernetics and Systems, 13, No.3, 1982, pp.201-212.

[155] Benningfield, L.M.: Circuit and Systems Theory. John Wiley, Chichester and New York, 1979, 575 pp.

[156] Benoit, B.M.: Fractals: Form, Chance, and Dimension. W.H.Freeman, San Francisco, 1977.

[157] Benseler, F., Hejl, P.M., Koeck, W.K.(eds.): Autopoiesis, Communications, and Society: The Theory of Autopoietic System in the Social Sciences. Campus Verlag, Frankfurt and New York, 1980, 229 pp.

[158] Bensoussan, A., Lions, J.L.(eds.): New Trends in Systems Analysis: International Symposium, Versailles 1976. Springer-Verlag, Berlin, FRG, and New York, 1977, 766 pp.

[159] Bensoussan, A., Lions, J.L.(eds.): International Symposium on Systems Optimization and Analysis. Springer-Verlag, Berlin, FRG, and New York, 1979, VIII+332 pp.

[160] Bensoussan, A., Lions, J.L.(eds.): Analysis and Optimization of Systems. (Proc. of a Conference in Versailles, France, Dec.1980). Springer-Verlag, Berlin, FRG, and New York, 1980, XIV+1000 pp.

[161] Bergeron, R.J.: Imaginary, Fluid, and Dancing Systems. Systems Trends, 3, No.9, Sept.1981, pp.8 - 13.

[162] Bernard, N.: Multiensembles, a Contribution to the Search of Fuzzy Sets Theory. In: Trappl, R.(ed.): Cybernetics and Systems Research. North-Holland, Amsterdam, and New York, 1982, pp.715-720.

[163] Bertoni, A., Haus, G., Mauri, G., Torelli, M.: Analysis and Compacting of Musical Texts. J. of Cybernetics, 8, No.3-4, 1978, pp.257-272.

[164] Best, D.P.: A Case Study in the Implementation of a Cybernetic Organisation Structure in a Medium Sized UK Company. In: Trappl, R.(ed.): Cybernetics and Systems Research. North-Holland, Amsterdam, and New York, 1982, pp.449-462.

[165] Best, D.P.: A Cybernetic Approach to the Application of Information Technology. In: Trappl, R.(ed.): Cybernetics and Systems Research, Vol.II. North-Holland, Amsterdam, and New York, 1984, pp.334-340.

[166] Bezdek, J.C., Harris, J.D.: Fuzzy Partitions and Relations: an Axiomatic Basis for Clustering. Fuzzy Sets and Systems, 1, No.2, April, 1978, pp.111-127.

[167] Bezdek, J.C.: Patterns Recognition with Fuzzy Objective Function Algorithms. Plenum Press, New York and London, 1981, XVI+256pp.

[168] Bibbero, R.J., Stern, D.M.: Microprocessor System, Interfacing and Applications. John Wiley, Chichester and New York, 1982, XIV+195pp.

[169] Binder, D.A.: Bayesian Cluster Analysis. Biometrika, 65, No.1, April 1978, pp.31-38.

[170] Binder, Z.: Stabilization, Co-ordination and Structuralization in the Integrated Automation of Complex Systems. Int. J. of Bio-Medical Computing, 7, 1976, pp.107-117.

[171] Biswas, A.K.: Mathematical Modelling and Environmental Decision-Making. Ecological Modelling, 1, No.1, May 1975,

pp.31-65.
[172] Blaquire, A., (et al.)(eds.): Dynamical Systems and
      Microphysics. (Seminar in Udine/Italy, Sept.1979).
      Springer-Verlag, Berlin, FRG, and New York, 1980, X+412
      pp.
[173] Blauberg, I.V., Sadovsky, V.N., Yudin, E.G.: Systems
      Theory: Philosophical and Methodological Problems.
      Progress Publishers, Moscow, 1977.
[174] Blauberg, I.V., Sadovsky, V.N., Yudin, E.G.: The Systemic
      Approach: Prerequisites, Problems and Difficulties.
      General Systems Yearbook, 25, 1980, pp.1-31.
[175] Blomberg, H., Ylinen, R.: Algebraic Theory for
      Multivariable Linear Systems. Academic Press, London, and
      New York, 1983.
[176] Blunden, M.: Military Technology and Systemic
      Interaction. In: Trappl, R.(ed.): Cybernetics and
      Systems Research. North-Holland, Amsterdam, and New York,
      1982, pp.485-488.
[177] Blunden, M.: Systems and Management - The Political
      Dimension. In: Trappl, R.(ed.): Cybernetics and Systems
      Research, Vol.II. North-Holland, Amsterdam, and New York,
      1984, pp.367-370.
[178] Bluth, B.J.: Parson's General Theory of Action. A
      Summary of the Basic Theory. National Behavior Systems,
      Granada Hills, Calif., 1982, IV + 132pp.
[179] Bobrow, D.G., Collins, A.(eds.): Representation and
      Understanding: Studies in Cognitive Science. Academic
      Press, London, and New York, 1975.
[180] Boden, M.: Artificial Intelligence and Natural Man.
      Basic Books, New York, 1977, 537 pp.
[181] Bogart, D.H.: Feedback, Feedforward, and Feedwithin:
      Strategic Information in Systems. Behavioral Science, 25,
      No.4, July 1980, pp.237-249.
[182] Bogdanski, Ch.: Basic Elements of Cybernetic Physics.
      In: Trappl, R.(ed.): Cybernetics and Systems Research.
      North-Holland, Amsterdam, and New York, 1982, pp.573-578.
[183] Bogner, S.: A Cybernetic Model of Cognitive Processes.
      In: Trappl, R.(ed.): Cybernetics and Systems Research,
      Vol.II. North-Holland, Amsterdam, and New York, 1984,
      pp.671-676.
[184] Bohm, D.: Wholeness and the Implicate Order. Routledge &
      Kegan Paul, London, 1980, XV+224pp.
[185] Bolter, M., Meyer, M., Probst, B.: A Statistical Scheme
      for Structural Analysis in Marine Ecosystems. Ecological
      Modelling, 9, No.2, March 1980.
[186] Boltyanskii, V.G.: Optimal Control of Discrete Systems.
      John Wiley, Chichester and New York, 1978, IX+392 pp.
[187] Boltyanskii, V.G.: Optimal Control of Discrete Systems.
      (Translated From Russian by Ron Hardin.) John Wiley,
      Chichester and New York, 1978, 392 pp.
[188] Boncina, R., Ribaric, M.: An Open System Composed of
      Non-Active Monotonous Parts in Passive Surroundings. In:
      Trappl, R.(ed.): Cybernetics and Systems Research,
      Vol.II. North-Holland, Amsterdam, and New York, 1984,
      pp.217-222.
[189] Bonczek, R.H., Holsapple C.W., Whinston, A.B.:
      Foundations of Decision Support Systems. Academic Press,
      London, and New York, 1981, XVIII+ 394 pp.
[190] Borodin, A., Munro, I.: The Computational Complexity of
      Algebraic and Numeric Problems. Elsevier / North-Holland,
      New York, 1975.
[191] Boroush, M.A., Chen, K., Christakis, A.N.(eds.):
      Technology Assessment: Creative Futures. North-Holland,
      Amsterdam, and New York, 1980.

[192] Bose, N.K.(ed.): Multidimensional Systems: Theory and Applications. IEEE, Piscataway, N.J., 1978, VII+295 pp.

[193] Bose, N.K.: Applied Multidimensional Systems Theory. Van Nostrand Reinhold, New York, 1981, 350 pp.

[194] Bosserman, R.W.: The Role of Mathematics in Systems Science. Behavioral Science, 26, No.4, October 1981, pp.388-393.

[195] Bosserman, R.W.: Internal Security Processes and Subsystems in Living Systems. In: Trappl, R.(ed.): Cybernetics and Systems Research. North-Holland, Amsterdam, and New York, 1982, pp.113-120.

[196] Bosserman, R.W.: Internal Security Processes in General Living Systems. Systems Science Institute, Univ. of Louisville, KY, 1982, 25pp. [IFSR-Depository]

[197] Botez, M.C.: Toward a Systemic Representation of General Systems: Open Multimodelling. In: Rose, J., Bilcio, C.(eds.): Modern Trends in Cybernetics and Systems (3 Vols.). Springer-Verlag, Berlin, FRG, and New York, 1977.

[198] Bottaci, L.: A Relational Scheme for the Abstract Specification of Production System Programs Used in Expert Systems. In: Trappl, R.(ed.): Cybernetics and Systems Research, Vol.II. North-Holland, Amsterdam, and New York, 1984, pp.799-804.

[199] Boulding, K.E.: The Universe as a General System: Fourth Annual Ludwig von Bertalanffy Memorial Lecture. Behavioral Science, 22, No.4, July 1977, pp.299-306.

[200] Boulding, K.E.: Ecodynamics: a New Theory of Societal Evolution. Sage, Beverly Hills, Ca., 1978.

[201] Boulding, K.E.: Human Knowledge as a Special System. Behavioral Science, 26, No.2, April 1981, pp.93-102.

[202] Bowen, B.D., Weisberg, H.F.: An Introduction to Data Analysis. W.H.Freeman, San Francisco, 1980, XVI+214 pp.

[203] Bowler, T.D.: General Systems Thinking: Its Scope and Applicability. North-Holland, Amsterdam, and New York, 1981, XII+234 pp.

[204] Boyce, W.E.(ed.): Case Studies in Mathematical Modeling. Pitman, London and Boston, 1981, XIII+386 pp.

[205] Boyd, G.M.: Cybernetic Aesthetics: Key Questions in the Design of Mutual Control in Education. In: Trappl, R.(ed.): Cybernetics and Systems Research, Vol.II. North-Holland, Amsterdam, and New York, 1984, pp.677-682.

[206] Bozic, S.M.: Digital and Kalman Filtering. John Wiley, Chichester and New York, 1979, VII+157 pp.

[207] Bozinovski, S.: A Self-Learning System Using Secondary Reinforcement. In: Trappl, R.(ed.): Cybernetics and Systems Research. North-Holland, Amsterdam, and New York, 1982, pp.397-402.

[208] Bozinovski, S.: A Representation of Pattern Classification Teaching. In: Trappl, R.(ed.): Cybernetics and Systems Research, Vol.II. North-Holland, Amsterdam, and New York, 1984, pp.775-780.

[209] Braae, M., Rutherford, D.A.: Selection of Parameters for a Fuzzy Logic Controller. Fuzzy Sets and Systems, 2, No.3, July 1979, pp.185-199.

[210] Bradley, R., (et al.): Case Studies in Mathematical Modelling: A Course Book for Engineers and Scientists. John Wiley, Chichester and New York, 1981, 256pp.

[211] Brams, S.J., Schotter, A., Schwodiauer, G.(eds.): Applied Game Theory. Physica-Verlag, Wuerzburg, Germany, 1979, 447 pp.

[212] Branin, F.H., Husseyin, K.(eds.): Problem Analysis in Science and Engineering. Academic Press, London, and New York, 1977, 514 pp.

[213] Brauer, W.(ed.): Net Theory and Applications. (Proc. of

the Advanced Course on General Net Theory of Processes and
Systems, Hamburg, 1979). Springer-Verlag, Berlin, FRG,
and New York, 1980.

[214] Brayton, R.K., Tong, C.H.: Stability of Dynamical
Systems: a Constructive Approach. IEEE Trans. on
Circuits and Systems, CAS-26, No.4, 1979.

[215] Breitenecker, F., Kaliman, J.: Simulation and Analysis of
Pathological Blood Pressure Behaviour after Treadmill Test
in Patients with Coarctation of the Aorta. In: Trappl,
R.(ed.): Cybernetics and Systems Research, Vol.II.
North-Holland, Amsterdam, and New York, 1984, pp.305-310.

[216] Brewer, J.N.: Structure and Cause and Effect Relations in
Social System Simulations. IEEE Trans. on Systems, Man,
and Cybernetics, SMC-7, No.6, June 1977, pp.468-474.

[217] Bridge, J.: Beginning Model Theory. Clarendon Press,
Oxford, 1977, 143 pp.

[218] Brigham, E.O.: The Fast Fourier Transform.
Prentice-Hall, Englewood Cliffs, New Jersey, 1974, XII+252
pp.

[219] Brightman, H.J.: Problem Solving: A Logical and Creative
Approach. Georgia State Univ., Atlanta, 1981.

[220] Brightman, H.L.: Problem Solving: A Logical and Creative
Approach. Coll. of Business Admin., Georgia State Univ.,
Atlanta, 1980, IX+242 pp.

[221] Brillinger, D.R.: Time Series: Data Analysis and Theory.
Holden-Day, San Francisco, 1980, XII+450pp.

[222] Brockett, P.L., Haaland, P.D., Levine, A.: Information
Theoretic Analysis of Questionnaire Data. IEEE Trans. on
Information Theory, IT-27, No.4, July 1981, pp.438-446.

[223] Broekstra, G.: Constraint Analysis and Structure
Identification II. Annals of Systems Research, 6, 1977,
pp.1-20.

[224] Broekstra, G.: On the Representation and Identification
of Structure Systems. Int. J. of Systems Science, 9,
No.11, Nov.1978, pp.1271-1293.

[225] Broekstra, G.: On the Foundations of GIT (General
Information Theory). Cybernetics and Systems, 11, No.1-2,
1980, pp.143-165.

[226] Broekstra, G.: C-Analysis of C-Structures:
Representation and Evaluation of Reconstruction Hypotheses
by Information Measures. Int. J. of General Systems, 7,
No.1, 1981, pp.33-62.

[227] Broekstra, G.: MAMA: Management by Matching. A
Consistency Model for Organizational Assessment and
Change. In: Trappl, R.(ed.): Cybernetics and Systems
Research, Vol.II. North-Holland, Amsterdam, and New York,
1984, pp.413-422.

[228] Brown, D.B.: Systems Analysis and Design for Safety.
Prentice-Hall, Englewood Cliffs, New Jersey, 1976, XIV+399
pp.

[229] Brown, D.J.H.: A Task-Free Concept Learning Model
Employing Generalization an Abstraction Techniques. J. of
Cybernetics, 9, No.4, 1979, pp.315-358.

[230] Brown, J.H.U.: Integration and Coordination of Metabolic
Processes: a Systems Approach to Endocrinology. Van
Nostrand Reinhold, New York, 1978, 248 pp.

[231] Bruckmann, G.(ed.): Input-Output Approaches in Global
Modelling. (Proc.of the Fifth IIASA Symposium on Global
Modelling). Pergamon Press, Oxford and New York, 1980,
IX+518 pp.

[232] Brugger, K., Hobitz, H.: A Simple Mathematical Model of
Biphasic Insulin Secretion. In: Trappl, R.(ed.):
Cybernetics and Systems Research. North-Holland,
Amsterdam, and New York, 1982, pp.299-304.

[233] Bruni, C., (et al.)(eds.): Systems Theory in Immunology. Springer-Verlag, Berlin, FRG, and New York, 1979, XI+273 pp. (Proc. of a conference in Rome,May,1978).

[234] Bryant, D.T., Niehaus, R.J.(eds.): Manpower Planning and Organization Design. Plenum Press, New York and London, 1978, 789 pp.

[235] Bubnicki, Z.: Global and Local Identification of Complex Systems with Cascade Structure. Systems Science, 1, No.1, 1975, pp.55-65.

[236] Bubnicki, Z.(ed.): Systems Science. (A Journal Published Quarterly.) Technical Univ. of Wroclaw, Poland, Wroclaw, Janiszewskiego 11/17..

[237] Buchberger, B., Roider, B.: Input/Output Codings and Transition Functions in Effective Systems. Int. J. of General Systems, 4, No.3, 1978, pp.201-210.

[238] Buchberger, E., Steinacker, I., Trappl, R., Trost, H., Leinfellner, E.: VIE-LANG: A German Language Understandig System. In: Trappl, R.(ed.): Cybernetics and Systems Research. North-Holland, Amsterdam, and New York, 1982, pp.851-856.

[239] Bunge, M.: General Systems and Holism. General Systems Yearbook, 22, 1977, pp.87-90.

[240] Bunge, M.: A Systems Concept of Society: Beyond Individualism and Holism. Theory and Decision, No.10, 1979, pp.13-20 (also General Systems Yearbook, 24, 1979, pp.27-44).

[241] Bunge, M.: Analogy between Systems. Int. J. of General Systems, 7, No.4, 1981, pp.221-224.

[242] Bunn, D.W., Thomas, H.(eds.): Formal Methods in Policy Analysis. Birkhaeuser Verlag, Basel and Stuttgart, 1978, 238 pp.

[243] Burghes, D.N., Wood, A.D.: Mathematical Models in the Social, Management and Life Sciences. Halstead, New York, 1980, 288pp.

[244] Burgin, M.S.: Products of Operators of Multidimensional Structured Models of Systems. Mathematical Social Sciences, 2, No.4, June 1982, pp.335-344.

[245] Burkhardt, H.: The Real World and Mathematics. Blackie, Glasgow, 1981, VII+190pp.

[246] Burks, A.W.: Models of Deterministic Systems. Mathematical Systems Theory, 3, 1975, pp.295-308.

[247] Burns, D.W.: The Synthesis of Forecasting Models in Decision Analysis. Birkhaeuser Verlag, Basel and Stuttgart, 1978, 238 pp.

[248] Burns, J.R., Ulgen, O.: A Sector Approach to the Formulation of Systems Dynamics. Int. J. of Systems Science, 9, No.6, 1978, pp.649-680.

[249] Burns, J.R., Ulgen, O., Beights, H.W.: An Algorithm for Converting Signed Digraphs to Forrester Schematics. IEEE Trans. on Systems, Man, and Cybernetics, SMC-9, No.3, March 1979, pp.115-124.

[250] Burns, J.R., Winstead, W.H.: An Input/Output Approach to the Structural Analysis of Digraphs. IEEE Trans. on Systems, Man, and Cybernetics, SMC-12, No.1, 1982, pp.15-24.

[251] Burton, D.J.: Methodology and Epistemology for Second-Order Cybernetics. Univ. of Calif. at Santa Cruz, 1979, 35pp. [IFSR-Depository]

[252] Busch, J.A., Busch, G.M.: Cybernetics IV: A System-Type Applicable to Human Groups. Univ. of Louisville, Louisville, Kentucky, 1981, 17pp. [IFSR-Depository]

[253] Butzer, K.W.: Civilizations: Organisms or Systems? American Scientist, 68, No.5, 1980, pp.517-523.

[254] Byerly, H.: Teleology and Evolutionary Theory Mechanisms

and Meanings.   Nature   and   Systems,   1, No.3, Sept.1979,
pp.157-176.

[255] Caianiello, E.R.,   Scarpetta,   G.,   Simoncelli,   G.:   A
Systemic Study of Monetary Systems. Int. J. of General
Systems, 8, No.2, 1982, pp.81-92.

[256] Caianiello, E.R.,   Di Giulio,   E.:   Energetics   Versus
Communication in the Nervous System. Cybernetics and
Systems, 13, No.2, 1982, pp.187-196.

[257] Callebaut, W.,   Van Bendegem, J.P.:   The   Distribution
Approach to Problem-Solving in Science: Prospects for
GSM. In: Trappl, R.(ed.):   Cybernetics   and   Systems
Research. North-Holland, Amsterdam, and New York, 1982,
pp.51-56.

[258] Camana, P.C., Hemami, H., Stockwell, C.W.:   Determination
of Feedback for Human Posture Control without Physical
Intervention. J. of Cybernetics, 7, No.3-4,   1977,
pp.199-226.

[259] Campbell, J.:   Grammatical   Man: Information, Entropy,
Language, and Life.   Simon and Schuster, New York, 1982,
320pp.

[260] Capocelli, R.M., Ricciardi, L.: A Cybernetic Approach to
Population Dynamics Modeling. J. of Cybernetics, 9, No.3,
1979, pp.297-312.

[261] Caputo, R.S.: Establishing a Rational Energy Policy for
Western Europe.   Austrian Society for Cybernetic Studies,
Vienna, 1982.

[262] Carlsson, C.: A System of Problems and How to Deal with
It. J. of Cybernetics, 10, No.4, 1980, pp.349-373.

[263] Carlsson, C.:   A System of Problems and a Way Treating it.
Prakseologia, 1980, No.2, pp.41-68. [Polish]

[264] Carlsson, C.:   An Approach to Handling Fuzzy Problem
Structures. Cybernetics and Systems, 14, No.1, 1983,
pp.33-54.

[265] Carter, C.F.,   Ford,   J.L.(eds.):   Uncertainty   and
Expectations in Economics: Essays in Honour of G.L.S.
Shackle. Basil Blackwell, Oxford, 1972.

[266] Carvallo, M.E.:   In Search of the Noetic Structure of
Systems. In: Trappl, R.(ed.): Cybernetics and Systems
Research. North-Holland, Amsterdam, and New York, 1982,
pp.31-38.

[267] Carvallo, M.E.:   Chance and the Necessity of the Third
Classical Paradigm. In: Trappl, R.(ed.): Cybernetics
and Systems Research, Vol.II.   North-Holland, Amsterdam,
and New York, 1984, pp.61-68.

[268] Caselles, A.:   A Method to Compare Theories in the Light
of General Systems Theory. In: Trappl, R.(ed.):
Cybernetics and Systems Research, Vol.II.   North-Holland,
Amsterdam, and New York, 1984, pp.27-32.

[269] Castelfranchi, C., Parisi, D., Stock, O.:   A Lexicon Based
Approach to Sentence Understanding. In: Trappl, R.(ed.):
Cybernetics and Systems Research.   North-Holland,
Amsterdam, and New York, 1982, pp.863-868.

[270] Casti, J.:   Illusion or Reality?:   the Mathematics of
Attainable Goals and Irreducible Uncertainties. IIASA,
Laxenburg, Austria, RM-76-042, 55 pp.

[271] Casti, J.:   Dynamical Systems and Their Applications:
Linear Theory. Academic Press, London, and New York,
1977, 240 pp.

[272] Casti, J.:   Connectivity, Complexity and Catastrophe in
Large-Scale Systems. John Wiley, Chichester and New York,
1979, XIII+203 pp.

[273] Casti, J., (et al.):   Lake Ecosystems: a Polyhedral
Dynamic Representation.   Ecological Modelling, 7, No.3,
Sept.1979, pp.223-237.

[274] Casti, J.: Topological Methods for Social and Behavioral Systems. Int. J. of General Systems, 8, No.4, 1982, pp.187-210.

[275] Cavallo, R.E., Islam, S.: General Systems Methodology and Large-Scale Systems. In: Oeren, T.I.(ed.): Cybernetics and Modelling and Simulation of Large Scale Systems. Int. Association for Cybernetics, Namur, 1976.

[276] Cavallo, R.E., Klir, G.J.: A Conceptual Foundation for Systems Problem Solving. Int. J. of Systems Science, 9, No.2, 1978.

[277] Cavallo, R.E.(ed.): The Role of Systems Methodology in Social Science Research. Martinus Nijhoff, Boston and The Hague, 1978.

[278] Cavallo, R.E.(ed.): Recent Developments in Systems Methodology for Social Science Research. Martinus Nijhoff, Boston and The Hague, 1979.

[279] Cavallo, R.E.(ed.): Supplementary Material to the SGSR Report on Systems Research Movement. SGSR, Louisville, Kentucky, 1979. [IFSR-Depository]

[280] Cavallo, R.E.(ed.): Sysems Research Movement: Characteristics, Accomplishments, and Current Developments. (A Report Sponsored by the Society for General Systems Research.) General Systems Bulletin, (Special Issue), IX, No.3, Summer 1979, 132 pp.

[281] Cavallo, R.E.(ed.): Systems Research Movement: Characteristics, Accomplishments and Current Developments. SGSR, Louisville, Kentucky, 1979, 132pp. [IFSR-Depository]

[282] Cavallo, R.E., Klir, G.J.: The Structure of Reconstructable Relations: a Comprehensive Study. J. of Cybernetics, 9, No.4, 1979, pp.399-413.

[283] Cavallo, R.E., Klir, G.J.: Reconstructability Analysis: Evaluation of Reconstruction Hypothesis. Int. J. of General Systems, 7, No.1, 1981, pp.7-32.

[284] Cavallo, R.E., Klir, G.J.: Reconstructability Analysis: Overview and Bibliography. Int. J. of General Systems, 7, No.1, 1981, pp.1-6.

[285] Cavallo, R.E., Klir, G.J.: Decision Making in Reconstructability Analysis. Int. J. of General Systems, 8, No.4, 1982, pp.243-256.

[286] Cavallo, R.E., Klir, G.J.: Reconstruction of Possibilistic Behavior Systems. Fuzzy Sets and Systems, 8, No.2, August 1982, pp.175-197.

[287] Cavallo, R.E.(ed.): Systems Methodology in Social Science Research: Recent Developments. Kluwer-Nijhoff, Boston and The Hague, 1982, X+194pp.

[288] Caws, P.: Coherence, System, and Structure. Idealistic Studies, 4, 1974, pp.2-17.

[289] Cellier, F.E.(ed.): Process in Modelling and Simulation. Academic Press, London, and New York, 1982, 488pp.

[290] Cerbone, G., Ricciardi, L., Sacerdote, L.: Mean Variance and Skewness of the First Passage Time for the Ornstein-Uhlenbeck Process. Cybernetics and Systems, 12, No.4, 1981, pp.395-429.

[291] Chaitin, G.J.: Algorithmic Information Theory. IBM Journal of Research and Development, 21, No.4, July 1977, pp.350-359.

[292] Chandrasekaran, B.: Natural and Social System Metaphors for Distributed Problem Solving. IEEE Trans. on Systems, Man, and Cybernetics, SMC-11, No.1, 1981, pp.1 - 5.

[293] Chang, L.C., Pradhan, D.K.: A Graph-Structural Approach for the Generalization of Data Management Systems. Information Sciences, 12, No.1, 1977, pp.1-18.

[294] Chang, Shi-Kou(ed.): Policy Analysis and Information

Systems, (A Journal Published Semi-Anually by the Univ. of
Ill. at Chicago Circle and Tankang College in Taipei,
Taiwan.) Univ. of Illinois At Chicago Circle, Dept. of
Information Engineering, Chicago, Ill., 60680, U.S.A..

[295] Chatterjee, K.: Comparison of Arbitration Procedures:
Models with Complete and Incomplete Information . IEEE
Trans. on Systems, Man, and Cybernetics, SMC-11, No.2,
1981, pp.101-109.

[296] Checkland, P.B.: The Origins and Nature of 'Hard' Systems
Thinking. J. of Applied Systems Analysis, 5, No.2, May
1978, pp.99-110.

[297] Checkland, P.B.: Techniques in 'Soft' Systems Practice.
Part 1: Systems Diagram - Some Tentative Guidelines.
Part 2: Building Conceptual Models. J. of Applied
Systems Analysis, 6, April 1979, Part 1: pp.33-40, Part
2: pp.41-49.

[298] Checkland, P.B.: The Shape of the Systems Movement. J.
of Applied Systems Analysis, 6, April 1979, pp.129- 135.

[299] Checkland, P.B.: The Systems Movement and the "Failure"
of Management Science. Cybernetics and Systems, 11, No.4,
1980, pp.317-324.

[300] Checkland, P.B.: Rethinking a Systems Approach. J. of
Applied Systems Analysis, 8, April 1981, pp.3-14.

[301] Checkland, P.B.: Systems Thinking, Systems Practice.
John Wiley, Chichester and New York, 1981, XIV+330 pp.

[302] Chen, C.N.(ed.): Pattern Recognition and Artificial
Intelligence. Academic Press, London, and New York, 1976,
621 pp.

[303] Chen, D.T.-W., Findler, N.V.: Toward Analogical Reasoning
in Problem Solving by Computers. J. of Cybernetics, 9,
No.4, 1979, pp.369-398.

[304] Chen, K.: Information Technology in Developing Countries.
Systems Research, 1, No.1, 1984, pp.83-86.

[305] Chen, K.: System Analysis in Different Social Settings.
Systems Research, 1, No.2, 1984, pp.117-126.

[306] Cheng, J.K., Huang, T.S.: A Subgraph Isomorphism
Algorithm Using Resolution. Pattern Recognition, 13,
No.5, 1981, pp.371-379.

[307] Chikan, A.: Analysis of a Model System. In: Trappl,
R.(ed.): Cybernetics and Systems Research.
North-Holland, Amsterdam, and New York, 1982, pp.535-542.

[308] Chomsky, N.: Rules and Representations. Columbia Univ.
Press, New York, 1980, VIII+300pp.

[309] Chow, G.C.: Econometric Analysis by Control Methods.
John Wiley, Chichester and New York, 1981, XVI+320 pages.

[310] Christensen, R.: Foundations of Inductive Reasoning.
Entropy, Lincoln, Mass., 1980, XII+364 pp.

[311] Christensen, R.: Entropy Minimax Sourcebook. Vol.1:
General Description; Vol.2: Philosophical Origins;
Vol.3: Computer Implementation; Vol.4: Applications.
Entropy, Lincoln, Mass., 1980/ 1981.

[312] Churchman, C.W.: The Systems Approach and its Enemies.
Basic Books, New York, 1979.

[313] Churchman, C.W.: Churchman's Conversations. Systems
Research, Vol.1, No.1, 1984, pp.3-4.

[314] Chytil, M.K.: Mathematical Methods as Cognitive
Problem-Solvers. Kybernetes, 9, No.3, 1980, pp.197-205.

[315] Ciborra, C., Migliarese, P., Romano, P.: A Methodological
Inquiry of Organizational Noise in Sociotechnical Systems.
Univ. della Calabria, Italy, 1982, 39pp.
[IFSR-Depository]

[316] Ciriani, T.A., Maione, U., Wallis, J.R.(eds.):
Mathematical Models for Surface Water Hydrology. (Proc.
of the Workshop on Mathematical Models in Hydrology, Pisa,

Italy, Dec.9-12, 1974) John Wiley, Chichester and New York, 1977, 440 pp.

[317] Ciupa, M.: A Cybernetic Approach to Packet Switched Data Communication Systems. In: Trappl, R.(ed.): Cybernetics and Systems Research. North-Holland, Amsterdam, and New York, 1982, pp.757-764.

[318] Clark, W.C., Jones, D.D., Holling, C.S.: Lessons for Ecology Policy Design: a Case Study of Ecosystem Management. Ecological Modelling, 7, No.1, June 1979, pp.1-53.

[319] Cliff, A.D., (et al.): Elements of Spatial Structure: a Quantitative Approach. Cambridge Univ. Press, Cambridge, Mass., 1975, XVII+258 pp.

[320] Cliff, A.D., Ord, J.K.: Spacial Processes: Models and Applications. Pion, London and Methuen, New York, 1981, XII+266 pp.

[321] Cliffe, M.J.: Cybernetics of Alcoholism. Cybernetics and Systems, 15, No.1-2, 1984, pp.197-220.

[322] Cliffe, M.J.: The Contribution of Cybernetics to the Science of Psychopathology. Kybernetes, 13, No.2, 1984, pp.93-98.

[323] Coblentz, A.M., Walter, J.R.(eds.): Systems Science in Health Care. Petrocelli, New York, 1977, 452 pp.

[324] Cochin, I.: Analysis and Design of Dynamic Systems. Harper and Row, New York, 1980, XVIII+796pp.

[325] Coffman, C.V., Fix, G.J.(eds.): Constructive Approaches to Mathematical Models. Academic Press, London, and New York, 1979, 458 pp.

[326] Cohen, J.: Non-Deterministic Algorithms. ACM Computing Surveys, 11, No.2, 1979, pp.79- 94.

[327] Cohen, P.R., Feigenbaum, E.A.(eds.): The Handbook of Artificial Intelligence (Vol.3). William Kaufmann, Los Altos, Calif., 1982, XVIII+640pp.

[328] Colby, J.A., McIntyre, C.D.: Mathematical Documentation for a Lotic Ecosystem Model. Oregon State Univ., Corvallis, Oregon, 1979, 54pp. [IFSR-Depository]

[329] Cole, I.D.: Naturalized Programming: A Method of Naturalizing Programming Languages. In: Trappl, R.(ed.): Cybernetics and Systems Research. North-Holland, Amsterdam, and New York, 1982, pp.745-750.

[330] Cole, I.D.: Naturalized System Development. In: Trappl, R.(ed.): Cybernetics and Systems Research, Vol.II. North-Holland, Amsterdam, and New York, 1984, pp.621-626.

[331] Collins, D.M.: Electron Density Images from Imperfect Data by Iterative Entropy Maximization. Nature, 298, 1982, pp.49-51.

[332] Collins, L.(ed.): Use of Models in the Social Sciences. Tavistock, London, 1974.

[333] Collot, F.: A New Mathematical Concept for General Systems Theory. In: Trappl, R.(ed.): Cybernetics and Systems Research. North-Holland, Amsterdam, and New York, 1982, pp.23-24.

[334] Colombano, S.P., MacElroy, R.D.: Complex Systems Tend to be Unstable. Is this Relevant to Ecology? In: Trappl, R.(ed.): Cybernetics and Systems Research. North-Holland, Amsterdam, and New York, 1982, pp.121-126.

[335] Comstock, F.L., Uyttenhove, H.J.J.: A Systems Approach to Grading of Flight Simulator Students. J. of Aircraft, Aug.1979.

[336] Conant, R.C.: Estimation of Entropy of a Binary Variable: Satisfying a Reliability Criterion. Kybernetik, 12, 1973, 1973, pp.55-57.

[337] Conant, R.C.: Faculty Observations of Variables 'and Relations. Int. J. of Systems Science, 10, No.4, 1979,

pp.471-475.
[338] Conant, R.C.:   Structural   Modelling   Using   a   Simple
      Information Measure.   Int.   J.   of   Systems   Science,   11,
      No.6, June 1980, pp.721-730.
[339] Conant, R.C.:   Detection   and   Analysis   of   Dependency
      Structures.   Int.   J.   of   General Systems,   7,   No.1,   1981,
      pp.81-92.
[340] Conant, R.C.:   Efficient   Proofs   of   Identities   in
      N-Dimensional   Information   Theory.   Cybernetica,   24,
      No,.3, 1981, pp.191-197.
[341] Conant,   R.C.:   Identifying   Relations   from   Partial
      Information.   Proc.   25th   Annual   SGSR   Meeting,   1981,
      pp.463-468.
[342] Conant, R.C.:   Set-Theoretic Structure Modelling.   Int.   J.
      of General Systems, 7, No.1, 1981.
[343] Conrad,   M.:   Adaptability   -   The   Significance   of
      Variability from Molecule to Ecosystem.   Plenum Press,   New
      York and London, 1983.
[344] Cook,   N.D.:   Stability and Flexibility: an   Analysis   of
      Natural Systems.   Pergamon   Press,   Oxford   and New York,
      1979, XI+246 pp.
[345] Cook, N.D.:   An Isomorphism of Control in Natural,   Social,
      and Cybernetic Systems.   J. of Cybernetics,   10,   No.1-3,
      1980, pp.29-40.
[346] Cordier, M.O., Rousset, M.C.:   Propagation: Another   Way
      for Matching   Patterns   in   KBS.   In: Trappl,   R.(ed.):
      Cybernetics and Systems Research, Vol.II.   North-Holland,
      Amsterdam, and New York, 1984, pp.787-792.
[347] Cornelis,   A.:   Anthropological   Transformations   in   the
      Dynamics   of   Scientific   Learning.   In: Trappl, R.(ed.):
      Cybernetics   and   Systems   Research.   North-Holland,
      Amsterdam, and New York, 1982, pp.371-378.
[348] Cornelis, A.:   Epistemological   Indicators   of   Scientific
      Identity.   In: Trappl, R.(ed.):   Cybernetics and Systems
      Research, Vol.II.   North-Holland, Amsterdam, and New York,
      1984, pp.683-690.
[349] Cosier,   R.A.,   Ruble,   T.L.,   Aplin,   J.C.:   An Evaluation   of
      the   Effectiveness   of   Dialectical   Inquiry   Systems.
      Management Science, 24, No.14, 1978, pp.1483- 1490.
[350] Cotterman, W.W.:   Systems   Analysis   and   Design: A
      Foundation   for   the 1980's.   Elsevier / North-Holland, New
      York, 1981.
[351] Couger, J.D., Colter, M.A., Knapp, R.W.:   Advanced   System
      Development/Feasibility   Techniques.   John   Wiley,
      Chichester and New York, 1982, XIII+506pp.
[352] Courtois, P.J.:   Decomposability: Queueing   and   Computer
      System Applications.   Academic   Press,   London,   and   New
      York, 1977, 201 pp.
[353] Cover, T.M., Shenhar, A.:   Compound Bayes   Predictors   for
      Sequences with   Apparent   Markov   Structure.   IEEE Trans. on
      Systems, Man, and   Cybernetics,   SMC-7   No.6,   June   1977,
      pp.421-424.
[354] Craine,   R.,   Havenner,   A.:   The   Optimal   Monetary
      Instrument: An   Empirical   Assessment.   J. of Cybernetics,
      7, No.1-2, 1977, pp.101-116.
[355] Cramer, F.:   Fundamental   Complexity: a   Concept   in
      Biological Science   and Beyond.   Interdisciplinary Science
      Reviews, 4, No.2, 1979, pp.132-139.
[356] Cremers,   A.B.,   Hibbard,   T.N.:   Orthogonality   of
      Information Structures.   Acta Informatica, 9, No.3, 1978,
      pp.243-261.
[357] Crimmins, E.T.:   The   Art   of   Abstracting.   ISI   Press,
      Philadelphia, 1982, XII + 150pp.
[358] Croskin, C.:   Ways of Knowing.   Cybernetica,   XXI,   No.3,

1978, pp.185-192.
[359] Cross, M., (et al.)(eds.): Modelling and Simulation in Practice. John Wiley, Chichester and New York, 1979, 358 pp.
[360] Csaki, P.: An Algebraic Approach to Some General Problems of Model Description. In: Trappl, R.(ed.): Cybernetics and Systems Research, Vol.II. North-Holland, Amsterdam, and New York, 1984, pp.47-52.
[361] Csanyi, V.: Considerations for the Scientific Formulation of Epistemology. L.Etoevoes Univ., Dept. of Behavior Genetics, Goed, Hungary, 1982, 22pp. [IFSR-Depository]
[362] Csanyi, V.: General Theory of Evolution. Akademiai Kiado, Budapest, 1982, 123pp.
[363] Csendes, T.: A Simulation Study on the Chemoton. Kybernetes, 13, No.2, 1984, pp.79-86.
[364] Cugno, F., Ferrero, M., Montrucchio, L.: Bellamy's Equalitarian Utopia: A System-Theory Approach. In: Trappl, R.(ed.): Cybernetics and Systems Research. North-Holland, Amsterdam, and New York, 1982, pp.521-526.
[365] Cull, P., Frank, W.: Flaws of Form. Int. J. of General Systems, 5, No.4, 1979, pp.201-212.
[366] Cumani, A.: On a possibilistic Approach to the Analysis of Fuzzy Feedback Systems. IEEE Trans. on Systems, Man, and Cybernetics, SMC-12, No.3, 1982, pp.417-422.
[367] Cushing, J.M.: Two Species Competition in a Periodic Environment. J. of Mathematical Biology, 10, No.4, 1980, pp.385-400.
[368] Czogala, E., Pedrycz, W.: On Identification of Fuzzy Systems and Its Application in Control Problems. Fuzzy Sets and Systems, 6, No.1, 1981, pp.73-83.
[369] Czogala, E., Pedrycz, W.: Control Problems in Fuzzy Systems. Fuzzy Sets and Systems, 7, No.3, May 1982, pp.257-273.
[370] Czogala, E., Pedrycz, W.: Fuzzy Rules Generation for Fuzzy Control. Cybernetics and Systems, 13, No.3, 1982, pp.275-293.
[371] Czogala, E., Zysno, P.: A Contribution to the Construction of the Gamma-Operator. In: Trappl, R.(ed.): Cybernetics and Systems Research, Vol.II. North-Holland, Amsterdam, and New York, 1984, pp.531-534.
[372] Dale, M.B.: Systems Analysis and Ecology. Ecology, 52, No.1, 1970, pp.2-16.
[373] Dalenoort, G.J.: Modularity of Behavior of Networks. In: Trappl, R.(ed.): Cybernetics and Systems Research. North-Holland, Amsterdam, and New York, 1982, 309-314.
[374] Dalenoort, G.J.: On the Representation of Control in Systems in General, and in Neural Networks in Particular. In: Trappl, R.(ed.): Cybernetics and Systems Research, Vol.II. North-Holland, Amsterdam, and New York, 1984, pp.251-256.
[375] Dammasch, I.E., Wagner, G.P.: On the Properties of Randomly Connected McCulloch-Pitts Networks: Differences between Input-Constant and Input-Variant Networks. Cybernetics and Systems, 15, No.1-2, 1984, pp.91-118.
[376] Darlington, J.L.: Interactive Validation of Program Specifications. In: Trappl, R.(ed.): Cybernetics and Systems Research. North-Holland, Amsterdam, and New York, 1982, pp.903-908.
[377] Darwish, M., Fantin, J., Grateloup, G.: On the Stability of Large-Scale Power Systems. J. of Cybernetics, 7, No.3-4, 1977, pp.159-168.
[378] Das, J.P., Kirby, J.R.: Simultaneous and Successive Cognitive Processes. Academic Press, London, and New York, 1979, XVII+248 pp.

[379] David, F.W., Nolle, H.: Experimental Modelling in Engineering. Butterworths, London, 1982, X+190pp.

[380] Davies, A.J.: The Finite Element Method: A First Approach. Clarendon Press, Oxford, 1980, XII+288 pp.

[381] Davio, M., Deschamps, J.-P., Thayse, A.: Discrete and Switching Functions. McGraw-Hill, New York, 1978, 729 pp.

[382] Davis, C., Demb, A., Espejo, R.: Organization for Program Management. John Wiley, Chichester and New York, 1979, XI+240 pp.

[383] Davis, R., Lenat, D.B.: Knowledge-Based Systems in Artificial Intelligence. McGraw-Hill, New York, 1981, XXII+490 pp.

[384] Day, R.H., Singh, I.: Economic Development as an Adaptive Process. Cambridge Univ. Press, Cambridge, Mass., 1977.

[385] Dayal, R.: An Integrated System of World Models. North-Holland, Amsterdam, and New York, 1981, XVIII + 398 pp.

[386] De Beaufrande,: Text, Discourse, and Process: Towards a Multidisciplinary Science of Texts. Ablex, Norwood, N.J., (Advances in Discourse Processes Series, Vol.4), 1980, XVI+352 pp.

[387] De Beaugrande, R.: Text, Discourse, and Process. Ablex, Norwood, N.J., 1980, XV+351 pp.

[388] De Cleris, M.: Certain System Concepts in Law and Politics. In: Trappl, R.(ed.): Cybernetics and Systems Research. North-Holland, Amsterdam, and New York, 1982, pp.507-512.

[389] De Delara, G.K., Stephenson, R.C.: Computer Simulation of Lateral Inhibition in the Retina. In: Trappl, R.(ed.): Cybernetics and Systems Research. North-Holland, Amsterdam, and New York, 1982, pp.287-292.

[390] De Greene, K.B.: Force Fields and Emergent Phenomena in Sociotechnical Macrosystems: Theories and Models. Behavioral Science, 23, No.1, Jan.1978, pp.1-14.

[391] De Greene, K.B.: The Adaptive Organization - Anticipation and Management of Crisis. John Wiley, Chichester and New York, 1982.

[392] De Jong, K.: Adaptive System Design: A Genetic Approach. IEEE Trans. on Systems, Man, and Cybernetics, SMC-10, No.9, 1980, pp.566-574.

[393] De La Sen, M.: Some Aspects of the Modeling of Multivariable Aperiodic Sampling Systems. In: Trappl, R.(ed.): Cybernetics and Systems Research. North-Holland, Amsterdam, and New York, 1982, pp.235-240.

[394] De La Sen, M.: Parameterizations in Single Input-Output Adaptive Control Systems to Avoid Delays in the Input Generation. In: Trappl, R.(ed.): Cybernetics and Systems Research, Vol.II. North-Holland, Amsterdam, and New York, 1984, pp.105-110.

[395] De Luca, A., Termini, S.: On the Convergence of Entropy Measures of a Fuzzy Set. Kybernetes, 6, No.3, 1977, pp.219-227.

[396] De Malherbe, R., De Malherbe, M.: Model Building in Ecology: a Hierarchical Approach. Kybernetes, 9, No.2, 1980, pp.141-150.

[397] De Mottoni, P., Schiaffino, A., Tesei, A.: Bifurcation of Regular Nonnegative Stationary Solutions for a Quasilinear Parabolic System. In: Trappl, R.(ed.): Cybernetics and Systems Research. North-Holland, Amsterdam, and New York, 1982, pp.345-350.

[398] De Raadt, J.D.R.: An Application of Ashby's Law to the Skewness of Income and Output Distributions. Systems Research, 1, No.2, 1984, pp.127-134.

[399] De Rosnay, J.: The Macroscope: a New World Scientific

System. Harper and Row, New York, 1980, XIX+247 pp.

[400] De Witt, C.T., Goudriaan, J.: Simulation of Ecological Processes (2nd Edition, Revised and Extended). Centre for Agricultural Publishing and Documentation, Wageningen, Neth., 1978, 175 pp.

[401] Dedons, A., (et al.): The Information Professional: Survey of an Emerging Field. Marcel Dekker, New York, 1981, XX+272pp.

[402] Degermendzhy, A.G.: The Reliability of Microevolution Process of the Stationary and Oscillatory Population in Open Systems. J. of Cybernetics, 8, No.2, 1978, pp.117-132.

[403] Dekker, L.(ed.): Simulation of Systems. North-Holland, Amsterdam, and New York, 1976.

[404] Delaney, W., Esposito, F., Vaccari, E.: A System Theory Based Simulation Language. In: Trappl, R.(ed.): Cybernetics and Systems Research, Vol.II. North-Holland, Amsterdam, and New York, 1984, pp.111-116.

[405] Della Riccia, G., De Santis, F.: The Approximation of Factor Scores in a Factor Analysis Algorithm. In: Trappl, R.(ed.): Cybernetics and Systems Research. North-Holland, Amsterdam, and New York, 1982, pp.209-214.

[406] Demarco, T.: Structural Analysis and System Specification. Yourdon, New York, 1978, 352 pp.

[407] Demillo, R.A., Lipton, R.J., Perlis, A.J.: Social Processes and Proofs of Theorems and Programs. ACM Communications, 22, No.5, 1979, pp.271- 280.

[408] Dempster, A.H., (et al.)(eds.): Deterministic and Stochastic Scheduling (NATO Advanced Study Institute). D.Reidel, Dordrecht, Holland, and Boston, 1982, XII+420pp.

[409] Denning, P.J., Dennis, J.B., Qualitz, J.E.: Machines, Languages, and Computation. Prentice-Hall, Englewood Cliffs, New Jersey, 1978, 601 pp.

[410] Dent, J.B., Blackie, M.J.: Systems Simulation in Agriculture. Applied Science Publishers, Barking, England, 1979, X+180 pp.

[411] Desouza, G.R.: System Methods for Socioeconomic and Environmental Impact Analysis. Lexington (Heath), Lexington, Mass., 1979, XII+164 pp.

[412] Dessauchoy, A.E.: Generalized Information Theory and Decomposability of Systems. Int. J. of General Systems, 9, No.1, 1982, pp.13-36.

[413] Destouches, J.-L.: Basic Concepts of System Theory. Cybernetics and Systems, 11, No.3, 1980, pp.195-214.

[414] Deutsch, K.W., Fritsch, B.: Toward a Theory of Simplification: Reduction of Complexity in Data Processing for World Models. Athenaeum, Koenigstein, 1980, 72 pp. [German]

[415] Devries, R.P., Van Hezewijk, R.: Systems Theory and the Philosophy of Science. Annals of Systems Research, 7, 1978, pp.91-125.

[416] Dhillon, B.S., Singh, S.: Engineering Reliability: New Techniques and Applications. John Wiley, Chichester and New York, 1981, XIX+329 pp.

[417] Dhrymes, P.J.: Mathematics for Econometrics. Springer-Verlag, Berlin, FRG, and New York, 1978.

[418] DiNola, A., Sessa, S.: On the Fuzziness of Solutions of Alpha-Fuzzy Relation Equations on Finite Spaces. Busefal, 11, 1982, pp.14-23.

[419] Dial, G.: A Note on Bayes Risk with Utilities. Cybernetics and Systems, 14, No.2-4, 1983, 159-164.

[420] Dial, G.: Entropy of Order Alpha and Type Beta and Shannon's Inequality. Cybernetics and Systems, 14, No.2-4, 1983, pp.165-172.

[421] Dijkstra, W., Van der Zouwen, J.: Testing Auxiliary Hypothesis Behind the Interview. Annals of Systems Research, 6, 1977, pp.49-63.

[422] Dimirivski, G.M., Gough, N.E., Barnett, S.: Categories in Systems and Control Theory. Int. J. of Systems Science, 8, No.10, Oct.1977, pp.1081-1090.

[423] Dimitrova, L.: Cybernetics and the Organization of Material Flow in the Industrial Enterprise. In: Trappl, R.(ed.): Cybernetics and Systems Research. North-Holland, Amsterdam, and New York, 1982, pp.431-436.

[424] Dirickx, Y.M.I., Jennergren, L.P.: Systems Analysis by Multilevel Methods: with Applications for Economics and Management. John Wiley, Chichester and New York, 1979, XI+217 pp.

[425] Ditleusen, O.: Uncertain Modelling. McGraw-Hill, New York, 1980.

[426] Dixon, N.R., Martin, T.B.(eds.): Automatic Speech and Speaker Recognition. IEEE, Piscataway, N.J., 1979, VI+433 pp.

[427] Dolby, J.L.: On the Notions of Ambiguity and Information Loss. Behavioral Science, 22, No.4, July 1977, pp.290-298.

[428] Dold, A., Eckmann, B.(eds.): Lecture Notes in Mathematics, Vol.453 Springer-Verlag, Berlin, FRG, and New York, 1975.

[429] Dombi, J., Zysno, P.: Comments on the Gamma-Model. In: Trappl, R.(ed.): Cybernetics and Systems Research. North-Holland, Amsterdam, and New York, 1982, pp.711-714.

[430] Dominique, J.: Logical Construction of Systems. Van Nostrand Reinhold, New York, 1981, XII+180 pp.

[431] Dominowski, R.L.: Research Methods. Prentice-Hall, Englewood Cliffs, New Jersey, 1980, VIII+390 pp.

[432] Domotor, Z.: Probabilistic and Causal Dependence Structures. Theory and Decision, 13, No.3, Sept.1981, pp.275-292.

[433] Dompere, K.K.: On Epistemology and Decision-Choice Rationality. In: Trappl, R.(ed.): Cybernetics and Systems Research. North-Holland, Amsterdam, and New York, 1982, pp.219-228.

[434] Donald, A.: Management, Information and Systems (Second Edition). Pergamon Press, Oxford and New York, 1979, XIII+253 pp.

[435] Dreben, B., Goldfarb, W.D.: The Decision Problem. Addison-Wesley, Reading, Mass., 1979, XIV+272 pp.

[436] Drechsler, F.S., Zamzeer, M.J.: Computer Integrated Manufacturing (CIM) in Irish Industry. In: Trappl, R.(ed.): Cybernetics and Systems Research, Vol.II. North-Holland, Amsterdam, and New York, 1984, pp.349-352.

[437] Dreyfus, S.E.: Dynamic Programming and the Calculus of Variations. Academic Press, London, and New York, 1965, 248 pp.

[438] Dreyfus, S.E., Law, A.M.: The Art and Theory of Dynamic Programming. Academic Press, London, and New York, 1977, 284 pp.

[439] Driankow, D., Stantchev, I.: An Approach to Multiobjectives Decision Making with Ordinal Information. In: Trappl, R.(ed.): Cybernetics and Systems Research. North-Holland, Amsterdam, and New York, 1982, pp.253-262.

[440] Drommenhoek, W., Sebus, J., Van Esch, G.J.: Biological Structures (translated from Dutch). University Park Press, Baltimore, 1980, 144 pp.

[441] Drozen, V.: Multilateral Interdependence. In: Trappl, R.(ed.): Cybernetics and Systems Research. North-Holland, Amsterdam, and New York, 1982, pp.65-68.

[442] Drozen, V.: Combinatory Spaces - An Approach to Pattern Analysis. Kybernetes, 13, No.1, 1984, pp.27-30.

[443] Dubois, D., Prade, H.: Fuzzy Sets and Systems: Theory and Applications. Academic Press, London, and New York, 1980, 416 pp.

[444] Dubois, D., Prade, H.: A Class of Fuzzy Measures Based on Triangular Norms. Int. J. of General Systems, 8, No.1, 1982, pp.43-62.

[445] Duda, R.O., Hart, P.E.: Pattern Classification and Scene Analysis. John Wiley, Chichester and New York, 1973.

[446] Duechting, W., Vogelsaenger, Th.: Recent Results in Modelling and Simulation of 3-D Tumor Growth. In: Trappl, R.(ed.): Cybernetics and Systems Research. North-Holland, Amsterdam, and New York, 1982, pp.293-298.

[447] Dunn, G., Everitt, B.S.: An Introduction to Mathematical Taxonomy. Cambridge Univ. Press, Cambridge, Mass., 1982, XII + 152pp.

[448] Duran, B.S., Odell, P.L.: Cluster Analysis: a Survey. Springer-Verlag, Berlin, FRG, and New York, 1974.

[449] Durkin, J.E.: Living Groups: Group Psychotherapy and General System Theory. Brunner/Mazel, New York, 1981, XXVII+370pp.

[450] Dutton, G.(ed.): Harvard Papers on Geographic Information Systems. Laboratory for Computer Graphics and Spatial Analysis, Harvard Univ., Cambridge, Mass., 1979.

[451] Dwyer, R.F., Kurz, L.: Sequential Partition Detectors. J. of Cybernetics, 8, No.2, 1978, pp.133-158.

[452] Dwyer, R.F., Kurz, L.: Sequential Partition Detectors with Dependent Sampling. J. of Cybernetics, 10, No.1-3, 1980, pp.211-232.

[453] Dym, C.L., Ivey, E.S.: Principles of Mathematical Modeling. Academic Press, London, and New York, 1980, XVI+262 pp.

[454] Economos, A.C.: Thoughts on a Systems View of Life. Cybernetica, 24, No.1, 1981, pp.33-44.

[455] Edmons, E.A.: Lattice Fuzzy Logics. Int. J. of Man-Machine Studies, 13, No.4, 1980, pp.455-465.

[456] Edwards, J.B., Owens, D.H.: Analysis and Control of Multipass Processes. Research Studies Press (A division of John Wiley), New York, 1982, XIX+298pp.

[457] Efstathiou, J., Rajkovic, V.: Multiattribute Decision Making Using a Fuzzy Heuristic Approach. IEEE Trans. on Systems, Man, and Cybernetics, SMC-9, No.6, 1969, pp.326-333.

[458] Egardt, B.: Stability of Adaptive Controllers. Springer-Verlag, Berlin, FRG, and New York, 1979, VI+158 pp.

[459] Egiazarian, K.O.: The Generalized Hadamard Transforms and Linear Filters. In: Trappl, R.(ed.): Cybernetics and Systems Research, Vol.II. North-Holland, Amsterdam, and New York, 1984, pp.131-136.

[460] Ehrenkrantz, E.: The Systems Approach to Design. McGraw-Hill, New York, 1980.

[461] Ehrig, H.: Universal Theory of Automata. B.G.Teubner, Stuttgart, 1974, 240 pp.

[462] Eigen, M., Schuster, P.: The Hypercycle: a Principle of Natural Selforganization. Springer-Verlag, Berlin, FRG, and New York, 1979.

[463] Eigen, M., Winkler, R.: Laws of the Game: How Principles of Nature Govern Chance. Alfred A.Knopf, New York, 1981, XV + 347 pp.

[464] Eilbert, R.F., Christensen, R.: Contrivedness: The Boundary Between Pattern Recognition and Numerology. Pattern Recognition, 15, No.3, 1982, pp.253-261.

[465] Eisen, M.: Mathematical Models in Cell Biology and Cancer Chemotherapy. Springer-Verlag, Berlin, FRG, and New York, 1979, IX+431 pp.

[466] Eisenstein, B.A., Vocarro, R.J.: Feature Extraction by System Identification. IEEE Trans. on Systems, Man, and Cybernetics, SMC-12, No.1, 1982, pp.42-50.

[467] El-Fattah, Y., Foulard, C.: Learning Systems: Decision, Simulation, and Control. Springer-Verlag, Berlin, FRG, and New York, 1978, VII+119 pp.

[468] El-Sherief, H., Sinha, N.K.: Identification and Modeling for Linear Multivariable Discrete-Time Systems: A Survey. J. of Cybernetics, 9, No.1, 1979, pp.43-72.

[469] El-Sherief, H., Sinha, N.K.: Identification of Multivariable Systems in the Transfer-Function Matrix Form. J. of Cybernetics, 9, No.2, 1979, pp.113-126.

[470] El-Sherief, H.: Identification and Control of Linear Discrete Multivariable Systems with Process Noise. Cybernetics and Systems, 14, No.1, 1983, pp.75-84.

[471] El-Sherief, H.: Recent Advances in Multivariable System Modeling and Identification Algorithms and their Applications. Systems Research, 1, No.1, 1984, pp.63-70.

[472] El-Shirbeeny, El-H.T., Elany, O.A.: A Sensitivity Augmented Conjugate Gradient Algorithm for Solving Constrained Control Design Problems. In: Trappl, R.(ed.): Cybernetics and Systems Research. North-Holland, Amsterdam, and New York, 1982, pp.169-174.

[473] Elandt-Johnson, R.C., Johnson, N.L.: Survival Models and Data Analysis. John Wiley, Chichester and New York, 1980, XVI+457 pp.

[474] Elias, D.: Procedures for Generating Reconstruction Hypotheses in the Reconstructability Analysis: APL Version. State University of New York, Binghamton, New York, 1979. [IFSR-Depository]

[475] Elison, D.I., Herschdorfer, I., Wilson, J.T.: Interactive Simulation on a Computer. Simulation, 38, No.5, May 1982, pp.161-175.

[476] Ellison, D.(ed.): Simulation Newsletter. Univ. of Lancaster, Centre in Simulation, Gillow House, Lancaster LA1 4YG, England.

[477] Elohim, J.L.: Further Steps in our Learning to Organize Educational Systems. In: Trappl, R.(ed.): Cybernetics and Systems Research. North-Holland, Amsterdam, and New York, 1982, pp.609-616.

[478] Elstob, C.M.: The Formal and the Physical. In: Trappl, R.(ed.): Cybernetics and Systems Research. North-Holland, Amsterdam, and New York, 1982, pp.25-30.

[479] Elstob, C.M.: Emergentism and Mind. In: Trappl, R.(ed.): Cybernetics and Systems Research, Vol.II. North-Holland, Amsterdam, and New York, 1984, pp.83-88.

[480] Elton, M.C.J., (et al.)(eds.): Evaluating New Telecommunication Services. Plenum Press, New York and London, 1978, 784 pp.

[481] Engell, S.: Variety, Information and Feedback. Kybernetes, 13, No.2, 1984, pp.73-78.

[482] England J.L.: Hierarchy Theory and Sociology. In: Trappl, R.(ed.): Cybernetics and Systems Research. North-Holland, Amsterdam, and New York, 1982, pp.97-104.

[483] England J.L., Warner, W.K.: An Ashby Hierarchy for Human Action. In: Trappl, R.(ed.): Cybernetics and Systems Research, Vol.II. North-Holland, Amsterdam, and New York, 1984, pp.665-670.

[484] Erdi, P.: System-Theoretical Approach to the Neural Organization: Feed-Forward Control of the Ontogenetic Development. In: Trappl, R.(ed.): Cybernetics and

Systems Research, Vol.II.   North-Holland, Amsterdam, and
New York, 1984, pp.229-236.

[485] Erickson, S.W.: A Systems View of the Foundations of
Education.   Systems Trends, 3, No.7, May 1981, pp.3-10.

[486] Ericson, R.F.(ed.): The General Systems Challenge. SGSR,
Louisville, Kentucky, 1978.

[487] Erlandson, R.F.:   The   Participant-Observer   Role   in
Systems-Methodologies. IEEE Trans. on Systems, Man, and
Cybernetics, SMC-10, No.1, 1980, pp.16-19.

[488] Erlandson, R.F.: The Satisfacting Process. A New Look.
IEEE Trans. on Systems, Man, and Cybernetics, SMC-11,
No.11, November 1981, pp.740-752.

[489] Espejo, R., Watt, J.: Management Information Systems: a
System for Design.   J. of Cybernetics, 9, No.3, 1979,
pp.259-283.

[490] Espejo, R.:   Cybernetic Praxis in Government: The
Management of Industry in Chile 1970-1973. Cybernetics
and Systems, 11, No.4, 1980, pp.325-338.

[491] Espejo, R.: Cybernetics in Management and Organization.
In: Trappl, R.(ed.):   Cybernetics - Theory and
Applications.   Hemisphere,   Washington,   D.C.,   1983,
pp.263-290.

[492] Espejo, R.:   Management and Information: The
Complementarity Control-Autonomy.   Cybernetics and
Systems, 14, No.1, 1983, pp.85-102.

[493] Estep, M.L.: A Systems Integration of Multidisciplinary
and Interdisciplinary Inquiry in Curriculum Design.   Univ.
of Texas at San Antonio, 1977, 20pp. [IFSR-Depository]

[494] Estep, M.L.:   Toward Alternative Methods of Systems
Analysis: The Case of Qualitative Knowing. In: Trappl,
R.(ed.): Cybernetics and Systems Research, Vol.II.
North-Holland, Amsterdam, and New York, 1984, pp.201-208.

[495] Eugene, J.: Aspects of the General Theory of Systems.
Maloine, Paris, 1981, 250 pages. [French]

[496] Eveland, S., Johnk, L.: The Use of the QBE Relational
Database System for a Gothic Ethymological Dictionary.
In: Trappl, R.(ed.): Cybernetics and Systems Research.
North-Holland, Amsterdam, and New York, 1982, pp.791-796.

[497] Everitt, B.S.: Unresolved Problems in Cluster Analysis.
Biometrics, 35, No.1, 1979, pp.169-181.

[498] Eyckhoff, P.(ed.):   Trends and Progress in Systems
Identification. Pergamon Press, Oxford and New York,
1981, XVI+402 pp.

[499] Eyckhoff, P.(ed.):   Trends and Progress in System
Identification. Pergamon Press, Oxford and New York,
1982, 410pp.

[500] Fairhurst, M.C.:   Cyclic Activity in Information
Processing Systems.   J. of Cybernetics, 7, No.1-2, 1977,
pp.37-48.

[501] Fairtlough, G.: How Systems Thinking Might Evolve. J. of
Applied Systems Analysis, 7, April 1980, pp.13 - 21.

[502] Fandel, G., Gal, T.(eds.):   Multiple Criteria Decision
Making Theory and Application.   (Proc. of a Conf. in
Hagen/Konigswinter, Germany, Aug.1979).   Springer-Verlag,
Berlin, FRG, and New York, 1980, XVI+570 pp.

[503] Fararo, T.J.: Social Activity and Social Structure: A
Contribution to the Theory of Social Systems. Cybernetics
and Systems, 12, No.1-2, 1981, pp.53-81.

[504] Faribault, M., Leon, J., Meissonnier, V., Memmi, D.,
Zarri, G.P.:   From Natural Language to a Canonical
Representation of the Corresponding Semantic
Relationships. In: Trappl, R.(ed.): Cybernetics and
Systems Research, Vol.II. North-Holland, Amsterdam, and
New York, 1984, pp.733-738.

[505] Farinas-del-Cerro, L.:   Prolegomena   for   Programming   in
      Modal   Logic.   In: Trappl,   R.(ed.):   Cybernetics   and
      Systems Research.   North-Holland, Amsterdam, and New York,
      1982, pp.917-920.
[506] Farkas, J.:   Rethinking of Theoretical Presuppositions   of
      Systems Theory.   Institute of Sociology, Hungarian Academy
      of Science, 1982, 35pp.   [IFSR-Depository]
[507] Farley, A.M.:   Issues in Knowledge-Based Problem   Solving.
      IEEE Trans.   on Systems,   Man,   and Cybernetics, SMC-10,
      No.8, August 1980, pp.446-459.
[508] Farreny, H.,   Prade, H.:   On the Best   Way   of   Designating
      Objects in Sentence Generation.   Kybernetes,   13, No.1,
      1984, pp.43-46.
[509] Faucheuz, C., Makridakis, S.:   Automation or   Autonomy   in
      Organizational Design.   Int.   J.   of General Systems, 5,
      No.4, 1979, pp.213-220.
[510] Favella, L.F.:   Mathematical   Foundations   of   ECG   Map
      Reconstruction.   Cybernetics   and   Systems,   11,   No.1-2,
      1980, pp.21-66.
[511] Favella, L.F.,   Reineri, M.T.,   Ricciardi,   L.,   Sacerdote,
      L.:   First   Passage   Time   Problems   and   Some   Related
      Computational Methods.   Cybernetics and Systems, 13, No.2,
      1982, pp.95-128.
[512] Favella, L.F., Morra, G.:   Application   of   Loeve-Karhunen
      Expansion to Acoustic   Diagnostic   Problems   for Internal
      Combustion Engines.   Cybernetics   and   Systems,   14,   No.1,
      1983, pp.55-74.
[513] Favella, L.F.:   Methodological and   Mathematical   Problems
      in ECG Analysis.   Cybernetics   and   Systems, 15, No.1-2,
      1984, pp.169-196.
[514] Fedanzo, A.J.:   The Origin of Information in Systems.   J.
      of   Social   and   Biological   Structures,   3,   No.1,   1980,
      pp.17-32.
[515] Fedra, K., Van Straten, G., Beck, M.B.:   Uncertainty   and
      Arbitrariness in   Ecosystem   Modelling: A Lake Modelling
      Example.   Ecological Modelling, 13,   Nos.1-2,   June   1981,
      pp.87-110.
[516] Feintuch, A., Saeks, R.:   System Theory: A Hilbert   Space
      Approach.   Academic   Press,   London,   and   New York, 1982,
      320pp.
[517] Feldman, J.A.,   Shields, P.C.:   Total   Complexity   and   the
      Inference of   Bent Programs.   Mathematical Systems Theory,
      10, No.3, 1976/1977, pp.181-191.
[518] Fernandez, E.B.,   Summers,   R.C.,   Wood,   C.:   Database
      Security and   Integrity.   Addison-Wesley, Reading, Mass.,
      1981, XIV + 320 pp.
[519] Ferracin, A.,   (et al.):   Self-Organizing Ability in Living
      Systems.   Bio-Systems, 10, 1978, pp.307-317.
[520] Fiacco, A.V., Kortanek, K.O.(eds.):   Extremal Methods   and
      Systems Analysis.   Springer-Verlag,   Berlin, FRG, and New
      York, 1980, XII+546 pp.
[521] Fick, G.,   Sprague,   R.H.Jr.(eds.):   Decision   Support
      Systems: Issues   and Challenges.   Pergamon Press, Oxford
      and New York, 1980, VIII+190 pp.
[522] Field, H.H.:   Science Without Numbers.   Basil   Blackwell,
      Oxford, 1980, XIII+130 pp.
[523] Findeisen, W.:   Control and Coordination   in   Hierarchical
      Systems (Vol.9 in the Wiley IIASA/ International Series on
      Applied Systems Analysis).   John Wiley, Chichester and New
      York, 1980, X+467 pp.
[524] Findler, N.V.(ed.):   Associative Networks: Representation
      and Use   of   Knowledge by Computers.   Academic   Press,
      London, and New York, 1979, XVIII+462 pp.
[525] Findler, N.V.:   Aspects of Computer Learning.   Cybernetics

and Systems, 11, No.1-2, 1980, pp.67-86.

[526] Findler, N.V.: A Multi-Level Learning Technique Using Production Systems. Cybernetics and Systems, 13, No.1, 1982, pp.25-30.

[527] Fitter, M.: Towards More 'Natural' Interactive Systems. Int. J. of Man-Machine Studies, 11, No.3, 1979, pp.339-350.

[528] Flavin, M.: Fundamental Concepts of Information Modelling. Yourdon, New York, 1981, 128 pp.

[529] Fleddermann, R.G.: Teaching the Systems Approach with a Group Thesis. Engineering Education, Feb.1980, pp.430-431.

[530] Fletcher, R.: Practical Methods in Optimization. Vol.1: Unconstrained Optimization; Vol.2: Constrained Optimization. John Wiley, Chichester and New York, Vol.1: 1980, 120 pp.; Vol.2: 1981, 224 pp.

[531] Floyd, R.W.: The Paradigms of Programming. ACM Communications, 22, No.8, Aug.1979, pp.455-460.

[532] Fogel, E., Huang, Y.F.: On the Value of Information in System Identification: Bounded Noise Case. Automatica, 18, No.2, March 1982, pp.229-238.

[533] Fontela, E., Rossier E.: Condensed Forms of Large Scale Models. Large Scale Systems, 1, No.4, 1980, pp.281-288.

[534] Foo, N.Y.: Stability Preservation under Homomorphisis. IEEE Trans. on Systems, Man, and Cybernetics, SMC-7, No.10, Oct.1977, pp.750-754.

[535] Foo, N.Y.: Closure Properties and Homomorphisms of Timevarying Systems. Mathematical Systems Theory, 12, No.1, 1978, pp.41-58.

[536] Fortuny-Amat, J., Talavage, J.: A Solution Procedure for the Hierarchical Coordination of Constrained Optimizing Systems. Int. J. of Systems Science, 12, No.2, 1981, pp.133-146.

[537] Fosko, P.D., Hancock, J.K.: Applications of General Systems Theory and Cybernetics to the Management of/ Research and Development. 1979, 20 pp. [IFSR-Depository]

[538] Fowler, T.B.: Brillouin and the Concept of Information. Int. J. of General Systems, 9, No.3, 1983, pp.143-156.

[539] Fox, M.S.: An Organizational View of Distributed Systems. IEEE Trans. on Systems, Man, and Cybernetics, SMC-11, No.1, 1981, pp.70-80.

[540] Frank, P.M.: Introduction to System Sensitivity Theory. Academic Press, London, and New York, 1978, 400 pp.

[541] Franksen, O.I., Flaster, P., Evans, F.J.: Qualitative Aspects of Large Scale Systems: Developing Design Rules Using APL. Springer-Verlag, Berlin, FRG, and New York, 1979, XII+120 pp.

[542] Franova, M.: Program Synthesis and Constructive Proofs Obtained by Beth's Tableaux. In: Trappl, R.(ed.): Cybernetics and Systems Research, Vol.II. North-Holland, Amsterdam, and New York, 1984, pp.715-720.

[543] Franta, W.R.: Process View of Simulation. Elsevier / North-Holland, New York, 1977, 195 pp.

[544] Frauenthal, J.C.: Mathematical Modelling in Epidemology. Springer-Verlag, Berlin, FRG, and New York, 1980, X+120 pp.

[545] Freeling, A.N.S.: Fuzzy Sets and Decision Analysis. IEEE Trans. on Systems, Man, and Cybernetics, SMC-10, No.7, July 1980, pp.341-354.

[546] Freuder, E.C.: Synthesizing Constraint Expressions. ACM Communications, 21, No.11, Nov.1978, pp.958-966.

[547] Fum, D., Guida, G., Tasso, C.: A Rule-Based Approach to Natural Language Text Representation and Comprehension. In: Trappl, R.(ed.): Cybernetics and Systems Research,

Vol.II. North-Holland, Amsterdam, and New York, 1984, pp.727-732.

[548] Furukawa, O., Ishizuchi, H.: Precoordination in Quality Control. Int. J. of General Systems, 9, No.4, 1983, pp.225-234.

[549] Furukawa, O., Ishizuchi, H.: Refinement of Quality Control Systems. Int. J. of General Systems, 9, No.3, 1983, pp.161-170.

[550] Furukawa, O., Ikeshoji, H., Ohmori, A.: A Methodology for Quality Goal-Seeking and Coordination, and the Practical Application. Systems Research, 1, No.1, 1984, pp.71-82.

[551] Gabasov, R., Kirilova, F.: The Qualitative Theory of Optimal Processes. (Translated From Russian by J. Casti). Marcel Dekker, New York, 1976, 640 pp.

[552] Gage, W.L.: Auditorium Management Information Systems. Cybernetics and Systems, 11, No.4, 1980, pp.369-380.

[553] Gage, W.L.: Flexibility in Network Planning. In: Trappl, R.(ed.): Cybernetics and Systems Research. North-Holland, Amsterdam, and New York, 1982, pp.463-470.

[554] Gage, W.L.: Evidence for Strategic Decisions. In: Trappl, R.(ed.): Cybernetics and Systems Research, Vol.II. North-Holland, Amsterdam, and New York, 1984, pp.173-178.

[555] Gaines, B.R., Kohout, L.J.: The Fuzzy Decade: a Bibliography of Fuzzy Systems and Closely Related Topics. Int. J. of Man-Machine Studies, 9, No.1, January 1977, pp.1-68.

[556] Gaines, B.R.: Fuzzy and Probability Uncertainty Logics. Information and Control, 38, No.2, Aug.1978, pp.154- 169.

[557] Gaines, B.R.: General Systems Research: Quo Vadis? General Systems Yearbook, 24, 1979, pp.1 - 9.

[558] Gaines, B.R.(ed.): General Systems. Yearbook of the Society for General Systems Research. SGSR, Louisville, Kentucky, Vol.24, 1979, IV+271 pp.

[559] Gaines, B.R.: Logical Foundations for Database Systems. Int. J. of Man-Machine Studies, 11, No.4, 1979, pp.481-500.

[560] Gaines, B.R.: Sequential Fuzzy System Identification. Fuzzy Sets and Systems, 2, No.1, Jan.1979, pp.15-24.

[561] Gaines, B.R.: Methodology in the Large: Modeling All There Is. Systems Research, 1, No.2, 1984.

[562] Galitsky, V.V.: On Modeling the Plant Community Dynamics. In: Trappl, R.(ed.): Cybernetics and Systems Research. North-Holland, Amsterdam, and New York, 1982, 677-682.

[563] Gall, J.: Systemantics: How Systems Work and Especially How They Fail. Quadrangle, New York, 1977.

[564] Gallopin, G.C.: The Abstract Concept of Environment. Int. J. of General Systems, 7, No.2, 1981, pp.139-150.

[565] Gams, M., Bratko, I.: A Circuit Analysis Program that Explains its Reasoning. In: Trappl, R.(ed.): Cybernetics and Systems Research, Vol.II. North-Holland, Amsterdam, and New York, 1984, pp.811-816.

[566] Ganascia, J.G.: Explanation Facilities for Diagnosis Systems. In: Trappl, R.(ed.): Cybernetics and Systems Research, Vol.II. North-Holland, Amsterdam, and New York, 1984, pp.805-810.

[567] Gane, C., Sarson, T.: Structural Systems Analysis: Tools and Techniques. Ist, New York, 1977, 373 pp.

[568] Garbade, K.: Time Variation in the Term Structure of Treasury Bills. J. of Cybernetics, 7, No.1-2, 1977, pp.117-132.

[569] Gardiner, P.C.: Decision Spaces. IEEE Trans. on Systems, Man, and Cybernetics, SMC-7, No.5, May 1977, pp.340-349.

[570] Gardner, M.R.: Predicting Novel Facts. Brit. J. for the

Philosophy of Science, 33, No.1, March 1982, pp.1-15.

[571] Garey, M.R., Johnson, D.S.: Computers and Intractability: a Guide to NP-Completeness. W.H.Freeman, San Francisco, 1979, 338 pp.

[572] Garfinkel, A.: Forms and Explanation. Yale Univ. Press, New Haven, Conn., 1981, XI+186 pp.

[573] Garfolo, B.T.: The Renal Dialysis Medical Information System. In: Trappl, R.(ed.): Cybernetics and Systems Research. North-Holland, Amsterdam, and New York, 1982, pp.645-650.

[574] Gasparski, W.W.(ed.): Problems of Design Methodology. Panstwowe Wydawnictwo Naukowe, Warsaw, 1977, 395 pp. [Polish, Russian, English]

[575] Gasparski, W.W.(ed.): Studies in Design Methodology in Poland (Special Issue of "Design Methods and Theories"). Polish Academy of Sciences, Warsaw, 1977, 103pp. [IFSR-Depository]

[576] Gasparski, W.W.: An Outline of the Logical Theory of Design. Praxiology, 1980, No.1, pp.95-112.

[577] Gasparski, W.W., Miller, D.(eds.): Design and System: Methodological Aspects, Vol.3. Polish Academy of Sciences, Warsaw, 1981, 174 pp. [Polish]

[578] Gasparski, W.W.: On Systems Theory and Systems Research and Science and Technology. Postepy Cybernetyki, 4, No.3, 1981, pp.7-21.

[579] Gasparski, W.W.(ed.): Papers from the Conference in Design Methods. Polish Academy of Sciences, Warsaw, 1982, 48pp. [IFSR-Depository]

[580] Gasparski, W.W.: Praxiology. Polish Academy of Sciences, Warsaw, 1982, 15pp. [IFSR-Depository]

[581] Gasparski, W.W.: Science, Technology, and Systems. Polish Academy of Sciences, Warsaw, 1982, 7pp. [IFSR-Depository]

[582] Gasparski, W.W.: Si Duo Dicunt Idem Non Est Idem: Or on Design Methodology as Seen from Neither an English nor a German Speaking Country. Polish Academy of Sciences, Warsaw, 1982, 15pp. [IFSR-Depository]

[583] Gasparski, W.W.: What Does it Mean a Notion of "System Modelling". Polish Academy of Sciences, Warsaw, 1982, 4pp. [IFSR-Depository]

[584] Gasparski, W.W.: Si duo dicunt idem non est idem: Or on Design Methodology as Seen from neither English nor German Speaking Country. In: Trappl, R.(ed.): Cybernetics and Systems Research, Vol.II. North-Holland, Amsterdam, and New York, 1984, pp.397-402.

[585] Gati, G.: Further Annotated Bibliography on the Isomorphism Disease. J. of Graph Theory, 3, No.2, 1979, pp.95-109.

[586] Gatlin, L.L.: Information Theory and the Living Systems. Columbia Univ. Press, New York, 1972, 210 pp.

[587] Gause, D.C., Rogers, G.: Cybernetics and Artificial Intelligence. In: Trappl, R.(ed.): Cybernetics - Theory and Applications. Hemisphere, Washington, D.C., 1983, pp.339-360.

[588] Gehrlein, W.V., Fishburn, P.C.: An Analysis of Simple Counting Methods for Ordering Incomplete Ordinal Data. Theory and Decision, 8, No.3, July 1977, pp.209-227.

[589] Gelfand, A.E., Walker, C.C.: A Systems-Theoretic Approach to the Management of Complex Organizations: Management by Consensus Level and its Interaction with Other Management Strategies. Behavioral Science, 25, No.4, July 1980, pp.250-260.

[590] Genesio, R., Sorrentino, N.: On the Model Order for Uncertain System Approximation. Systems Science, 3, No.3,

1977, pp.215-225.
[591] Gengoux, K.G., David, P.K.: Citizen Participation: A New Factor in Social Systems. In: Trappl, R.(ed.): Cybernetics and Systems Research. North-Holland, Amsterdam, and New York, 1982, pp.495-500.
[592] Gentile, S.: A Discrete Model for the Study of a Lake. Applied Mathematical Modelling, 3, June 1979, pp.193-198.
[593] George, F.H.: Cybernetics and the Environment. Paul Elek, London, 1977.
[594] George, F.H.: The Foundations of Cybernetics. Gordon and Breach, New York, and London, 1977, XIV+286 pp.
[595] George, F.H.: Philosophical Foundations of Cybernetics. Abacus Press, Tunbridge Wells, England, 1979, IX+157 pp.
[596] George, F.H.: The Science of Philosophy. Gordon and Breach, New York, and London, 1981, IX + 326 pp.
[597] Georgiev, A.A.: Nonparametric Mathematical Model from Individual Human Growth Curve. In: Trappl, R.(ed.): Cybernetics and Systems Research, Vol.II. North-Holland, Amsterdam, and New York, 1984, pp.277-280.
[598] Gerardy, R.: Probabilistic Finite State System Identification. Int. J. of General Systems, 8, No.4, 1982, pp.229-242.
[599] Gerardy, R.: Experiments with Some Methods for the Identification of Finite-State Systems. Int. J. of General Systems, 9, No.4, 1983, pp.197-204.
[600] Gericke, M., Straube, P.: Changes of Regional Passenger-Transport Demands by Society-Controlled Activities - A Fuzzy Simulation Approach. In: Trappl, R.(ed.): Cybernetics and Systems Research, Vol.II. North-Holland, Amsterdam, and New York, 1984, pp.573-580.
[601] Getis, A., Boots, B.: Models of Spatial Processes. Cambridge Univ. Press, Cambridge, Mass., 1978, XVI+198 pp.
[602] Getz, W.M.(ed.): Mathematical Modelling in Biology and Ecology. Springer-Verlag, Berlin, FRG, and New York, 1980, VII+356pp.
[603] Geyer, R.F., Van der Zouwen, J.(eds.): Sociocybernetics, Vol.II. Martinus Nijhoff, Boston and The Hague, 1978.
[604] Geyer, R.F.: Alienation Theories: a General Systems Approach Pergamon Press, Oxford and New York, 1980, XIX+201 pp.
[605] Gheorghe, A., Popovici, A., Stoica, M.: Energy Management Planning Tools for the Transition towards a Medium Developed Economy. In: Trappl, R.(ed.): Cybernetics and Systems Research. North-Holland, Amsterdam, and New York, 1982, pp.689-694.
[606] Ghiselli, E.E., Campbell, J.P., Zedeck, S.: Measurement Theory for the Behavioral Sciences. W.H.Freeman, San Francisco, 1981, XVIII+484 pp.
[607] Ghosal, A.: Applied Cybernetics: its Relevance to Operations Research. Gordon and Breach, New York, and London, 1978, 180 pp.
[608] Ghosal, A.: Relevance of Cybernetics in the Study of Developing Countries. In: Trappl, R.(ed.): Cybernetics and Systems Research, Vol.II. North-Holland, Amsterdam, and New York, 1984, pp.483-490.
[609] Giere, R.N.: Propensity and Necessity. Synthese, 40, No.3, 1979, pp.439-451.
[610] Gigley, H.: Artificial Intelligence Meets Brain Theory. An Integrated Approach to Simulation Modelling of Natural Language Processing. In: Trappl, R.(ed.): Cybernetics and Systems Research. North-Holland, Amsterdam, and New York, 1982, pp.937-942.
[611] Gihman, I.I., Skorohod, A.V.: Controlled Stochastic Processes. Springer-Verlag, Berlin, FRG, and New York,

1979, VIII+238 pp.
[612] Gill, A.: Applied Algebra for the Computer Sciences. Prentice-Hall, Englewood Cliffs, New Jersey, 1976.
[613] Gill, P., Murray, W., Wright, M.H.: Practical Optimization. Academic Press, London, and New York, 1982, 420pp.
[614] Gilmore, R.: Catastrophe Theory for Scientists and Engineers. John Wiley, Chichester and New York, 1981, XX+666pp.
[615] Gini, G., Gini, M., Somalvico, M.: Deterministic and Nondeterministic Programming in Robot Systems. Cybernetics and Systems, 12, No.4, 1981, pp.345-362.
[616] Ginzberg, M.J.: A Prescription Model for System Implementation. Systems, Objectives, Solutions, 1, No.1, Jan.1981, pp.33-46.
[617] Gladun, V.P., Vaschenko, N.D.: Adaptive Problem-Solving Systems. Kybernetes, 9, No.3, 1980, pp.181-188.
[618] Glanville, R.: Consciousness: And so on. J. of Cybernetics, 10, No.4, 1980, pp.301-312.
[619] Glanville, R.: The Model's Dimensions: A Form for Argument. Int. J. of Man-Machine Studies, 13, No.3, 1980, pp.305-322.
[620] Glanville, R.: Inside Every White Box there are Two Black Boxes Trying to Get Out. Behavioral Science, 27, No.1, January 1982, pp.1-11.
[621] Glanville, R.: Distinguished and Exact Lies. In: Trappl, R.(ed.): Cybernetics and Systems Research, Vol.II. North-Holland, Amsterdam, and New York, 1984, pp.655-664.
[622] Glaser, F.B., Greenberg, S.W., Barrett, M.: A System Approach to Alcohol Treatment. Addiction Research Foundation, Toronto, 1978, XXII+304 pp.
[623] Glass, A.L., Holyoak, K.J., Santa, J.L.: Cognition. Addison-Wesley, Reading, Mass., 1979, XX+522 pp.
[624] Globus, G.G:: Is There a Ghost in the Machine after all? Dept. of Psychiatry and Human Behavior, Univ.of Calif. at Irvine, 1979, 16pp. [IFSR-Depository]
[625] Gloria-Bottini, F., Bottini, E., Ginzburg, L.R., Rowe, R.E.: Selective Polymorphism of the Erythrocyte Acid Phosphate Locus in Humans. In: Trappl, R.(ed.): Cybernetics and Systems Research, Vol.II. North-Holland, Amsterdam, and New York, 1984, pp.299-304.
[626] Glushkov, V.M., (et al.): Algebra, Languages Programming. Naukova Dumka, Kiev, 1978, 319 pp. [Russian]
[627] Goffman, W., Warren, K.S.: Scientific Information Systems and the Principle of Selectivity. Praeger, New York, 1980, X+192 pp.
[628] Goh, B.S.: Stability, Vulnerability and Persistence of Complex Ecosystems. Ecological Modelling, 1, No.2, July 1975, pp.105-116.
[629] Gokhale, D.V., Kullback, S.: The Information in Contingency Tables. Marcel Dekker, New York, 1978.
[630] Gold, H.J.: Mathematical Modeling of Biological Systems: an Introductory Guidebook. John Wiley, Chichester and New York, 1977, 357 pp.
[631] Golemanov, L.A., Hakkala, L.: On the Coordinability of Hierarchical Systems. Systems Science, 3, No.3, 1977, pp.285-297.
[632] Golledge, R.G., Rayner, J.N.(eds.): Proximity and Preference: Problems in the Multidimensional Analysis of Large Data Sets. Univ. of Minnesota Press, Minneapolis, 1982, XI+312pp.
[633] Gollmann, D.: On the Identification of Certain Non-Linear Networks of Automata. In: Trappl, R.(ed.): Cybernetics

and Systems Research. North-Holland, Amsterdam, and New York, 1982, pp.147-150.

[634] Gollmann, D.: On the Theory of Bilinear Shift Registers. In: Trappl, R.(ed.): Cybernetics and Systems Research, Vol.II. North-Holland, Amsterdam, and New York, 1984, pp.125-130.

[635] Golubitsky, M.: An Introduction to Catastrophe Theory and its Applications. SIAM Reviews, 20, No.2, 1978, pp.352-387 (also General Systems Yearbook, 24, 1979, pp.65-100).

[636] Gomez, P.: Top-Down Versus Bottom-Up Organizational Design: a Cybernetic Perspective. J. of Enterprise Management, 1, No.3, 1978, pp.229-239.

[637] Gomez, P., Probst, G.J.B.: Centralization Versus Decentralization in Business Organisations: Cybernetic Rules for Effective Management. Cybernetics and Systems, 11, No.4, 1980, pp.381-400.

[638] Gomez, P.: Systems-Methodology in Action: Organic Problem Solving in a Publishing Company. J. of Applied Systems Analysis, 9, April 1982, pp.67-85.

[639] Gonzales, R.C., Wintz, P.: Digital Image Processing. Addison-Wesley, Reading, Mass., 1977.

[640] Gonzales, R.C., Thomason, M.G.: Syntactic Pattern Recognition. Addison-Wesley, Reading, Mass., 1978, XIX+283 pp.

[641] Goodman, L.A.: Analyzing Qualitative / Categorical Data. Abt Books, Cambridge, Mass., 1979, VIII+471 pp.

[642] Goodman, L.A., Kruskal, W.H.: Measures of Association for Cross Classification. Springer-Verlag, Berlin, FRG, and New York, 1979, X+146 pp.

[643] Gopal, K., Aggarwal, K.K., Gupta, J.S.: An Event Expansion Algorithm for Reliability Evaluation in Complex Systems. Int. J. of Systems Science, 10, No.4, 1979, pp.363-371.

[644] Gorbatov, V.A.: Theory of Partially Ordered Systems. Sovietskoye Radio, Moscow, 1976, 336 pp. [Russian]

[645] Gottinger, H.W.(ed.): Systems Approaches and Environmental Problems. Vanderhoek-Ruprecht, Goettingen, 1974.

[646] Gottinger, H.W.: Complexity and Catastrophe: Applications of Dynamic System Theory. In: Rose, J., Bilcio, C.(eds.): Modern Trends in Cybernetics and Systems (3 Vols.). Springer-Verlag, Berlin, FRG, and New York, 1977, pp.13-26.

[647] Gottinger, H.W.: Toward an Algebraic Theory of Complexity in Dynamic Systems. J. of Cybernetics, 7, No.1-2, 1977, pp.69-100 (also General Systems Yearbook, 22, pp.73-83).

[648] Gottinger, H.W.: Structural Characteristics Economical Models: a Study of Complexity. Policy Analysis and Information Systems, 2, No.2, 1979, pp.11-30.

[649] Gould, P.: Q-Analysis or a Language of Structure: An Introduction for Social Scientists, Geographers and Planners. Int. J. of Man-Machine Systems, 13, No.2, 1980, pp.169-199.

[650] Gourlay, A.R., McLean, J.M., Shepherd, P.: Identification and Analysis of the Subsystems Structure of Models. Applied Mathematical Modelling, 1, No.5, June 1977, pp.245-252.

[651] Grabowski, J., Janiak, A.: Sequencing Problem with Resource Constraints. In: Trappl, R.(ed.): Cybernetics and Systems Research, Vol.II. North-Holland, Amsterdam, and New York, 1984, pp.329-334.

[652] Graedel, T.E., McGill, R.: Graphical Presentation of Results from Scientific Computer Models. Science, 215,

No.4537, 1982, pp1191-1198.
[653] Graham, J.H., Saridis, G.N.: Linguistic Decision
Structures for Hierarchical Systems. IEEE Trans. on
Systems, Man, and Cybernetics, SMC-12, No.3, 1982,
pp.325-333.
[654] Grams, R.R.: Systems-Analysis Workbook. Charles
C.Thomas, Springfield, Ill., 1972, 61 pp.
[655] Granero-Porati, M.I., Kron-Morelli, R., Porati, A.:
Random Ecological Systems with Structure:
Stability-Complexity Relationship. Bulletin of
Mathematical Biology, 44, No.1, 1982, pp.103-117.
[656] Grant, D,P., Gasparski, W.W.(eds.): Design Methods and
Theories: Vol.15, No.2. California Polytechnic State
Univ., San Luis Opisbo, Calif., 1982, 48pp.
[IFSR-Depository]
[657] Grant, D,P., Gasparski, W.W.(eds.): Design Methods and
Theories: Vol.15, No.3. California Polytechnic State
Univ., San Luis Opisbo, Calif., 1982, 46pp.
[IFSR-Depository]
[658] Grauer, M., Lewandovski, A., Schrattenholzer, L.: The
Generation of Efficient Energy Supply Strategies using
Multi-Criteria Optimization. In: Trappl, R.(ed.):
Cybernetics and Systems Research. North-Holland,
Amsterdam, and New York, 1982, pp.683-688.
[659] Green, M.A.: Solar Cells: Operating Principles,
Technology, and System Application. Prentice-Hall,
Englewood Cliffs, New Jersey, 1982, XIV+274 pp.
[660] Greenberg, H.J., Maybee, J.S.(eds.): Computer-Assisted
Analysis and Model Simplification. Academic Press,
London, and New York, (Symposium in Boulder, Colorado,
March 80), 1981, XII+260 pp.
[661] Greenberg, H.J., Maybee, J.S.(eds.): Computer-Assisted
Analysis and Model Simplification. Academic Press,
London, and New York, 1982, 563pp.
[662] Greene, D.H., Knuth, D.E.: Mathematics for the Analysis
of Algorithms. Birkhaeuser Verlag, Basel and Stuttgart,
1981, 108pp.
[663] Greenspan, D.: A Completely Arithmetic Formulation of
Classical and Special Relativistic Physics. Int. J. of
General Systems, 4, No.2, 1978, pp.105-112.
[664] Greenspan, D.: Arithmetic Applied Mathematics. Pergamon
Press, Oxford and New York, 1980, VIII+166 pp.
[665] Greenspan, D.: Discrete Modeling in the Microcosm and in
the Macrocosm. Int. J. of General Systems, 6, No.1, 1980,
pp.25-46.
[666] Greenspan, D.: A New Computer Approach to the Modeling of
Atoms, Ions and Molecules. (Part I: Hydrogen; Part II:
The Oxygen Molecule). J. of Computational and Applied
Mathematics, 7, 1981, (I: No.1, pp.41-49; II: No.2,
pp.129-133).
[667] Gregory, S.A.: Large Technological Projects. Cybernetics
and Systems, 11, No.4, 1980, pp.401-424.
[668] Greibach, S.A.: Theory of Program Structures: Schemes,
Semantics, Verification. Springer-Verlag, Berlin, FRG,
and New York, 1975.
[669] Grenander, U.: Regular Structures: Lectures in Pattern
Theory. Vol.3. Springer-Verlag, Berlin, FRG, and New
York, 1981, 576 pp.
[670] Grindley, K.: Systematics: a New Approach to Systems
Analysis. Petrocelli, New York, 1977.
[671] Groen, J.C.F.L.: A Constraint Analytic Measure for the
Distance between Neighbouring Structures. In: Trappl,
R.(ed.): Cybernetics and Systems Research.
North-Holland, Amsterdam, and New York, 1982, pp.57-64.

[672] Grogono, P., Nelson, S.H.: Problem Solving and Computer Programming. Addison-Wesley, Reading, Mass., 1982, XVI-284pp.

[673] Grollmann, J.: Description of Feedback Systems via Approximation by Delayed Systems. In: Trappl, R.(ed.): Cybernetics and Systems Research. North-Holland, Amsterdam, and New York, 1982, pp.157-162.

[674] Gross, D., Harris, C.M.: Fundamentals of Queueing Theory. John Wiley, Chichester and New York, 1974.

[675] Grubbstrom, R.W., Lundquist, J.: Theory of Relatively Closed Systems and Applications. Profil, Linkoping, Sweden, 1975, 131 pp.

[676] Grubbstrom, R.W., Lundquist, J.: The Axsater Integrated Production-Inventory System Interpreted in Terms of the Theory of Relatively Closed Systems. J. of Cybernetics, 7, 1977, pp.49-67.

[677] Gruber, H.: Modularization and Abstracting; a Discussion on the Basis of Sequential Systems. In: Trappl, R.(ed.): Cybernetics and Systems Research, Vol.II. North-Holland, Amsterdam, and New York, 1984, pp.155-160.

[678] Gruska, J.(ed.): Mathematical Foundations of Computer Science. Springer-Verlag, Berlin, FRG, and New York, 1977.

[679] Guida, G., Mandriolo, D., Somalvica, M.: An Integrated Model of Problem Solver. Information Sciences, 13, No.1, 1977, pp.11- 33.

[680] Guida, G., Somalvico, M.: Interactivity and Incrementality in Natural Language Understanding Systems. Cybernetics and Systems, 12, No.4, 1981, pp.363-383.

[681] Guida, G., Tasso, C.: Natural Language Access to Online Data Bases. In: Trappl, R.(ed.): Cybernetics and Systems Research. North-Holland, Amsterdam, and New York, 1982, pp.891-896.

[682] Guidorzi, R.P.: Multistructural Model Selection. In: Trappl, R.(ed.): Cybernetics and Systems Research. North-Holland, Amsterdam, and New York, 1982, pp.135-140.

[683] Gumovski, I., Mira, C.: Recurrences and Discrete Dynamic Systems. Springer-Verlag, Berlin, FRG, and New York, 1980, VI+272 pp.

[684] Gupta, M.M., Ragade, R.K., Yager, R.R.(eds.): Advances in Fuzzy Set Theory and Applications. North-Holland, Amsterdam, and New York, 1969.

[685] Gupta, M.M., Saridis, G.N., Gaines, B.R.(eds.): Fuzzy Automata and Decision Processes. Elsevier / North-Holland, New York, 1977.

[686] Gupta, M.M., Ragade, R.K., Yager, R.R.(eds.): Advances in Fuzzy Set Theory and Applications. North-Holland, Amsterdam, and New York, 1979, XV+753 pp.

[687] Gupta, M.M., Nikiforuk, P.N.: On the Design of an Adaptive Controller for Uncertain Plants via Liapunov Signal Synthesis Method. J. of Cybernetics, 9, No.1, 1979, pp.73-98.

[688] Gupta, M.M.: Hierarchical Dynamic Optimization for Linear Discrete Systems. J. of Cybernetics, 10, No.1-3, 1980, pp.41-76.

[689] Gupta, M.M.: Feedback Control Applications of Fuzzy Set Theory: A Survey. Proc. 8th World IFAC Congress, Kyoto, Japan, Vol.5, pp.1-7.

[690] Gurdial,, Taneja, I.J.: A Note on Fano Inequality Involving Distortion. J. of Cybernetics, 7, No.3-4, 1977, pp.249-256.

[691] Gurel, O., Rossler, O.E.(eds.): Bifurcation Theory and Applications in Scientific Disciplines. New York Academy of Sciences, New York, 1979, 708 pp.

[692] Gutenbaum, J., Niezgodka, M.(eds.): Applications of Systems Theory to Economics, Management, and Technology. Polish Sci. Publ., Warsaw, 1980, 631pp.

[693] Gutierrez, L.T., Fey, W.R.: Ecosystem Succession. M.I.T.Press, Cambridge, Mass., 1980.

[694] Gvishiani, J.M., (et al.)(eds.): Systems Research: Methodological Problems - 1979 Yearbook. Nauka, Moscow, 1980, 384 pp. [Russian]

[695] Gyarfas, F., Popper, M.: CODEX: Prototypes Driven Backward and Forward Chaining Computer-Based Diagnostic Expert System. In: Trappl, R.(ed.): Cybernetics and Systems Research, Vol.II. North-Holland, Amsterdam, and New York, 1984, pp.821-824.

[696] Gyula, P.: Common Laws of Sciences and Systems. Akademiai Kiado, Budapest, 1973. [Hungarian]

[697] Haack, S.: Do We Need 'Fuzzy Logic'? Int. J. of Man-Machine Studies, 11, No.4, July 1979, pp.437-445.

[698] Hafele, W., Kirchmayer, L.K.(eds.): Modelling of Large Scale Energy Systems. (IIASA Proceedings Series, Vol.12). Pergamon Press, Oxford and New York, 1981, XVI+490 pp.

[699] Hagg, C.: Possibility and Cost in Decision Analysis. Fuzzy Sets and Systems, 1, No.2, April 1978, pp.81-86.

[700] Haimes, Y.Y.: Hierarchical Analysis of Water Resources Systems: Modeling and Optimization of Large-Scale Systems. McGraw-Hill, New York, 1977, XVIII+478 pp.

[701] Haimes, Y.Y.: Hierarchical Holographic Modeling. IEEE Trans. on Systems, Man, and Cybernetics, SMC-11, No.9, Sept.1981, pp.606-617.

[702] Hajek, P., Havranek, T.: On Generation of Inductive Hypotheses. Int. J. of Man-Machine Studies, 9, No.4, July 1977, pp.415-434.

[703] Hajek, P., Havranek, T.: Mechanizing Hypothesis Formationmathematical Foundations of a General Theory. Springer-Verlag, Berlin, FRG, and New York, 1978.

[704] Hajek, P.(ed.): Special Issue of the International Journal of Man-Machine Studies on the Guha Method of Mechanized Hypothesis Formation. Int. J. of Man-Machine Studies, 10, No.1, Jan.1978, pp.1-93.

[705] Haken, H.(ed.): Synergetics - A Workshop. Springer-Verlag, Berlin, FRG, and New York, 1977, X+274 pp.

[706] Haken, H.: Synergetics. Springer-Verlag, Berlin, FRG, and New York, 1977.

[707] Haken, H.(ed.): Pattern Formation by Dynamic Systems and Pattern Recognition. Springer-Verlag, Berlin, FRG, and New York, 1979, VII+306 pp.

[708] Haken, H.(ed.): Dynamics of Synergetic Systems. Springer-Verlag, Berlin, FRG, and New York, 1980, VIII+272pp.

[709] Hakkala, L.: An Infeasible Gradient-Type Coordination Algorithm for Dynamical Systems. Int. J. of Systems Science, 8 No.5, May 1977, Pp.489-496.

[710] Haley, K.B.(ed.): Applied Operatons Research in Fishing. (Vol.10 in NATO Conference Series.). Plenum Press, New York and London, 1981, XVI+490 pp.

[711] Halfon, E., Reggiani, M.G.: Adequacy of Ecosystem Models. Ecological Modelling, 4, No.1, Jan.1978, pp.29-50.

[712] Halfon, E.(ed.): Theoretical Systems Ecology: Advances and Case Studies. Academic Press, London, and New York, 1979, XVI+516 pp.

[713] Hall, C.A.S., Day, J.W.(eds.): Ecosystem Modeling in Theory and Practice: an Introduction with Case Studies. John Wiley, Chichester and New York, 1977.

[714] Hall, E.L.: Computer Image Processing and Recognition.

Academic Press, London, and New York, 1979, 584 pp.

[715] Hallam, T.G., Simberloff, D.S.: On the Intrinsic Structure of Differential Equation Models of Cosystems. Ecological Modelling, 3, No.3, Aug.1977, pp.167-182.

[716] Halldin, S.(ed.): Comparison of Forest Water and Energy Exchange Models. (Proc. of an IUFRO Workshop, Uppsala, Sweden, Sept.24-30, 1978). ISEM, Copenhagen and Elsevier, Amsterdam, 1978.

[717] Halmes, A., (et al.)(eds.): Topics in Systems Theory: Publication in Honour of Prof.Hans Blomberg. Acta Polytechnica Scandinavica, Mathematics and Computer Science Series, No.31, Helsinki, 1979, 191 pp.

[718] Hanassab, S., Fatmi, H.A.: Parallel Processors for Cybernetic Systems. Cybernetics and Systems, 11, No.1-2, 1980, pp.179-192.

[719] Hanassab, S., Fatmi, H.A.: An Interpolator Based Controller for Cybernetic Systems. In: Trappl, R.(ed.): Cybernetics and Systems Research. North-Holland, Amsterdam, and New York, 1982, pp.165-168.

[720] Hanken, A.F.G.: Cybernetics and Society: An Analysis of Social Systems. Heyden, Philadelphia, 1981.

[721] Hanken, A.F.G., Reuver, H.A.: Social Systems and Learning Systems. Martinus Nijhoff, Boston and The Hague, 1981, X+246 pp.

[722] Hanson, O.J.: The Role of Systems Analysis in Designing Systems and Methods of Training Using Systems Techniques. In: Gasparski, W.W.(ed.): Problems of Design Methodology. Panstwowe Wydawnictwo Naukowe, Warsaw, 1977.

[723] Hanson, O.J.: Assessing and Modifying an Advanced Business Systems Analysis Course Using Feedback from Past Students. In: Trappl, R.(ed.): Cybernetics and Systems Research. North-Holland, Amsterdam, and New York, 1982, pp.603-608.

[724] Hanson, O.J.: Review and Re-Design of an Advanced Systems Analysis Course. In: Trappl, R.(ed.): Cybernetics and Systems Research, Vol.II. North-Holland, Amsterdam, and New York, 1984, pp.371-380.

[725] Haralik, R.M., Davis, L.S., Rosenfeld, A., Milgram, D.L.: Reduction Operation for Constraint Satisfaction. Information Sciences, 14, No.3, June 1978, pp.199-219.

[726] Haralik, R.M.: The Characterization of Binary Relation Homomorphisms. Int. J. of General Systems, 4, No.2, 1978, pp.113-121.

[727] Harris, L.R.: Using the Data Base as a Semantic Component to Aid in the Parsing of Natural Language Data Base Queries. J. of Cybernetics, 10, No.1-3, 1980, pp.77-96.

[728] Harrison, M.A.: Introduction to Formal Language Theory. Addison-Wesley, Reading, Mass., 1978, 594 pp.

[729] Hartl, R.F., Mehlmann, A.: Optimal Seducing Policies for Dynamic Continuous Lovers Under Risk of Being Killed a Rival. Cybernetics and Systems, 15, No.1-2, 1984, pp.119-126.

[730] Hartley, R., (et al.)(eds.): Recent Developments in Markov Decision Processes. Academic Press, London, and New York, 1980, XIV+334 pp.

[731] Hartnett, W.E.(ed.): Systems: Approaches, Theories, Applications. D.Reidel, Dordrecht, Holland, and Boston, 1977.

[732] Haruna, K., Komoda, N.: Structural Sensitivity Analysis and its Applications. Large Scale Systems, 1, No.2, 1980.

[733] Hassell, M.P.: The Dynamics of Arthropod Predator-Prey Systems. Princeton University Press, Princeton, N.J., 1978, 237 pp.

[734] Hastings, N.A.J., Mello, J.M.C.: Decision Networks. John

Wiley, Chichester and New York, 1978, IX+196 pp.
[735] Hattiangadi, J.N.: The Structure of Problems. Philosophy
of Social Sciences, Part 1: 8, No.4, 1978; Part 2: 9,
No.1, 1979.
[736] Hatze, H.: Myocybernetics - A New Development in
Neuromuscular Control. J. of Cybernetics, 10, No.4, 1980,
pp.341-348.
[737] Havranek, T., Chyba, M., Pokorny, D.: Processing
Sociological Data by the Guha Method: an Example. Int.
J. of Man-Machine Studies, 9, No.4, July 1977, pp.439-447.
[738] Havranek, T.: An Alternative Approach to Missing
Information in the Guha Method. Kybernetika, 16, No.2,
1980, pp.145-1555.
[739] Hawkes, N.(ed.): International Seminar on Trends in
Mathematical Modelling. Springer-Verlag, Berlin, FRG, and
New York, 1971, 288 pp.
[740] Hawkes, N.(ed.): International Seminar on Trends in
Mathematical Modelling: a UNESCO Report.
Springer-Verlag, Berlin, FRG, and New York, 1973, VI+288
pp.
[741] Hawkins, D.M.(ed.): Topics in Applied Multivariate
Analysis (2 volumes). Nat.Res.Inst. for Mathematical
Sciences, Pretoria, SA, January 1981, XVI+466 pp.
[742] Haydon, P.G., Holden, A.V., Winlow, W.: Coherent and
Incoherent Computation in Identified Molluscan Neurones.
In: Trappl, R.(ed.): Cybernetics and Systems Research.
North-Holland, Amsterdam, and New York, 1982, pp.265-270.
[743] Hayes, P.: Trends in Artificial Intelligence. Int. J. of
Man-Machine Studies, 10, No.3, May 1978, pp.295-299.
[744] Hazlett, B.A.(ed.): Quantitative Methods in the Study of
Animal Behavior. Academic Press, London, and New York,
1977, X+222 pp.
[745] Hazony, Y.: Problem Solving with Interactive Graphics.
Perspectives in Computing, 1, No.1, Feb.1981, pp.23-28.
[746] Heim, R., Palm, G.: Theoretical Approaches to Complex
Systems. Springer-Verlag, Berlin, FRG, and New York,
1978.
[747] Hellerman, L.: Symbolic Initiative and its Application to
Computers. Int. J. of General Systems, 8, No.3, 1982,
pp.161-168.
[748] Hellwig, Z.H., Fischer, M.M.: Basic Problems of Modern
Econometrics and Some Problems of its Acceptability and
Interpretation. Austrian Society for Cybernetic Studies,
Vienna, 1983.
[749] Hendel, R.J.: The Grue Paradox: An Information-Theoretic
Solution. Yeshiva Univ., Manhattan, NY, 1979, 24pp.
[IFSR-Depository]
[750] Henize, J.: Modelling Large Scale Socio-Economic Systems:
Can We Begin to Do Better. Large Scale Systems, 1, No.2,
1980, pp.89-105.
[751] Herman, G.T.(ed.): Image Reconstruction From Projections:
Implementation and Applications. Springer-Verlag, Berlin,
FRG, and New York, 1979, XII+286 pp.
[752] Hermann, R., Martin, C.F.: Applications of Algebraic
Geometry to Systems Theory, Part 1 IEEE Trans. on
Automatic Control, AC-22 No.1, Feb.1977, pp.19-25.
[753] Hickin, J., Sinha, N.K.: Applications of Projective
Reduction Methods to Estimation and Control. J. of
Cybernetics, 8, No.2, 1978, pp.159-182.
[754] Higashi, M., Klir, G.J.: Measures of Uncertainty and
Information Based on Possibility Distributions. Int. J.
of General Systems, 9, No.1, 1982, pp.43-58.
[755] Higashi, M., Klir, G.J.: On Measures of Fuzziness and
Fuzzy Completeness. Int. J. of General Systems, 8, No.3,

        1982, pp.169-180.
[756] Higashi, M.,   Klir,   G.J.:   On the Notion of Distance
        Representing Information Closeness: Possibility and
        Probability Distributions. Int. J. of General Systems, 9,
        No.2, 1983, pp.103-116.
[757] Hirata, H., Fukao, T.: A Model of Mass and Energy Flow in
        Ecosystems.   Mathematical   Biosciences,   33,   1977,
        pp.321-334.
[758] Hirata, H.:   A Model of Hierarchical Ecosystems with
        Migration. Bulletin of Mathematical Biology, 42, 1980,
        pp.119-130.
[759] Hirata, H.:   A Model of Hierarchical Ecosystems with
        Utility Efficiency Mass and its Stability. Int. J. of
        Systems Science, 11, No.4, April 1980, pp.487-493.
[760] Hirzinger, G.:   Robot-Teaching via Force-Torque-Sensors.
        In: Trappl, R.(ed.):   Cybernetics and Systems Research.
        North-Holland, Amsterdam, and New York, 1982, pp.955-964.
[761] Hisdal, E.:   Conditional Possibilities, Independence and
        Noninteraction Fuzzy Sets and Systems, 1, No.4, 1978,
        pp.283-297.
[762] Hisdal, E.:   Generalized   Fuzzy   Set   Systems   and
        Particularization. Fuzzy Sets and Systems, 4, No.3, 1980,
        pp.275-291.
[763] Hoaglin, D.C., (et al.): Data for Decisions: Information
        Strategies for Policymakers. Abt Books, Cambridge, Mass.,
        1982, XXII+318pp.
[764] Hoehle, U.:   A Remark on Entropies with Respect to
        Plausibility Measures. In: Trappl, R.(ed.): Cybernetics
        and Systems Research.   North-Holland, Amsterdam, and New
        York, 1982, pp.735-738.
[765] Hoernig, K.M.: Aspects of Automatic Program Construction.
        In: Trappl, R.(ed.):   Cybernetics and Systems Research.
        North-Holland, Amsterdam, and New York, 1982, pp.909-916.
[766] Hoffer, J.A.,   Kennedy,   M.H.:   Factors Affecting the
        Comparative Advantages of Sequential and Batched Decision
        Making. J. of Cybernetics, 10, No.1-3, 1980, pp.97-116.
[767] Hofferbert, R.I.,   Schaefer,   G.F.:   The Application of
        General Systems Methodology to the Comparative Study of
        Public Policy. Int. J. of General Systems, 8, No.2, 1982,
        pp.93-108.
[768] Hofmann, Th.R.:   Qualitative Terms for Quantity.   In:
        Trappl, R.(ed.):   Cybernetics and Systems Research.
        North-Holland, Amsterdam, and New York, 1982, pp.403-408.
[769] Hogeweg, P.: Locally Synchronised Developmental Systems:
        Concrete Advantages of Discrete Event Formalism. Int. J.
        of General Systems, 6, No.2, 1980, pp.57-73.
[770] Holden, A.V.: Cybernetic Approach to Global and Local
        Mode of Action of General Anestetics. J. of Cybernetics,
        9, No.2, 1979, pp.143-150.
[771] Holden, A.V., Muhamad, M.A.: Chaotic Activity in Neural
        Systems. In: Trappl, R.(ed.): Cybernetics and Systems
        Research, Vol.II. North-Holland, Amsterdam, and New York,
        1984, pp.245-250.
[772] Holler, M.J.:   Markets   as   Self-Policing   Quality
        Enforcement Systems. In: Trappl, R.(ed.): Cybernetics
        and Systems Research, Vol.II. North-Holland, Amsterdam,
        and New York, 1984, pp.477-482.
[773] Holling, C.S.: Resilience and Stability of Ecological
        Systems. Annual Review of Ecology and Systematics, 4,
        1973, pp.1-23.
[774] Holling, C.S.(ed.): Adaptive Environmental Assessment and
        Management. John Wiley, Chichester and New York, 1978,
        XVIII+377 pp.
[775] Holst, P.: Computer Simulation 1951-1976: An Index to

the Literature. Mansell, London, 1979, 438 pp.

[776] Honkasalo, A.: Design of Work Tasks and Education of Labor Force. In: Trappl, R.(ed.): Cybernetics and Systems Research. North-Holland, Amsterdam, and New York, 1982, pp.595-602.

[777] Honkasalo, A.: Entropic Processes and Economic Systems. In: Trappl, R.(ed.): Cybernetics and Systems Research, Vol.II. North-Holland, Amsterdam, and New York, 1984, pp.429-434.

[778] Hooda, D.S., Tuteja, R.K.: Two Generalized Measures of "Useful" Information. Information Sciences, 23, No.1, Feb.1981, pp.11-24.

[779] Hooker, C.A.(ed.): Physical Theory as Logico-Operational Structure. D.Reidel, Dordrecht, Holland, and Boston, 1979, XVIII+336 pp.

[780] Hopcroft, J.E., Ulman, J.D.: Introduction to Automata Theory, Languages, and Computation. Addison-Wesley, Reading, Mass., 1979, 418 pp.

[781] Horacek, H., Buchberger, E.: On Generation of Anaphora and Gapping in German. In: Trappl, R.(ed.): Cybernetics and Systems Research, Vol.II. North-Holland, Amsterdam, and New York, 1984, pp.759-766.

[782] Horn, W., Buchstaller, W., Retti, J.: Interactive Computer-Aided Planning: Estimating Health-Care Resource Requirements. Cybernetics and Systems, 11, No.1-2, 1980, pp.167-178.

[783] Horn, W., Retti, J., Buchstaller, W., Trappl, R.: Health Care in Austria: 1980-2000. Austrian Society for Cybernetic Studies, Vienna, 1981. [German]

[784] Horn, W., Trappl, R.: Fast Walsh vs.Fast Fourier Transform: A Comparison of Time-Efficiency. In: Trappl, R., Klir, G.J., Pichler, F.(eds.): Progress in Cybernetics and Systems Research, Vol.VIII. Hemisphere, Washington, D.C., 1982, pp.329-332.

[785] Horn, W., Buchstaller, W., Trappl, R.: The Structure of Manifestations in a Medical Consultation System. In: Trappl, R.(ed.): Cybernetics and Systems Research. North-Holland, Amsterdam, and New York, 1982, pp.927-932.

[786] Horn, W.: An Artificial Intelligence System for Medical Decision Support. Austrian Society for Cybernetic Studies, Vienna, 1983. [German]

[787] Horn, W., Trappl, R., Ulrich, D., Chroust, G.: A Frame-Based Real-Time Graphic Interaction System. In: Trappl, R.(ed.): Cybernetics and Systems Research, Vol.II. North-Holland, Amsterdam, and New York, 1984, pp.825-830.

[788] Horowitz, E., Sahni, S.: Fundamentals of Data Structures. Computer Science Press, Woodland Hills, Calif., 1976.

[789] Horvath, T.: Method and Systematic Reflections. Ultimate Reality and Meaning, 3, No.2, 1980, pp.144-163.

[790] Hoskins, R.F.: Generalised Functions. John Wiley, Chichester and New York, 1979, XII+351 pp.

[791] Hossforf, H., Van Amerongen, C.: Model Analysis of Structures. Van Nostrand Reinhold, New York, 1974, 235 pp.

[792] Hostetter, G.H.: Fundamentals of Network Analysis. Harper and Row, New York, 1979, XX+474 pp.

[793] Hoye, R.E., Bryant, D.T.: Current Status of Hospital Management Information Systems. Systems Research, 1, No.1, 1984, pp.55-62.

[794] Hu, N.C.: Multi-Dimensional FFT Computation. In: Trappl, R.(ed.): Cybernetics and Systems Research, Vol.II. North-Holland, Amsterdam, and New York, 1984, pp.137-142.

[795] Huber, O.: Nontransitive Multidimensional Preferences: Theoretical Analysis of a Model. Theory and Decision, 10, 1979, pp.147-165.

[796] Hull, R., Sadek, K.E., Willmer, M.A.P.: Communication and Information: Their Relationship to Computer-Based Information Systems. In: Trappl, R.(ed.): Cybernetics and Systems Research. North-Holland, Amsterdam, and New York, 1982, pp.827-832.

[797] Hunt, E.B.: Artificial Intelligence. Academic Press, London, and New York, 1975, 468 pp.

[798] Hunt, L.R.: Controllability of General Nonlinear Systems. Mathematical Systems Theory, 12, 1979, pp.361- 370.

[799] Hutt, A.T.F.: A Relational Data Base Management System. John Wiley, Chichester and New York, 1979, XVIII+226 pp.

[800] Huxham, C.S., Bennett, P.G.: Hypergames as a Soft Systems Methodology. In: Trappl, R.(ed.): Cybernetics and Systems Research. North-Holland, Amsterdam, and New York, 1982, pp.229-234.

[801] Hwang, C.L., Masud, A.S.: Multiple Objective Decision Making: Methods and Applications (a State-of-the-Art Survey). Springer-Verlag, Berlin, FRG, and New York, 1979, XII+352 pp.

[802] Hwang, C.L.: Decomposition of Binary Sequences of Finite Periods and a Class of Error Correcting Codes. Digital Processes, 6, No.1, 1980, pp.1 - 20.

[803] Hwang, C.L., Yoon, K.: Multiple Attribute Decision Making: Methods and Applications: A State-of-the-Art Survey. Springer-Verlag, Berlin, FRG, and New York, 1981, XI+259pp.

[804] Hwang, C.L., Tillman, F.A., Lee, M.H.: System-Reliability Evaluation Techniques for Complex/Large Systems: A Review. IEEE Trans. on Reliability, R-30, No.5, December 1981, pp.416-423.

[805] Hyde, M.O.: Computers that Think?: The Search for Artificial Intelligence. Enslow, Hillside, N.J., 1982, XII+202pp.

[806] Hyvarinen, L.P.: Information Theory for Systems Engineers. Springer-Verlag, Berlin, FRG, and New York, 1970, 197 pp.

[807] Ibidapo-Obe, O.: An Approximate Differential Structure for Error Covariance in Nonlinear Dynamical Systems. J. of Cybernetics, 7, No.1-2, 1977, pp.15-22.

[808] Ihara, J.: A Structural Analysis of Criteria for Selecting Model Variables. IEEE Trans. on Systems, Man, and Cybernetics, SMC-10, No.8, August 1980, pp.460-466.

[809] Iijima, J.: Serial Decomposition of General Systems. In: Trappl, R.(ed.): Cybernetics and Systems Research. North-Holland, Amsterdam, and New York, 1982, pp.73-78.

[810] Ingemarsson, I.: Toward a Theory of Unknown Functions. IEEE Trans. on Information Theory, IT-24, No.2, March 1978, pp.238-240.

[811] Inmon, W.H.: Effective Data Base Design. Prentice-Hall, Englewood Cliffs, New Jersey, 1981, XI+228pp.

[812] Innis, G.S.(ed.): New Directions in the Analysis of Ecological Systems. Simulation Council, La Jolla, Ca., 1977, 248 pp.

[813] Innocent, P.R.: Toward Self-Adaptive Interface System. Int. J. of Man-Machine Studies, 16, No.3, April 1982, pp.287-299.

[814] Iri, M., Fujishige, S.: Use of Matroid Theory in Operations Research, Circuits and Systems Theory. Int. J. of Systems Science, 12, No.1, 1981, pp.27-54.

[815] Iserman, R.(ed.): Identification and System Parameter Estimation. Pergamon Press, Oxford and New York, 1980.

[816] Iserman, R.(ed.): System Identification (Papers from an IFAC Symposium in Darmstadt, W.Germany , Sept.1979). Pergamon Press, Oxford and New York, 1981.

[817] Ivan, I.T., Arhire, R.I.: System of Indicators for Estimating the Software Performances. In: Trappl, R.(ed.): Cybernetics and Systems Research. North-Holland, Amsterdam, and New York, 1982, pp.765-772.

[818] Jaburek, W.J.: Teleconferencing via Videotex. In: Trappl, R.(ed.): Cybernetics and Systems Research, Vol.II. North-Holland, Amsterdam, and New York, 1984, pp.593-598.

[819] Jacak, W., Tchon, K.: Events and Systems. Systems Science, 1, No.2, 1975, pp.77-88.

[820] Jacak, W., Tchon, K.: On Goal-Oriented Systems. Systems Science, 3, No.3, 1977, pp.307-313.

[821] Jacak, W., Rezenblit, J.: Modelling of Evenistic Systems. Postepy Cybernetyki, 4, No.3, 1981, pp.65-73.

[822] Jachymczyk, W., Stachowicz, M.: Karhunen-Loewe Transform Approximation for Certain Class of Random Processes. In: Trappl, R.(ed.): Cybernetics and Systems Research. North-Holland, Amsterdam, and New York, 1982, pp.175-178.

[823] Jack, W., Tchon, K.: On the Eventistic Approach to the Notion of System. Technical Univ. of Wroclaw, Poland, 1977, 13pp. [IFSR-Depository]

[824] Jacob, F.: The Possible and the Actual. Univ. of Washington Press, Seattle, Wash., 1982, 72pp.

[825] Jacobs, B.E.: On Generalized Computational Complexity. J. of Symbolic Logic, 42, No.1, March 1977, pp.47-58.

[826] Jacoby, S.L.S., Kowalik, J.S.: Mathematical Modeling with Computers. Prentice-Hall, Englewood Cliffs, New Jersey, 1980, XI+292 pp.

[827] Jahoda, M.: Wholes and Parts, Meaning and Mechanism. Nature, 296, 1982, pp.497-498.

[828] Jain, V., Christakis, A.N.: A Systemic Assessment and Application of the Ekistic Typology. Ekistics, Nov./Dec.1980.

[829] Jakubowski, R., Krol, J.: General Algorithms of Complex Dynamic Systems Simulation. Systems Science, 1, No.1, 1975, pp.17-27.

[830] Jakubowski, R.: Syntactic Characterization of Machine Parts Shapes. Cybernetics and Systems, 13, No.1, 1982, pp.1-24.

[831] James, D.J.G., McDonald, J.J.: Case Studies in Mathematical Modelling. John Wiley, Chichester and New York, 1981, 224pp.

[832] Janko, J.: Systems Approach in Natural Science Reflecting Orientation Towards Practice: and Episode from the History of Czech Science in the Middle of the 19th Century. Teorie Rozvoje Vedy, 3, No.4, 1979, pp.39-48.

[833] Janssens, L., Hoebeke, L., Michiels, H.: The Application of Cybernetic Principles for the Training of Operator Crews of Nuclear Power Plants. In: Trappl, R.(ed.): Cybernetics and Systems Research, Vol.II. North-Holland, Amsterdam, and New York, 1984, pp.387-390.

[834] Jantch, E.: The Self-Organizing Universe. Pergamon Press, Oxford and New York, 1980 XVII+343 pp.

[835] Jantch, E.(ed.): The Evolutionary Vision: Toward a Unifying Paradigm of Physical, Biological and Sociocultural Evolution (AAAS Selected Symposium). Westview Press, Boulder, Colorado, 1981.

[836] Jantzen, J.: A Generalized Fuzzy Relational Matrix. Int. J. of Man-Machine Studies, 13, No.4, 1980, pp.413-421.

[837] Jarisch, W., Detwiler, J.S.: Stochastic Modeling of Fetal Heart Rate Variability with Applications to Clinical

Risk-Assessment   and   Detection   of   Fetal   Respiratory
Movements-Preliminary Results.   Cybernetics   and   Systems,
11, No.3, 1980, pp.215-244.
[838] Jarisch, W., Allen, R., Hsu, K.: Signal Analysis of
Visual Evoked Potentials. In: Trappl, R.(ed.):
Cybernetics and Systems Research. North-Holland,
Amsterdam, and New York, 1982, pp.281-286.
[839] Jaron, J.: Goal-Oriented Cybernetical Systems. Systems
Science, 1, No.1, 1975, pp.5-16.
[840] Jaron, J.: Relations and Topologies in the Goal-Space of
a System. Systems Science, 3, No.2, 1977, pp.101-118.
[841] Jean, R.V.: A Methodological Paradigm in Plant Biology.
Rimuoski, Quebec, 1982, 31pp. [IFSR-Depository]
[842] Jeffers, J.N.R.: An Introduction to Systems Analysis:
with Ecological Applications. Edward Arnold, London,
1978, X+198 pp.
[843] Jennergren, L.P.: The Multilevel Approach: a Systems
Analysis Methodology. J. of Systems Engineering, 4, No.2,
Jan.1976, pp.97-106.
[844] Jensen, H.L., Horvitz, J.S.: A Theoretical Framework for
Quantifying Legal Decisions. Jurimetrics Journal, 20,
No.2, 1979, pp.121-139.
[845] Jezek, J.: Universal Algebra and Theory of Models. SNTL,
Prague, 1976, 226 pp. [Czech]
[846] Joergensen, S.E.(ed.): State of the Art in Ecological
Modelling. Int. Society for Ecological Modelling,
Copenhagen, 1979, 891 pp.
[847] Joerres, R., Martin, W., Brinkmann, K.: Identification of
the Temperature Masking of the Circadian System of Euglena
Gracilis. In: Trappl, R.(ed.): Cybernetics and Systems
Research, Vol.II. North-Holland, Amsterdam, and New York,
1984, pp.293-298.
[848] Johnson, D.T., Schubert, L.K.: A Planning Control
Strategy that Allows for the Cost of Planning. In:
Trappl, R.(ed.): Cybernetics and Systems Research.
North-Holland, Amsterdam, and New York, 1982, pp.965-972.
[849] Johnson, L.: A Theoretical Framework for Human
Communication. In: Trappl, R.(ed.): Cybernetics and
Systems Research. North-Holland, Amsterdam, and New York,
1982, pp.823-826.
[850] Johnson, L., Hartley, R.T.: Conversation Theory and
Cognitive Coherence Theories. In: Trappl, R.(ed.):
Cybernetics and Systems Research, Vol.II. North-Holland,
Amsterdam, and New York, 1984, pp.647-650.
[851] Johnson, P.E., Hassebrock, F.: Validating Computer
Simulation Models of Expert Reasoning. In: Trappl,
R.(ed.): Cybernetics and Systems Research.
North-Holland, Amsterdam, and New York, 1982, pp.921-926.
[852] Johnson, W.A., Shankar, P.: Applications of Systems
Theory to Ecology. George Washington Univ., Maryland,
1979, 30pp. [IFSR-Depository]
[853] Jones, A.H.: Cybernetic Design Features for Computerised
Business Information Systems. In: Trappl, R.(ed.):
Cybernetics and Systems Research. North-Holland,
Amsterdam, and New York, 1982, pp.815-822.
[854] Jones, B.: Determination of Reconstruction Families.
Int. J. of General Systems, 8, No.4, 1982, pp.225-228.
[855] Jones, B.: General Systems Theory and Algorithm Theory.
Int. J. of General Systems, 9, No.3, 1983, pp.157-160.
[856] Jones, D.D.: Catastrophe Theory Applied to Ecological
Systems. Simulation, 29, No.1, July 1977, pp.1-15.
[857] Jones, L.M.: Defining Systems Boundaries in Practice:
Some Proposals and Guidelines. J. of Applied Systems
Analysis, 9, April 1982, pp.41-55.

[858] Jorgenson, S.E., Harleman, D.R.F.(eds.): Hydrophysical and Ecological Modelling of Deep Lakes and Reservoirs. IIASA, Laxenburg, Austria, 1978, IX+38 pp.

[859] Jorgenson, S.E.(ed.): Hydrophysical and Ecological Models of Shallow Lakes (Summary of an IIASA Workshop). IIASA, Laxenburg, Austria, 1979, VIII+159 pp.

[860] Jorgenson, S.E.(ed.): State of the Art in Ecological Modelling. Int. Society for Ecological Modelling, Copenhagen, 1979, 891pp.

[861] Jumarie, G.: Some Technical Applications of Relativistic Information Fuzzy Sets, Linguistics, Relativistic Sets and Communication. Cybernetica, 20, No.2, 1977, pp.91-128.

[862] Jumarie, G.: A Relativistic Approach to Modelling Dynamic Systems Involving Human Factors. Int. J. of Systems Science, 10, No.1, 1979, pp.89-112.

[863] Jumarie, G.: A New Approach to the Transinformation of Random Experiments Involving Fuzzy Observations. In: Trappl, R.(ed.): Cybernetics and Systems Research, Vol.II. North-Holland, Amsterdam, and New York, 1984, pp.567-572.

[864] Jutila, S.T.: Social Gaming and Control by Dislocations. Int. J. of General Systems, 6, No.3, 1980.

[865] Kahne, S.: Introducing Systems Concepts to All University Students. Engineering Education, Feb.1980, pp.427-429.

[866] Kahneman, D., Slovic, P., Tversky, A.(eds.): Judgement Under Uncertainty: Heuristics and Biases. Cambridge Univ. Press, Cambridge, Mass., 1982, XIV + 556 pp.

[867] Kaindl, H.: Dynamic Control of the Quiescence Search in Computer Chess. In: Trappl, R.(ed.): Cybernetics and Systems Research. North-Holland, Amsterdam, and New York, 1982, pp.973-978.

[868] Kaiser, J.S.: Organizational Behavior: A Vector of Single-Valued Transformations within a Determinate Machine. J. of Cybernetics, 10, No.1-3, 1980, pp.117-134.

[869] Kak, S.: On Efficiency of Chemical Homeostasis - an Information-Theoretic Approach. IEEE Trans. on Systems, Man, and Cybernetics, SMC-9, No.3, March 1979, pp.160-163.

[870] Kalaba, R.E., Tesfatsion, L.: Two Solution Techniques for Adaptive Reinvestment: A Small Sample Comparison. J. of Cybernetics, 8, No.1, 1978, pp.101-112.

[871] Kalata, P., Priemer, R.: On System Identification with and without Certainty. J. of Cybernetics, 8, No.1, 1978, pp.31-50.

[872] Kalman, R.E.: On Partial Realizations, Transfer Functions, and Canonical Forms. Acta Polytechnica Scandinavica, Mathematics and Computer Science Series, No.31, Helinski, 1979, pp.9-32.

[873] Kandel, A., Byatt, W.J.: Fuzzy Sets, Fuzzy Algebra, and Fuzzy Statistics. IEEE Proc., 66, No.12, 1978, pp.1619-1639.

[874] Kandel, A., Lee, S.C.: Fuzzy Switching and Automata: Theory and Applications. Crane Russak, New York, and Edward Arnold, London, 1979, X+303 pp.

[875] Kandel, A., Byatt, W.J.: Fuzzy Processes. Fuzzy Sets and Systems, 4, No.2, 1980, pp.117-152.

[876] Kania, A.: On Stability of Formal Fuzziness Systems. Information Sciences, 22, No.1, 1980, pp.21-68.

[877] Kannellakis, P.C.: On the Computational Complexity of Cardinality Constraints in Relational Databases. Information Processing Letters, 11, No.2, Oct.1980, pp.98-101.

[878] Kantor, J.R.: System Structure and Scientific Psychology. Psychological Record, 23, 1973, pp.451-458.

[879] Kaposi, A., Rzevski, G.: Research into the Assessment of

Complexity. In: Gasparski, W.W., Miller, D.(eds.): Design and System: Methodological Aspects, Vol.3. Polish Academy of Sciences, Warsaw, 1981, pp.49-71.

[880] Kapur, J.N.: On Maximum-Entropy Complexity Measures. Int. J. of General Systems, 9, No.2, 1983, pp.95-102.

[881] Karlin, S., Taylor, H.M.: A Second Course in Stochastic Processes. Academic Press, London, and New York, 1981, XVIII+542 pp.

[882] Kashyap, R.L.: A Bayesian Comparison of Different Classes of Dynamic Models Using Empirical Data. IEEE Trans. on Automatic Control, CA-22, No.5, Oct.1977, pp.715-727.

[883] Katz, M.J.: Questions of Uniqueness and Resolution in Reconstruction from Projections. Springer-Verlag, Berlin, FRG, and New York, 1978, 175pp.

[884] Katz, M.J., Goffman, W.: Preformation of Autogenetic Patterns. Philosophy of Science, 48, No.3, Sept.1981, pp.438-453.

[885] Kauffman, L.H.: Network Synthesis and Varela's Calculus. Int. J. of General Systems, 4, No.3, 1978, pp.179-188.

[886] Kaufmann, A., Grouchko, D., Cruon, R.: Mathematical Models for the Study of the Reliability of Systems. Academic Press, London, and New York, 1977, 221 pp.

[887] Kaufmann, G.: Visual Imagery and Its Relation to Problem Solving. Columbia Univ. Press, New York, 1980.

[888] Kaushik, M.L.: Burst-Error-Correcting Codes with Weight-Constraints Under a New Metric. J. of Cybernetics, 8, No.2, 1978, pp.183-202.

[889] Kaushik, M.L.: Single Error and Burst Error Correcting Codes Through a New Metric. J. of Cybernetics, 9, No.1, 1979, pp.1-16.

[890] Kavakoglu, I.(ed.): Mathematical Modelling of Energy Systems. Sijthoff and Nordhoff, Alphen aan den Rijn (The Nederlands), 1981, XIV + 476pp.

[891] Kawamura, K., Da Silvera e Silva, W.: On Developing the Microelectronics Industry: A Systems Approach. Systems Research, 1, No.1, 1984, pp.45-54.

[892] Kaynak, E.: A Holistic Approach in Marketing Planning. In: Trappl, R.(ed.): Cybernetics and Systems Research. North-Holland, Amsterdam, and New York, 1982, pp.561-566.

[893] Kedem, B.: Binary Time Series. Marcel Dekker, New York, 1980.

[894] Keen, G.P.W., Morton, M.S.S.: Decision Support Systems. Addison-Wesley, Reading, Mass., 1978, 264 pp.

[895] Keen, G.P.W., Morton, M.S.S.: Decision Support Systems: an Organizational Perspective. Addison-Wesley, Reading, Mass., 1978, 264 pp.

[896] Keeny, B.P.: Ecosystemic Epistemology: an Alternative Paradigm for Diagnosis. Family Process, 18, No.2, 1979, pp.117-130.

[897] Kefalas, A.G.St.: Management by Wisdom: On the Art and Science of Anticipatory Management. In: Trappl, R.(ed.): Cybernetics and Systems Research. North-Holland, Amsterdam, and New York, 1982, pp.421-426.

[898] Kekes, J.: The Centrality of Problem Solving. Inquiry, 22, No.4, 1979, pp.405-421.

[899] Kellerman, W., Pichet, J.: Computer-Aided Identification of System Structure: An Attempt at Implementation. In: Gasparski, W.W., Miller, D.(eds.): Design and System: Methodological Aspects, Vol.3. Polish Academy of Sciences, Warsaw, 1981, pp.73-93. [Polish]

[900] Kellermayr, K.H.: Performance Simulation of an Optical Bus Area Network. In: Trappl, R.(ed.): Cybernetics and Systems Research, Vol.II. North-Holland, Amsterdam, and New York, 1984, pp.605-610.

[901] Kennedy, P.: A Guide to Econometrics. M.I.T.Press, Cambridge, Mass., 1979, XII+176 pp.

[902] Kent, W.: Data and Reality. North-Holland, Amsterdam, and New York, 1978, XV+211 pp.

[903] Keravnou, E.T., Johnson, L.: Design of Expert Systems from the Perspective of Conversation Theory Methodology. In: Trappl, R.(ed.): Cybernetics and Systems Research, Vol.II. North-Holland, Amsterdam, and New York, 1984, pp.651-654.

[904] Kerefky, P., Ruda, M.: Program Optimization and Manipulation on User Level. In: Trappl, R.(ed.): Cybernetics and Systems Research. North-Holland, Amsterdam, and New York, 1982, pp.797-802.

[905] Kerefky, P., Ratko, I., Ruda, M.: Patient Registers on Microcomputers. In: Trappl, R.(ed.): Cybernetics and Systems Research, Vol.II. North-Holland, Amsterdam, and New York, 1984, pp.519-524.

[906] Kerin, T., Slavtchev, A., Nedeltchev, N., Kjurktchiev, T.: A Method for Performance Assessment of Medical Radioisotope Equipment. In: Trappl, R.(ed.): Cybernetics and Systems Research, Vol.II. North-Holland, Amsterdam, and New York, 1984, pp.311-318.

[907] Kerman, C.E.: Creative Tension: the Life and Thought of Kenneth Boulding. Univ. of Michigan Press, Ann Arbor, Mich., 1974, XV+380 pp.

[908] Kerner, H., Pitrik, R.: Functional versus Imperative Programming - Complement or Contradiction? In: Trappl, R.(ed.): Cybernetics and Systems Research, Vol.II. North-Holland, Amsterdam, and New York, 1984, pp.627-634.

[909] Keyes, R.W.: Fundamental Limits in Digital Information Processing. IEEE Proc., 69, No.2, Feb.1981, pp.267-278.

[910] Kickert, W.J.M.: An Example of Linguistic Modelling. Annals of Systems Research, 7, 1978, pp.37-62.

[911] Kickert, W.J.M.: Fuzzy Theories on Decision-Making. (Vol.3 in Frontiers in Systems Research Book Series.) Martinus Nijhoff, Boston and The Hague, 1978, 182 pp.

[912] Kickert, W.J.M., Van Gigch, J.P.: A Metasystem Approach to Organizational Decision-Making. Management Science, 25, No.12, 1979, pp.1217- 1231.

[913] Kickert, W.J.M.: A Formal Approach to Organization Design. Eindhoven Univ. of Technology, Eindhoven, Netherlands, 1980, 21pp. [IFSR-Depository]

[914] Kickert, W.J.M.: Organization of Decision-Making: a Systems-Theoretical Approach. North-Holland, Amsterdam, and New York, 1980, XIV+278 pp.

[915] Kieras, D.E.: Finite Automata and S-R Models. J. of Mathematical Psychology, 13, No.2, April 1976, pp.127-147.

[916] Kilmann, R.H.: Social Systems Design: Normative Theory and the Maps Technology. North-Holland, Amsterdam, and New York, 1977, XV+327 pp.

[917] Kindler, E.: Classification of Simulation Programming Languages: I. Declaration of Necessary System Conceptions. Elektronische Informationsverarbeitung u.Kybernetik, 14, No.10, 1978, pp.519-526.

[918] Kindler, E.: Classification of Simulation Programming Languages: II. Description of Types and Individual Typology. Elektronische Informationsverarbeitung u.Kybernetik, 14, No.11, 1978, pp.575-584

[919] Kindler, E.: Dynamic Systems and Theory of Simulation. Kybernetika, 15, No.2, 1979, pp.78-87.

[920] Kindler, E.: Simula 67 and Systems Analysis. J. of Information Processing, 15, No.10, 1979, pp.513-524.

[921] Kindler, E.: Simulation Programming Languages. SNTL, Prague, 1980, 277 pp. [Czech]

[922] Kindler, E.: A Formalization of Some Simulation Language Concepts. Int. J. of General Systems, 6, No.4, 1981, pp.183-190.

[923] Kindler, E., Chochol, S., Prokop, C.: Systems of Material Flow. Int. J. of General Systems, 9, No.4, 1983, pp.217-224.

[924] Kindler, J.(ed.): Proceedings of a Workshop on Modelling of Water Demands. IIASA, Laxenburg, Austria, 1978, X+156 pp.

[925] King, L.T.: Problem Solving in a Project Environment. John Wiley, Chichester and New York, 1981, XI + 204 pp.

[926] King, L.T.: Problem Solving in a Project Environment: A Consulting Process. John Wiley, Chichester and New York, 1981, 216 pp.

[927] Kinston, W.: Improving Health Care Institutions: An Action Research Approach to Organisation of Complex Systems. In: Trappl, R.(ed.): Cybernetics and Systems Research. North-Holland, Amsterdam, and New York, 1982, pp.633-640.

[928] Kinston, W.: Linking System Research and Delivery Systems. The Example of Health Care. In: Trappl, R.(ed.): Cybernetics and Systems Research, Vol.II. North-Holland, Amsterdam, and New York, 1984, pp.505-512.

[929] Kintala, C.M.R.: Amounts of Nondeterminism in Finite Automata. Acta Informatica, 13, No.2, 1980, pp.199-204.

[930] Kittler, J.: A Comment on Fast Monte Carlo Integration of P.D.F. Estimators. J. of Cybernetics, 8, No.3-4, 1978, pp.253-256.

[931] Kittler, J., Fu, K.S., Pau, L.F.(eds.): Pattern Recognition Theory and Applications. D.Reidel, Dordrecht, Holland, and Boston, 1982, X+576pp.

[932] Kleijnen, J.P.C.: Design and Analysis of Simualtion: Practical Statistical Techniques. Department of Business and Economics, Katholieke, 1976, 34pp. [IFSR-Depository]

[933] Klement, E.P.: Proceedings of the International Seminar on Fuzzy Set Theory. Johannes Kepler Univ., Linz, Austria, September 24-29, 1979.

[934] Klement, E.P.: On the Cardinality of Fuzzy Sets. In: Trappl, R.(ed.): Cybernetics and Systems Research. North-Holland, Amsterdam, and New York, 1982, pp.701-704.

[935] Klement, E.P., Puri, M.L., Ralescu, D.: Law of Large Numbers and Central Limit Theorem for Fuzzy Random Variables. In: Trappl, R.(ed.): Cybernetics and Systems Research, Vol.II. North-Holland, Amsterdam, and New York, 1984, pp.525-530.

[936] Kleyle, R., De Korvin, A.: Switching Mechanisms in a Generalized Information System. Cybernetics and Systems, 15, No.1-2, 1984, pp.145-168.

[937] Klir, G.J., Rogers, G., Gesyps, R.G.(eds.): Basic and Applied General Systems Research: A Bibliography. School of Advanced Technology, SUNY-Binghampton, Binghampton, NY, 1977, 250 pp. [IFSR-Depository]

[938] Klir, G.J., Uyttenhove, H.J.J.: On the Problem of Computer-Aided Structure Identification: Some Experimental Observations and Resulting Guidelines. Int. J. of Man-Machine Studies, 9, No.6, 1977, pp.593-628.

[939] Klir, G.J.: Pattern Discovery in Activity Arrays. In: Hartnett, W.E.(ed.): Systems: Approaches, Theories, Applications. D.Reidel, Dordrecht, Holland, and Boston, 1977, pp.121-158.

[940] Klir, G.J.(ed.): Applied General Systems Research. Plenum Press, New York and London, 1978.

[941] Klir, G.J., Braten, S., Casti, J.(eds.): Frontiers in Systems Research: Implications for the Social Sciences (a

New Book Series) . Martinus Nijhoff, Boston and The Hague.

[942] Klir, G.J.: The General Systems Research Movement. In: Sharif, N., Adulbhan, P.(eds.): Proceedings of the Int. Conference on Systems Modelling in Developing Countries. Asian Institute of Technology, Bangkok, 1978, pp.23-70.

[943] Klir, G.J.(ed.): The North-Holland Series in General Systems Research. Elsevier / North-Holland, New York.

[944] Klir, G.J.: Architecture of Structure Systems: a Basis for the Reconstructability Analysis. Acta Polytechnica Scandinavica, Mathematics and Computer Science Series, No.31, Helsinki, 1979, pp.33-43.

[945] Klir, G.J.: Computer-Aided Systems Modelling. In: Halfon, E.(ed.): Theoretical Systems Ecology: Advances and Case Studies. Academic Press, London, and New York, 1979, pp.291-323.

[946] Klir, G.J.: General Systems Problem Solving Methodology. In: Zeigler, B.P., (et al.)(eds.): Methodology in Systems Modelling and Simulation. North-Holland, Amsterdam, and New York, 1979, pp.3-28.

[947] Klir, G.J., Uyttenhove, H.J.J.: Procedures for Generating Reconstruction Hypotheses in the Reconstructability Analysis. Int. J. of General Systems, 5, No.4, 1979, pp.231-246.

[948] Klir, G.J.: On Systems Methodology and Inductive Reasoning: The Issue of Parts and Wholes. General Systems Yearbook, 26, 1982, pp.29-36.

[949] Klir, G.J.: General Systems Concepts. In: Trappl, R.(ed.): Cybernetics - Theory and Applications. Hemisphere, Washington, D.C., 1983, pp.91-120.

[950] Klir, G.J.: Possibilistic Information Theory. In: Trappl, R.(ed.): Cybernetics and Systems Research, Vol.II. North-Holland, Amsterdam, and New York, 1984, pp.3-8.

[951] Kloeden, P.E.: Fuzzy Dynamical Systems. Fuzzy Sets and Systems, 7, No.3, May 1982, pp.275-296.

[952] Knappe, U., Heymann, J.: Microcomputer-Based System for Industrial Energy Control. In: Trappl, R.(ed.): Cybernetics and Systems Research. North-Holland, Amsterdam, and New York, 1982, pp.665-670.

[953] Kobayashi, H.: Modelling and Analysis: an Introduction to System Performance Evaluation Methodology. Addison-Wesley, Reading, Mass., 1978, XVII+446 pp.

[954] Kobayashi, I.: Information and Information Processing Structure. Information Systems, 1, No.2, April 1975, pp.39-49.

[955] Kobsa, A.: Knowledge Representation: The Representation of Knowledge in the Computer. Austrian Society for Cybernetic Studies, Vienna, 1982. [German]

[956] Kobsa, A.: On Regarding AI Programs as Theories. In: Trappl, R.(ed.): Cybernetics and Systems Research. North-Holland, Amsterdam, and New York, 1982, pp.933-9636.

[957] Kobsa, A.: Knowledge Representation: A Survey of its Mechanisms, a Sketch of its Semantics. Cybernetics and Systems, 15, No.1-2, 1984, pp.41-90.

[958] Kobsa, A., Trost, H.: Representing Belief Models in Semantic Networks. In: Trappl, R.(ed.): Cybernetics and Systems Research, Vol.II. North-Holland, Amsterdam, and New York, 1984, pp.753-758.

[959] Koczy, L.T.: Interactive Sigma-Algebras and Fuzzy Objects of Type N. J. of Cybernetics, 8, No.3-4, 1978, pp.273-290.

[960] Kodratoff, Y., Costa, E.: Non-Canonical Simplification.

In: Trappl, R.(ed.): Cybernetics and Systems Research, Vol.II. North-Holland, Amsterdam, and New York, 1984, pp.705-714.

[961] Koenderink, J.J., Van Doorn, A.J.: Perception of Solid Shape and Spatial Lay-Out through Photometric Invariants. In: Trappl, R.(ed.): Cybernetics and Systems Research. North-Holland, Amsterdam, and New York, 1982, pp.943-948.

[962] Koenderink, J.J., Doorn Van, A.J.: On Spotting Three-Dimensional Shape: Rigid Movement, Bending and Stretching by Eye: Implications for Robot Vision. In: Trappl, R.(ed.): Cybernetics and Systems Research, Vol.II. North-Holland, Amsterdam, and New York, 1984, pp.767-774.

[963] Koffman, E.B.: Problem Solving and Structured Programming in PASCAL. Addison-Wesley, Reading, Mass., 1981.

[964] Kohonen, T.: Associative Memory: a Systems-Theoretical Approach. Springer-Verlag, Berlin, FRG, and New York, 1977, 176 pp.

[965] Kohout, L.J., Pinkava, V.: The Algebraic Structure of the Spencer-Brown and Varela Calculi. Int. J. of General Systems, 6, No.3, 1980.

[966] Kohout, L.J., Bandler, W., Trayner, C., Anderson, J.: Construction of an Expert Therapy Adviser as a Special Case of a General System Protection Design. In: Trappl, R.(ed.): Cybernetics and Systems Research, Vol.II. North-Holland, Amsterdam, and New York, 1984, pp.97-104.

[967] Kontur, I.F., Ambrus, S.Z.: Runoff Simulation with a Random Walk Model. In: Trappl, R.(ed.): Cybernetics and Systems Research, Vol.II. North-Holland, Amsterdam, and New York, 1984, pp.209-216.

[968] Koppelaar, H., Kruijt, D.: Regular Career Systems II. Annals of Systems Research, 6, 1977, pp.35-47.

[969] Koppelaar, H.(ed.): Systematica. Utrecht, The Netherlands, 1981.

[970] Korhonen, P.: A Stepwise Procedure for Multivariate Clustering. Univ. of Helsinki, Doctoral Thesis, 1979.

[971] Kornai, J.: Descriptive-Explanatory Theoretical Models of the Socialist Economy: Review of a Research Direction. Systems Research, 1, No.2, 1984, pp.135-144.

[972] Kornwachs, K., Lucadou, W.v.: Pragmatic Information and Nonclassical Systems. In: Trappl, R.(ed.): Cybernetics and Systems Research. North-Holland, Amsterdam, and New York, 1982, pp.191-198.

[973] Kornwachs, K., Lucadou, W.v.: Some Notes on Information and Interaction. In: Trappl, R.(ed.): Cybernetics and Systems Research, Vol.II. North-Holland, Amsterdam, and New York, 1984, pp.9-14.

[974] Kovalenko, Y.I., Borisyuk, R.M., Borisyuk, G.N., Kirillov, A.B., Kryukov, V.I.: Short-Term Memory as a Metastable State. II. Simulation Model. In: Trappl, R.(ed.): Cybernetics and Systems Research, Vol.II. North-Holland, Amsterdam, and New York, 1984, pp.267-272.

[975] Kovalevsky, V.A.: Image Pattern Recognition. Springer-Verlag, Berlin, FRG, and New York, 1980, XVII+242pp.

[976] Kowalaski, R.: Logic of Problem Solving. North-Holland, Amsterdam, and New York, 1979, 287 pp.

[977] Koyayashi, H.: Modeling and Analysis: an Introduction to System Performance Evaluation Methodology. Addison-Wesley, Reading, Mass., 1978.

[978] Kozen, D.: Positive First-Order Logic Is NP-Complete. IBM Journal of Research and Development, 25, No.4, July 1981, pp.327-332.

[979] Kraft, R.W.: Symbols, Systems, Science & Survival.

Vantage Press, New York, 1975, 246 pp.

[980] Krallmann, H.:   Enlarging   the   Systems   Paradigm:
Historical View   from   1973 until 1983.   Systems Research,
1, No.3, 1984, pp.167-190.

[981] Kramer, J.T.A., De Smit, J.: Systems Thinking: Concepts
and Notions.   Martinus   Nijhoff,   Boston   and   The Hague,
1974.

[982] Kriegel, U.,   Mende,   W.,   Grauer,   M.:   An   Evolutionary
Analysis of World Energy Consumption and World Population.
In: Trappl,   R.(ed.):   Cybernetics and Systems Research,
Vol.II.   North-Holland, Amsterdam,   and   New   York,   1984,
pp.435-442.

[983] Krippendorff, K.(ed.):   Communication   and   Control   in
Society.   Gordon   and   Breach, New York, and London, 1979,
529 pp.

[984] Krippendorff, K.:   On   the   Algorithm   for   Identifying
Structures in   Multi-Variate   Data.   Int.   J.   of General
Systems, 7, No.1, 1981, pp.63-80.

[985] Krishana, V.S.:   An   Introduction   to   Category   Theory.
North-Holland, Amsterdam, and New York, 1981, X+173 pp.

[986] Krone, R.M.:   Systems   Analysis   and   Policy   Sciences:
Theory and Practice.   John Wiley, Chichester and New York,
1980, XXVII+216pp.

[987] Kronsjo, L.I.:   Algorithms: Their   Complexity   and
Efficiency.   John   Wiley,   Chichester   and New York, 1979,
XV+361 pp.

[988] Kryukov, V.I.:   Short-Term Memory as a   Metastable   State.
I.   Master   Equation   Approach.   In: Trappl,   R.(ed.):
Cybernetics and Systems Research, Vol.II.   North-Holland,
Amsterdam, and New York, 1984, pp.261-266.

[989] Krzanik, L.:   Evolutionary Projects: Dynamic Optimization
of Delivery   Step   Structure.   In: Trappl,   R.(ed.):
Cybernetics and   Systems Research, Vol.II.   North-Holland,
Amsterdam, and New York, 1984, pp.323-328.

[990] Kubat, L.,   Zeman, J.(eds.):   Entropy   and   Information   in
Science and   Philosophy.   Elsevier   /   North-Holland, New
York, 1975.

[991] Kubinski, T.:   An   Outline   of   the   Logical   Theory   of
Questions.   Akademie-Verlag, Berlin, 1980.

[992] Kucera, V.:   Discrete   Linear   Control: the   Polynomial
Equation Approach.   John   Wiley, Chichester and New York,
1979, 206 pp.

[993] Kuhlenbeck, H., Gerlach, J.:   The   Human   Brain   and   its
Universe.   Vol.I: The   World of Natural Sciences and its
Phenomenology.   Vol.II: The Brain and its Mind.   Karger,
Basel, 1982, Vol.1: XIV+282 pp.;   Vol.2:   XIV+374 pp.

[994] Kuhn, H.W.,   Szego,   G.P.(eds.):   Mathematical   Systems
Theory and   Economics.   Springer-Verlag, Berlin, FRG, and
New York, 1969.

[995] Kuyk, W.:   Complementarity   in   Mathematics.   D.Reidel,
Dordrecht, Holland, and Boston, 1977.

[996] Kwakernaak, H.:   Fuzzy   Random   Variables.   Part   1:
Definitions   and   Theorems.   Part   2:   Algorithms   and
Examples for the   Discrete   Case.   Information   Sciences,
Part 1:   15 No.1,   1978,   pp.1-15;   Part 2:   17, No.   3,
1979, pp.253-278.

[997] La Patra, J.W.:   Applying the Systems   Approach   to   Urban
Development.   McGraw-Hill, New York, 1977.

[998] La Porte, T.R.(ed.):   Organized Complexity: Challenge   to
Politics   and   Policy.   Princeton   University   Press,
Princeton, N.J., 1975.

[999] LaSalle, J.P.:   The Stability of Dynamical Systems.   SIAM,
Philadelphia, 1976, V+76 pp.

[1000] Labos, E.:   Periodic and Non-Periodic Motions in Different

Classes of Formal Neuronal Networks and Chaotic Spike
Generators. In: Trappl, R.(ed.): Cybernetics and
Systems Research, Vol.II. North-Holland, Amsterdam, and
New York, 1984, pp.237-244.

[1001] Labudde, R.A.: Discrete Hamiltonian Mechanics. Int. J.
of General Systems, 6, No.1, 1980, pp.3-12.

[1002] Ladanyi, O.G.: The Eighties - A World in Motion, A
Challenge for Systems Engineering. In: Trappl, R.(ed.):
Cybernetics and Systems Research. North-Holland,
Amsterdam, and New York, 1982, pp.427-430.

[1003] Ladanyi, O.G.: Modelling Economic Movements Using
Published International Statistics. In: Trappl, R.(ed.):
Cybernetics and Systems Research, Vol.II. North-Holland,
Amsterdam, and New York, 1984, pp.391-396.

[1004] Laing, R.: Machines as Organisms: an Exploration of the
Relevance of Recent Results. Bio-Systems, 11, No.2,3,
1979, pp.20-215.

[1005] Lakshmivarahan, S., Rajasethupathy, K.S.: Considerations
for Fuzzifying Formal Languages and Synthesis of Fuzzy
Grammars. J. of Cybernetics, 8, No.1, 1978, pp.83-100.

[1006] Lamport, L.: Time, Clock and the Ordering of Events in
Distributed Systems. ACM Communications, 21, No.7, July
1976, pp.558-565.

[1007] Landau, Y.D.: Adaptive Control: the Model Reference
Approach. Marcel Dekker, New York, 1979, XXII+406 pp.

[1008] Landweber, L.H., Robertson, E.L.: Properties of
Conflict-Free and Persistent Petri Nets. ACM Journal, 25,
No.3, July 1978, pp.352-364.

[1009] Langefors, L., Samuelson, K.: Information and Data in
Systems. Petrocelli, New York, 1976.

[1010] Lasker, G.E.(ed.): Applied Systems and Cybernetics (6
Vols). Pergamon Press, Oxford and New York, 1981.

[1011] Lasker, G.E.(ed.): The Relation between Major World
Problems and Systems Learning. Intersystems, Seaside,
Ca., 1983.

[1012] Lasota, A., Mackey, M.C.: The Extinction of Slowly
Evolving Dynamical Systems. J. of Mathematical Biology,
10, No.4, 1980, pp.333-345.

[1013] Laszlo, C.A.: Champ - A System and Computer Program
Package for Health Hazard Appraisal. Cybernetics and
Systems, 12, No.1-2, 1981, pp.37-42.

[1014] Laszlo, E.: GST: Prospects and Principles. In: Rose,
J., Bilcio, C.(eds.): Modern Trends in Cybernetics and
Systems (3 Vols.). Springer-Verlag, Berlin, FRG, and New
York, 1977, pp.131-152.

[1015] Laszlo, E.: Some Reflections on Systems Theory's Critics.
Nature and Systems, 2, No.1, March 1980, pp.49-53.

[1016] Laszlo, E.: Cybernetics and Global Planning. In:
Trappl, R.(ed.): Cybernetics - Theory and Applications.
Hemisphere, Washington, D.C., 1983, pp.373-396.

[1017] Lauenroth, H.-G.: Algorithmic Model Systems and
Innovation Strategies for Automation of Industrial
Production and Information Systems. In: Trappl, R.(ed.):
Cybernetics and Systems Research, Vol.II. North-Holland,
Amsterdam, and New York, 1984, pp.319-322.

[1018] Launer, R.L., Siegel, A.F.(eds.): Modern Data Analysis.
Academic Press, London, and New York, 1982, 224pp.

[1019] Lavorel, P.M., Gigley, H.: How we Name or Misname
Objects. In: Trappl, R.(ed.): Cybernetics and Systems
Research. North-Holland, Amsterdam, and New York, 1982,
pp.351-356.

[1020] Lazarevic, B.(ed.): Global and Large-Scale System Models.
(Proc. of a Seminar in Dubrovnik, Yugoslavia, August 1978)
Springer-Verlag, Berlin, FRG, and New York, 1979, VIII+232

pp.
[1021] Le Bonniec, G.P.: The Development of Model Reasoning. Academic Press, London, and New York, 1980.
[1022] Le Moigne, J.-L.: Analyse de Systeme: Bibliographie et Commentaires (Systems Analysis: Bibliography and Comments). Univ. De Droit, Aix-en-Provence, France, Aix-en-Provence, France, 1978, 418 pp. [French]
[1023] Le Moigne, J.-L.: Analyse de Systeme. Univ. De Droit, Aix-en-Provence, France, 1979, 418pp. [IFSR-Depository]
[1024] LeMoigne, J.-L.: Modeling Complexity: A French History and its Prospects. Human Systems Management, 2, No.3, Oct.1981, pp.143-155.
[1025] Leal, A., Pearl, J.: An Interactive Program for Conversational Elicitation of Decision Structures. IEEE Trans. on Systems, Man, and Cybernetics, SMC-7, No.5, May 1977, pp.368-376.
[1026] Lee, E.B., Neftci, S., Olbrot, A.: Canonical Forms for Time Delay Systems. IEEE Trans. on Automatic Control, AC-27, No.1, Feb.1982, pp.128-132.
[1027] Lee, E.T.: Applications of Fuzzy Set Theory to Image Sciences. J. of Cybernetics, 10, No.1-3, 1980, pp.127-136.
[1028] Lee, R.M.: Expert vs. Management Support Systems: Semantic Issues. Cybernetics and Systems, 14, No.2-4, 1983, 139-158.
[1029] Leeming, A.M.C.: A Systems Study of the Design of a User Interface. In: Trappl, R.(ed.): Cybernetics and Systems Research, Vol.II. North-Holland, Amsterdam, and New York, 1984, pp.615-620.
[1030] Lefebre, V.A.: Algebra of Conscience: A Comparative Analysis of Wester and Soviet Ethical Systems. D.Reidel, Dordrecht, Holland, and Boston, 1982, XXVII+194pp.
[1031] Lefebre, V.A.: Second Order Cybernetics in the Soviet Union and the West. In: Trappl, R.(ed.): Power, Autonomy, Utopia: New Approaches Towards Complex Systems. Plenum Press, New York and London, 1985.
[1032] Legget, R.W., Williams, L.R.: A Reliability Index for Models. Ecological Modelling, 13, No.4, Sept.1981, pp.303-312.
[1033] Leigh, J.R.: Functional Analysis and Linear Control Theory. Academic Press, London, and New York, 1981, 176pp.
[1034] Leinfellner, E.: Causality and Language: Phonetics / Phonology, Syntax, Semantics. Austrian Society for Cybernetic Studies, Vienna, 1980. [German]
[1035] Leinhardt, S.(ed.): Special Issue of Mathematical Sociology on Social Network Research. Mathematical Sociology on Social Network Research, 5, No.1, 1977, pp.1-147.
[1036] Lendaris, G.G.: Structural Modeling - A Tutorial Guide. IEEE Trans. on Systems, Man, and Cybernetics, SMC-10, No.12, Dec.1980, pp.807-840.
[1037] Lesmo, L., Saitta, L., Torasso, P.: An Interpreter of Fuzzy Production Rules. In: Trappl, R.(ed.): Cybernetics and Systems Research, Vol.II. North-Holland, Amsterdam, and New York, 1984, pp.793-798.
[1038] Lester, F.K., Garofalo, J.(eds.): Mathematical Problem Solving. Franklin Institute Press, Philadelphia, 1982, XII + 140pp.
[1039] Leung, Y.: A Fuzzy Set Analysis of Sociometric Structure. J. of Mathematical Sociology, 7, No.2, 1980, pp.159-180.
[1040] Leung-yan-cheung, S.K., Cover, T.M.: Some Equivalences between Shannon Entropy and Kolmogorov Complexity. IEEE Trans. on Information Theory, IT-24, No.3, May 1978,

pp.331-338.
[1041] Levien, R.E.: Introduction to IIASA: Applying Systems
Analysis in an International Setting. Behavioral Science,
24, No.3, 1979, pp.155-168.
[1042] Levine, R.D., Tribus,(eds.): The Maximum Entropy
Formalism. M.I.T.Press, Cambridge, Mass., 1979, 489 pp.
[1043] Levitan, G.L.: Use of Deductive Inference in Situation
Control Systems. Engineering Cybernetics, 1977.
[1044] Levy, L.S.: Discrete Structures of Computer Science.
John Wiley, Chichester and New York, 1980, 310 pp.
[1045] Lewicka-Strzalecka, A.: The Systems Approach as a
Methodological Attitude. An Empirical Analysis. In:
Trappl, R.(ed.): Cybernetics and Systems Research,
Vol.II. North-Holland, Amsterdam, and New York, 1984,
pp.53-60.
[1046] Lewin, D.: Theory and Design of Digital Computer Systems.
John Wiley, Chichester and New York, 1980 (2nd Edition) ,
VII+472 pp.
[1047] Lewis, H.R., Papadimitriou, C.H.: The Efficiency of
Algorithms. Scientific American, 238, No.1, Jan.1978,
pp.96-109.
[1048] Lieberherr, K.J., Specker, E.: Complexity of Partial
Satisfaction. ACM Journal, 28, No.2, April 1981,
pp.411-421.
[1049] Lilienfeld, R.: The Rise of Systems Theory: an
Ideological Analysis. John Wiley, Chichester and New
York, 1978.
[1050] Linger, R.C., Mills, H.D., Witt, B.I.: Structured
Programming: Theory and Practice. Addison-Wesley,
Reading, Mass., 1979, 402 pp.
[1051] Linkens, D.A:(ed.): Biological Systems Modelling and
Control. IEEE, Piscataway, N.J., 1979, 350pp.
[1052] Linstone, H.A.: The Multiple Perspective Concept with
Applications to Technology and Other Decision Areas.
Technological Forecasting and Social Chance, 20, 1981,
pp.275-325.
[1053] Linstone, H.A.: Cybernetics and the Future. In: Trappl,
R.(ed.): Cybernetics - Theory and Applications.
Hemisphere, Washington, D.C., 1983, pp.397-412.
[1054] Lions, J.I.: Some Methods in the Mathematical Analysis of
Systems and Their Control. Science Press, Beijing, and
Gordon and Breach, New York, 1981, XXVI+542pp.
[1055] Lions, J.L., Systems and T: Some Methods in the
Mathematical Analysis of Systems and Their Control.
Science Press, Beijing, and Gordon and Breach, New York,
1981, XII+542pp.
[1056] Lipset, D.: Gregory Bateson: The Legacy of a Scientist.
Beacon Press, Boston, 1982, XII + 364pp.
[1057] Lipski, W.: Information Systems with Incomplete
Information. Proc. 3rd Int.Symposium on Automata Theory,
Languages and Programs, 1977.
[1058] Lipski, W.: On the Logic of Incomplete Information. In:
Gruska, J.(ed.): Mathematical Foundations of Computer
Science. Springer-Verlag, Berlin, FRG, and New York,
1977, pp.374-381.
[1059] Locker, A., Coulter, N.A.: A New Look at the Description
and Prescription of Systems. Behavioral Science, 22,
No.3, May 1977, pp.197-206.
[1060] Loefgren, L.: Describability and Learnability of Complex
Systems. In: Oeren, T.I.(ed.): Cybernetics and
Modelling and Simulation of Large Scale Systems. Int.
Association for Cybernetics, Namur, 1976, pp.41-56.
[1061] Loefgren, L.: Knowledge of Evolution and Evolution of
Knowledge. 1981, pp.129-151.

[1062] Loefgren, L.: Excerpts from the Autology of Learning. In: Trappl, R.(ed.): Cybernetics and Systems Research. North-Holland, Amsterdam, and New York, 1982, pp.357-362.

[1063] Loo, S.G.: Measures of Fuzziness. Cybernetica, XX, No.3, 1977, pp.201-210.

[1064] Loo, S.G.: Fuzzy Relations in the Social and Behavioral Science. J. of Cybernetics, 8, No.1, 1978, pp.1-16.

[1065] Lopes, M.A., Veloso, P.A.S.: Operations on Problems and Their Solution Spaces. In: Trappl, R.(ed.): Cybernetics and Systems Research. North-Holland, Amsterdam, and New York, 1982, pp.379-384.

[1066] Lowen, W.: Creep-Creep: Out of Human Scale. Human Systems Management, 2, No.3, Oct.1981, pp.217-220.

[1067] Lowen, W.: Dichotomies of the Mind: A Systems Science Model of the Mind and Personality. John Wiley, Chichester and New York, 1982, XII + 314pp.

[1068] Lozinskii, E.L.: Construction of Relations in Relational Databases. ACM Trans. on Database Systems, 5, No.2, 1980, pp.208-224.

[1069] Lozinskii, E.L., Nirenburg, S.: Locality in Natural Language Processing. In: Trappl, R.(ed.): Cybernetics and Systems Research. North-Holland, Amsterdam, and New York, 1982, pp.875-880.

[1070] Lucas, R.G., Hardin, R.C.: Political Determinants of Information System Configuration. In: Trappl, R.(ed.): Cybernetics and Systems Research, Vol.II. North-Holland, Amsterdam, and New York, 1984, pp.353-358.

[1071] Luenberger, D.G.: Introduction to Dynamic Systems: Theory, Models, and Applications. John Wiley, Chichester and New York, 1979, 446 pp.

[1072] Luhman, N.: Trust and Power. John Wiley, Chichester and New York, 1979.

[1073] Luhman, N.: The World Society as a Social System. Int. J. of General Systems, 8, No.3, 1982, pp.131-138.

[1074] Lundeberg, M., Goldkuhl, G., Nilsson, A.: Information Systems Development: A Systematic Approach. Prentice-Hall, Englewood Cliffs, New Jersey, 1981, XII+337 pp.

[1075] Lusk, E.J., Wright, H.: A Discussion of External Validity in Systemic Research. J. of Applied Systems Analysis, 9, April 1982, pp.57-66.

[1076] M'Pherson, P.K.: Systems Engineering: An Approach to Whole-System Design. Radio and Electronic Engineer, 50, No.11/12, 1980, pp.545-558.

[1077] M'Pherson, P.K.: A Framework for Systems Engineering Design. Radio and Electronic Engineer, 51, No.2, 1981, pp.59-93.

[1078] MacDonald, B.A., Andreae, J.H.: The Competence of a Multiple Context Learning System. Int. J. of General Systems, 7, No.2, 1981, pp.123-138.

[1079] MacDonald, N.: Time Lags in Biological Models. Springer-Verlag, Berlin, FRG, and New York, 1979, VI+112 pp.

[1080] Machlup, F.: Knowledge: Its Creation, Distribution and Economic Significance. Vol.2: The Branches of Learning. Princeton University Press, Princeton, N.J., 1982, XII + 208pp.

[1081] Machtey, M.: An Introduction to the General Theory of Algorithms. Elsevier / North-Holland, New York, 1978, 272 pp.

[1082] Maciejowski, J.M.: The Modelling of Systems with Small Observation Sets. Springer-Verlag, Berlin, FRG, and New York, 1978, VI+242 pp.

[1083] Mackenzie, K.D.: A Process Based Measure for the Degree

of Hierarchy in a Group. J. of Enterprise Management, 1,
No.3, 1978, pp.153-184.

[1084] Mackenzie, K.D., Bello, J.A.: Leadership as a Task
Process Uncertainty Control Process. Human Systems
Management, 2, No.3, Oct.1981, pp.199-213.

[1085] Mackinnon, A.J., Wearing, A.J.: Complexity and
Decision-Making. Behavioral Science, 25, No.4, July 1980,
pp.285-296.

[1086] Madan, S., Jain, J.L.: Cybernetic Modeling of Queueing
Complex. J. of Cybernetics, 9, No.2, 1979, pp.151-159.

[1087] Madan, S.: Basic Concepts on the Theoretical Treatment of
Cybernetic Queueing Systems. J. of Cybernetics, 10,
No.1-3, 1980, pp.239-248.

[1088] Mager, P.P.: Biorhythms, System Organization and
Bioactive Compounds. Chronobiologia, VII, No.1, 1980,
pp.55-79.

[1089] Mahmoud, M.S., Singh, M.G.: Large Scale Systems
Modelling. Pergamon Press, Oxford and New York, 1982,
XII+326pp.

[1090] Majewska-Tronovicz, J.: Selected Praxiological
Bibliography. Praxiology, 1980, No 1, pp.157-197.

[1091] Majone, G., Quade, E.(eds.): Pitfalls of Analysis. John
Wiley, Chichester and New York, 1980, VIII+213 pp.

[1092] Majster, M.E.: Extended Directed Graphs, a Formalism for
Structured Data and Data Structures. Acta Informatica, 8,
No.1, 1977, pp.37-59.

[1093] Mak, K.L.: Dynamics and Control of Integrated
Production-Inventory-Shipping Systems. Cybernetics and
Systems, 12, No.3, 1981, pp.205-224.

[1094] Maker, P.P: Input-Output Relations in Multivariate
Biosystems. In: Unger, C.K., Stocker, G.(eds.):
Biophysical Ecology and Ecosystem Research.
Akademie-Verlag, Berlin, 1981, pp.218-226.

[1095] Makridakis, S., Wheelwright, S.C.: Forecasting: Methods
and Applications. John Wiley, Chichester and New York,
1978, XVI+713 pp.

[1096] Malec, J.: Scenarios as a Tool for Dynamic Scene
Representation. In: Trappl, R.(ed.): Cybernetics and
Systems Research, Vol.II. North-Holland, Amsterdam, and
New York, 1984, pp.781-786.

[1097] Malik,, F., Probst, G.J.B.: Evolutionary Management.
Cybernetics and Systems, 13, No.2, 1982, pp.153-174.

[1098] Mamdani, E.H.: Application of Fuzzy Logic to Approximate
Reasoning Using Linguistic Synthesis. IEEE Trans. on
Computers, C-26, No.12, Dec.1977, pp.1182-1191.

[1099] Mamdani, E.M., Gaines, B.R.(eds.): Fuzzy Reasoning and
Its Applications. Academic Press, London, and New York,
1981, XVIII + 381pp.

[1100] Manescu, M.: Economic Cybernetics. Heyden, Philadelphia,
1980.

[1101] Mann, K.H.: Ecology of Coastal Waters: A Systems
Approach (Studies in Ecology, Volume 8). Blackwell,
Oxford, 1982, 312pp.

[1102] Manohar Raho, M.J.: Control Systems and Quantitative
Economic Policy. In: Trappl, R.(ed.): Cybernetics and
Systems Research, Vol.II. North-Holland, Amsterdam, and
New York, 1984, pp.451-456.

[1103] Marchuk, G.I.(ed.): Modelling and Optimization of Complex
Systems. (Proc. of a Conference in Novosibirsk, USSR,
July 1978) Springer-Verlag, Berlin, FRG, and New York,
1979, VI+294 pp.

[1104] Margolis, D.L.: Bond-graphs for Some Classic Dynamic
Systems. Simulation, 35, No.3, 1980, pp.81-87.

[1105] Mark, J.W., Dasiewicz, P.P.: Application of Iterative

Algorithms to Adaptive Predicitive Coding. J. of
Cybernetics, 7, No.3-4, 1977, pp.279-316.

[1106] Marko, H.: Methods of System Theory. Springer-Verlag,
Berlin, FRG, and New York, 1977, 220 pp. [German]

[1107] Marmarelis, P.Z., Marmarelis, V.Z.: Analysis of
Physiological Systems. Plenum Press, New York and London,
1978.

[1108] Marshall, J.E.: Control of Time-Delay Systems. IEEE,
Piscataway, N.J., 1979, 237pp..

[1109] Martin, J.N.T.: On Mapping Real Systems. J. of Applied
Systems Analysis, 7, April 1980, pp.151-156.

[1110] Martin, N.F.G., England, J.W.: Mathematical Theory of
Entropy. Addison-Wesley, Reading, Mass., 1981, XXIV+258
pp.

[1111] Martin, P.E.: Simplification and Control of Information
Patterns - Further Developments. In: Trappl, R.(ed.):
Cybernetics and Systems Research. North-Holland,
Amsterdam, and New York, 1982, pp.589-594.

[1112] Maruyama, M.: Heterogenistics: an Epistemological
Restructuring of Biological and Social Sciences.
Cybernetica, XX, No.1, 1977, pp.70-86.

[1113] Maryanski, F.I.: Digital Computer Simulation. Hayden,
Rochelle Park, New Jersey, 1980, 328 pp.

[1114] Mason, R.O., Swanson, E.B.: Measurement for Management
Decision. Addison-Wesley, Reading, Mass., 1981, XIV+550
pp.

[1115] Masuda, Y.: The Information Society as Past-Industrial
Society. Inst. for the Information Society, Tokio, Japan,
1981, XIV+176pp (U.S. Distributor: World Future Society,
Bookservice, Maryland.

[1116] Matejko, A.: The Structural Criteria of Social System
Maturity. In: Niemeyer, H.(ed.): Soziale
Beziehungsgeflechte. Dunker & Humbolt, Berlin, 1980,
pp.57-76.

[1117] Matevosyan, A.K.: The Invers Problem of Karhunen-Loeve.
In: Trappl, R.(ed.): Cybernetics and Systems Research,
Vol.II. North-Holland, Amsterdam, and New York, 1984,
pp.143-148.

[1118] Matis, J.H., Patten, B.C., White, G.C.(eds.):
Compartmental Analysis of Ecosystems Models.
International Cooperative Publ. House, Burtonsville,
Maryland, 1979,.

[1119] Matsuda, T., Takatsu, S.: Algebraic Properties of
Satisficing Decision. Information Sciences, 17, No.3,
1979, pp.221- 237.

[1120] Matsuno, K.: Compartmentalization of Self-Reproducing
Machineries: Multiplication of Microsystems with
Self-Instructing Polymerization of Amino Acids. Origins
of Life, 10, 1980, pp.361-370.

[1121] Matsuno, K.: Disequilibrium Dynamics of Autonomous
Systems and Their Structural Transformation. Int. J. of
General Systems, 6, No.2, 1980, pp.75-82.

[1122] Matsuno, K.: Operational Description Microsystems
Formation in Probiological Molecular Evolution. Origins
of Life, 10, 1980, pp.39-45.

[1123] Matsuno, K.: Dynamics of Relations, Sequential
Connectedness, and Feedback Loops in Open Systems. IEEE
Trans. on Systems, Man, and Cybernetics, SMC-11, No.4,
April 1981, pp.310-312.

[1124] Matsuno, K.: Material Self-Assembly as a Physiochemical
Process. Biosystems, 13, 1981, pp.237-241.

[1125] Matsuno, K.: Self-Sustaining Multiplication and
Reproduction of Microsystems in Protobiogenesis.
Biosystems, 14, 1981, pp.163-170.

[1126] Mattessich, R.: Instrumental Reasoning and Systems Methodology: an Epistemology of the Applied and Social Sciences. D.Reidel, Dordrecht, Holland, and Boston, 1978, 330 pp.

[1127] Mattessich, R.: Axiomatic Representation of the Systems Framework: Similarities and Differences Between Mario Bunges's World of the Systems and My Own Systems Methodology. Cybernetics and Systems, 13, No.1, 1982, pp.51-75.

[1128] Maturana, H.R., Varela, F.J.: Autopoiesis and Cognition: The Realization of the Living. D.Reidel, Dordrecht, Holland, and Boston, 1980, XXX+142 pp.

[1129] Maurer, H.: The Austrian Approach to Videotex. In: Trappl, R.(ed.): Cybernetics and Systems Research, Vol.II. North-Holland, Amsterdam, and New York, 1984, pp.589-592.

[1130] May, R.M.: Will a Large Complex System be Stable? Nature, 238, 1972, pp.413-414.

[1131] Maybeck, P.S.: Stochastic Models, Estimation, and Control. Academic Press, London, and New York, 1979, 432 pp.

[1132] McAleer, W.E.: Systems: A Concept for Business and Management. J. of Applied Systems Analysis, 9, April 1982, pp.99-129.

[1133] McClamroch, N.H.: State Models of Dynamic Systems. Springer-Verlag, Berlin, FRG, and New York, 1980, VIII+248pp.

[1134] McClamroch, N.H.: State Models of Dynamic Systems: A Case Study Approach. Springer-Verlag, Berlin, FRG, and New York, 1980, VIII+248 pp.

[1135] McCorkell, C., Swanick, B.H.: A Hierarchical Adaptive Control System. Cybernetica, XX, No.3, 1977, pp.173-190.

[1136] McDowall, D., (et al.): Interrupted Time Series Analysis. Sage, Beverly Hills, Ca., 1980, 96 pp.

[1137] McGowan, C.L., Kelly, J.R.: A Review of Decomposition and Design Methodologies. In: Rees, R.K.D.(ed.): Software Reliability (2 Vols). Infotech International, Maidenhead, Berkshire, UK, 1977.

[1138] McIntosh, J.E.A., McIntosh, R.P.: Mathematical Modelling and Computers in Endocrinology. Springer-Verlag, Berlin, FRG, and New York, 1980, XII+358pp.

[1139] McIntyre, C.D., Colby, J.A.: A Hierarchical Model of Lotic Ecosystems. Oregon State Univ., Corvallis, Oregon, 1979, 24pp. [IFSR-Depository]

[1140] McIntyre, C.D., White, C.: Addendum to Internal Report 165: Mathematical Documentation for a Lotic Ecosystem Model. Oregon State Univ., Corvallis, Oregon, 1979, 32pp. [IFSR-Depository]

[1141] McLean, J.M.: Methods of System Analysis and Model Building. In: Trappl, R.(ed.): Cybernetics - Theory and Applications. Hemisphere, Washington, D.C., 1983, pp.121-140.

[1142] McPherson, P.K.: On Understanding Modelling and Improving Human Systems. J. of Applied Systems Analysis, 7, April, 1980, pp.131-150.

[1143] Meadows, D.L.: Techniques for Implementing Computer-Based Policy Models. In: Trappl, R.(ed.): Power, Autonomy, Utopia: New Approaches Towards Complex Systems. Plenum Press, New York and London, 1985.

[1144] Medgyessy, P.: Decomposition of Superpositions of Density Functions and Discrete Distributions. Adam Hilger, Bristol, 1977, 308 pp.

[1145] Medina, B.F.: Structured System Analysis: A New Technique. Gordon and Breach, New York, and London, 1981,

XIV+82 pp.
[1146] Mees, A.L.: Dynamics of Feedback Systems. John Wiley, Chichester and New York, 1981, X+214 pp.
[1147] Mehlhorn, K.: Nearly Optimal Binary Search Trees. Acta Informatica, 5, No.4, 1975, pp.287-295.
[1148] Mel'cuk, I.A.: Cybernetics and Linguistics: Some Reasons for as Well as Some Consequences of Bringing them Together. Austrian Society for Cybernetic Studies, Vienna, 1977.
[1149] Mel'cuk, I.A.: Cybernetics and Linguistics. In: Trappl, R.(ed.): Cybernetics - Theory and Applications. Hemisphere, Washington, D.C., 1983, pp.323-338.
[1150] Meldman, M.J., (et al.): Riss: a Relational Data Base Management System for Minicomputers. Van Nostrand Reinhold, New York, 1978, IX+113 pp.
[1151] Melkumian, A.V.: Distortion Measures in Image Data Compression and Transmission Systems. In: Trappl, R.(ed.): Cybernetics and Systems Research, Vol.II. North-Holland, Amsterdam, and New York, 1984, pp.149-154.
[1152] Mergenthaler, W.: The Total Repair Cost in a Defective, Coherent Binary System. Cybernetics and Systems, 13, No.3, 1982, pp.219-243.
[1153] Mesarovic, M.D., Pestel, E.(eds.): Multilevel Computer Model of World Development System. IIASA, Laxenburg, Austria, 1974.
[1154] Metz, H.A.J.: State Space Models for Animal Behaviour. Annals of Systems Research, 6, 1977, pp.65-109.
[1155] Metzer, J.R., Barnes, B.H.: Decision Table Languages and Systems. Academic Press, London, and New York, 1977.
[1156] Meyer, A.R.: Weak Monadic Second Order Theory of Successor Is Not Elementary Recursive. In: Dold, A., Eckmann, B.(eds.): Lecture Notes in Mathematics, Vol.453 Springer-Verlag, Berlin, FRG, and New York, 1975.
[1157] Meyer, J.F.: Closed-Form Solutions of Performability. IEEE Trans. on Computers, C-31, No.7, July 1982, pp.648-657.
[1158] Michalski, R.S.: Pattern Recognition as Rule-Guided Inductive Inference. IEEE Trans. on Pattern Analysis and Machine Intelligence, PAMI-2, No.4, 1980, pp.349-361.
[1159] Michel, A.N., Miller, R.K.: Qualitative Analysis of Large Scale Systems. Academic Press, London, and New York, 1977, 289 pp.
[1160] Michie, D.(ed.): Expert Systems in the Microelectronic Age. Edinburgh Univ. Press, Edinburgh, 1979.
[1161] Michie, D.: Machine Intelligence and Related Topics: An Information Scientist's Weekend Book. Gordon and Breach, New York, and London, 1982, XI+316pp.
[1162] Mihram, G.A.: Simulation: Methodology for Decision Theorists. In: White, D.J., Brown, K.C.(eds.): Role and Effectiveness of Theories of Decision in Practice. Hodder / Stoughton, London, 1975, pp.320-327.
[1163] Milgram, M.: Barycentric Entropy of Mankov Systems. Cybernetics and Systems, 12, No.1-2, 1981, pp.141-178.
[1164] Millendorfer, J.: Growth Reducing Factors in Complex Systems. In: Trappl, R.(ed.): Cybernetics and Systems Research. North-Holland, Amsterdam, and New York, 1982, pp.543-548.
[1165] Millendorfer, J.: Systemanalytical Approaches to Long Economic Waves. In: Trappl, R.(ed.): Cybernetics and Systems Research, Vol.II. North-Holland, Amsterdam, and New York, 1984, pp.443-450.
[1166] Miller, G.L.: Graph Isomorphism, General Remarks. J. of Computer and Systems Sciences, 18, No.2, April 1979, pp.128-142.

[1167] Miller, J.G.: Living Systems. McGraw-Hill, New York, 1978, XII+1102 pp.
[1168] Miller, M.C., (et al.)(eds.): Mathematical Models in Medical Diagnosis. Praeger, New York, 1981, XVIII+188pp.
[1169] Miller, R.B.: Archetypes of Man-Computer Problem Solving. Ergonomics, 12, 1969, pp.559-581.
[1170] Milsum, J.H., Laszlo, C.A.: Cybernetics and Health Care. In: Trappl, R.(ed.): Cybernetics - Theory and Applications. Hemisphere, Washington, D.C., 1983, pp.235-262.
[1171] Mingers, J.C.: Towards an Appropriate Social Theory for Applied Systems Thinking: Critical Theory and Soft Systems Methodology. J. of Applied Systems Analysis, 7, April 1980, pp.41-49.
[1172] Mitroff, I.I., Kilmann, R.H.: Methodological Approaches to Social Science. Jossey-Bass, San Francisco, 1978, XVIII+150 pp.
[1173] Mitroff, I.I., Williams, J.G., Flynn, R.: On the Strength of Belief of Dialectical Belief Systems. Int. J. of General Systems, 4, No.3, 1978, pp.189-200.
[1174] Mitsch, W.J., (et al.)(eds.): Energetics and Systems. Ann Arbor Science, Ann Arbor, Mich., 1982, X + 132pp.
[1175] Mittermeir, R.: CML-Graphs - A Notation for Systems Development. In: Trappl, R.(ed.): Cybernetics and Systems Research. North-Holland, Amsterdam, and New York, 1982, pp.803-810.
[1176] Mizumoto, M.: Note on the Arithmetic Rule by Zadeh for Fuzzy Conditional Inference. Cybernetics and Systems, 12, No.3, 1981, pp.247-306.
[1177] Mock, T.J., Grove, H.D.: Measurement, Accounting, and Organizational Information. John Wiley, Chichester and New York, 1979, XVIII+238 pp.
[1178] Mohler, R., Ruberti, R.(eds.): Theory and Applications of Variable Structure Systems. Academic Press, London, and New York, 1972.
[1179] Mohler, R., Ruberti, A.(eds.): Recent Developments in Variable Structure Systems, Economics and Biology. Springer-Verlag, Berlin, FRG, and New York, 1978, VI+326 pp.
[1180] Mohler, R., Kolodziej, W.J.: An Overview of Bilinear System Theory and Applications. IEEE Trans. on Systems, Man, and Cybernetics, SMC-10, No.10, 1980, pp.683-688.
[1181] Molina, F.W.: A Survey of Resource Directive Decomposition in Mathematical Programming. ACM Computing Surveys, 11, No.2, 1979, pp.95- 104.
[1182] Molloy, K.J.: Insights into Complex Organisational Problem Solving. In: Trappl, R.(ed.): Cybernetics and Systems Research. North-Holland, Amsterdam, and New York, 1982, pp.527-534.
[1183] Monahan, G.E.(ed.): A Survey of Partially Observable Markov Decision Processes Theory, Models, and Algorithms. Management Science, 28, No.1, Jan.1982, pp.1-16.
[1184] Moraga, C.: Introduction to Linear p-Adic Invariant Systems. In: Trappl, R.(ed.): Cybernetics and Systems Research, Vol.II. North-Holland, Amsterdam, and New York, 1984, pp.121-124.
[1185] Morein, R.: The Use of the Calculus of Variations and Functionals in the Analysis of Threshold Logic Unit Networks. Cybernetics and Systems, 12, No.1-2, 1981, pp.43-52.
[1186] Morin, E.: Complexity. Social Science Forum, XXVI, No.4, 1974, pp.555-582.
[1187] Morley, D.A.: Mathematical Modelling in Water and Wastewater Treatment. Applied Science Publishers,

Barking, England, 1979, XV+366 pp.
[1188] Morris, C.N., Rolph, J.E.: Introduction to Data Analysis and Statistical Inference. Prentice-Hall, Englewood Cliffs, New Jersey, 1981, XX+389 pp.
[1189] Morris, H.M.: Critical Stability in a Conserved Nutrient Ecosystem. Int. J. of Systems Science, 8, No.6, 1977, pp.651- 664.
[1190] Moser, J.: Hidden Symmetrics in Dynamical Systems. American Scientist, 67, No.6, 1979, pp.689- 695.
[1191] Moss, Brian: Ecology of Fresh Waters. Halstead, New York, 1980, 332 pp.
[1192] Moylan, P.J., Hill, D.J.: Stability Criteria for Large-Scale Systems. IEEE Trans. on Automatic Control, AC-23, No.2, April 1978, pp.143-149.
[1193] Mulej, M.: Dialectical Systems Theory. Rozvojni Center Celje, Celje, Yugoslavia, 1979, XII+218 pp. [Slovene]
[1194] Mulej, M.: Dialectical Systems Theory and Work Simplification. In: Trappl, R.(ed.): Cybernetics and Systems Research. North-Holland, Amsterdam, and New York, 1982, pp.579-588.
[1195] Mulej, M., Pirc, V.: Method "Wholistic Creativity of Many (WCM)" Applied to Collect and Organize Associates' Ideas about New Products and Programs. In: Trappl, R.(ed.): Cybernetics and Systems Research, Vol.II. North-Holland, Amsterdam, and New York, 1984, pp.341-348.
[1196] Mumford, E.: Participative Systems Design: Structure and Method. Systems, Objectives, Solutions, 1, No.1, January 1981, pp.5-19.
[1197] Murdick, R.G.: MIS: Concepts and Design. Prentice-Hall, Englewood Cliffs, New Jersey, 1980, XIV+610 pp.
[1198] Murphy, R.E.: Adaptive Processes in Economic Systems. Academic Press, London, and New York, 1965.
[1199] Nachane, D.M.: Optimization Methods in Multi-Level Systems. In: Trappl, R.(ed.): Cybernetics and Systems Research, Vol.II. North-Holland, Amsterdam, and New York, 1984, pp.69-76.
[1200] Nagel, E.: Teleology Revisited. J. of Philosophy, LXXIV, No.5, May 1977, pp.261-301.
[1201] Nahmias, S.: Fuzzy Variables. Fuzzy Sets and Systems, 1, No.2, April 1978, pp.97-110.
[1202] Nash, P.: Systems Modelling and Optimization. IEEE, Piscataway, N.J., 1981, 224pp.
[1203] Naughton, J.: Craft Knowledge in Systems Research: The Rules of the Game. Open Univ., Milton Keynes, England, 1979, 21pp. [IFSR-Depository]
[1204] Naylor, A.W.: On Decomposition Theory: Generalized Dependence. IEEE Trans. on Systems, Man, and Cybernetics, SMC-11, No.10, October 1981, pp.699-713.
[1205] Neal, M.: The Construction of Personality in the Novel. In: Trappl, R.(ed.): Cybernetics and Systems Research. North-Holland, Amsterdam, and New York, 1982, pp.363-366.
[1206] Neck, R.: Interactions between the Political and the Economic System in Austria. In: Trappl, R.(ed.): Cybernetics and Systems Research. North-Holland, Amsterdam, and New York, 1982, pp.513-520.
[1207] Neck, R.: A System-Theoretic Analysis of a Simple Macroeconomic Model. In: Trappl, R.(ed.): Cybernetics and Systems Research, Vol.II. North-Holland, Amsterdam, and New York, 1984, pp.457-464.
[1208] Negoita, C., Zadeh, L.A., Zimmermann, H.J.(eds.): Fuzzy Sets and Systems: a New International Journal. North-Holland, Amsterdam, and New York, 1977.
[1209] Negoita, C.: Management Applications of Systems Theory. Birkhaeuser Verlag, Basel and Stuttgart, 1979, 155 pp.

[1210] Negoita, C.:   Fuzzy   Systems.   Abacus   Press,   Tunbridge
       Wells, England, 1980.
[1211] Negoita, C.V.,   Roman,   R.:   On the Logic of Discrete
       Systems Dynamics.   Kybernetes, 9, No.3, 1980, pp.189-192.
[1212] Negoita, C.V.:   Fuzzy Systems.   Abacus   Press,   Tunbridge
       Wells, England, 1981, VIII+111 pp.
[1213] Nelson, R.J.:   Structure of Complex Systems. In: Suppes,
       P., Asquith, P.D.(eds.): PSA 1976. Philosophy of Science
       Asscociation, East Lansing, Mich., 1977, pp.523-542.
[1214] Netchiporenko, V.I.:   Structure Analysis of Systems.
       Sovietskoye Radio, Moscow, 1977, 214 pp. [Russian]
[1215] Neuwirth, E.: Fuzzy Similiarity Relations and Distance
       Coefficients in Thesauri.   In: Trappl,   R.(ed.):
       Cybernetics and   Systems   Research.   North-Holland,
       Amsterdam, and New York, 1982, pp.811-814.
[1216] Nevitt, B.:   ABC of Prophecy: Understanding   the
       Environment. Canadian Futures, Toronto, 1980, 92 pp.
[1217] Newell, A.,   Simon,   H.A.:   Human   Problem Solving.
       Prentice-Hall, Englewood Cliffs, New Jersey, 1972.
[1218] Newell, A.:   The   Knowledge   Level.   Artificial
       Intelligence, 18, No.1, Jan.1982, pp.87-127.
[1219] Newton-Smith, W.H.: The Structure of Time.   Routledge &
       Kegan Paul, London, 1980, XII+262 pp.
[1220] Nguyen, H.T.: On Conditional Possibility Distributions.
       Fuzzy Sets and Systems, 1, No.4, Oct.1978, pp.299-309.
[1221] Nicholson, H.:   Structure of Interconnected   Systems.
       IEEE, Piscataway, N.J., 1978, 258pp.
[1222] Nicholson, H.(ed.):   Modelling of Dynamical Systems.
       Peter Peregrinus, London and New York, Vol.I: 1980,
       XI+227 pp.; Vol.II: 1981, XIV+264 pp.
[1223] Nicholson, H.(ed.):   Modelling of Dynamical Systems.
       IEEE, Piscataway, N.J., Vol.1: 1980, 256 pp; Vol.2:
       1981, 288 pp.
[1224] Nicolis, G.,   Prigogine,   I.:   Self-Organization   in
       Non-Equilibrium Systems.   John Wiley, Chichester and New
       York, 1977.
[1225] Nicolis, J.S.:   Inadequate   Communication   between
       Self-Organizing   Systems   and   Desynchronization   of
       Physiological Rhythms.   General Systems Yearbook, 22,
       1977, pp.119-136.
[1226] Nicolis,   J.S.,   Protonotarios: Bifurcation   in
       Non-Negotiable Games: A Paradigm for Self-Organization in
       Cognitive Systems.   Int. J. of Bio-Medical Computing, 10,
       1979, pp.417-447.
[1227] Niemeyer, H.(ed.):   Soziale Beziehungsgeflechte. Dunker &
       Humbolt, Berlin, 1980. [German]
[1228] Nigro, A.: Language Behavioral Feedback. Cybernetica,
       20, No.2, 1977, pp.141-146.
[1229] Nijssen, G.M.(ed.): Modelling in Data Base Management
       Systems. Elsevier / North-Holland, New York, 1976.
[1230] Nilsson, N.J.: Principles of Artificial Intelligence.
       Tioga Publishing Co., Palo Alto, Calif., 1980, XVI+476 pp.
[1231] Ninneman, J.: The Survival of U.S.Business: A General
       Systems Approach. Systems Trends, 3, No.3, March 1981,
       pp.3 - 12.
[1232] Nishimura, H.:   Realization of Dependency   Structures.
       Bulletin of Mathematical Biology, 39, No.4,   1977,
       pp.499-503.
[1233] Nitecki, Z., Robinson, C.(eds.):   Global   Theory   of
       Dynamical Systems. Springer-Verlag, Berlin, FRG, and New
       York, 1980, X+500 pp.
[1234] Nobile, A.G., Ricciardi, L., Sacerdote, L.: On a Class of
       Discrete Models for Regulated Growth with Intrinsic Lower
       Bounds. In: Trappl, R.(ed.): Cybernetics and Systems

Research, Vol.II. North-Holland, Amsterdam, and New York, 1984, pp.281-292.

[1235] Noguchi, S., Tanaka, A., Sugawara, K.: Properties of Analogue Neural Networks. J. of Cybernetics, 10, No.4, 1980, pp.265-282.

[1236] Norman, D.A.(ed.): Perspectives in Cognitive Science. (Papers from a Meeting in La Jolla, Cal., Aug.1979). Ablex, Norwood, N.J., 1981, 304 pp.

[1237] Norman, M.: Software Package for Economic Modelling. IIASA, Laxenburg, Austria, 1977, XII+127 pp., Kybernetes, 7, No.1, 1978, pp.13-18.

[1238] Nowak, S.: Methodology of Sociological Research. D.Reidel, Dordrecht, Holland, and Boston, 1977.

[1239] Nowakowska, M.: Methodological Problems of Measurement of Fuzzy Concepts in the Social Sciences. Behavioral Science, 22, No.2, March 1977, pp.107-115.

[1240] Nowakowska, M.: The Logical Structure of the Development of a Scientific Discipline. Austrian Society for Cybernetic Studies, Vienna, 1977. [German]

[1241] Nowakowska, M.: Theories of Research: Modelling Approaches. Panstwowe Wydawnictwo Naukowe, Warsaw, 1977. [Polish]

[1242] Nowakowska, M.: Instability, Monopolies, and Alienation in Science. Austrian Society for Cybernetic Studies, Vienna, 1979.

[1243] Nowakowska, M.: Formal Semiotics and Multidimensional Semiotic Systems. Cybernetics and Systems, 12, No.1-2, 1981, pp.83-102.

[1244] Nowakowska, M.: New Theoretical, Methodological, and Empirical Possibilities in Decision Theory. Cybernetics and Systems, 12, No.4, 1981, pp.311-343.

[1245] Nowakowska, M.: Cybernetics in the Social Sciences. In: Trappl, R.(ed.): Cybernetics - Theory and Applications. Hemisphere, Washington, D.C., 1983, pp.177-234.

[1246] Nowotny, H.: Not Quite Human: Science and Utopia. In: Trappl, R.(ed.): Power, Autonomy, Utopia: New Approaches Towards Complex Systems. Plenum Press, New York and London, 1985.

[1247] Nugyen, V.V., Wood, E.F.: Review and Unification of Linear Identifiability Concepts. SIAM Reviews, 24, No.1, 1982, pp.34-51.

[1248] Nurmi, H.: On Strategies of Cybernetic Model-Building. Kybernetes, 7, No.1, 1978, pp.13-18.

[1249] Nussbaumer, H.J.: Fast Fourier Transform and Convolution Algorithms. Springer-Verlag, Berlin, FRG, and New York, 1980, 330 pp.

[1250] O'Donovan, T.M.: GSPS: Simulation Made Simple. John Wiley, Chichester and New York, 1979, XII+127 pp.

[1251] O'Muircheartaigh, C.A., Payne, C.(eds.): The Analysis of Survey Data. Vol.1: Exploring Data Structures. Vol.2: Model Fitting. John Wiley, Chichester and New York, 1977.

[1252] Oden, A.S.: An Overview of Mathematical Modeling of the Behavior of Hydrocarbon Reservoirs. SIAM Reviews, 24, No.3, July 1982, pp.263-273.

[1253] Odum, H.T., Odum, E.C.: Energy Basis for Man and Nature. McGraw-Hill, New York, 1976.

[1254] Oeren, T.I.(ed.): Cybernetics and Modelling and Simulation of Large Scale Systems. Int. Association for Cybernetics, Namur, 1976.

[1255] Oeren, T.I.: Rationale for Large Scale System Simulation Software Based on Cybernetic and General Systems Theories. In: Oeren, T.I.(ed.): Cybernetics and Modelling and Simulation of Large Scale Systems. Int. Association for Cybernetics, Namur, 1976, pp.149-179.

[1256] Oeren, T.I., (et al.)(eds.):   Simulation with Discrete
       Models. IEEE, Piscataway, N.J., 1980, XI+258 pp.
[1257] Ogborn, J.M.,   Johnson,   L.:   Conversation   Theory.
       Kybernetes, 13, No.1, 1984, pp.7-16.
[1258] Oguntade, O.O.:   Implementing a Pragmatic Theory of
       Humanistic Systems.   Int. J. of General Systems, 8, No.1,
       1982, pp.33-42.
[1259] Ohlsson, S.: On the Automated Learning of Problem Solving
       Rules. In: Trappl, R.(ed.):   Cybernetics and Systems
       Research. North-Holland, Amsterdam, and New York, 1982,
       pp.979-984.
[1260] Okeda, M., Siljak, D.D.:   Overlapping Decompositions,
       Expansions, and Contractions of Dynamic Systems. Large
       Scale Systems, 1, No.1, Feb.1980, pp.29-38.
[1261] Okuda, T.H., Asai, K.: A Formulation of Fuzzy Decision
       Problems with Fuzzy Information Using Probability Measures
       of Fuzzy Events.   Information and Control, 38, No.2,
       Aug.1978, pp.135-147.
[1262] Oldershaw, R.L.:   Conceptual   Foundations   of   the
       Self-Similar Hierarchical Cosmology.   Int. J. of General
       Systems, 7, No.2, 1981, pp.151-158.
[1263] Oldershaw, R.L.:   On the Number of Levels in the
       Self-Similar Hierarchical Cosmology.   Int. J. of General
       Systems, 7, No.2, 1981, pp.159-164.
[1264] Oldershaw, R.L.: Empirical and Theoretical Support for
       Self-Similarity between Atomic and Stellar Systems.   Int.
       J. of General Systems, 8, No.1, 1982, pp.1-6.
[1265] Oldershaw, R.L.:   New Evidence for the Principle of
       Self-Similarity. Int. J. of General Systems, 9, No.1,
       1982, pp.37-42.
[1266] Ollero, A., Freire, E.: The Structure of Relations in
       Personnel Management.   Fuzzy Sets and Systems, 5, No.2,
       March 1981, pp.115-125.
[1267] Optner, S.L.O.(ed.): Systems Analysis.   Penguin Books,
       London, 1973.
[1268] Orava, P.J.:   On the Concepts of Input-Output Model,
       Consolity, and State in the Theory of Dynamical Systems
       and Control. Acta Polytechnica Scandinavica, Mathematics
       and Computer Science Series, No.31, Helsinki, 1979,
       pp.120-127.
[1269] Oren, T.I.: Simulation - As it Has Been, Is, and Should
       Be. Simulation, 29, No.5, Nov.1977, pp.161-164.
[1270] Oren, T.I., Zeigler, B.P.:   Concepts for Advanced
       Simulation Methodologies.   Simulation, 32, No.3, March
       1979, pp.69-82.
[1271] Orlovsky, S.A.: Decision-Making with a Fuzzy Preference
       Relation. Fuzzy Sets and Systems, 1, No.3, July 1978,
       pp.155-167.
[1272] Orlovsky, S.A.: Effective Alternatives for Multiple Fuzzy
       Preference Relations. In: Trappl, R.(ed.):   Cybernetics
       and Systems Research.   North-Holland, Amsterdam, and New
       York, 1982, pp.185-190.
[1273] Otnes, R.K., Enochson, L.: Applied Time Series Analysis.
       Vol.1: Basic Techniques. John Wiley, Chichester and New
       York, 1978, XIV+449 pp.
[1274] Overton, W.S.: A Strategy of Model Construction.   In:
       Hall, C.A.S., Day, J.W.(eds.): Ecosystem Modeling in
       Theory and Practice: an Introduction with Case Studies.
       John Wiley, Chichester and New York, 1977.
[1275] Overton, W.S., White, C.: Evolution of a Hydrology Model
       - an Exercise in Modelling Strategy Int. J. of General
       Systems, 4, No.2, 1978, pp.89-104.
[1276] Overton, W.S., White, C.: On Constructing a Hierarchical
       Model in the Flex Paradigm Illustrated by Structural

Aspects of a Hydrology Model. Int. J. of General Systems, 6, No.4, 1981, pp.191-216.

[1277] Owens, D.H.: Feedback and Multivariate Systems. IEEE, Piscataway, N.J., 320 pp.

[1278] Palm, G.: Rules for Synaptic Changes and their Relevance for the Storage of Information in the Brain. In: Trappl, R.(ed.): Cybernetics and Systems Research. North-Holland, Amsterdam, and New York, 1982, pp.277-280.

[1279] Palvoelgyi, L.: Network-Modelling of Learning. In: Trappl, R.(ed.): Cybernetics and Systems Research. North-Holland, Amsterdam, and New York, 1982, pp.409-414.

[1280] Pappis, C.P., Mandani, E.H.: A Fuzzy Logic Controller for a Traffic Junction. IEEE Trans. on Systems, Man, and Cybernetics, SMC-7, No.10, Oct.1977, pp.707-717.

[1281] Pardo, L.: Information Energy of a Fuzzy Event and a Partition of Fuzzy Events. In: Trappl, R.(ed.): Cybernetics and Systems Research, Vol.II. North-Holland, Amsterdam, and New York, 1984, pp.541-544.

[1282] Paritsis, N.C., Stewart, D.J.: Adaptional Problems in Natural Intelligent Systems with Changes in Environmental Variety. In: Trappl, R.(ed.): Cybernetics and Systems Research. North-Holland, Amsterdam, and New York, 1982, pp.415-420.

[1283] Parks, M.S., Steinberg, E.: Dichotic Property and Teleogenesis. Kybernetes, 7, No.4, 1978, pp.259-264.

[1284] Pask, G.: Consciousness. J. of Cybernetics, 9, No.3, 1979, pp.211-258.

[1285] Pask, G.: Development in Conversational Theory - Part 1 Int. J. of Man-Machine Studies, 13, No.4, 1980, pp.357-411.

[1286] Pask, G.: The Originality of Cybernetics and the Cybernetics of Originality. In: Trappl, R.(ed.): Cybernetics and Systems Research. North-Holland, Amsterdam, and New York, 1982, pp.367-370.

[1287] Pask, G.: Cybernetics in Psychology and Education. In: Trappl, R.(ed.): Cybernetics - Theory and Applications. Hemisphere, Washington, D.C., 1983, pp.159-176.

[1288] Pask, G.: The Architecture of Knowledge and Knowledge of Architecture. In: Trappl, R.(ed.): Cybernetics and Systems Research, Vol.II. North-Holland, Amsterdam, and New York, 1984, pp.641-646.

[1289] Patel, R.V., Munro, N.: Multivariate System Theory and Design. Pergamon Press, Oxford and New York, 1982, XII+374pp.

[1290] Patten, B.C., (et al.): Propagation of Cause in Ecosystems. In: Patten, B.C.(ed.): Systems Analysis in Ecology. Academic Press, London, and New York, 1976.

[1291] Patten, B.C.(ed.): Systems Analysis in Ecology. Academic Press, London, and New York, 1976.

[1292] Patten, B.C.: Systems Approach to the Concept of Environment. Ohio Journal of Science, 78, No.4,1978, pp.206-222.

[1293] Patten, B.C., Auble, G.T.: Systems Approach to the Concept of Niche. Synthese, 43, 1980, pp.155-181.

[1294] Patten, B.C., Auble, G.T.: System Theory and the Ecological Niche. American Naturalist, 117, No.6, June 1981, pp.893-922.

[1295] Patten, B.C., Odum, E.P.: The Cybernetic Nature of Ecosystem. American Naturalist, 118, 1981, pp.886-895.

[1296] Pau, L.F.: Failure Diagnosis and Performance Monitoring. Marcel Dekker, New York, 1982.

[1297] Paulre, B.E.(ed.): Systems Dynamics and the Analysis of Chance. North-Holland, Amsterdam, and New York, 1981, XII + 382pp.

[1298] Peacocke, C.:   Holistic   Explanation.   Clarendon   Press,
       Oxford, 1980.
[1299] Pearce, J.G.:   Telecommunications   Switching.   Plenum
       Press, New York and London, 1981, IX+338 pp.
[1300] Pearl, J.:   On the   Complexity   of   Inexact   Computations.
       Information   Processing   Letters,   4,   No.3,   Dec.1975,
       pp.77-81.
[1301] Pearl, J.:   On   the   Storage   Economy   of   Inferential
       Question-Answer Systems.   IEEE Trans. on Systems, Man, and
       Cybernetics, SMC-5, No.6, Nov.1975, pp.595-602.
[1302] Pearl, J.:   A Framework for Processing   Value   Judgements.
       IEEE Trans. on Systems, Man, and Cybernetics, SMC-7, No.5,
       May 1977, pp.349-357.
[1303] Pearl, J.:   On   Summarizing   Data   Using   Probabilistic
       Assertions.   IEEE   Trans.   on   Information   Theory, IT-23,
       No.4, July 1977, pp.459-465.
[1304] Pearl, J.:   An   Economic   Basis   for   Certain   Methods   of
       Evaluating   Probabilistic   Forecasts.   Int.   J.   of
       Man-Machine Studies, 10, No.2, March 1978, pp.175-183.
[1305] Pearlman, W.:   Beyond   Gestalt.   Kybernetes,   13,   No.1,
       1984, pp.17-20.
[1306] Peay, E.R.:   Structural Models   with   Qualitative   Values.
       J. of   Mathematical   Sociology, 8, No.2, 1982, pp.161-192.
[1307] Pedretti,   A.:   Epistemology,   Semantics,   and
       Self-Reference.   J.   of Cybernetics,   10,   No.4,   1980,
       pp.313-340.
[1308] Pedretti, A.:   The   Cybernetics   of   Language.   Princelet
       Editions, London, 1981.
[1309] Pedretti, A.:   Where Coincidence   Coincides   -   Coining   a
       Notion of   Cybernetic   Description?   In: Trappl, R.(ed.):
       Cybernetics   and   Systems   Research.   North-Holland,
       Amsterdam, and New York, 1982, pp.79-84.
[1310] Pedrycz, W.:   Fuzzy Relational Equations   with   Triangular
       Norms and their Resolutions. Busefal, 11, 1982, pp.24-32.
[1311] Pedrycz, W.:   Some   Applicational   Aspects   of   Fuzzy
       Relational Equations   in   Systems   Analysis.   Int.   J. of
       General Systems, 9, No.3, 1983, pp.125-132.
[1312] Pedrycz, W.:   A   Model   of   Decision-Making   in   a   Fuzzy
       Environment.   Kybernetes, 13, No.2, 1984, pp.99-102.
[1313] Pedrycz, W.:   Construction   of   Fuzzy   Relational   Models.
       In: Trappl,   R.(ed.):   Cybernetics and Systems Research,
       Vol.II.   North-Holland, Amsterdam,   and   New   York,   1984,
       pp.545-550.
[1314] Peery, N.S.:   General   Systems   Theory   Approaches   to
       Organizations: Some   Problems   in   Application.   J.   of
       Management Studies, 12, No.3, Oct.1975, 266-275.
[1315] Peixoto, M.M.(ed.):   Dynamical Systems.   Academic   Press,
       London, and New York, 1973, XIII+745 pp.
[1316] Pelzmann, L., Stueckler, H.:   Learned   Helplessness   -   An
       Attribution-Model for   Unemployed   Workers?   In: Trappl,
       R.(ed.):   Cybernetics   and   Systems   Research.
       North-Holland, Amsterdam,   and New York, 1982, pp.391-396.
[1317] Perlis, A.J.,   Sayward,   F.,   Shaw,   M.(eds.):   Software
       Metrics: An   Analysis   and   Evaluation.   M.I.T.Press,
       Cambridge, Mass., 1981, XI+404 pp.
[1318] Perlis, J.H.,   Ignizio,   J.P.:   Stability   Analysis: An
       Approach to   the Evaluation of System Design.   Cybernetics
       and Systems, 11, No.1-2, 1980, pp.87-103.
[1319] Pernici, B.:   Problem Solving in Game   Playing: Computer
       Chess. Cybernetics and Systems, 13, No.1, 1982, pp.31-49.
[1320] Peschel, M.,   Mende,   W.,   Voigt,   M.:   Application   of
       Polyoptimization   to   Processes   of   Evolution.   Austrian
       Society for Cybernetic Studies, Vienna, 1979.   [German]
[1321] Peschel, M., Bocklisch, S.F.,   Meyer,   W.,   Straube,   P.,

Richardt, J.: Data- and Behaviour Analysis with Classification Models. Austrian Society for Cybernetic Studies, Vienna, 1981. [German]

[1322] Peschel, M., Mende, W.: Ecological Approach to Systems Analysis Based on Volterra Equations. Cybernetics and Systems, 13, No.2, 1982, pp.175-186.

[1323] Peschel, M., Mende, W.: Ecological Approach to Systems Analysis Based on Volterra Equations. In: Trappl, R.(ed.): Cybernetics and Systems Research. North-Holland, Amsterdam, and New York, 1982, pp.127-134.

[1324] Peschel, M.: Engineering Cybernetics. In: Trappl, R.(ed.): Cybernetics - Theory and Applications. Hemisphere, Washington, D.C., 1983, pp.291-322.

[1325] Peschel, M., Breitenecker, F.: Socio-Economic Consequences of the Volterra Approach for Nonlinear Systems. In: Trappl, R.(ed.): Cybernetics and Systems Research, Vol.II. North-Holland, Amsterdam, and New York, 1984, pp.423-428.

[1326] Peterson, J.L.: Petri Nets. ACM Computing Surveys, 9 No.3, Sept.1977, pp.223-252.

[1327] Peterson, J.L.: Petri Net Theory and the Modeling of Systems. Prentice-Hall, Englewood Cliffs, New Jersey, 1981, X+290 pp.

[1328] Petkoff, B.: A Cybernetic Model of Scientific Research and Cognition. In: Trappl, R.(ed.): Cybernetics and Systems Research, Vol.II. North-Holland, Amsterdam, and New York, 1984, pp.721-726.

[1329] Petrov, B.N., (et al.): Some Problems in the Theory of Searchless Self-Adaptive Systems, I. Engineering Cybernetics, 14, 1976, pp.132-140.

[1330] Petrovski, A., Yashin, A.: Medico-Demographic Models in Health Care Systems. In: Trappl, R.(ed.): Cybernetics and Systems Research. North-Holland, Amsterdam, and New York, 1982, pp.629-632.

[1331] Pfeilsticker, A.: The Systems Approach and Fuzzy Set Theory Bridging the Gap between Mathematical and Language-Oriented Economics. Fuzzy Sets and Systems, 6, No.3, Nov.1981, pp.209-233.

[1332] Phillips, D.C.: Holistic Thought in Social Science. Stanford Univ. Press, Stanford, Ca., 1976, 149 pp.

[1333] Phillips, R.J., Beaumont, M.J., Richardson, D.: AESOP: an Architectural Relational Data Base. Computer-Aided Design, 11, No.4, July 1979, pp.217-226.

[1334] Pichler, F., Ottendorfer: Decomposition of General Dynamic Systems. Systems Science, 5, No.1, 1979, pp.5-18.

[1335] Pichler, F., Hanika, F.de P.(eds.): Progress in Cybernetics and Systems Research, Vol.VII. Hemisphere, Washington, D.C., 1980, 393 pp.

[1336] Pichler, F., Trappl, R.(eds.): Progress in Cybernetics and Systems Research, Vol.VI. Hemisphere, Washington, D.C., 1982, 398 pp.

[1337] Pichler, F.: Dynamical Systems Theory. In: Trappl, R.(ed.): Cybernetics - Theory and Applications. Hemisphere, Washington, D.C., 1983, pp.43-56.

[1338] Pichler, F., Hellwagner, H.(eds.): Robotics and Image Processing. Austrian Society for Cybernetic Studies, Vienna, 1983. [German]

[1339] Pichler, F.: General Systems Algorithms for Mathematical Systems Theory. In: Trappl, R.(ed.): Cybernetics and Systems Research, Vol.II. North-Holland, Amsterdam, and New York, 1984, pp.161-164.

[1340] Pierce, J.R., Posner, E.C.: Introduction to Communication Science and Systems. Plenum Press, New York and London, 1980, XVI+390pp.

[1341] Pinske, M.S.: Information and Information Stability of Random Variables and Processes. Holden-Day, San Francisco, 1964.

[1342] Pipino, L.L., Van Gigch, J.P.: Potential Impact of Fuzzy Sets on the Social Sciences. Cybernetics and Systems, 12, No.1-2, 1981, pp.21-35.

[1343] Pippenger, N.: Information Theory and the Complexity of Boolean Functions. Mathematical Systems Theory, 10, No.2, 1977, pp.127-167.

[1344] Pippenger, N.: Complexity Theory. Scientific American, 238, No.6, 1978, pp.114- 124.

[1345] Pla Lopez, R.: Systemic Transition from Ideological Learning to Scientific Learning. In: Trappl, R.(ed.): Cybernetics and Systems Research, Vol.II. North-Holland, Amsterdam, and New York, 1984, pp.691-696.

[1346] Plotkin, A.A.: Hierarchical Systems of Subsets. Automation and Remote Control, 42, No.5, Part 2, May 1981, pp.670-675.

[1347] Poch, F.A.: Some Applications of Fuzzy Sets to Statistics . Instituto Nacional De Estadistica, Madrid, 1979, 115 pp. [Spanish]

[1348] Pogonowski, J.: Set-Theoretical Approach to General Systems Theory. In: Trappl, R.(ed.): Cybernetics and Systems Research. North-Holland, Amsterdam, and New York, 1982, pp.15-18.

[1349] Porenta, G., Riederer, P.: A Mathematical Model of the Dopaminergic Synapse: Stability and Sensitivity Analyses, and Simulation of Parkinson's Disease and Aging Processes. Cybernetics and Systems, 13, No.3, 1982, pp.257-274.

[1350] Porter, B.: Synthesis of Dynamical Systems. Nelson, London, 1969.

[1351] Pospelov, D.A.: Large Systems: Situational Control . Znaniye Press, Moscow, 1975. [Russian]

[1352] Pospelov, D.A.: Logical-Linguistic Models in Control Systems. Energoizdat, Moscow, 1981, 231 pp. [Russian]

[1353] Poston, T., Stewart, I.: Catastrophe Theory and its Applications. Pitman, London and Boston, 1978, 491 pp.

[1354] Pot, John S.: Scientific Relevance and the Rehabilitation of the Goal Concept. Uitgewejij Stabo/All Round B.V., Groningen, 1980.

[1355] Potter, J.M., Anderson, B.D.O.: Partial Prior Information and Decision-Making. IEEE Trans. on Systems, Man, and Cybernetics, SMC-10, No.3, March 1980, pp.125-133.

[1356] Pratt, W.K.: Digital Image Processing. John Wiley, Chichester and New York, 1978, X+750 pp.

[1357] Presern, S.: Plan Generating System for a Tactile Sensor. In: Trappl, R.(ed.): Cybernetics and Systems Research. North-Holland, Amsterdam, and New York, 1982, pp.949-954.

[1358] Pressman, R.S.: Software Engineering: A Practitioner's Approach. McGraw-Hill, New York, 1982, XVI+352pp.

[1359] Prevost, P.: "Soft" Systems Methodology, Functionalism and the Social Sciences. J. of Applied Systems Analysis, 5, No.1, Nov.1976, pp.65-73.

[1360] Prigogine, I.: From Being to Becoming - Time and Complexity in the Physical Sciences. W.H.Freeman, San Francisco, 1982.

[1361] Primas, H.: Theory Reduction and Non-Boolean Theories. J. of Mathematical Biology, 4, No.3, 1977, pp.281-301.

[1362] Pritsker, A.A.B.: Modeling and Analysis Using Q-Gert Networks. John Wiley, Chichester and New York, 1977.

[1363] Probst, G.J.B., Gomez, P.: New ('second order') Cybernetics against Mismanagement. In: Trappl, R.(ed.): Cybernetics and Systems Research. North-Holland, Amsterdam, and New York, 1982, pp.437-448.

[1364] Probst, G.J.B., Guentert, B.: The Necessity of Sensitivity Models for Managers, Planners and Politicians in Health Care Systems. In: Trappl, R.(ed.): Cybernetics and Systems Research. North-Holland, Amsterdam, and New York, 1982, pp.617-628.

[1365] Prokop, A.: Systems Analysis and Synthesis in Biology and Biotechnology. Int. J. of General Systems, 8, No.1, 1982, pp.7-32.

[1366] Protzen, J.-P., Gasparski, W.W.(eds.): Design Methods and Theories: Vol.10, No.2. Univ. of California at Berkeley, Berkeley, California, 1981, 102pp. [IFSR-Depository]

[1367] Pszczolowski, T.: Praxiology - The Theory with Past and Future. Praxiology, 1, 1980, pp.3-18.

[1368] Pugh, R.E.: Evaluation of Policy Simulation Models: .a Conceptual Approach to Case Study. Information Resources Press, Washington, D.C., 1977.

[1369] Putt, A.M.(ed.): General Systems Theory Applied to Nursing. Little, Brown and Company, Boston, 1978, 195 pp.

[1370] Quade, E., (et al.): The State-of-the-Art Questionnaire on Applied Systems Analysis: a Report on the Response. IIASA, Laxenburg, Austria, 1976, IX+64 pp.

[1371] Quine, W.V.: Theories and Things. Harvard University Press, Cambridge, Mass., 1981, XII + 220pp.

[1372] Rabin, M.O.: Complexity of Computations. ACM Communications, 20, No.9, Sept.1977, pp.625-633.

[1373] Radecki, T.: Level Fuzzy Sets. J. of Cybernetics, 7, No.3-4, 1977, pp.189-198.

[1374] Ragade, R.K.: Fuzzy Interpretative Modeling. J. of Cybernetics, 6, Nos.3-4, 1976, pp.189- 211.

[1375] Ragade, R.K.(ed.): General Systems Yearbook. SGSR, Louisville, Kentucky, 25, 1980, VIII+197pp.

[1376] Ragade, R.R.: General Systems: Yearbook of the Society for General Systems Research. SGSR, Louisville, Kentucky, Vol.26, 1981, VI+270pp.

[1377] Raghaven, V.V., Yu, C.T.: A Comparison of the Stability Characteristics of Some Graph Theoretic Clustering Methods. IEEE Trans. on Pattern Analysis and Machine Intelligence, PAMI-3, No.4, July 1981, pp.393-402.

[1378] Ralescu, D.: Toward a General Theory of Fuzzy Variables. J. of Mathematical Analysis and Applications, 86, No.1, March 1982, pp.176-193.

[1379] Randers, J.(ed.): Elements of the System Dynamic Method. M.I.T.Press, Cambridge, Mass., 1980, 488 pp.

[1380] Rao, G.P., Rutherford, D.A.: Approximate Reconstruction of Mapping Functions from Linguistic Descriptions in Problems of Fuzzy Logic Applied to Systems Control. Cybernetics and Systems, 12, No.3, 1981, pp.225-236.

[1381] Rao, G.V.: Complex Digital Control Systems. Van Nostrand Reinhold, New York, 1979, XXII+516 pp.

[1382] Rapoport, A., Stein, W.E., Burkheiner, G.J.: Response Models for Detection of Change. D.Reidel, Dordrecht, Holland, and Boston, 1979, VIII+200 pp.

[1383] Rapoport, A.: Cybernetics as a Link between two Theories of Cognition. In: Trappl, R.(ed.): Cybernetics and Systems Research. North-Holland, Amsterdam, and New York, 1982, pp.3-14.

[1384] Rastogi, P.N.: Cybernetic Study of Societal Systems (Part I): Theory and Methodology. Kybernetes, 6, No.2, 1977, pp.95-105.

[1385] Rastogi, P.N.: The Behaviour of Societal Systems. Indian Institute of Advanced Study, Simla, India, 1978.

[1386] Rastogi, P.N.: An Introduction to Social and Management Cybernetics. Affiliated East-West Press, New Dehli / Madras, 1979, 255 pp.

[1387] Rasvan, V.: Some System Theory Ideas Connected with the Stability Problem. J. of Cybernetics, 8, No.2, 1978, pp.203-215.

[1388] Ratko, I., Csukas, M., Vaszary, P.: Computer Registration of Patients Waiting for Cardiac Operation. In: Trappl, R.(ed.): Cybernetics and Systems Research. North-Holland, Amsterdam, and New York, 1982, pp.651-654.

[1389] Ratko, I.: On Evaluating of Logical Expressions in Programming Languages. In: Trappl, R.(ed.): Cybernetics and Systems Research, Vol.II. North-Holland, Amsterdam, and New York, 1984, pp.611-614.

[1390] Rauch, W.D.: Automatic Refereeing and Excerpting. Austrian Society for Cybernetic Studies, Vienna, 1977. [German]

[1391] Rauch, W.D.: Automatic Extracting as an Interactive Process. J. of Cybernetics, 9, No.2, 1979, pp.103-112.

[1392] Ray, W.H.: Advanced Process Control. McGraw-Hill, New York, 1981, XIII+376pp.

[1393] Read, R.C., Corneil, D.G.: The Graph Isomorphism Disease. J. of Graph Theory, 1, No.4, Winter 1977, pp.339-364.

[1394] Reckmeyer, W.J.(ed.): General Systems Research and Design: Precursors and Futures. SGSR, Louisville, Kentucky, 1981, 623pp.

[1395] Reddy, N.P.: A Thermodynamic Theory of Knowledge. Cybernetics and Systems, 13, No.3, 1982, pp.213-217.

[1396] Redhead, M.: Models in Physics. Brit. J. for the Philosophy of Science, 31, No.2, 1980, pp.145-163.

[1397] Rees, R.K.D.(ed.): Software Reliability (2 Vols). Infotech International, Maidenhead, Berkshire, UK, 1977.

[1398] Reichle, D.E.(ed.): Dynamic Properties of Forest Ecosystems. Cambridge Univ. Press, Cambridge, Mass., 1981, XXVI+684 pp.

[1399] Reisig, G.H.R.: Statistical Data Analysis in the Information Domain. Kybernetes, 6, No.2, 1977, pp.107-123.

[1400] Rescher, N.: Methodological Pragmatism: a Systems-Theoretical Approach to the Theory of Knowledge. New York Univ. Press, New York, 1977, 315 pp.

[1401] Rescher, N.: Cognitive Systematization: a Systems-Theoretic Approach to a Coherent Theory of Knowledge. Basil Blackwell, Oxford, 1979, XII+211 pp.

[1402] Rescher, N.: Induction. Basil Blackwell, Oxford, 1980, XII+225 pp.

[1403] Rescher, N.: Empirical Inquiry. Athlone-Press, London, 1982, XII+291pp.

[1404] Resconi, G.: Logical Analysis of Dynamical Systems-Applications to Life Sciences. In: Trappl, R.(ed.): Cybernetics and Systems Research. North-Holland, Amsterdam, and New York, 1982, pp.315-320.

[1405] Resconi, G., Aleo, F.: Analysis of the Purine Biosynthesis by Mathematical Logic. Cybernetics and Systems, 14, No.2-4, 1983, pp.103-138.

[1406] Resconi, G.: A New Formalization of the General Systems by Logical and Algebraic Structures. In: Trappl, R.(ed.): Cybernetics and Systems Research, Vol.II. North-Holland, Amsterdam, and New York, 1984, pp.33-38.

[1407] Retti, J.: SISSY - A Fast Interactive Simulation System. Austrian Society for Cybernetic Studies, Vienna, 1979. [German]

[1408] Retti, J.: A Framework for Interactive Engineering of Simulation Models of Time-Oriented Systems. In: Trappl, R.(ed.): Cybernetics and Systems Research. North-Holland, Amsterdam, and New York, 1982, pp.199-204.

[1409] Retti, J.: Frame-Oriented Knowledge Representation for

Modelling and Simulation. Austrian Society for Cybernetic Studies, Vienna, 1984. [German]

[1410] Retti, J.: MOSES: A Schema-Based Knowledge Representation for Modelling and Simulation of Dynamical Systems. In: Trappl, R.(ed.): Cybernetics and Systems Research, Vol.II. North-Holland, Amsterdam, and New York, 1984, pp.817-820.

[1411] Reusch, B., Szwillus, G.: Continuous Systems with Digital Behavior. In: Trappl, R.(ed.): Cybernetics and Systems Research. North-Holland, Amsterdam, and New York, 1982, pp.151-156.

[1412] Reusch, B., Szwillus, G.: Characterization of Continuous Systems that are Digital. Cybernetics and Systems, 14, No.2-4, 1983, pp.253-292.

[1413] Reuver, H.A.: Learning within the Context of General Systems Theory. In: Trappl, R., Pask, G.(eds.): Progress in Cybernetics and Systems Research, Vol.IV. Hemisphere, Washington, D.C., 1978, pp.113-121.

[1414] Revell, N.: Systematic Methodologies for the Design of Small Database Systems. In: Trappl, R.(ed.): Cybernetics and Systems Research, Vol.II. North-Holland, Amsterdam, and New York, 1984, pp.599-604.

[1415] Reynolds, J.C.: Reasoning About Arrays. ACM Communications, 22, No.5, 1979, pp.290-299.

[1416] Ricci, F.J., Wilson, G.: Computer Simulation of a Cybernetic Model of Man in his Cultural Environment. J. of Cybernetics, 8, No.3-4, 1978, pp.325-350.

[1417] Ricciardi, L.: Cybernetics in Biology and Medicine: An Introduction. In: Trappl, R.(ed.): Cybernetics and Systems Research. North-Holland, Amsterdam, and New York, 1982, pp.263-264.

[1418] Ricciardi, L.: Cybernetics and Biology: Classic Attempts at Neuronal Modeling. In: Trappl, R.(ed.): Cybernetics - Theory and Applications. Hemisphere, Washington, D.C., 1983, pp.141-158.

[1419] Rifkin, J.: Entropy: a New World View. Viking Press, New York, 1980, XI+305 pp.

[1420] Ringle, M.(ed.): Philosophical Perspectives in Artificial Intelligence. Humanities Press, Atlantic Highlands, N.J., 1980, XII+244 pp.

[1421] Rissanen, J., Langdon, G.G.: Universal Modeling and Coding. IEEE Trans. on Information Theory, IT-27, No.1, Jan.1981, pp.12-23.

[1422] Rivett, P.: Model Building for Decision Analysis. John Wiley, Chichester and New York, 1980, XI+172 pp.

[1423] Robb, F.: Cybernetics in Management Thinking. Systems Research, 1984, Vol.1, No.1, 1984, pp.5-24.

[1424] Roberts, F.S.: Discrete Mathematical Models. Prentice-Hall, Englewood Cliffs, New Jersey, 1976.

[1425] Roberts, F.S.: Measurement Theory with Applications to Decision-Making, Utility, and the Social Sciences. Addison-Wesley, Reading, Mass., 1979, XXII+420 pp.

[1426] Roberts, G.F., Di Cesare, F.: An Input/Output Approach to the Structural Analysis of Digraphs. IEEE Trans. on Systems, Man, and Cybernetics, SMC-12, No.1, 1982, pp.3-14.

[1427] Roberts, P.C.: Modelling Large Systems. Taylor & Francis, London, 1978, 124 pp.

[1428] Roberts, P.C.: Modelling Large Systems: Limits to Growth Revisited. John Wiley, Chichester and New York, 1978, 120 pp.

[1429] Roberts, S.L.: Systems Approach to Assessing Behavioral Problems of Critical Care Patients. Heart and Lung, 4, No.4, July/August 1975, pp.593-598.

[1430] Robertshaw, J.E., Mecca, S.J., Rerick, M.N.: Problem Solving: a Systems Approach. McGraw-Hill, New York, 1979, XIII+272 pp.
[1431] Robinson, A.: Non-Standard Analysis. North-Holland, Amsterdam, and New York, 1974.
[1432] Robinson, D.F., Foulds, L.R.: Digraphs: Theory and Techniques. Gordon and Breach, New York, and London, 1980, XV+256 pp.
[1433] Robinson, E.A., Silva, M.T.: Digital Foundations of Time Series Analysis. Vol.1: The Box-Jenkins Approach. Holden-Day, San Francisco, 1979, VIII+450 pp.
[1434] Roedding, W.: Different Approaches to the Aggregation of Preferences. Cybernetics and Systems, 13, No.3, 1982, pp.245-256.
[1435] Roffel, B., Rijndorp, J.E.: Process Dynamics, Control and Protection. Ann Arbor Science, Ann Arbor, Mich., 1982, XVI + 416pp.
[1436] Rose, J., Bilcio, C.(eds.): Modern Trends in Cybernetics and Systems (3 Vols.). Springer-Verlag, Berlin, FRG, and New York, 1977.
[1437] Rose, J.(ed.): Current Topics in Cybernetics and Systems (Proc.Fourth Mt.Congress on Cybernetics and Systems, Amsterdam, Aug.21-25, 1978). Springer-Verlag, Berlin, FRG, and New York, 1978, XI+409 pp.
[1438] Rose, J.(ed.): Robotica (Int.Journal of Information, Education and Research in Robotics and Artificial Intelligence). Cambridge Univ. Press, Cambridge, Mass., Vol.1: 1983; Vol.2: 1984.
[1439] Rosen, R.: Dynamical Similarity and the Theory of Biological Transformations. Bulletin of Mathematical Biology, 40, No.5, 1978, pp.549-579.
[1440] Rosen, R.: Feedforwards and Global Systems Failure. J. of Theoretical Biology, 74, 1978, pp.579-590.
[1441] Rosen, R.: Fundamentals of Measurement and Representation of Natural Systems. North-Holland, Amsterdam, and New York, 1978, XVII+211 pp..
[1442] Rosen, R.: On Anticipatory Systems. J. of Social and Biological Structures, 1, 1978, pp.155-180.
[1443] Rosen, R.: Anticipatory Systems in Retrospect and Prospect. General Systems Yearbook, 24, 1979, pp.11-23.
[1444] Rosen, R.: Protein Folding: a Prototype for Control of Complex Systems. Int. J. of Systems Science, 11, No.5, 1980, pp.527-540.
[1445] Rosen, R.(ed.): Progress in Theoretical Biology, Vol.6. Academic Press, London, and New York, 1981, XIV+214pp.
[1446] Rosen, R.: The Physics of Complex Systems. In: Trappl, R.(ed.): Power, Autonomy, Utopia: New Approaches Towards Complex Systems. Plenum Press, New York and London, 1985.
[1447] Rosenberg, R.M.: Analytical Dynamics of Discrete Systems. Plenum Press, New York and London, 1977, 424 pp.
[1448] Rosenkrantz, R.D.: Inference, Method and Decision. D.Reidel, Dordrecht, Holland, and Boston, 1977, 262 pp.
[1449] Ross, S.D.: Learning and Discovery. Gordon and Breach, New York, and London, 1981, IX + 138pp.
[1450] Rota, G.C.: Finite Operator Calculus. Academic Press, London, and New York, 1975.
[1451] Roth, G., Schwegler, H.(eds.): Self-Organizing Systems: An Interdisciplinary Approach. Campus Verlag, Frankfurt and New York, 1981, 187pp.
[1452] Rowe, W.D.: An Anatomy of Risk. John Wiley, Chichester and New York, 1977.
[1453] Royce, J.R., Buss, A.R.: The Role of General Systems and Information Theory in Multifactor Individuality Theory. Canadian Psychological Review, 17, No.1, 1976, pp.1-21

(also General Systems Yearbook, 24, 1979, pp.185-205).

[1454] Rozenberg, G., Salomaa, A.: The Mathematical Theory of LSystems. Academic Press, London, and New York, 1979, XVI+352 pp.

[1455] Ruben, B.D., Kim, J.Y.(eds.): General Systems Theory and Human Communication. Hayden, Rochelle Park, New Jersey, 1975.

[1456] Rubinstein, R.Y.: Simulation and Monte Carlo Method. John Wiley, Chichester and New York, 1981, 278pp.

[1457] Rudall, B.H., Secker, J.A.: Developing Automated Teaching Systems. In: Trappl, R.(ed.): Cybernetics and Systems Research. North-Holland, Amsterdam, and New York, 1982, pp.385-390.

[1458] Rudall, R.: Computers and Cybernetics. Heyden, Philadelphia, 1981.

[1459] Rudawitz, L.M., Freeman, P.: Client-Centered Design: Concepts and Experience. Systems, Objectives, Solutions, 1, No.1, January 1981, pp.21-32.

[1460] Rudwick, B.H.: Solving Management Problems: a Systems Approach to Planning and Control. John Wiley, Chichester and New York, 1979, XIV+496 pp.

[1461] Ruffner, M.A.H.: Human Factors in Systems Analysis. IEEE Trans. on Systems, Man, and Cybernetics, SMC-11, No.7, July 1981, pp.509-514.

[1462] Rugh, W.J.: Nonlinear Systems Theory: The Volterra/Wiener Approach. John Hopkins University Press, Baltimore, 1981, XIV + 325pp.

[1463] Ruotsalainen, P., Nokso-Koivisto, P.: Modeling and Simulating the Resource Allocation of the Regional Hospital System with Different Service Levels. In: Trappl, R.(ed.): Cybernetics and Systems Research. North-Holland, Amsterdam, and New York, 1982, pp.641-644.

[1464] Rus, T.: Data Structures and Operating Systems. John Wiley, Chichester and New York, 1979, XI+364 pp.

[1465] Rushinek, A., Rushinek, S.F.: An Analysis of Computer Software and Its Importance to the Accounting Information System. In: Trappl, R.(ed.): Cybernetics and Systems Research, Vol.II. North-Holland, Amsterdam, and New York, 1984, pp.635-640.

[1466] Ryabinin, I.: Reliability of Engineering Systems: Principles and Analysis. Mir Publishers, Moscow, 1976, 326 pp.

[1467] Ryan, R.M.: Conceptualizing the Integration of Social Welfare Delivery Systems: A Matrix Organization Approach. In: Trappl, R.(ed.): Cybernetics and Systems Research. North-Holland, Amsterdam, and New York, 1982, pp.501-506.

[1468] Rzevski, G.: Some Philosophical Aspects of System Design. In: Trappl, R.(ed.): Cybernetics and Systems Research. North-Holland, Amsterdam, and New York, 1982, pp.39-44.

[1469] Sachs, W., Calhoun, G.: Systemic Inactivism: A Modern Laissez-Faire Thinking. Human Systems Management, 2, No.3, Oct.1981, pp.191-198.

[1470] Sadovskiy, L.Y., (et al.): Questions Regarding Simulation of Hierarchical Systems. Engineering Cybernetics, 1977, No.2, pp.14-22.

[1471] Sadovsky, V.N.: The System Principle and Leibnitz's Theoretical Heritage. In: ß: Logik, Erkenntnistheorie, Wissenschaftstheorie, Metaphysik, Theologie. Franz Steiner Verlag, Wiesbaden, 1980, pp.136-144.

[1472] Saeks, R.(ed.): Large-Scale Dynamical Systems. Western Periodicals, North Hollywood, Ca., 1976, 314 pp.

[1473] Sagal, P.T.: Epistemology of Economics. J. for General Philosophy of Science, 8, No.1, 1977, , pp.144-162.

[1474] Sage, A.P.: Methodology for Large Scale Systems.

McGraw-Hill, New York, 1977.

[1475] Sage, A.P.: Behavioral and Organizational Considerations in the Design of Information Systems and Process of Planning and Decision Support. IEEE Trans. on Systems, Man, and Cybernetics, SMC-11, No.9, Sept.1981, pp.640-678.

[1476] Sage, A.P.: Sensitivity Analysis in Systems for Planning and Decision Support. J. of the Franklin Institute, 312, 1981, pp.265-291.

[1477] Sahal, D.: Principles of Regulation and Control. Kybernetes, 7, No.1, 1978, pp.19-24.

[1478] Sahal, D.: Systemic Similitude. Int. J. of Systems Science, 9, No.12, 1978, pp.1351-1357.

[1479] Sahal, D.: A Theory of Evolution of Technology. Int. J. of Systems Science, 10, No.3, 1979, pp.259-274.

[1480] Sahal, D.: A Unified Theory of Self-Organization. J. of Cybernetics, 9, No.2, 1979, pp.127-142.

[1481] Sahal, D.: Patterns of Technological Innovations. Addison-Wesley, Reading, Mass., 1981, XVI + 381.

[1482] Sain, M.K.: Introduction to Algebraic System Theory. Academic Press, London, and New York, 1981, 368 pp.

[1483] Sakawa, M., Yano, H.: An Interactive Goal Attainment Method for Multiobjective Nonconvex Problems. In: Trappl, R.(ed.): Cybernetics and Systems Research, Vol.II. North-Holland, Amsterdam, and New York, 1984, pp.189-194.

[1484] Salzberg, P.M., Seibert, P.: A Unified Theory of Attraction in General Systems. Int. J. of General Systems, 6, No.3, 1980.

[1485] Sampson, J.R.: Adaptive Information Processing: An Introductory Survey. Springer-Verlag, Berlin, FRG, and New York, 1976, 214pp.

[1486] Samuelson, K., Borko, H., Amey, G.X.: Information Systems and Networks. North-Holland, Amsterdam, and New York, 1977, X+148 pp.

[1487] Sandblom, C.-L.: Optimization of Economic Policy Using Lagged Controls. J. of Cybernetics, 7, No.3-4, 1977, pp.257-268.

[1488] Sandblom, C.-L.: Simulation Results on the Evaluation of Economic Policies. J. of Cybernetics, 10, No.1-3, 1980, pp.161-176.

[1489] Sanderson, A.C., Wong, A.K.C.: Pattern Trajectory Analysis of Nonstationary Multivariate Data. IEEE Trans. on Systems, Man, and Cybernetics, SMC-10, No.7, July 1980, pp.364-392.

[1490] Sanderson, J.G.: A Relational Theory of Computing. Springer-Verlag, Berlin, FRG, and New York, 1980, VI+147 pp.

[1491] Sanderson, M.: Successful Problem Management. John Wiley, Chichester and New York, 1979, VII+227 pp.

[1492] Santis, F.de, Pagliuca, A.: On Factor Analysis and Fisher's Linear Discriminant Analysis. Cybernetics and Systems, 13, No.1, 1982, pp.77-91.

[1493] Santos, E.S.: Probabilistic Machines and Languages. J. of Cybernetics, 9, No.2, 1979, pp.185-204.

[1494] Saridis, G.N.: Self-Organizing Control of Stochastic Systems. Marcel Dekker, New York, 1977, 448 pp.

[1495] Sastry, D., Gauvrit, M.: A Two-Level Algorithm for the Simultaneous State and Parameter Identification. Int. J. of Systems Science, 9, No.12, 1978, pp.1397-1418.

[1496] Sauer, C.H., Chandy, K.M.: Computer Systems Performance Modeling. Prentice-Hall, Englewood Cliffs, New Jersey, 1981, XIII+352 pp.

[1497] Saunders, P.T.: An Introduction to Catastrophe Theory. Cambridge Univ. Press, Cambridge, Mass., 1980, XII+144pp.

[1498] Savage, G.J., Kesavan, H.K.:   The Graph-Theoretic Field Model-I - Modelling and Formulations. J. of the Franklin Institute, 307, No.2, Feb.1979, pp.107-147.

[1499] Sawaragi, Y., Soeda, T., Omatu, S.: Modeling, Estimation, and Their Applications for Distributed Parameter Systems. Springer-Verlag, Berlin, FRG, and New York, 1978, VI+269 pp.

[1500] Scandura, J.M.: Problem Solving: a Structural Process Approach with Instructional Implications. Academic Press, London, and New York, 1977, X+591 pp.

[1501] Scandura, J.M.: Theoretical Foundations of Instruction: A Systems Alternative to Cognitive Psychology. J. of Structural Learning, 6, 1980, pp.347-394.

[1502] Scavia, D., Robertson, A.(eds.): Perspectives on Lake Eco-System Modeling. Ann Arbor Science, Ann Arbor, Mich., 1979, 326 pp.

[1503] Scavia, D.: An Ecological Model of Lake Ontario. Ecological Modelling, 8, Jan.1980, pp.49-78.

[1504] Schal, R.K.: A Collection of Thoughts of General Systems Theorists Concerning the Process of Human Transformation. Systems Trends, 4, No.6, June 1982, pp.4-12.

[1505] Schellenberg, E.: Project Management, Key to Successful Computer Projects. In: Trappl, R.(ed.): Cybernetics and Systems Research. North-Holland, Amsterdam, and New York, 1982, pp.567-572.

[1506] Schellhorn, J.P.: Model Following in Multiple Control: Exact and Approximate. In: Trappl, R.(ed.): Cybernetics and Systems Research. North-Holland, Amsterdam, and New York, 1982, pp.91-96.

[1507] Schieve, W.C., Allen, P.M.(eds.): Self-Organization and Dissipative Structures: Applications in the Physical and Social Sciences. Univ. of Texas Press, 1982, XII+362pp.

[1508] Schiffman, Y.M.: A Systems Approach for Estimating Energy Savings in Communities. In: Trappl, R.(ed.): Cybernetics and Systems Research. North-Holland, Amsterdam, and New York, 1982, pp.671-676.

[1509] Schilderinck, J.H.F.: Regression and Factor Analysis in Econometrics. Martinus Nijhoff, Boston and The Hague, 1977, 239 pp.

[1510] Schimanovich, W.: An Axiomatic Theory of Winning Mathematical Games and Economic Actions. Cybernetics and Systems, 12, No.1-2, 1981, pp.1-19.

[1511] Schkolnick, M.: A Clustering Algorithm for Hierarchical Structures. ACM Trans. on Database Systems, 2, No.1, March 1977, pp.27-44.

[1512] Schmutzer, M.E.A., Bandler, W.: Hi & Low - In & Out: Approaches to Social Status. J. of Cybernetics, 10, No.4, 1980, pp.283-300.

[1513] Schnakenberg, J.: Thermodynamic Network Analysis of Biological Systems (Second Edition). Springer-Verlag, Berlin, FRG, and New York, 1981, X+150pp.

[1514] Scholz, C.: The Architecture of Hierarchy. Kybernetes, 11, 1982, pp.175-181.

[1515] Schwartz, D.G.: Isomorphisms of Spencer-Brown's Laws of Form and Varela's Calculus for Self-Reference. Int. J. of General Systems, 6, No.4, 1981, pp.239-256.

[1516] Schwarzenback, J., Gill, K.F.: Systems Modelling and Control. John Wiley, Chichester and New York, 1978, 230 pp.

[1517] Scoltock, J.: A Survey of the Literature on Cluster Analysis. Computer Journal, 25, No.1, Feb.1982, pp.130-134.

[1518] Scott, C.H., Jefferson, T.R.: On the Analysis of an Information Theoretic Model of Spacial Interaction.

Information Sciences, 20, No.1, 1980, pp.1-12.

[1519] Scott, W.A., Osgood, D.W., Peterson, C.: Cognitive Structure: Theory and Measurement of Individual Differences. John Wiley, Chichester and New York, 1979, 252 pp.

[1520] Scott, W.R.: Organizations: Rational, Natural and Open Systems. Prentice-Hall, Englewood Cliffs, New Jersey, 1981, XVIII+383 pp.

[1521] Segel, L.A.(ed.): Mathematical Models in Molecular and Cellular Biology. Cambridge Univ. Press, Cambridge, Mass., 1981, X+758pp.

[1522] Sellers, P.H.: Combinatorial Complexes: A Mathematical Theory of Algorithms. D.Reidel, Dordrecht, Holland, and Boston, 1979, XV+185 pp.

[1523] Sembugamoorthy, V.: PLAS, a Paradigmatic Language Acquisition System. In: Trappl, R.(ed.): Cybernetics and Systems Research. North-Holland, Amsterdam, and New York, 1982, pp.857-862.

[1524] Sengupta, J.K.: Bilinear Models in Stochastic Programming. J. of Cybernetics, 9, No.2, 1979, pp.161-168.

[1525] Sengupta, J.K.: Testing and Validation Problems in Stochastic Linear Programming. J. of Cybernetics, 9, No.1, 1979, pp.17-42.

[1526] Sengupta, J.K.: Minimax Solutions in Stochastic Programming. Cybernetics and Systems, 11, No.1-2, 1980, pp.1-19.

[1527] Seo, F., Sakawa, M.: Fuzzy Extension of Multiattribute Utility Analysis for Collective Choice. In: Trappl, R.(ed.): Cybernetics and Systems Research, Vol.II. North-Holland, Amsterdam, and New York, 1984, pp.195-200.

[1528] Sernadas, A.: SYSTEMATICS: Its Syntax and Semantics as a Query Language. Computer Journal, 24, No.1, February 1981, pp.56-61.

[1529] Seton, F.: The Evalution of Economic Flows and Systemic Efficiency. In: Trappl, R.(ed.): Cybernetics and Systems Research. North-Holland, Amsterdam, and New York, 1982, pp.471-476.

[1530] Shaeffer, D.L.: A Model Evaluation Methodology Applicable to Environmental Assessment Models. Ecological Modelling, 8, Jan.1980, pp.275-295.

[1531] Shahane, A.H.: Interdisciplinary Models of Water Systems. Ecological Modelling, 2, No.2, June 1976, pp.117-145.

[1532] Shakun, M.F.: Formalizing Conflict Resolution in Policy Making. Int. J. of General Systems, 7, No.3, 1981, pp.207-216.

[1533] Shakun, M.F.: Policy Making and Meaning as Design of Purposeful Systems. Int. J. of General Systems, 7, No.4, 1981, pp.235-252.

[1534] Shakun, M.F., Sudit, E.F.: Effectiveness, Productivity and Design of Purposeful Systems: The Profit-Making Case. Int. J. of General Systems, 9, No.4, 1983, pp.205-216.

[1535] Shapiro, L.G., Haralik, R.M.: Structural Descriptions and Inexact Matching. IEEE Trans. on Pattern Analysis and Machine Intelligence, PAMI-3, No.5, Sept.1981, pp.504-519.

[1536] Sharif, N., Adulbhan, P.(eds.): Proceedings of the Int. Conference on Systems Modelling in Developing Countries. Asian Institute of Technology, Bangkok, 1978.

[1537] Sharif, N., Adulbhan, P.(eds.): Systems Models for Decision Making. Asian Institute of Technology, Bangkok, 1978, 433 pp.

[1538] Sharma, B.D., Dass, B.K.: Adjacent-Error Correcting Binary Perfect Codes. J. of Cybernetics, 7, No.1-2, 1977, pp.9-14.

[1539] Sharma, B.D., Gupta, S.N.: Block-Wise Error-Protecting Burst Codes with Weight Constraints. J. of Cybernetics, 7, No.1-2, 1977, pp.1-8.

[1540] Sharma, B.D., Kaur, B.: Phased Error Correcting Perfect Codes. J. of Cybernetics, 7, No.3-4, 1977, pp.227-248.

[1541] Shaw, M.L.G.: On Becoming a Personal Scientist. Academic Press, London, and New York, 1980, XVI+332 pp.

[1542] Shaw, M.L.G., Gaines, B.R.: The Personal Scientist in the Community of Science. SGSR, Louisville, Kentucky, 1981, pp.59-68.

[1543] Shepard, R.N.: Multidimensional Scaling, Tree-fitting, and Clustering. Science, 210, 24 October 1980, pp.390-398.

[1544] Shepp, L.A., Slepian, D., Wyner, A.D.: On Prediction of Moving-Averaging Processes. Bell System Technical Journal, 59, No.3, 1980, pp.367-415.

[1545] Sherlock, R.A.: Analysis of the Behaviour of Kauffman Binary Networks: Part 1: State Space Description and Distribution of Limit Cycle Lengths; Part 2: the State Cycle Fraction for Networks of Different Connectivities. Bulletin of Mathematical Biology, 41, No.5, 1979; Part 1: pp.687-705; Part 2: pp.707-724.

[1546] Shieh, L., Yates, R.E., Navarro, J.M.: Representation of Continuous Time State Equations by Discrete-Time State Equations. IEEE Trans. on Systems, Man, and Cybernetics, SMC-8, No.6, June 1978, pp.485-492.

[1547] Shigan, E.N.(ed.): Systems Modeling in Health Care. IIASA, Laxenburg, Austria, 1977.

[1548] Shmulyan, B.L.: Interaction of Entropy Systems. Automation and Remote Control, 42, No.9, Part 2, Sept.1981, pp.1213-1220.

[1549] Shoesmith, D.J., Smiley, T.J.: Multiple-Conclusion Logic. Cambridge Univ.Press, Cambridge, Mass., 1978, XIII+396 pp.

[1550] Shore, J.E.: Cross-Entropy Minimization Given Fully-Decomposable Subset and Aggregate Constraints. Naval Research Lab.Memo., Report 4430 (Feb.13, 1981), 22 pp.

[1551] Shore, J.E., Johnson, R.W.: Properties of Cross-Entropy Minimization. IEEE Trans. on Information Theory, IT-27, No.4, July 1981, pp.472-482.

[1552] Shugart, H.H., O'Neill, R.V.(eds.): Systems Ecology. Dowden, Hutchinson and Ross, Stroudsburg, Penn., 1979, XIV+370 pp.

[1553] Sierocki, I., Straka, G., Pichler, F.: Invariance-Relations in General Dynamic Systems. Austrian Society for Cybernetic Studies, Vienna, 1983. [German]

[1554] Siewiorek, D.P., Swarz, R.S.: The Theory and Practice of Reliable System Design. Digital Press, Bedford, Mass., 1982, XXIV+772pp.

[1555] Siljak, D.D.: Large-Scale Dynamic Systems. North-Holland, Amsterdam, and New York, 1978, XVI+416 pp.

[1556] Silvern, L.C.: Systems Engineering Applied to Training. Education and Training Consultants Co., Los Angeles, Box 49899, Ca.90049, 1972, 170 pp.

[1557] Silvern, L.C.: Systems Engineering of Education I: the Evolution of Systems Thinking in Education. Education and Training Consultants Co., Los Angeles, Box 49899, Ca.90049, 1975, 128 pp.(Third Edition).

[1558] Silvert, W.: The Formulation and Evaluation of Predictions. Int. J. of General Systems, 7, No.3, 1981, pp.189-206.

[1559] Silviu, G.: Information Theory with Applications. McGraw-Hill, New York, 1977, 439 pp.

[1560] Simionescu, C.L.: On a Model Neurology. In: Trappl,

R.(ed.): Cybernetics and Systems Research. North-Holland, Amsterdam, and New York, 1982, pp.339-344.

[1561] Simon, H.A.: How Complex Are Complex Systems? In: Suppes, P., Asquith, P.D.(eds.): PSA 1976. Philosophy of Science Asscociation, East Lansing, Mich., 1977, pp.507-522.

[1562] Simon, H.A.: Models of Discovery and Other Topics in the Methods of Science. D.Reidel, Dordrecht, Holland, and Boston, 1977, XX+464 pp.

[1563] Simon, H.A.: Models of Bounded Rationality (2 Volumes). M.I.T.Press, Cambridge, Mass., 1982, XXIV+478 pp.

[1564] Simon, J.C.(ed.): Computer Oriented Learning Processes. Noordhoff, Leiden, Holland, , 1976.

[1565] Simon, J.C., Haralik, R.M.(eds.): Digital Image Processing. D.Reidel, Dordrecht, Holland, and Boston, 1981, VIII+596pp.

[1566] Singh, M.G., Titli, A.(eds.): Handbook of Large Scale Systems Engineering Applications. North-Holland, Amsterdam, and New York, 1979, X+564 pp.

[1567] Singh, R.: Systems Approaches to the Study of Industrial Relations. J. of Applied Systems Analysis, 5, No.1, Nov 1976, pp.49-63.

[1568] Sinha, D.K.(ed.): Catastrophe Theory and Applications. John Wiley, Chichester and New York, 1982, XII+158pp.

[1569] Sinha, K.K., Law, S.S.Y.: Adaptive Nuclear Reactor Control without Explicit Identification. J. of Cybernetics, 7, No.1-2, 1977, pp.23-36.

[1570] Skala, H.J.: On Many-Valued Logics, Fuzzy Sets, Fuzzy Logics and Their Applications. Fuzzy Sets and Systems, 1, No.2, April 1978, pp.129-149.

[1571] Sklenar, J.: Apparatus for Formal Description of Discrete Systems. Int. J. of General Systems, 7, No.4, 1981, pp.225-234.

[1572] Skvoretz, J.: Information Diffusion in Formally Structured Populations: An Information Processing Approach. J. of Cybernetics, 8, No.1, 1978, pp.51-82.

[1573] Slawski, C.: A Systems Model of Negotiation: Arabs Versus Israeli. Calif. State Univ., Long Beach, Calif., 1979, 53pp. [IFSR-Depository]

[1574] Slawski, C.: Developing the Learning Relationship: In Theory and Practice. Calif. State Univ., Long Beach, Calif., 1979, 50pp. [IFSR-Depository]

[1575] Slawski, C.: Identity Bargaining: A Policy Systems Research Model of Career Development. Calif. State Univ., Long Beach, Calif., 1979, 45pp. [IFSR-Depository]

[1576] Slawski, C.: Institutionalizing Innovation in an Organization: A Model and Case Study. Calif. State Univ., Long Beach, Calif., 1979, 61pp. [IFSR-Depository]

[1577] Slawski, C.: Love, Power and Conflict: A Systems Model of Interparty Negotiation. Calif. State Univ., Long Beach, Calif., 1979, 50pp. [IFSR-Depository]

[1578] Slawski, C.: Mapping Joint Action. Calif. State Univ., Long Beach, Calif., 1979, 9pp. [IFSR-Depository]

[1579] Slawski, C.: Social Actualization: A Theory of Community Living Arrangements and a Practical Utopia. Calif. State Univ., Long Beach, Calif., 1979, 36pp. [IFSR-Depository]

[1580] Slawski, C.: The Interpersonal Conflict Episode: A Systems Model. Calif. State Univ., Long Beach, Calif., 1979, 23pp. [IFSR-Depository]

[1581] Slawski, C.: Towards a Synthesis of Social Psychological "Theories". Calif. State Univ., Long Beach, Calif., 1979, 29pp. [IFSR-Depository]

[1582] Slyadz, N.N., Borisov, A.N.: Analysis of Fuzzy Initial Information in Decision-Making Models. In: Trappl,

R.(ed.):   Cybernetics   and   Systems   Research.
North-Holland, Amsterdam, and New York, 1982, pp.739-744.
[1583] Smale, S.: The Mathematics of Time: Essays on Dynamical
Systems, Economic Processes and Related Topics.
Springer-Verlag, Berlin, FRG, and New York, 1980, VI+152
pp.
[1584] Small, M.G.:   Systemic and Global Learning.   Systems
Trends, 3, No.7, May 1981, pp.10-16.
[1585] Smetanic, Y.S., Pozolov, R.V.: On the Algorithm for
Determining the Primary Structure of Biopolymers.
Bulletin of Mathematical Biology, 41, 1979, pp.1-20.
[1586] Smets, P.: The Degree of Belief in a Fuzzy Event.
Information Sciences, 25, No.1, 1981, pp.1-19.
[1587] Smith, A.W.:   Mangagement Systems Analyses and
Applications. Dryden-Press, New York, 1982.
[1588] Smith, C.S.: A Search for Structure: Selected Essays on
Science, Art and History. M.I.T.Press, Cambridge, Mass.,
1981, X+410 pp.
[1589] Smith, J.M.:   Evolution and the Theory of Games.
Cambridge Univ.Press, Cambridge, Mass., 1982, 218pp.
[1590] Smith, M.:   Models in Ecology.   Cambridge Univ.Press,
Cambridge, Mass., 1978, 157 pp.
[1591] Smith, P.M.: An Exploration of Shared Knowledge about
Procedures Used in Fault Diagnosis Tasks. In: Trappl,
R.(ed.): Cybernetics and Systems Research, Vol.II.
North-Holland, Amsterdam, and New York, 1984, pp.697-704.
[1592] Smith, R.B.: How to Plan, Design & Implement a Bad
System. Petrocelli, New York, 1981, X+157 pp.
[1593] Smith, R.G., Davis, R.: Frameworks for Cooperation in
Distributed Problem Solving. IEEE Trans. on Systems, Man,
and Cybernetics, SMC-11, No.1, 1981, pp.61-70.
[1594] Smithson, M.: Models for Fuzzy Nominal Data. Theory and
Decision, 14, No.1, March 1982, pp.51-74.
[1595] Smyth, D.S., Checkland, P.B.: Using a Systems Approach:
the Structure of Root Definitions. J. of Applied Systems
Analysis, 5, No.1, Nov.1976, pp.75-83.
[1596] Soete, W.:   Structural Methods for Decomposition of
Large-Scale Systems. Regelungstechnik, 28, No.2, February
1980, pp.37-44. [German]
[1597] Soi, I.M.,   Aggarwal,   K.K.:   An   Overview   of
Computer-Orientied Learning Processes.   Regional
Engineering College, Kurukshetra, India, 1980, 18pp.
[IFSR-Depository]
[1598] Sonde, B.S.: Introduction to System Design Using
Integrated Circuits. John Wiley, Chichester and New York,
XII + 261 pp.
[1599] Sonnier, I.L.: Holistic Education: Teaching of Science
in the Affective Domain. Philosophical Library, New York,
1982, XX+126pp.
[1600] Soong, T.T.: Probabilistic Modeling and Analysis in
Science and Engineering. John Wiley, Chichester and New
York, 1981, XIV + 384 pp.
[1601] Southwood, T.R.E.: Ecological Methods (2nd Edition).
Chapman and Hall, London, 1978, 540 pp.
[1602] Spagovich, V.G.: Theory of Adaptive Systems. Nauka,
Moscow, 1976, 319pp. [Russian]
[1603] Spedding, C.R.W.: Agricultural Systems. Applied Science
Publishers, Barking, England, 1979, X+175 pp.
[1604] Sperb, R.P.: Maximum Principles and their Applications.
Academic Press, London, and New York, 1981, X+224pp.
[1605] Spoehr, K.T.,   Lehmkule,   S.W.:   Visual Information
Processing. W.H.Freeman, San Francisco, 1982, XII+298pp.
[1606] Spriet, J.A.,   Vansteenkiste,   G.C.:   Computer-Aided
Modelling and Simulation. Academic Press, London, and New

York, 1982, X+490pp.

[1607] Squire, E.: Introducing Systems Design. Addison-Wesley, Reading, Mass., 1980, XVII+346 pp.

[1608] Srinivasan, S.K., Subramanian, R.: Probabilistic Analysis of Redundant Systems. Springer-Verlag, Berlin, FRG, and New York, 1980, XVI+262 pp.

[1609] Srivastava, M.S., Khatri, C.G.: An Introduction to Multivariate Statistics. North-Holland, Amsterdam, and New York, 1979, XVI+350 pp.

[1610] Starr, M.K., Zeleny, M.(eds.): Multiple Criteria Decision Making. North-Holland, Amsterdam, and New York, 1977.

[1611] Stefanski, J.: A Heuristic Search for the Best Hierarchical Structure. Kybernetes, 10, No.2, 1981, pp.91-98.

[1612] Stegmuller, W.: The Structure and Dynamics of Theories. Springer-Verlag, Berlin, FRG, and New York, 1976, XVII+284 pp.

[1613] Stegmuller, W.: The Structuralist View of Theories. Springer-Verlag, Berlin, FRG, and New York, 1979.

[1614] Steinacker, I.: Computer-Aided Research in Learning. Austrian Society for Cybernetic Studies, Vienna, 1978. [German]

[1615] Steinacker, I., Trost, H., Leinfellner, E.: Disambiguation in German. In: Trappl, R.(ed.): Cybernetics and Systems Research. North-Holland, Amsterdam, and New York, 1982, pp.869-874.

[1616] Steinacker, I., Trappl, R., Horn, W.: Future, Impacts, and Future Impacts of Artificial Intelligence: A Bibliography. Austrian Society for Cybernetic Studies, Vienna, 1983.

[1617] Steinacker, I., Trost, H.: Focussing as a Method for Disambiguation and Anaphora Resolution. In: Trappl, R.(ed.): Cybernetics and Systems Research, Vol.II. North-Holland, Amsterdam, and New York, 1984, pp.747-752.

[1618] Steinberg, J.B., Conant, R.C.: An Informational Analysis of the Inter-Male Behavior of the Grasshopper Chortophaga Viridifaciata. Animal Behavior, 22, 1974, pp.617-627.

[1619] Stemmer, N.: The Reliability of Inductive Inferences and Our Innate Capacities. J. for General Philosophy of Science, IX, No.1, 1978, pp.93-105.

[1620] Steward, D.V.: Systems Analysis and Management: Structure, Strategy and Design. Petrocelli, New York, 1981, 292 pp.

[1621] Stockmeyer, L.J., Chandra, A.K.: Intrinsically Difficult Problems. Scientific American, 239, No.5, May 1979, pp.140-159.

[1622] Stoklosa, J.: Transformations of Some Dynamical Systems-Networks of (a,k)-Machines Generating Cyclic Sequences. In: Trappl, R.(ed.): Cybernetics and Systems Research. North-Holland, Amsterdam, and New York, 1982, pp.179-184.

[1623] Stoklosa, J.: On the Reachability Problem for Some Dynamical Systems. In: Trappl, R.(ed.): Cybernetics and Systems Research, Vol.II. North-Holland, Amsterdam, and New York, 1984, pp.117-120.

[1624] Stone, L.D.: Theory of Optimal Search. Academic Press, London, and New York, 1975.

[1625] Stoy, J.E.: Denotational Semantics: the Scott-Strachey Approach to Programming Languages. M.I.T.Press, Cambridge, Mass., 1977, 414 pp.

[1626] Straszak, A., Owsinski, J.W.(eds.): New Trends in Mathematical Modelling. Polish Academy of Sciences, Warsaw, 1978, 368 pp.

[1627] Strauss, A.: An Introduction to Optimal Control Theory.

Springer-Verlag, Berlin, FRG, and New York, 1968.

[1628] Strejc, V.: State Space Theory of Discrete Linear Control. John Wiley, Chichester and New York, 1981, 426 pp.

[1629] Strombach, W.: Wholeness, Gestalt, System: On the Meaning of these Concepts in German Language. Int. J. of General Systems, 9, No.2, 1983, pp.65-72.

[1630] Stucki, P.(ed.): Advances in Digital Image Processing. Plenum Press, New York and London, 1979, IX+332 pp.

[1631] Subrahmaniam, K.: Multivariate Analysis: a Selected and Abstracted Bibliography, 1957-1972 Marcel Dekker, New York, 1973.

[1632] Subrahmanian, E., Cannon, R.L.: A Generator for Models of Discrete-Event Systems. Simulation, 36, No.3, 1981, pp.93-101.

[1633] Sugeno, M., Terano, T.: A Model of Learning Based on Fuzzy Information. Kybernetes, 6, No.3, 1977, pp.157-166.

[1634] Sugiyama, K.S., Tagawa, S., Tuda, M.: Methods for Visual Understanding of Hierarchical System Structures. IEEE Trans. on Systems, Man, and Cybernetics, SMC-11, No.2, 1981, pp.109-125.

[1635] Sugiyama, K.S., Toda, M.: Visual Q-Analysis (I): A Case of Future Computer Systems Development in Japan. Cybernetics and Systems, 14, No.2-4, 1983, pp.185-228.

[1636] Sultan, L.H.: Some Considerations on Systems Organization for Fuzzy Information Processing. In: Trappl, R.(ed.): Cybernetics and Systems Research, Vol.II. North-Holland, Amsterdam, and New York, 1984, pp.551-556.

[1637] Suppe, F.: The Structure of Scientific Theories (Second Ed.) Univ.of Illinois Press, Urbana, Ill., 1977.

[1638] Suppes, P.: Some Remarks About Complexity. In: Suppes, P., Asquith, P.D.(eds.): PSA 1976. Philosophy of Science Asscociation, East Lansing, Mich., 1977, pp.543-547.

[1639] Suppes, P., Asquith, P.D.(eds.): PSA 1976. Philosophy of Science Asscociation, East Lansing, Mich., 1977.

[1640] Sussman, G.J., Steele, G.L.: CONSTRAINTS - A Language for Expressing Almost Hierarchical Descriptions. Artificial Intelligence, 14, No.1, August 1980, pp.1-39.

[1641] Sussmann, H.J., Zahler, R.S.: Catastrophe Theory as Applied to the Social and Biological Sciences: a Critique. Synthese, 37, No.2, Feb.1978, pp.117-216.

[1642] Sutherland, J.W.: Societal Systems: Methodology, Modelling and Management. North-Holland, Amsterdam, and New York, 1978, XV+336 pp.

[1643] Sutherland, J.W., Yee, K.G.: Towards a System-Theoretical Decision Logic. Theory and Decision, 10, 1979, pp.31-59.

[1644] Suzuki, R., Kawato, M., Tatsumi, H.: Mathematical and Phenomenological Studies of Biological Rhythms. In: Trappl, R.(ed.): Cybernetics and Systems Research. North-Holland, Amsterdam, and New York, 1982, pp.327-332.

[1645] Svoboda, A., White, D.E.: Advanced Logical Circuit Design Techniques. Garland Stpm Press, New York, 1979.

[1646] Swets, J.A., Pickett, R.M.: Evaluation of Diagnostic Systems: Methods from Signal Detection Theory. Academic Press, London, and New York, 1982, XIV+256pp.

[1647] Sworder, D.D., Archer, S.M.: Influence of Sensor Failures on Log Regulators. J. of Cybernetics, 7, No.3-4, 1977, pp.269-278.

[1648] Sydow, A.(ed.): Systems and Simulation 1980 Akademie-Verlag, Berlin, 1980, 496pp.

[1649] Syski, R.: Random Processes: a First Look. Marcel Dekker, New York, 1979, XVIII+290 pp.

[1650] Szaniawksi, K.: Science as an Information-Seeking Process. Postepy Cybernetyki, 4, No.3, 1981, pp.23-32.

[1651] Szanto, E.: Generation of Technological Information. In: Trappl, R.(ed.): Cybernetics and Systems Research. North-Holland, Amsterdam, and New York, 1982, pp.549-554.

[1652] Szuecs, E.: Similitude and Modelling. Elsevier / North-Holland, New York, 1980, 335 pp.

[1653] Tacker, E.C., Sanders, C.W.: Decentralized Structures for State Estimation on Large Scale Systems. Large Scale Systems, 1, No.1, Feb.1980, pp.39-49.

[1654] Takahara, Y.: A System Theoretic Consideration of the World Model. In: Mesarovic, M.D., Pestel, E.(eds.): Multilevel Computer Model of World Development System. IIASA, Laxenburg, Austria, 1974.

[1655] Takahara, Y., Ikeshoji, H.: Characterization of the Causality of General Dynamic Systems. Int. J. of Systems Science, 9, No.6, 1978, pp.639-648.

[1656] Takahara, Y., Yakano, B., Kijima, K.: A Unified Theory of Decision Principles. Int. J. of Systems Science, 11, No.11, 1980, pp.1295-1314.

[1657] Takahara, Y., Nakano, B.: A Characterization of Interactions. Int. J. of General Systems, 7, No.2, 1981, pp.109-122.

[1658] Takahara, Y., Nakano, B., Kijima, K.: A Structure of Rational Decision Principles. Int. J. of General Systems, 7, No.3, 1981, pp.175-188.

[1659] Takahara, Y., Nakano, B., Kubota, H.: A Stationarization Functor and its Concrete Realizations. Int. J. of General Systems, 9, No.3, 1983, pp.133-142.

[1660] Takatsu, S.: Multiple-Objective Satisficing Decision Problems. Kybernetes, 13, No.1, 1984, pp.21-26.

[1661] Takatsu, S.: Organizational Equilibrium under Uncertainty. Kybernetes, 13, No.2, 1984, pp.87-92.

[1662] Takatsuji, M.: An Information-Theoretical Approach to a System of Interacting Elements. Biological Cybernetics, 17, No.4, 1975, pp.207-210.

[1663] Takeuchi, A.: Evolutionary Automata - Comparison of Automation Behavior and Restle's Learning Model. Information Sciences, 20, No.2, March 1980, pp.91-99.

[1664] Takeuchi, A., (et al.): The Foundations of Multivariate Analysis: A Unified Approach by Means of Projection onto Linear Subspeces. Halstead, New York, 1982, XII+458 pp.

[1665] Tamburrini, S., Termini, S.: Do Cybernetics, Systems Science and Fuzzy Sets Share Some Epistemological Problems? Systems Trends, 4, No.4, 1982, pp.13-18.

[1666] Tamura, H., Halfon, E.: Identification of a Dynamic Lake Model by the Method of Data Handling: An Application to Lake Ontario. Ecological Modelling, 11, No.2, 1980, pp.81-100.

[1667] Tancig, P.: SOVA - General Purpose Software Environment for NLU-Systems. In: Trappl, R.(ed.): Cybernetics and Systems Research. North-Holland, Amsterdam, and New York, 1982, pp.845-850.

[1668] Tani, S.T.: A Perspective on Modelling in Decision Analysis. Management Science, 24, No.14, 1978, pp.1500-1506.

[1669] Tao, K.M., Hsia, P.: On the Structure of Directed Graphs with Applications: A Rapprochement with Systems Theory (Part I). Int. J. of General Systems, 8, No.3, 1982, pp.147-160.

[1670] Tao, K.M., Hsia, P.: On the Structure of Directed Graphs with Applications: A Rapprochement with Systems Theory (Part II). Int. J. of General Systems, 8, No.4, 1982, pp.211-224.

[1671] Taschdjian, E.: The Cybernetics of Stressed Systems. In: Trappl, R.(ed.): Cybernetics and Systems Research,

Vol.II. North-Holland, Amsterdam, and New York, 1984, pp.77-82.

[1672] Taylor, A.M.: A Systems Approach to the Political Organization of Space. Social Science Information, 14, 1975, p.35.

[1673] Taylor, J.R.: An Introduction to Error Analysis. University Science Books, Mil Valley, Calif., 1982, XVI + 270pp.

[1674] Taylor, W.R.: Using Systems Theory to Organize Confusion. Family Process, 18, No.4, 1979, pp.479-488.

[1675] Tazaki, E., Amagasa, M.: Structural Modeling in a Class of Systems Using Fuzzy Sets Theory. Fuzzy Sets and Systems, 2, No.1, Jan.1979, pp.87-103.

[1676] Tazaki, E., Amagasa, M.: Structural Modeling with Consensus for Societal Systems Using Fuzzy Set Theory. Proc.8th World IFAC Congress, Kyoto, Japapn, 1981, Vol.5, pp.35-40.

[1677] Tchon, K.: A 'Transversal' Approach to Evenistic Systems. Postepy Cybernetyki, 4, No.3, 1981, pp.77-79.

[1678] Tchon, K., Wojciechowska, J.: Towards a Formal Theory of Description. Int. J. of General Systems, 6, No.4, 1981, pp.217-224.

[1679] Tchon, K.: On Some Operations Preserving Generic Properties of Systems. Int. J. of General Systems, 9, No.2, 1983, pp.89-94.

[1680] Tchon, K.: Towards a Global Analysis of Systems. Int. J. of General Systems, 9, No.3, 1983, pp.171-176.

[1681] Techo, R.: Data Communications: an Introduction to Concepts and Design. Plenum Press, New York and London, 1980, X+293 pp.

[1682] Teodorescu, D.: Cybernetical Actions and Markov Processes. J. of Cybernetics, 9, No.2, 1979, pp.169-184.

[1683] Tesfatsion, L.: A New Approach to Filtering and Adaptive Control: Optimality Results J. of Cybernetics, 7, No.1-2, 1977, pp.133-146.

[1684] Thissen, W.: Investigations into World 3 Model: Lessons for Understanding Complicated Models. IEEE Trans. on Systems, Man, and Cybernetics, SMC-8, No.3, March 1978, pp.183-193.

[1685] Thomas, A.: Generating Tension for Constructive Change: The Use and Abuse of Systems Models. Cybernetics and Systems, 11, No.4, 1980, pp.339-354.

[1686] Thompson, J.M., Hunt, G.W.: The Instability of Evolving Systems. Interdisciplinary Science Reviews, 2, No.3, 1977, pp.240-262.

[1687] Thowsen, A.: Identifiability of Dynamic Systems. Int. J. of Systems Science, 9, No.7, July 1978, pp.813-825.

[1688] Thulasiraman, K., Swamy, M.N.S.: Graphs, Networks, and Algorithms. John Wiley, Chichester and New York, 1981, XX+592pp.

[1689] Tighe, T.J.: Modern Learning Theory: Foundations and Fundamental Issues. Oxford Univ. Press, Oxford and New York, 1982, X+422pp.

[1690] Tilti, A., Singh, M.G.(eds.): Large Scale System Theory and Applications (Proc. of a Symposium in Toulouse, France, June 1980). Pergamon Press, Oxford and New York, 1981, XXXIV+614pp.

[1691] Tjoa, A.M., Wagner, R.R.: A System-Theoretical Approach to Database Systems. Kybernetes, 9, No.4, 1980, pp.257-264.

[1692] Tjoa, A.M., Wagner, R.R.: The Universal Relation as Logical View of Databases. Austrian Society for Cybernetic Studies, Vienna, 1981. [German]

[1693] Toda, M.: Man, Robot, and Society: Models and

Speculations. Martinus Nijhoff, Boston and The Hague, 1982, XVIII+235pp.

[1694] Toda, M., Sugiyama, K.S.: Visual Q-Analysis (II): Strategy Analysis for Future Computer Systems Development in Japan. Cybernetics and Systems, 14, No.2-4, 1983, pp.229-252.

[1695] Toern, A.A.: Simulation Graphs: A General Tool for Modeling Simulation Designs. Simulation, 37, No.6, Dec.1981, pp.187-194.

[1696] Toffoli, T.: Computation and Construction Universality of Reversible Cellular Automata. J. of Computer and Systems Sciences, 15, 1977, pp.213-231.

[1697] Tomlinson, R.: A Systems Approach to Planning in Organizations - Developing a Collaborative Research Study. Cybernetics and Systems, 11, No.4, 1980, pp.355-368.

[1698] Tompa, F.W.: A Practical Example of the Specification of Abstract Data Types. Acta Informatica, 13, No.3, 1980, pp.205-224.

[1699] Tonella, G.: Pluralistic Modelling for Development Planning. In: Trappl, R.(ed.): Cybernetics and Systems Research, Vol.II. North-Holland, Amsterdam, and New York, 1984, pp.491-498.

[1700] Tong, R.M.: Synthesis of Fuzzy Models for Industrial Processes - Some Recent Results. Int. J. of General Systems, 4, No.3, 1978, pp.143-162.

[1701] Tong, R.M., Bonisonne, P.P.: A Linguistic Approach to Decision Making with Fuzzy Sets. IEEE Trans. on Systems, Man, and Cybernetics, SMC-10, No.11, 1980, pp.716-723.

[1702] Tong, R.M.: Some Properties of Fuzzy Feedback Systems. IEEE Trans. on Systems, Man, and Cybernetics, SMC-10, No.6, June 1980, pp.327-330.

[1703] Tong, R.M.: The Evaluation of Fuzzy Models Derived from Experimental Data. Fuzzy Sets and Systems, 4, No.1, July 1980, pp.1-12.

[1704] Topolski, J.: Methodology of History. D.Reidel, Dordrecht, Holland, and Boston, 1976, X+690 pp.

[1705] Tou, J.T.(ed.): Advances in Information Systems Science, Vol.6. Plenum Press, New York and London, 1976.

[1706] Tou, J.T.: Knowledge Engineering. Int. J. of Computer and Information Sciences, 9, No.4, 1980, pp.275-285.

[1707] Toulotte, J.M., Parsy, J.P.: A Method for Decomposing Interpreted Petri Nets and its Utilization. Digital Processes, 5, No.3-4, 1979, pp.223-234.

[1708] Toutenburg, H.: Prior Information in Linear Models. John Wiley, Chichester and New York, 1982, 215pp.

[1709] Tracy, L.: A Dynamic Living-Systems Model of Work Motivation. Systems Research, 1, No.3, 1984, pp.191-204.

[1710] Trappl, R., Von Luetzow, A., Wuendsch, L., Bornschein, H.: A Second Order Model of the Optic Generator Potential and its Relation to Stevens' Power Function. Pfluegers Archiv, 372, 1977, pp.165-168.

[1711] Trappl, R.: Non-Computerized Formal Methods with Computer Efficiency in Medicine. Proc.Int.Cong.on Computing in Medicine, Online Conf., Uxbridge, 1977.

[1712] Trappl, R., Klir, G.J., Ricciardi, L.(eds.): Progress in Cybernetics and Systems Research, Vol.III. Hemisphere, Washington, D.C., 1978.

[1713] Trappl, R., Pask, G.(eds.): Progress in Cybernetics and Systems Research, Vol.IV. Hemisphere, Washington, D.C., 1978.

[1714] Trappl, R.: Simple Forecasting Procedures in Health Care. In: Trappl, R., Pask, G.(eds.): Progress in Cybernetics and Systems Research, Vol.IV. Hemisphere, Washington, D.C., 1978.

[1715] Trappl, R.: Computer Psychotherapy: Is it Acceptable, Feasible, Advisable?. Cybernetics and Systems, 12, No.4, 1981, pp.385-394.

[1716] Trappl, R.: The Soft Side of Innovation: Medicine - a Hard Case. Univ.of Vienna, Dept.of Medical Cybernetics and AI, Report 81-01, 1981.

[1717] Trappl, R., Horn, W.: A Health Hazard Appraisal Program for Austria. In: Trappl, R., Ricciardi, L., Pask, G.(eds.): Progress in Cybernetics and Systems Research, Vol.IX. Hemisphere, Washington, D.C., 1982, pp.509-518.

[1718] Trappl, R.(ed.): Cybernetics and Systems Research. North-Holland, Amsterdam, and New York, 1982, XVIII+986 pp.

[1719] Trappl, R., Leinfellner, E., Steinacker, I., Trost, H.: Ontology on the Computer. Hoelder-Pichler-Tempsky, Wien, Akten des 6.Int.Wittgenstein Symposiums, 1982.

[1720] Trappl, R., Ricciardi, L., Pask, G.(eds.): Progress in Cybernetics and Systems Research, Vol.IX. Hemisphere, Washington, D.C., 1982, 560 pp.

[1721] Trappl, R., Ricciardi, L., Pask, G.(eds.): Progress in Cybernetics and Systems Research, Vol.VII. Hemisphere, Washington, D.C., 1982, 544 pp.

[1722] Trappl, R., Klir, G.J., Pichler, F.(eds.): Progress in Cybernetics and Systems Research, Vol.VIII. Hemisphere, Washington, D.C., 1982.

[1723] Trappl, R., Hanika, F.de P., Tomlinson, R.(eds.): Progress in Cybernetics and Systems Research, Vol.X. Hemisphere, Washington, D.C., 1982, 480 pp.

[1724] Trappl, R., Findler, N.V., Horn, W.(eds.): Progress in Cybernetics and Systems Research, Vol.XI. Hemisphere, Washington, D.C., 1982, 650 pp.

[1725] Trappl, R.(ed.): Cybernetics - Theory and Applications. Hemisphere, Washington, D.C., 1983.

[1726] Trappl, R.(ed.): Cybernetics and Systems Research, Vol.II. North-Holland, Amsterdam, and New York, 1984.

[1727] Trappl, R.: Impacts of Artificial Intelligence. In: Trappl, R.(ed.): Cybernetics and Systems Research, Vol.II. North-Holland, Amsterdam, and New York, 1984, pp.831-838.

[1728] Trappl, R.: Artificial Intelligence: A One-Hour Course. In: Trappl, R.(ed.): The Impacts of Artificial Intelligence. Science, Technology, Military, Economy, Society, Culture, and Politics. North-Holland, Amsterdam, and New York, 1985.

[1729] Trappl, R., Horn, W., Klir, G.J.(eds.): Basic and Applied General Systems Research: A Bibliography, 1977-1984. Hemisphere, Washington, D.C., 1985.

[1730] Trappl, R.: Impacts of Artificial Intelligence: An Overview. In: Trappl, R.(ed.): The Impacts of Artificial Intelligence. Science, Technology, Military, Economy, Society, Culture, and Politics. North-Holland, Amsterdam, and New York, 1985.

[1731] Trappl, R.(ed.): Power, Autonomy, Utopia: New Approaches Towards Complex Systems. Plenum Press, New York and London, 1985.

[1732] Trappl, R.: Reducing International Tension through Artificial Intelligence: A Proposal for 3 Projects. In: Trappl, R.(ed.): Power, Autonomy, Utopia: New Approaches Towards Complex Systems. Plenum Press, New York and London, 1985.

[1733] Trappl, R.(ed.): The Impacts of Artificial Intelligence. Science, Technology, Military, Economy, Society, Culture, and Politics. North-Holland, Amsterdam, and New York, 1985.

[1734] Trattnig, W.: Denotational Semantics of a Software Specification Language Based on the Data Flow Model. In: Trappl, R.(ed.): Cybernetics and Systems Research. North-Holland, Amsterdam, and New York, 1982, pp.781-790.

[1735] Tremblay, J.P., Sorenson, P.G.: An Introduction to Data Structures with Applications. McGraw-Hill, New York, 1976.

[1736] Tripp, R.S., Rainey, L.B., Pearson, J.M.: The Use of Cybernetics in Organizational Design and Development: An Illustration within Air Force Logistics Command. Cybernetics and Systems, 14, No.2-4, 1983, pp.293-314.

[1737] Troncale, L.: Linkage Propositions between Fifty Principal Systems Concepts. In: Klir, G.J.(ed.): Applied General Systems Research. Plenum Press, New York and London, 1978, pp.29-52.

[1738] Troncale, L.: Origins of Hierarchical Levels: An Emergent Evolutionary Process Based on Systems Concepts. In: Ericson, R.F.(ed.): The General Systems Challenge. SGSR, Louisville, Kentucky, 1978, pp.84-94.

[1739] Troncale, L.: Are Levels of Complexity in Bio-Systems Real? Applications of Clustering Theory to Modeling Systems Emergence. In: Lasker, G.E.(ed.): Applied Systems and Cybernetics (6 Vols). Pergamon Press, Oxford and New York, 1981, pp.1020-1026.

[1740] Troncale, L.: On a Possible Discrimination between Bioevolution and a Theory of Systems Emergence. In: Reckmeyer, W.J.(ed.): General Systems Research and Design: Precursors and Futures. SGSR, Louisville, Kentucky, 1981, pp.225-234.

[1741] Troncale, L.(ed.): A General Survey of Systems Methodology. Intersystems, Seaside, Ca., 1982.

[1742] Troncale, L.(ed.): A General Survey of Systems Methodology. (Proc.of the Twenty-Sixth Annual Meeting of the SGSR). SGSR, Louisville, Kentucky, January 5-9, 1982, 3 volumes, 1144pp.

[1743] Troncale, L.: Linkage Propositions between Principal Systems Concepts. In: Troncale, L.(ed.): A General Survey of Systems Methodology. Intersystems, Seaside, Ca., 1982, pp.27-38.

[1744] Troncale, L.: Some Key, Unanswered Questions about Hierarchies. In: Troncale, L.(ed.): A General Survey of Systems Methodology. Intersystems, Seaside, Ca., 1982, pp.77-81.

[1745] Troncale, L.: Testing Hierarchy Models with Data Using Computerized, Empirical Data Bases. In: Troncale, L.(ed.): A General Survey of Systems Methodology. Intersystems, Seaside, Ca., 1982, pp.90-102.

[1746] Troncale, L., Voorhees, B.: Towards a Formalization of Systems Linkage Propositions. In: Lasker, G.E.(ed.): The Relation between Major World Problems and Systems Learning. Intersystems, Seaside, Ca., 1983, pp.341-349.

[1747] Troncale, L.: A Hybrid Systems Method: Tests for Hierarchy & Links between Isomorphs. In: Trappl, R.(ed.): Cybernetics and Systems Research, Vol.II. North-Holland, Amsterdam, and New York, 1984, pp.39-46.

[1748] Troncale, L.: Future of General Systems Science: Obstaches, Potentials, Case Studies. Systems Research, 1, No.4, 1985.

[1749] Troncale, L.: Knowing Natural Systems Enables Better Design of Man-Made Systems: The Linkage Proposition Model. In: Trappl, R.(ed.): Power, Autonomy, Utopia: New Approaches Towards Complex Systems. Plenum Press, New York and London, 1985.

[1750] Troncale, L.: Systems Theory Applied to the Biological

and Medical Sciences. In: Banathy, B.H.(ed.): Proceedings of the 29th Annual Meeting. SGSR, Louisville, Kentucky, 1985, 16 pp.

[1751] Trost, H.: Computer Simulation of Neurotic Processes. Austrian Society for Cybernetic Studies, Vienna, 1978. [German]

[1752] Trost, H.: Outline of a Natural-Language Understanding System. Austrian Society for Cybernetic Studies, Vienna, 1980. [German]

[1753] Trost, H., Buchberger, E., Steinacker, I., Trappl, R.: VIE-LANG: A German Language Dialogue System. Cybernetics and Systems, 14, No.2-4, 1983, pp.343-356.

[1754] Trost, H.: Declarative Knowledge Representation: An Overview and an Application in the Field of Natural Language Understanding. Austrian Society for Cybernetic Studies, Vienna, 1984. [German]

[1755] Trussel, P., Brandt, A., Knapp, S.: Using Nursing Research: Discovery, Analysis, and Interpretation. Nursing Resources, Wakefield, Mass., 1981.

[1756] Trussel, P.: Constructs from General Systems Theory to a Health History. In: Trappl, R.(ed.): Cybernetics and Systems Research, Vol.II. North-Holland, Amsterdam, and New York, 1984, pp.513-518.

[1757] Tsai, W.H., Fu, K.S.: Attributed Grammar - A Tool for Combining Syntactic and Statistical Approaches to Pattern Recognition. IEEE Trans. on Systems, Man, and Cybernetics, SMC-10, No.12, Dec.1980, pp.873-885.

[1758] Tsetlin, M.L.: Automata Theory and Modelling of Biological Systems. Academic Press, London, and New York, 1980.

[1759] Tsokos, C.P., Thrall, R.M.(eds.): Decision Information. Academic Press, London, and New York, 1979, X+528 pp.

[1760] Tsypkin, Y.Z.: The Theory of Adaptive and Learning Systems. In: Trappl, R.(ed.): Cybernetics - Theory and Applications. Hemisphere, Washington, D.C., 1983, pp.57-90.

[1761] Turksen, I.B., Yao, D.D.W.: Bounds for Fuzzy Inference. In: Trappl, R.(ed.): Cybernetics and Systems Research. North-Holland, Amsterdam, and New York, 1982, pp.729-734.

[1762] Turney, P.: On Laws of Form. Toronto, Ontario, 1982, 25pp. [IFSR-Depository]

[1763] Tussey, C.C., Wolf, D.J.: The Kentucky Longevity Information System. Cybernetics and Systems, 13, No.2, 1982, pp.129-152.

[1764] Tzafestas, S.G.: Optimization of System Reliability: a Survey of Problems and Techniques. Int. J. of Systems Science, 11, No.4, April 1980, pp.455-486.

[1765] Uccellini, L.W., Petersen, R.A.: Applying Discrete Model Concepts to the Computation of Atmospheric Trajectories. Int. J. of General Systems, 6, No.1, 1980, pp.13-24.

[1766] Ulanowicz, R.E.: The Empirical Modelling of an Ecosystem. Ecological Modelling, 4, No.1, Jan.1978, pp.29-40.

[1767] Ulrich, W.: The Design of Problem Solving Systems. Management Science, 23, No.10, June 1977, pp.1099-1108.

[1768] Ulrich, W.: A Critique of Pure Cybernetic Reason: The Chilean Experience with Cybernetics. J. of Applied Systems Analysis, 8, April 1981, pp.33-59.

[1769] Umnov, A., Albegov, M.: An Approach to Distributed Modeling. Behavioral Science, 26, No.4, Oct.1981, pp.354-365.

[1770] Umpleby, S.: Methods for Making Social Organisation Adaptive. In: Trappl, R.(ed.): Power, Autonomy, Utopia: New Approaches Towards Complex Systems. Plenum Press, New York and London, 1985.

[1771] Unger, C.K., Stocker, G.(eds.): Biophysical Ecology and Ecosystem Research. Akademie-Verlag, Berlin, 1981.

[1772] Unterberger, H.: Automatic Recognition of Voiced Signals using Articulation Models. In: Trappl, R.(ed.): Cybernetics and Systems Research. North-Holland, Amsterdam, and New York, 1982 , pp.745-750 751-756.

[1773] Urban, H.B.: The Concept of Development for a Systems Perspective. In: Baltes, P.B.(ed.): Life Span Development and Behavior, Vol.I. Academic Press, London, and New York, 1978, pp.45-83.

[1774] Utkin, V.I.: Variable Structure Systems with Sliding Modes. IEEE Trans. on Automatic Control, AC-22, No.2, April 1977, pp.212-222.

[1775] Uttenhove, H.J.: SAPS (Systems Approach Problem Solver): An Introduction and Guide. Computing and Systems Consultants, Binghampton, NY, 1981.

[1776] Uyeno, D.H.: PASSIM: A Discrete-Event Simulation Package for PASCAL. Simulation, 35, No.6, 1980, pp.183-190.

[1777] Uyttenhove, H.J.J.: On the Two-dimensionality in the Behavioral System Identification Problem (Part I). Int. J. of Man-Machine Studies, 12, No.4, 1980, pp.325-340.

[1778] Uyttenhove, H.J.J., Gomez, P.: The Systems Approach Problem Solver and a Stockmarket Example. SGSR, Louisville, Kentucky, 1981, pp.341-348.

[1779] Vaccari, E., Delaney, W.: The Role of System Modelling in Natural Language Processing. In: Trappl, R.(ed.): Cybernetics and Systems Research, Vol.II. North-Holland, Amsterdam, and New York, 1984, pp.739-746.

[1780] Vagnucci, A.H., (et al.): Concatenation and Homogeneity of Biological Time Series. IEEE Trans. on Systems, Man, and Cybernetics, SMC-7, No.6, June 1977, pp.483-491.

[1781] Valdes-Perez, R.E., Conant, R.C.: Information Loss Due to Data Quantization in Reconstructability Analysis. Int. J. of General Systems, 9, No.4, 1983, pp.235-248.

[1782] Vallee, R.: Evolution of Dynamical Linear System with Random Initial Conditions. In: Trappl, R.(ed.): Cybernetics and Systems Research. North-Holland, Amsterdam, and New York, 1982, pp.163-164.

[1783] Vallee, R.: "Eigen-Elements" for Observing and Interacting Subjects. In: Trappl, R.(ed.): Cybernetics and Systems Research, Vol.II. North-Holland, Amsterdam, and New York, 1984, pp.89-92.

[1784] Vallet, C., Ferre, M., Moulin, Th., Chastang, J.: Use of a Partially Self-Referent Formalism for Description of Natural Systems. In: Trappl, R.(ed.): Cybernetics and Systems Research. North-Holland, Amsterdam, and New York, 1982, pp.85-90.

[1785] Van Amerongen, J., (et al.): An Autopilot for Ships Designed with Fuzzy Sets. In: Van Navta, L.(ed.): Digitial Computer Applications to Process Control. North-Holland, Amsterdam, and New York, 1977, pp.479-487.

[1786] Van Campenhout, J.M., Cover, T.M.: Maximum Entropy and Conditional Probability. IEEE Trans. on Information Theory, IT-27, No.4, July 1981, pp.483-489.

[1787] Van Gigch, J.P.: A Methodological Comparison of the Science, Systems and Metasystem Paradigms. Int. J. of Man-Machine Studies, 11, No.5, 1979, pp.651-663.

[1788] Van Gigch, J.P., Kramer, N.J.T.A.: A Taxonomy of Systems Science. Int. J. of Man-Machine Analysis, 14, No.2, 1981, pp.179-191.

[1789] Van Navta, L.(ed.): Digitial Computer Applications to Process Control. North-Holland, Amsterdam, and New York, 1977.

[1790] Van Rootselaar, B.(ed.): Annals of Systems Research,

Vol.7. Martinus Nijhoff, Boston and The Hague, 1978.

[1791] Van Steenkiste, G.C.(ed.): Modeling, Identification and Control in Environmental Systems. North-Holland, Amsterdam, and New York, 1978.

[1792] Van Straaten,, (et al.)(eds.): Proceedings of the Second Joint MTA/IIASA Task Force Meeting on Lake Balaton Modelling Veszprem (Hungary). 1980, 2 Volumes, 233pp and 281pp.

[1793] Van den Bogard, W., Kleijnen, J.P.C.: Minimizing Wafting Times Using Priority Classes: A Case Study in Response Surface Methodology. Katholieke Hogeschool, Tilburg, Netherlands, 1977, 15pp. [IFSR-Depository]

[1794] Van der Ven, A.H.G.S.: Introduction to Scaling. John Wiley, Chichester and New York, 1980, XI+301 pp.

[1795] Van der Zanden, J.: Measurement Theory and Signal Theory: Part I Investigations on the Foundations. Univ.of Technology, Delft, Netherlands, 1981, 18pp. [IFSR-Depository]

[1796] Van der Zouwen, J.: Inventory of Systems Research in the Netherlands. Vrije Universiteit, Amsterdam, Netherlands, 1977, 15pp. [IFSR-Depository]

[1797] Van der Zouwen, J.: Inventory of Systems Research on the Netherlands. Vrije Universiteit, Amsterdam, Netherlands, 1979, 20pp. [IFSR-Depository]

[1798] Varela, F.J., Goguen, J.A.: The Arithmetic of Closure. J. of Cybernetics, 8, No.3-4, 1978, pp.291- 324.

[1799] Varela, F.J.: Principles of Biological Autonomy. (Second Vol. in the North-Holland Series in General Systems Research, Edited by G.J. Klir.) North-Holland, Amsterdam, and New York, 1979, XX+306 pp.

[1800] Varela, F.J.: Steps to a Cybernetics of Autonomy. In: Trappl, R.(ed.): Power, Autonomy, Utopia: New Approaches Towards Complex Systems. Plenum Press, New York and London, 1985.

[1801] Varshavskiy, V.I., (et al.): Asynchronous Processes II. Composition and Matching. Engineering Cybernetics, 18, No.5, 1980, pp.92-97.

[1802] Varshavskiy, V.I., (et al.): Asynchronous Processes. I. Definition and Interpretation. Engineering Cybernetics, 19, No.4, July-August 1980, pp.93-97.

[1803] Vavrousek, J.: Multiple Structure of Systems. Ekonomicko-matematicky obzor, 1980, No.4, pp.380-397. [Czech]

[1804] Vavrousek, J.: An Approach to Modelling in Empirical Sciences. 1981, 31pp. [IFSR-Depository]

[1805] Vavrousek, J.: Multiple Structure of Systems. 1981, 34 pp. [IFSR-Depository]

[1806] Veelenturf, L.P.J.: An Automata-Theoretical Approach to Developing Learning Neural Networks. Cybernetics and Systems, 12, No.1-2, 1981, pp.179-202.

[1807] Veloso, P.A.S., Martins, R.C.B.: On Reducibilities Among General Problems. In: Trappl, R.(ed.): Cybernetics and Systems Research, Vol.II. North-Holland, Amsterdam, and New York, 1984, pp.21-26.

[1808] Vemuri, V.: Modeling of Complex Systems: an Introduction. Academic Press, London, and New York, 1978, 464 pp.

[1809] Venkatesan, M.: An Alternate Approach to Transitive Coupling in ISM. IEEE Trans. on Systems, Man, and Cybernetics, SMC-9, No.3, March 1979, pp.125-130.

[1810] Ventriglia, F.: Kinetic Theory of Neural Systems: Memory Effects. In: Trappl, R.(ed.): Cybernetics and Systems Research. North-Holland, Amsterdam, and New York, 1982, pp.271-276.

[1811] Ventriglia, F.: Kinetic Theory of Neural Systems: An Attempt Towards a Simulation of Cerebellar Cortex Activity. In: Trappl, R.(ed.): Cybernetics and Systems Research, Vol.II. North-Holland, Amsterdam, and New York, 1984, pp.223-228.

[1812] Verhagen, C.J.D.M.: Some General Remarks About Pattern Recognition; its Definitions; its Relation with Other Disciplines; a Literature Survey. Pattern Recognition, 7, No.3, Sept.1975, pp.109-116.

[1813] Veroy, B.S.: Large Scale Network Architecture Synthesis: Interactive Strategy. Int. J. of General Systems, 9, No.2, 1983, pp.73-88.

[1814] Veroy, B.S.: Overall Design of Reliable Centralized Voice/Data Communication Network. Int. J. of General Systems, 9, No.4, 1983, pp.183-196.

[1815] Vickers, G.: Education in Systems Thinking. J. of Applied Systems Analysis, 7, April 1980, pp.3-10.

[1816] Viliums, E.R.: Methods to Obtain Interlevel Consistency of Decisions under Uncertainty. In: Trappl, R.(ed.): Cybernetics and Systems Research. North-Holland, Amsterdam, and New York, 1982, pp.241-248.

[1817] Viliums, E.R., Sukur, L.Y.: Practical Aspects of Alternatives Evaluating and Decision Making Under Uncertainty and Multiple Objectives. In: Trappl, R.(ed.): Cybernetics and Systems Research, Vol.II. North-Holland, Amsterdam, and New York, 1984, pp.165-172.

[1818] Vincent, T.L., Grantham, W.J.: Optimality in Parametric Systems. John Wiley, Chichester and New York, 1981, XVIII+244pp.

[1819] Vitanyi, P.M.B.: Lindenmayer Systems: Structure, Languages, and Growth Functions. Mathematisch Centrum, Amsterdam, 1980, 209pp.

[1820] Von Bertalanffy, L.: Systems View of Man (edited by Paul La Violette). Westview Press, Boulder, Colorado, 1980.

[1821] Von Glasersfeld, E.: Steps in the Construction of "Others" and "Reality": A Study of Self-Regulation. In: Trappl, R.(ed.): Power, Autonomy, Utopia: New Approaches Towards Complex Systems. Plenum Press, New York and London, 1985.

[1822] Von Luetzow, A., Porenta, G., Trappl, R.: A System Analysis of Information Transmission of Cat Retinal Horizontal Cells. In: Trappl, R., Ricciardi, L., Pask, G.(eds.): Progress in Cybernetics and Systems Research, Vol.IX. Hemisphere, Washington, D.C., 1982, pp.127-134.

[1823] Von Plato, J.: Probability and Determinism. Philosophy of Science, 49, No.1, March 1982, pp.51-66.

[1824] Von Wright, G.H.: The Logic of Preference. Edinburgh Univ.Press, Edinburgh, 1963.

[1825] Voyat, G.E.: Piaget Systematized. Erlbaum, Hillsdale, N.J., 1982, XX + 214pp.

[1826] Wade, J.W.: Architecture, Problems, and Purposes: Architectural Design as a Basic Problem-Solving Process. John Wiley, Chichester and New York, 1977, XII+350 pp.

[1827] Waldhauer, F.D.: Feedback. John Wiley, Chichester and New York, 1982.

[1828] Walichiewicz, L.: Decomposition of Linguistic Rules in the Design of a Multi-Dimensional Fuzzy Control Algorithm. In: Trappl, R.(ed.): Cybernetics and Systems Research, Vol.II. North-Holland, Amsterdam, and New York, 1984, pp.557-562.

[1829] Walker, B.J., Blake, I.F.: Computer Security and Protection Structures. Dowden, Hutchinson and Ross, Stroudsburg, Penn., 1977.

[1830] Walker, C.C., Gelfand, A.G.: A System Theoretic Approach

to the Management of Complex Organizations: Management by
Exception, Priority, and Input Span in a Class of
Fixed-Structure Models. Behavioral Science, 24, No.2,
March 1979, pp.112-120.

[1831] Walker, J.A.: Dynamical Systems and Evolution Equations:
Theory and Applications. Plenum Press, New York and
London, 1980, VIII+236 pp.

[1832] Walker, V.R.: Criteria for Object Perception. J. of
Cybernetics, 7, No.3-4, 1977, pp.169-188.

[1833] Wallace, M.R.: Applications of Ultrastability.
Kybernetes, 8, No.3, 1979, pp.185-191.

[1834] Wallenmaier, T.E.: Towards a Quantitative Concept of
Teleological Systems. Nature and Systems, 1, No.3,
Sept.1979, pp.147-155.

[1835] Waller, R.J.: Complexity and the Boundaries of Human
Policy Making. Int. J. of General Systems, 9, No.1, 1982,
pp.1-12.

[1836] Wang, P., Chang, S.K.(eds.): Fuzzy Systems: Theory and
Application to Policy Analysis and Information Systems.
Plenum Press, New York and London, 1980, IX+413pp.

[1837] Wang, Z., Dang, Y.: Application of Fuzzy Decision Making
and Fuzzy Linear Programming in Production and Planning of
Animal Husbandry System in Farming Region. In: Trappl,
R.(ed.): Cybernetics and Systems Research, Vol.II.
North-Holland, Amsterdam, and New York, 1984, pp.563-566.

[1838] Wang, Z.: M-Fuzzy Numbers and Random Variables. In:
Trappl, R.(ed.): Cybernetics and Systems Research,
Vol.II. North-Holland, Amsterdam, and New York, 1984,
pp.535-538.

[1839] Warfield, J.N.: Crossing Theory and Hierarchy Mapping.
IEEE Trans. on Systems, Man, and Cybernetics, SMC-7, No.7,
July 1977, pp.505-523.

[1840] Warfield, J.N.: Some Principles of Knowledge
Organization. IEEE Trans. on Systems, Man, and
Cybernetics, SMC-9, No.6, 1979, pp.317-327.

[1841] Warfield, J.N.: Complementary Relations and Map Reading.
IEEE Trans. on Systems, Man, and Cybernetics, SMC-10,
No.6, June 1980, pp.285-291.

[1842] Warnier, J.-D.: Logical Construction of Systems. Van
Nostrand Reinhold, New York, 1982.

[1843] Wasinowski, R.: Choosing among Alternative Strategies.
In: Trappl, R.(ed.): Cybernetics and Systems Research.
North-Holland, Amsterdam, and New York, 1982, pp.249-252.

[1844] Wasniowski, R.: The Use of Computer-Communication Systems
in Futures Research. In: Trappl, R.(ed.): Cybernetics
and Systems Research. North-Holland, Amsterdam, and New
York, 1982, pp.833-836.

[1845] Watanabe, S.: A Generalized Fuzzy-Set Theory. IEEE
Trans. on Systems, Man, and Cybernetics, SMC-8, No.10,
Oct.1978, pp.756-760.

[1846] Watanabe, S.: Pattern Recognition as a Quest for Minimum
Entropy. Pattern Recognition, 13, No.5, 1981, pp.381-387.

[1847] Waterman, D.A., Hayes-Roth, F.(eds.): Pattern-Directed
Inference Systems. Academic Press, London, and New York,
1978, 672 pp.

[1848] Watson, S.R., Weiss, J.J., Donnell, M.L.: Fuzzy Decision
Analysis. IEEE Trans. on Systems, Man, and Cybernetics,
SMC-9, No.1, Jan.1969, pp.1-9.

[1849] Watt, K.E.F., Molloy, L.F., Varshney, C.K., Weeks, D.,
Wirosardjono, S.: The Unsteady State - Environmental
Problems, Growth, and Culture. University Press of Hawai,
Honolulu, 1977.

[1850] Webber, B.L., Nilsson, N.J.(eds.): Readings in Artificial
Intelligence. W.H.Freeman, San Francisco, 1982, 557pp.

[1851] Wechler, W.: The Concept of Fuzziness in Automata and Language Theory. Akademie-Verlag, Berlin, 1978.

[1852] Wedde, H.F.: An Exercise in Flexible Formal Modeling under Realistic Assumptions. Systems Research, 1, No.2, 1984, pp.105-116.

[1853] Weinberg, G.M., Weinberg, D.: On the Design of Stable Systems. John Wiley, Chichester and New York, 1979, XVI+353 pp.

[1854] Weingartner, P.: Analogy Among Systems. Dialectica, 35, Nos.3-4, 1979, pp.355-378.

[1855] Weir, M.: Design for a Goal-Directed System. In: Trappl, R.(ed.): Cybernetics and Systems Research. North-Holland, Amsterdam, and New York, 1982, pp.45-50.

[1856] Wellstead, P.E.: Introduction to Physical System Modelling. Academic Press, London, and New York, 1979, X+280pp.

[1857] Wendt, S.: The Programmed Action Module: an Element for System Modelling. Digital Processes, 5, No.3-4, 1979, pp.213- 222.

[1858] Wenger, M.S:: Thoughts on a Cybernetic View of Social Organization. In: Trappl, R.(ed.): Cybernetics and Systems Research, Vol.II. North-Holland, Amsterdam, and New York, 1984, pp.471-476.

[1859] West, B.J.: Mathematical Models as a Tool for the Social Sciences. Gordon and Breach, New York, and London, 1980, 120 pp.

[1860] White, A.M.(ed.): Interdisciplinary Teaching. Jossey-Bass, San Francisco, 1981, 113pp.

[1861] White, C., Sage, A.P.: A Multiple Objective Optimization-Based Approach to Choicemaking. IEEE Trans. on Systems, Man, and Cybernetics, SMC-10, No.6, June 1980, pp.315-326.

[1862] White, D.J., Brown, K.C.(eds.): Role and Effectiveness of Theories of Decision in Practice. Hodder / Stoughton, London, 1975.

[1863] Wilder, R.L.: Mathematics as a Cultural System. Pergamon Press, Oxford and New York, 1981, XII+182pp.

[1864] Wildhelm, J.: Objectives and Multiobjective Decision Making Under Uncertainty. Springer-Verlag, Berlin, FRG, and New York, 1975, IV+111 pp.

[1865] Williams, P.M.: Bayesian Conditionalisation and the Principle of Minimum Information. Brit. J. for the Philosophy of Science, 31, 1980, pp.131-144.

[1866] Windeknecht, T.G., D'Angelo, H.: System Theoretic Implications of Numerical Methods Applied to the Solution of Ordinary Differential Equations. IEEE Trans. on Systems, Man, and Cybernetics, SMC-7, No.11, 1977, pp.805-810.

[1867] Winkel, D., Prosser F.: The Art of Digital Design: An Introduction to Top-Down Design. Prentice-Hall, Englewood Cliffs, New Jersey, 1980, XIV + 498 pp.

[1868] Winograd, T.: Language as a Cognitive Process (Vol.1: Syntax). Addison-Wesley, Reading, Mass., 1983, XIV+640 pp.

[1869] Winston, P.H.: Artificial Intelligence. Addison-Wesley, Reading, Mass., 1977.

[1870] Winston, P.H.: Learning and Reasoning by Analogy. ACM Communications, 23, No.12, 1980, pp.689-703.

[1871] Winston, P.H., Brown, R.H.(eds.): Artificial Intelligence: An MIT Perspective (2 Vols.). M.I.T.Press, Cambridge, Mass., Vol.1: XVI+492 pp.; Vol.2: XVI+486 pp., 1982.

[1872] Wisniewski, T., Wasinowski, R.: An Interactive Model for Analysing Climate Change due to Increasing Carbon Dioxide.

In: Trappl, R.(ed.): Cybernetics and Systems Research. North-Holland, Amsterdam, and New York, 1982, pp.695-700.

[1873] Witten, I.H.: Probabilistic Behaviour-Structure Transformations Using Transitive Moore Models. Int. J. of General Systems, 6, No.3, 1980.

[1874] Witten, I.H.: Some Recent Results on Nondeterministic Modelling of Behaviour Sequences. SGSR, Louisville, Kentucky, 1981, pp.265-274.

[1875] Witten, I.H.: Principles of Computer Speech. Academic Press, London, and New York, 1982, XII+286 pp.

[1876] Wnuk, A.J.: The Block-Oriented Interactive Simulator BORIS. In: Trappl, R.(ed.): Cybernetics and Systems Research. North-Holland, Amsterdam, and New York, 1982, pp.773-780.

[1877] Wolaver, T.G.: Effect of Systems Structure, Connectivity, and Recipient Control on the Sensitivity Characteristics of a Model Ecosystem. Int. J. of Systems Science, 11, No.3, March 1980, pp.291-303.

[1878] Wolkowski, Z.W.: Hierarchical Perception of Time and Periodicity. In: Trappl, R.(ed.): Cybernetics and Systems Research. North-Holland, Amsterdam, and New York, 1982, pp.321-326.

[1879] Wood-Harper, A.T., Fitzgerald, G.: A Taxonomy of Current Approaches to Systems Analysis. Computer Journal, 25, No.1, Feb.1982, pp.12-16.

[1880] Wrathal, C.P.: Computer Acronyms, Abbreviations, etc. Petrocelli, New York, 1981, IX+483 pp.

[1881] Wu, W.T.: Heuristic Approach to Chinese-Character Search. Kybernetes, 13, No.1, 1984, pp.39-42.

[1882] Wuensch, G.: System Theorie. Geest & Portig, Leipzig, 1975. [German]

[1883] Wulf, W.A., (et al.): Fundamental Structures of Computer Science. Addison-Wesley, Reading, Mass., 1981, XVIII+621 pp.

[1884] Wyner, A.D.: A Definition of Conditional Mutual Information for Arbitrary Ensembles. Information and Control, 38, No.1, July 1978, pp.51-59.

[1885] Wyner, A.D.: Fundamental Limits in Information Theory. IEEE Proc., 69, No.2, February 1981, pp.239-251.

[1886] Xuemou, W.: Pansystems Analysis and Fuzzy Sets. Busefal, No.10, 1982, pp.7-16.

[1887] Yager, R.R.: Fuzzy Decision Making Including Unequal Objectives. Fuzzy Sets and Systems, 1, No.2, April 1978, pp.87-95.

[1888] Yager, R.R.: Linguistic Model and Fuzzy Truths. Int. J. of Man-Machine Studies, 10, No.5, Sept.1978, pp.483-494.

[1889] Yager, R.R.: Validation of Fuzzy-Linguistic Models. J. of Cybernetics, 8, No.1, 1978, pp.17-30.

[1890] Yager, R.R.: On the Measure of Fuzziness and Negation. Part 1: Membership in the Unit Interval. Int. J. of General Systems, 5, No.4, 1979, pp.221-230.

[1891] Yager, R.R.: A Foundation for a Theory of Possibility. J. of Cybernetics, 10, No.1-3, 1980, pp.177-204.

[1892] Yager, R.R.: A Linguistic Variable for Importance of Fuzzy Sets. J. of Cybernetics, 10, No.1-3, 1980, pp.249-260.

[1893] Yager, R.R.: An Approach to Inference in Approximate Reasoning. Int. J. of Man-Machine Studies, 1980, pp.323-338.

[1894] Yager, R.R.: Aspects of Possibilistic Uncertainty. Int. J. of Man-Machine Studies, 12, No.3, April 1980, pp.283-298.

[1895] Yager, R.R.: Fuzzy Sets, Probabilities, and Decision. J. of Cybernetics, 10, No.1-3, 1980, pp.1-18.

[1896] Yager, R.R.: Fuzzy Subsets of Type II in Decisions. J. of Cybernetics, 10, No.1-3, 1980, pp.137-160.

[1897] Yager, R.R.: Fuzzy Thinking as Quick and Efficient. Cybernetica, 23, No.4, 1980, pp.265-298.

[1898] Yager, R.R.: On a General Class of Fuzzy Connectives. Fuzzy Sets and Systems, 4, No.3, 1980, pp.235-242.

[1899] Yager, R.R.: Approximate Reasoning and Possibilistic Models in Classification. Int. J. of Computer and Information Sciences, 10, No.2, April 1981, pp.141-175.

[1900] Yager, R.R.: Prototypical Values for Fuzzy Subsets. Kybernetes, 10, No.2, 1981, pp.135-139.

[1901] Yager, R.R.: Some Properties of Fuzzy Relationships. Cybernetics and Systems, 12, Nos.1-2, 1981, pp.123-140.

[1902] Yager, R.R.: Fuzzy Prediction Based on Regression Models. Information Sciences, 26, No.1, 1982, pp.45-63.

[1903] Yager, R.R.(ed.): Fuzzy Set and Possibility Theory. Pergamon Press, Oxford and New York, 1982, XIV+633 pp.

[1904] Yager, R.R.: Lingustic Hedges: Their Relation to Context and Their Experimental Realization. Cybernetics and Systems, 13, No.4, 1982, pp.357-374.

[1905] Yager, R.R.: Measuring Tranquility and Anxiety in Decision Making: An Application of Fuzzy Sets. Int. J. of General Systems, 8, No.3, 1982, pp.139-146.

[1906] Yager, R.R.: Some Procedures for Selecting Fuzzy Set-Theoretic Operators. Int. J. of General Systems, 8, No.2, 1982, pp.115-124.

[1907] Yager, R.R.: Entropy and Specificity in a Mathematical Theory of Evidence. Int. J. of General Systems, 9, No.4, 1983, pp.249-260.

[1908] Yager, R.R.: Membership in a Compound Fuzzy Subset. Cybernetics and Systems, 14, No.2-4, 1983, pp.173-184.

[1909] Yager, R.R.: On Different Classes of Linguistic Variables Defined via Fuzzy Subsets. Kybernetes, 13, No.2, 1984, pp.103-110.

[1910] Yakowitz, S.J.: Computational Probability and Simulation. Addison-Wesley, Reading, Mass., 1977.

[1911] Yashin, A.: A New Proff and New Results in Conditional Gaussian Estimation Procedures. In: Trappl, R.(ed.): Cybernetics and Systems Research. North-Holland, Amsterdam, and New York, 1982, pp.205-208.

[1912] Yeh, R.T.(ed.): Current Trends in Programming Methodology: Vol. IV - Data Structuring. Prentice-Hall, Englewood Cliffs, New Jersey, 1978, 321 pp.

[1913] Yezhkova, I.V., Pospelov, D.A.: Decision Making on Fuzzy Grounds. Engineering Cybernetics, 1977, No.6.

[1914] Young, T.Y., Liu, P.S., Rondon, R.J.: Statistical Pattern Classification with Binary Variables. IEEE Trans. on Pattern Analysis and Machine Intelligence, PAMI-3, No.2, 1981, pp.155-163.

[1915] Zarri, G.P.: The RESEDA Project: Using AI Techniques in Order to Solve the Problem of Information Incompleteness. In: Trappl, R.(ed.): Cybernetics and Systems Research. North-Holland, Amsterdam, and New York, 1982, pp.897-902.

[1916] Zeeman, E.C.: Catastrophe Theory: Selected Papers 1972-1977. Addison-Wesley, Reading, Mass., 1977, 675 pp.

[1917] Zeigler, B.P.: Canonical Realization of General Time Systems. Information Sciences, 12, No.2, 1977, pp.179-186.

[1918] Zeigler, B.P.: Structuring the Organization of Partial Models. Int. J. of General Systems, 4, No.2, 1978, pp.81-88.

[1919] Zeigler, B.P., (et al.)(eds.): Methodology in Systems Modelling and Simulation. North-Holland, Amsterdam, and New York, 1979, XV+537 pp.

[1920] Zeigler, B.P.: Structuring Principles for Multifacet System Modelling. In: Zeigler, B.P., (et al.)(eds.): Methodology in Systems Modelling and Simulation. North-Holland, Amsterdam, and New York, 1979, pp.83-135.

[1921] Zeleny, M.(ed.): Autopoiesis, Dissipative Structures, and Spontaneous Social Orders. Westview Press, Boulder, Colorado, 1980, XXI+149 pp.

[1922] Zeleny, M.(ed.): Autopoiesis: A Theory of Living Organization. North-Holland, Amsterdam, and New York, 1981, XVII+314 pp.

[1923] Zeleny, M.: On the Squandering of Resources and Profits Via Linear Programming. Interfaces, 11, No.5, Oct. 1981, pp.101-107.

[1924] Zeleny, M.: Multiple Criteria Decision Making. McGraw-Hill, New York, 1982, XXII + 561pp.

[1925] Zhukovin, V.E.: The Multicriteria Decision Making with Vector Fuzzy Preference Realtion. In: Trappl, R.(ed.): Cybernetics and Systems Research, Vol.II. North-Holland, Amsterdam, and New York, 1984, pp.179-182.

[1926] Zimmermann, H.J., Zysno, P.: Latent Connectives in Human Decision Making. Fuzzy Sets and Systems, 4, No.1, July 1980, pp.37-51.

[1927] Zimmermann, H.J.: Testability and Measuring of Mathematical Models. Mathematical Modelling, 1980, pp.123-139.

[1928] Zionts, S.(ed.): Multiple Criteria Problem Solving. Springer-Verlag, Berlin, FRG, and New York, 1978.

[1929] Zorbas, Y.G., Aripov, K.N.: Medical Monitoring of Man's Reliability during Cosmic Flights. In: Trappl, R.(ed.): Cybernetics and Systems Research. North-Holland, Amsterdam, and New York, 1982, pp.305-308.

[1930] Zouwen, J.v.d., Dijkstra, W.: Modeling Interaction Processes in Interviews; a Contribution of Systems Methodology. In: Trappl, R.(ed.): Cybernetics and Systems Research. North-Holland, Amsterdam, and New York, 1982, pp.105-112.

[1931] Zwart, P.J.: About Time. North-Holland, Amsterdam, and New York, 1976.

[1932] Zwick, M.: The Cusp Catastrophe and the Laws of Dialectics. Nature and Systems, 1, No.3, Sept. 1979, pp.177-187.

[1933] General Systems Research: A Science, a Methodology, a Technology (Proceedings of the 1979 North American Meeting). SGSR, Louisville, Kentucky, 1979, 288 pp.

[1934] Proceedings of the First International Symposium on Policy Analysis and Information Systems. Duke University, Durham, N.C., June 28-30, 1979.

[1935] Logik, Erkenntnistheorie, Wissenschaftstheorie, Metaphysik, Theologie. Franz Steiner Verlag, Wiesbaden, 1980.

[1936] Non-Monotonic Logic (Special Issue). Artificial Intelligence, 13, Nos. 1+2, April 1980, pp.1-174.

[1937] Praxiology (a Yearbook published by the Department of Praxiology, Insitute of Philosophy and Sociology). Polish Academy of Sciences, Warsaw, (since 1980 in English).

[1938] Design Studies. A New Journal Covering the Study, Research and Techniques of All Design (Published Quarterly in Co-operation with the Design Research Society). IPC Science and Technology Press, Guildford, Surrey, England.

[1939] Metamedicine. A New International Journal for Philosophy and Methodology of Medicine. D.Reidel, Dordrecht, Holland, and Boston, 1981.

[1940] Proceedings of the Workshop on Applications of Adaptive Systems Theory. Medicine. Yale University, May 1981, 197

pp.

[1941] Second Special Issue on the Guha Method of Mechanized
       Hypothesis Formation. Int. J. of Man-Machine Studies, 15,
       No.3, October 1981, pp.251-358.

[1942] Special English Issue of the Polish Journal "Progress in
       Cybernetics" Devoted to Papers Presented at the Second
       Spring School in Jadwisin, Poland, April 13-19, 1980.
       Postepy Cybernetyki, 4, No.3, 1981, pp.1-133.

[1943] Special Issue on Simulation Modelling and Statistical
       Computing. ACM Communications, 24, No.4, April 1981,
       pp.171-273.

[1944] Systems Research: Methodological Problems - 1982
       Yearbook. Nauka, Moscow, 1982, 402 pp. [Russian]